Springer Series in

OPTICAL SCIENCES 170

founded by H.K.V. Lotsch

Springer Series in
OPTICAL SCIENCES

The Springer Series in Optical Sciences, under the leadership of Editor-in-Chief William T. Rhodes, Georgia Institute of Technology, USA, provides an expanding selection of research monographs in all major areas of optics: lasers and quantum optics, ultra fast phenomena, optical spectroscopy techniques, optoelectronics, quantum information, information optics, applied laser technology, industrial applications, and other topics of contemporary interest.
With this broad coverage of topics, the series is of use to all research scientists and engineers who need up-to-date reference books.
The editors encourage prospective authors to correspond with them in advance of submitting a manuscript. Submission of manuscripts should be made to the Editor-in-Chief or one of the Editors. See also www.springer.com/series/624

For further volumes:
http://www.springer.com/series/624

Zhigang Chen • Roberto Morandotti
Editors

Nonlinear Photonics and Novel Optical Phenomena

Springer

Editors
Zhigang Chen
Physics and Astronomy
San Francisco State University
San Francisco, CA 94132, USA

and

Applied Physics School
Nankai University
Tianjin 300457, China

Roberto Morandotti
INRS-EMT
Université du Québec
Varennes J3X 1S2, Canada

ISSN 0342-4111 ISSN 1556-1534 (electronic)
ISBN 978-1-4614-3537-2 ISBN 978-1-4614-3538-9 (eBook)
DOI 10.1007/978-1-4614-3538-9
Springer New York Heidelberg Dordrecht London

Library of Congress Control Number: 2012938482

Printed on acid-free paper

Springer is part of Springer Science+Business Media (www.springer.com)

Preface

Nonlinear optics and photonics is a broad field of history as venerable as that of lasers, yet the field has been advanced dramatically with many discoveries and emerging new frontiers in recent years. This book contains a total of 13 chapters from more than 50 coauthors affiliated with leading groups in the field of nonlinear optics and photonics from different institutions well distributed globally (*USA, Australia, China, Japan, Israel, Greece, Italy, Germany, France, Spain, United Kingdom, and Canada*). The content of the book covers both theoretical and experimental studies on novel phenomena in a variety of optical materials and photonic systems.

In Chap. 1, the authors present an overview on a new type of optical beams, namely, self-accelerating Airy beams. Research activities on such nondiffracting and self-bending optical beams have surged in recent years due to fundamental interest and many proposed applications of Airy beams. In this chapter, the authors discuss the generation and control of Airy beams as well as the recent developments in the area. In Chap. 2, the authors present a review on integrated photonics based on recently developed high-index doped-silica glass. This material combines the optimal linear properties of single-mode fibers with the typical nonlinear properties of other materials, such as semiconductors and nonlinear glasses. This novel glass material may lead to the new and important possibility such as integration of complex spiral guiding structures onto chip-size areas.

The next three chapters deal with linear and nonlinear spatial beam dynamics in photonic lattices and waveguide arrays. In the past decade, the field of discrete optics has grown rapidly and tremendously because of numerous new findings in discrete optical systems that have a tight link to many other branches of physics. In Chap. 3, the authors report their experimental work on wave transport in amorphous photonic structures, and show that the concept of band-gaps and defect states has greater importance and wider implications than those traditionally conceived in the context of crystalline structures. In Chap. 4, the authors present their theoretical and experimental results on all-optically controlled spatial reshaping and localization of multicolor light beams in nonlinear waveguide arrays. In addition, they review briefly the recent developments on polychromatic light control in photonic lattices and waveguide arrays. In Chap. 5, the authors provide a brief overview on spatial beam dynamics mediated by the interplay between beam diffraction or lattice coupling and self-focusing/defocusing hybrid nonlinearity. Such hybrid nonlinearity plays a key role in unusual nonlinear beam dynamics in both continuous and discrete regimes.

Chapter 6 is about the theory of polariton solitons in semiconductor microcavities. The authors review the physics behind the formation of localized states of exciton-polaritons, the polariton solitons. In particular, the authors show that pump momenta associated with a positive or negative effective polariton mass can give rise to the formation of dark or bright solitons in semiconductor microresonators. In Chap. 7, the authors review a series of experiments on the study of localized structures in semiconductor optical devices, including localized single addressable optical vortices observed in a system formed by two face-to-face VCSELs. Chapters 8–10 report on various intriguing nonlinear phenomena in different systems, including scale-free optics, for which diffraction and evanescent wave formation of subwavelength beams are circumvented by using nonlinearity in nanodisordered ferroelectrics; spatial dispersive shock waves, in which nonlinearity enhances diffraction as observed in both local and nonlocal media; and finally wavelength-scale plasma gratings in air as a result of nonlinear interaction of intense ultrashort filaments.

Chapters 11 and 12 focus on Terahertz waves. In Chap. 11, the authors discuss the emerging field of ultrafast nonlinear optics in the terahertz regime, and introduce THz nonlinear spectroscopy through the absorption-bleaching phenomenon in thin-film semiconductor. In Chap. 12, the authors discuss the generation of Terahertz radiation via Purcell-enhanced nonlinear frequency mixing, which relies on the dramatic enhancements of the conversion efficiency of an arbitrary difference-frequency down-conversion process. Finally, in Chap. 13 the authors discuss the emerging realm of photonic interband transition nanophotonics, where exciting possibilities linked to the dynamical modulation of photonic structures for applications in optical isolation and tunable resonance are presented.

Most of the chapters included in this book are based on invited presentations from the Second International Workshop on *Nonlinear Optics and Novel Phenomena* held at TEDA Applied Physics School of Nankai University, Tianjin, China in the summer of 2010. The editors wish to take this opportunity to acknowledge the support from local organizers at Nankai University, and to thank all contributors for their hard effort and patience to bring this book into reality.

San Francisco, CA, USA Zhigang Chen
Varennes, Canada Roberto Morandotti

David Duchesne INRS-EMT, Varennes, QC, Canada

Massachusetts Institute of Technology, Cambridge, MA, USA

Nikolaos K. Efremidis Department of Applied Mathematics, University of Crete, Heraklion, Crete, Greece

O.A. Egorov Institute of Condensed Matter Theory and Solid State Optics, Friedrich Schiller Universität Jena, Jena, Germany

Shanhui Fan Department of Electrical Engineering, Ginzton Laboratory, Stanford University, Stanford, CA, USA

Marcello Ferrera INRS-EMT, Varennes, QC, Canada

School of Physics and Astronomy, University of St Andrews, Fife, UK

Jason W. Fleischer Department of Electrical Engineering, Princeton University, Princeton, NJ, USA

P. Genevet CNRS, Institut Nonlineaire de Nice (UMR6618), Universite de Nice-Sophia Antipolis, Valbonne, France

M. Giudici CNRS, Institut Nonlineaire de Nice (UMR6618), Universite de Nice-Sophia Antipolis, Valbonne, France

X. Hachair CNRS, Institut Nonlineaire de Nice (UMR6618), Universite de Nice-Sophia Antipolis, Valbonne, France

F.A. Hegmann Department of Physics, University of Alberta, Edmonton, AB, Canada

Yi Hu Department of Physics and Astronomy, San Francisco State University, San Francisco, CA, USA

The Key Laboratory of Weak-Light Nonlinear Photonics, Ministry of Education and TEDA Applied Physics School, Nankai University, Tianjin, China

INRS-EMT, University of Quebec, Varennes, QC, Canada

Shu Jia Department of Electrical Engineering, Princeton University, Princeton, NJ, USA

Yuri S. Kivshar Nonlinear Physics Centre, Research School of Physics and Engineering, Australian National University, Canberra, ACT, Australia

F. Lederer Institute of Condensed Matter Theory and Solid State Optics, Friedrich Schiller Universität Jena, Jena, Germany

Brent E. Little Infinera Corporation, Sunnyvale, CA, USA

Jia Liu State Key Laboratory of Precision Spectroscopy, East China Normal University, Shanghai, China

Contributors

Jose Azaña INRS-EMT, Varennes, QC, Canada

S. Barland CNRS, Institut Nonlineaire de Nice (UMR6618), Universite de Nice-Sophia Antipolis, Valbonne, France

Christopher Barsi Department of Electrical Engineering, Princeton University, Princeton, NJ, USA

F. Blanchard Institute for Integrated Cell-Material Sciences (iCeMS), Kyoto University, Kyoto-shi, Japan

J. Bravo-Abad Departamento de Fisica Teorica de la Materia Condensada, Universidad Autonoma de Madri, Madrid, Spain

E. Caboche CNRS, Institut Nonlineaire de Nice (UMR6618), Universite de Nice-Sophia Antipolis, Valbonne, France

C. Conti Dipartimento di Fisica, Universit'a di Roma Sapienza and ISC-CNR, Rome, Italy

Zhigang Chen Department of Physics and Astronomy, San Francisco State University, San Francisco, CA, USA

The Key Laboratory of Weak-Light Nonlinear Photonics, Ministry of Education and TEDA Applied Physics School, Nankai University, Tianjin, China

Demetrios N. Christodoulides CREOL/College of Optics, University of Central Florida, Orlando, FL, USA

Sai T. Chu Infinera Corporation, Sunnyvale, CA, USA

City University of Hong Kong, Hong Kong, China

E. DelRe Dipartimento di Fisica, Universitá di Roma Sapienza and IPCF-CNR, Roma, Italy

Contents

Sheng Liu The Key Laboratory of Space Applied Physics and Chemistry, Ministry of Education and Shaanxi Key Laboratory of Optical Information Technology, School of Science, Northwestern Polytechnical University, Xi'an, China

Cibo Lou The Key Laboratory of Weak-Light Nonlinear Photonics, Ministry of Education and TEDA Applied Physics School, Nankai University, Tianjin, China

Roberto Morandotti INRS-EMT, Université du Québec, Varennes, J3X 1S2, Canada

David J. Moss IPOS/CUDOS, School of Physics, University of Sydney, Sydney, NSW, Australia

Dragomir N. Neshev Nonlinear Physics Centre, Research School of Physics and Engineering, Australian National University, Canberra, ACT, Australia

T. Ozaki INRS-EMT, Varennes, QC, Canada

Yong-Woo Park INRS-EMT, Varennes, QC, Canada

Alessia Pasquazi INRS-EMT, Varennes, QC, Canada

Marco Peccianti INRS-EMT, Varennes, QC, Canada

F. Pedaci CNRS, Institut Nonlineaire de Nice (UMR6618), Universite de Nice-Sophia Antipolis, Valbonne, France

L. Razzari IIT, Istituto Italiano di Tecnologia, Genoa, Italy

INRS-EMT, Varennes, QC, Canada

Mikael Rechtsman Physics Department, Solid State Institute, Technion, Haifa, Israel

M. Reid Department of Physics, University of Northern British Columbia, Prince George, BC, Canada

Mordechai Segev Physics Department, Solid State Institute, Technion, Haifa, Israel

G. Sharma INRS-EMT, Université du Québec, Montreal, QC, Canada

Georgios A. Siviloglou CREOL/College of Optics, University of Central Florida, Orlando, FL, USA

D.V. Skryabin Department of Physics, Centre for Photonics and Photonic Materials, University of Bath, Bath, UK

M. Soljačić Physics Department, MIT, Cambridge, MA, USA

F.H. Su Institute of Solid State Physics, Hefei, China

Andrey A. Sukhorukov Nonlinear Physics Centre and Centre, Research School of Physics and Engineering, Australian National University, Canberra, ACT, Australia

Alexander Szameit Physics Department, Solid State Institute, Technion, Haifa, Israel

J.R. Tredicce CNRS, Institut Nonlineaire de Nice (UMR6618), Universite de Nice-Sophia Antipolis, Valbonne, France

Wenjie Wan Department of Electrical Engineering, Princeton University, Princeton, NJ, USA

Jingjun Xu The Key Laboratory of Weak-Light Nonlinear Photonics, Ministry of Education and TEDA Applied Physics School, Nankai University, Tianjin, China

Zongfu Yu Department of Electrical Engineering, Ginzton Laboratory, Stanford University, Stanford, CA, USA

Heping Zeng State Key Laboratory of Precision Spectroscopy, East China Normal University, Shanghai, China

Peng Zhang Department of Physics and Astronomy, San Francisco State University, San Francisco, CA 94132, USA

Jianlin Zhao The Key Laboratory of Space Applied Physics and Chemistry, Ministry of Education and Shaanxi Key Laboratory of Optical Information Technology, School of Science, Northwestern Polytechnical University, Xi'an, China

Chapter 1
Self-accelerating Airy Beams: Generation, Control, and Applications

Yi Hu, Georgios A. Siviloglou, Peng Zhang, Nikolaos K. Efremidis, Demetrios N. Christodoulides, and Zhigang Chen

1.1 Introduction

More than three decades ago, Berry and Balazs made an important prediction within the context of quantum mechanics: they proposed theoretically that the Schrödinger equation describing a free particle can exhibit a nonspreading Airy wave packet solution [1]. Perhaps, the most remarkable feature of this Airy packet is its ability to *freely accelerate* even in the absence of any external potential. As first noted in Berry's paper, this Airy packet happens to be unique, e.g., it is the only nontrivial solution (apart from a plane wave) that remains invariant with time

Y. Hu (✉)
Department of Physics and Astronomy, San Francisco State University,
San Francisco, CA 94132, USA

The Key Laboratory of Weak-Light Nonlinear Photonics, Ministry of Education and TEDA Applied Physics School, Nankai University, Tianjin 300457, China

INRS-EMT, University of Quebec, 1650 Blvd. Lionel Boulet, Varennes, QC, Canada J3X 1S2,
e-mail: hynankai@gmail.com

G.A. Siviloglou • D.N. Christodoulides
CREOL/College of Optics, University of Central Florida, Orlando, FL 32816, USA

P. Zhang
Department of Physics and Astronomy, San Francisco State University,
San Francisco, CA 94132, USA

N.K. Efremidis
Department of Applied Mathematics, University of Crete, Heraklion, Crete 71409, Greece

Z. Chen
Department of Physics and Astronomy, San Francisco State University,
San Francisco, CA 94132, USA

The Key Laboratory of Weak-Light Nonlinear Photonics, Ministry of Education and TEDA Applied Physics School, Nankai University, Tianjin 300457, China

Z. Chen and R. Morandotti (eds.), *Nonlinear Photonics and Novel Optical Phenomena*, Springer Series in Optical Sciences 170, DOI 10.1007/978-1-4614-3538-9_1,

in one-dimensional (1D) domain [1, 2]. However, this work has hibernated in the literature for decades, and it never led to experimental realization of any self-accelerating Airy wave packet.

Over the years, nonspreading or nondiffracting wave configurations have been systematically investigated in higher-dimensions (2D and 3D), particularly in the areas of optics and atom physics [3–8]. What makes the analogy between these two seemingly different disciplines possible is the mathematical correspondence between the quantum mechanical Schrödinger equation and the paraxial equation of diffraction [9]. In terms of experimental realization, optics has thus far provided a fertile ground in which the properties of such nonspreading localized waves can be directly observed and studied in detail. Perhaps, the best known example of such 2D diffraction-free optical waves is the so-called *Bessel beams* as first suggested and observed by Durnin et al. [3, 4]. This work has since sparked considerable theoretical and experimental activity and paved the way toward the discovery of other interesting nondiffracting solutions [5–7]. It should be noted that even though at first sight these propagation-invariant beams may appear dissimilar, they in fact share common characteristics. In particular, they are all generated from an appropriate conical superposition of plane waves [3–7]. Even more importantly, all these solutions are known to convey infinite power: a direct outcome of their nondiffracting nature. Of course, in practice, all these nonspreading beams are normally truncated by an aperture (because of lack of space and power) and as such they tend to diffract slightly during propagation [10]. Yet, if the geometrical size of the limiting aperture greatly exceeds the spatial features of the ideal propagation-invariant fields, the diffraction process is considerably "slowed down" over the intended propagation distance and hence for all practical purposes these beams are called "diffraction-free" [11]. It should be emphasized that no localized 1D propagation-invariant beam can be synthesized through conical superposition.

Recently, a specific type of nondiffracting beams, namely, self-accelerating *Airy beams* [12, 13] has attracted a great deal of interest due to their unique properties and many potential applications such as in optical micromanipulation [14–16], plasma guidance [17, 18], vacuum electron acceleration [19, 20], and routing surface plasmon polaritons [21]. In contradistinction with the Bessel beams, the Airy beams do not rely on simple conical superposition of plane waves, and they possess the properties of self-acceleration in addition to nondiffraction and self-healing. For the past few years, tremendous research work has been devoted to the study of Airy beams, from theoretical predictions to experimental observations, from linear control to nonlinear self-trapping, and from fundamental aspects to demonstrations of proposed applications.

In this chapter, we provide an overview on generation of linear and nonlinear control of Airy beams and recent developments in this area. In just a few years, driven by both fundamental interest and application potential, the number of research papers dealing with optical Airy beams has risen dramatically. Thus, we discuss only a selection of our published papers and mention a few others significant to the field. This overview is by no means all-inclusive, nor is it meant to be.

1.2 Generation of Optical Self-accelerating Airy Beams

1.2.1 Ideal Infinite-Energy Airy Beams

We begin our analysis by considering the (1 + 1)D paraxial equation of diffraction that governs the propagation dynamics of the electric field envelope φ associated with planar optical beams:

$$i\frac{\partial \varphi}{\partial \xi} + \frac{1}{2}\frac{\partial^2 \varphi}{\partial s^2} = 0. \tag{1.1}$$

In (1.1), $s = x/x_0$ represents a dimensionless transverse coordinate, x_0 is an arbitrary transverse scale, $\xi = z/kx_0^2$ is a normalized propagation distance (with respect to the Rayleigh range), and $k = 2\pi n/\lambda_0$ is the wave number of the optical wave. Incidentally, this same equation is also known to govern pulse propagation in dispersive media.

As first shown in [1], (1.1) admits the following Airy nondispersive solution:

$$\varphi(s, \xi) = \text{Ai}\left[s - \left(\frac{\xi}{2}\right)^2\right] \exp\left(is\frac{\xi}{2} - \frac{i\xi^3}{12}\right). \tag{1.2}$$

Clearly, at the origin $\varphi(s, 0) = \text{Ai}(s)$, (1.2) shows that the intensity profile of this wave remains invariant during propagation while it experiences constant *transverse acceleration*. The term $(\xi/2)^2$ in (1.2) describes this ballistic trajectory. Figure 1.1 depicts the diffraction-free propagation of such an accelerating Airy wave packet as a function of distance ξ. An alternative interpretation of this interesting result was given by Greenberger through the principle of equivalence [22]. More specifically, he remarked that a stationary Airy wave packet associated with a quantum mechanical particle in a constant gravitational field will be perceived as accelerating upwards by a free-falling

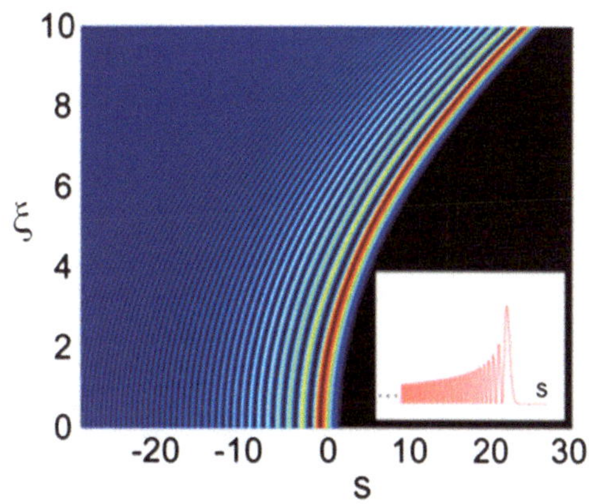

Fig. 1.1 Propagation dynamics of a diffraction-free Airy wave. The corresponding input intensity of the beam is shown in the *inset*

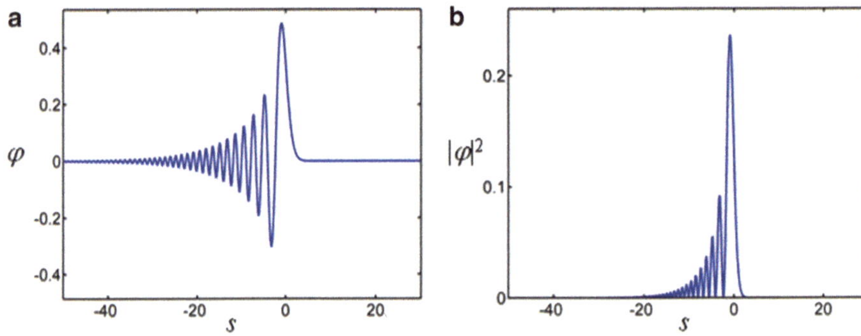

Fig. 1.2 Normalized field profile (**a**) and intensity profile (**b**) of a finite-energy Airy beam when $\alpha = 0.1$

observer in whose frame of inertia gravitational forces are absent. As also indicated in [1], this accelerating behavior is by no means in conflict with Ehrenfest's theorem which describes the motion of the center of gravity of a wave packet [1, 9]. This is simply because the Airy beam is not square integrable ($\int \mathrm{Ai}^2(x)\mathrm{d}x \to \infty$), and thus, its center of mass cannot be defined [1, 23]. Note that, ideally, the Airy beam would have infinite energy, thus it keeps the transverse acceleration and diffraction-free propagation no matter how far it travels, much like a free-falling object that always keeps the gravitational acceleration in absence of friction or obstacles.

1.2.2 Truncated 1D and 2D Airy Beams: Theory

Infinite-energy Airy beams are impossible in practice. One possible way to realize such beams is to introduce an exponential aperture function, i.e., let

$$\varphi(s,0) = \mathrm{Ai}(s)\,\exp(\alpha s) \tag{1.3}$$

at the input of the system ($\xi = 0$). In (1.3) the decay factor $\alpha > 0$ is a positive quantity so as to ensure containment of the infinite Airy tail and can thus enable the physical realization of such beams. Note that the positive branch of the Airy function decays very rapidly and thus the convergence of the function in (1.3) is guaranteed. Figure 1.2a depicts the field profile of such a truncated Airy beam at $\xi = 0$, whereas Fig. 1.2b plots its corresponding intensity.

Of particular interest is the Fourier spectrum of this beam which in the normalized k-space is given by:

$$\Phi_0(k) = \exp(-\alpha k^2)\,\exp\left(\frac{i}{3}(k^3 - 3\alpha^2 k - i\alpha^3)\right). \tag{1.4}$$

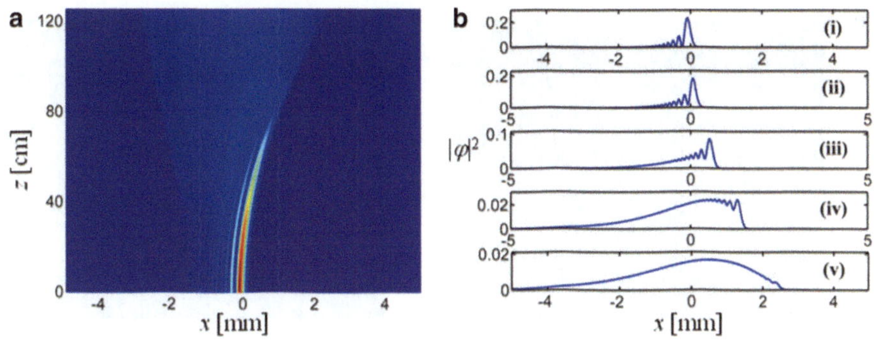

Fig. 1.3 (**a**) Propagation dynamics of a finite-energy Airy beam as a function of distance, (**b**) cross sections of the normalized beam intensity at (i) $z = 0$ cm, (ii) 31.4 cm, (iii) 62.8 cm, (iv) 94.3 cm, and (v) 125.7 cm

From (1.4) it becomes directly evident that the wave packet power spectrum is Gaussian. From Parseval's theorem, the total power of this finite-energy Airy wave packet can be directly obtained and is given by:

$$\int_{-\infty}^{\infty} \mathrm{d}s|\varphi(s,\xi=0)|^2 = \sqrt{\frac{1}{8\pi\alpha}}\exp\left(\frac{2\alpha^3}{3}\right). \tag{1.5}$$

By directly solving (1.1) under the initial conditions of (1.3), we find that the Airy beam evolves according to:

$$\varphi(s,\xi) = \mathrm{Ai}\left[s - \left(\frac{\xi}{2}\right)^2 + i\alpha\xi\right]\exp\left(\alpha s - \frac{\alpha\xi^2}{2} - \frac{i\xi^3}{12} + i\alpha^2\frac{\xi}{2} + is\frac{\xi}{2}\right). \tag{1.6}$$

Note that in the limit $\alpha = 0$ our solution reduces to the nondispersive wave packet shown in (1.2). Figure 1.3a shows the propagation of such a planar Airy beam up to a distance of 1.25 m when $x_0 = 100$ μm and the decay parameter is $\alpha = 0.1$. The corresponding cross sections of the intensity profiles at various distances are shown in Fig. 1.3b. For these parameters, the intensity FWHM of the first lobe of this beam is 171 μm. We note that for a Gaussian beam of this same width, its Rayleigh range would have been 13.25 cm at a wavelength of $\lambda_0 = 0.5$ μm. For this example, the intensity features of this beam remain essentially invariant up to 75 cm as clearly seen in Fig. 1.3. Evidently this wave endures because of the quasi-diffraction free character of the Airy wave packet. We emphasize that for this same distance, the front lobe of the beam would have expanded by at least six times. As illustrated in Fig. 1.3b, the beam starts to deteriorate first from the tail as a result of truncation. The last feature to disappear (around 100 cm) is the front lobe. After a certain distance (in this case 120 cm), the beam intensity becomes Gaussian-like, i.e., as expected from its Gaussian power spectrum in the Fraunhofer limit.

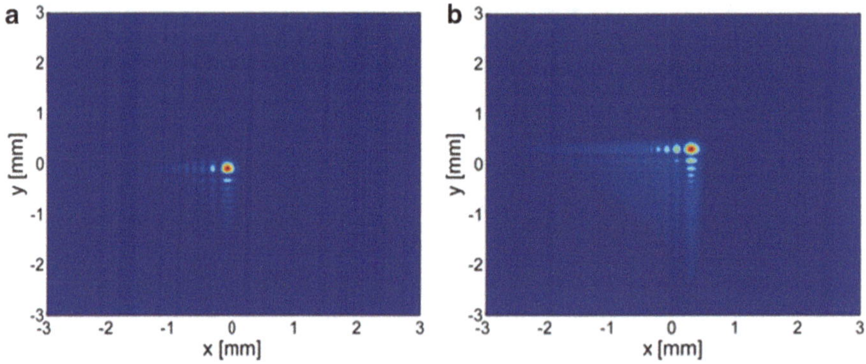

Fig. 1.4 A two-dimensional finite-energy Airy beam at the input $z = 0$ (**a**) and after propagating to $z = 50$ cm (**b**)

Even more importantly, in spite of its truncation (necessary for its realization), the Airy wave packet still exhibits its most exotic feature, i.e., its trend to freely accelerate. This characteristic is rather peculiar given the fact that it may occur in free space, e.g., in the absence of any index gradients such as from prisms or layered media. This behavior is reflected in the term $s - (\xi/2)^2$ that appears in the argument of the Airy function in (1.6). These acceleration dynamics can be clearly seen in Fig. 1.3a, where the beam's parabolic trajectory becomes evident. For the example discussed here, the beam will shift by 880 μm at $z = 75.4$ cm [12].

These results can be readily generalized into 2D domain, i.e., when the initial field envelope is given by

$$\varphi(x, y, z = 0) = \mathrm{Ai}\left(\frac{x}{x_0}\right)\mathrm{Ai}\left(\frac{y}{y_0}\right)\exp\left(\frac{x}{w_1}\right)\exp\left(\frac{y}{w_2}\right). \tag{1.7}$$

The intensity profile of such a 2D beam at $z = 0$ and $z = 50$ cm is shown in Fig. 1.4a, b when $x_0 = y_0 = 100$ μm and $w_1 = w_2 = 1$ mm. In this case, the 2D Airy beam remains almost invariant up to a distance of $z = 50$ cm along the longitudinal direction while accelerating in the same manner along the 45° axis in the x–y transverse direction. This again suggests that experimental realization of finite-energy Airy beams is possible simply by truncation.

1.2.3 Truncated 1D and 2D Airy Beams: Experiment

From (1.4), one can readily deduce that the angular Fourier spectrum of the truncated Airy beam is Gaussian and involves a cubic phase (k^3) resulting from the Fourier transform of the Airy function itself. This particular form of the spectrum has important implications in terms of experimental synthesis of

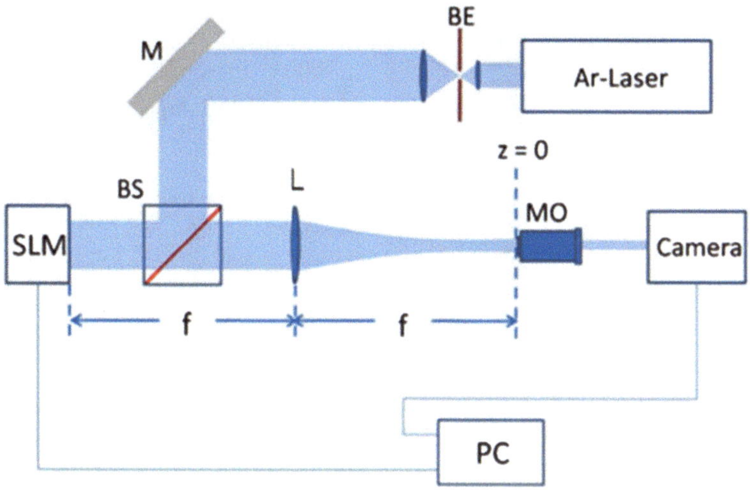

Fig. 1.5 Experimental setup for generation of truncated Airy beams. *SLM* spatial light modulator, *BE* beam expander, *MO* microscope objective

the truncated version of Airy packets. As a result, this wave can be generated from a broad Gaussian beam through a Fourier transformation provided that a cubic phase is imposed. Experimentally, such a cubic phase can be readily realized with Gaussian laser beam by using a spatial light modulator (SLM).

A typical experimental setup for Airy beam generation is illustrated in Fig. 1.5. An air-cooled Argon-ion continuous-wave laser operating at 488 nm emits a linearly polarized fundamental Gaussian beam that is subsequently collimated to a width of 6.7 mm (FWHM). This broad Gaussian beam is then reflected from the front facet of a computer-controlled liquid crystal SLM. The SLM is used to impose the cubic-phase modulation (from -20π to 20π in 2 cm) that is necessary to produce the Airy beam. In order to generate a 1D (or 2D) Airy beam, a converging cylindrical (or circular) lens with a focal length of $f = 1.2$ m is placed at a distance f in front of the SLM phase array. After the SLM, the Fourier transform of the phase-modulated Gaussian beam is then obtained at a distance $d = f = 1.2$ m behind the lens. The Airy beam produced is then imaged on a carefully aligned CCD camera through a 5× microscope objective. The propagation of the Airy beam is then monitored by translating the imaging apparatus. Figure 1.6a, b shows the phase masks used to generate the 1D and 2D Airy beams, respectively.

Experimental results of a 1D Airy beam propagation in free-space are shown in Fig. 1.7, where Fig. 1.7a depicts the intensity profile of the 1D exponentially truncated Airy beam at the origin ($z = 0$). In our experiment, $x_0 = 53$ μm and $\alpha = 0.11$. Figure 1.7b, c shows the corresponding intensity profiles of this Airy packet at $z = 10$, 20 cm, respectively. As expected, the beam remains almost diffraction-free while its main lobe tends to quadratically accelerate. Our measurements show that the spatial FWHM width of the main lobe (containing

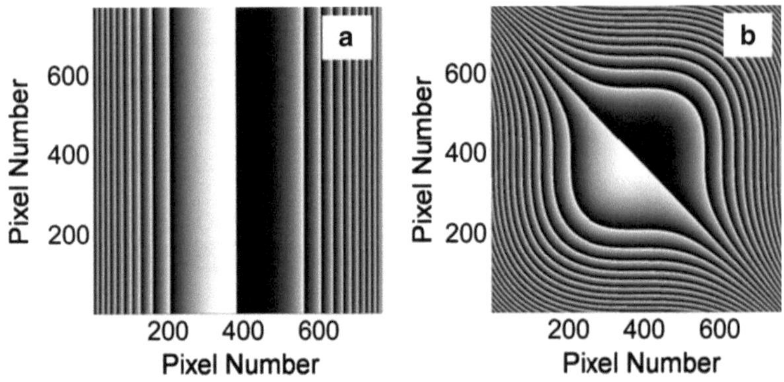

Fig. 1.6 Phase masks used to generate (**a**) 1D and (**b**) 2D-Airy beams. The cubic phase is "wrapped" between $[0, 2\pi]$. In the gray-scale pattern, *black* corresponds to 0 and *white* to 2π radians

in this case more than 70% of the total beam energy) remains almost invariant up to a distance of approximately 25 cm and retains its original value of ≈90 μm. It is worth noting that this occurs in free space and is by no means a result of any optical nonlinearity [24]. Figure 1.7d–f depicts the corresponding expected theoretical behavior of this same Airy packet at these same distances in good agreement with the experiment. Note that a Gaussian beam of this size would have diffracted at least 6–7 times in this same distance, shown in Fig. 1.7g, h.

Our experiment also demonstrated the transverse acceleration of the local intensity maxima, shown in Fig. 1.8. This parabolic-like trajectory is a result of acceleration and is well described by the theoretical relation $x_d = \lambda_0^2 z^2/(16\pi^2 x_0^3)$, as long as the beam remains quasi-diffraction free and before diffraction effects take over. The solid line in Fig. 1.8 corresponds to the latter analytical expression. As these results indicate, after 30 cm of propagation the beam experiences a deflection of 820 μm comparable to the total size of the packet (≈first 10 lobes of the Airy beam). Again, we emphasize that the acceleration observed here refers to the local intensity features of the packet. In all cases, the center of gravity $\langle x \rangle$ of this wave remains invariant [1, 25] since $\mathrm{d}\langle x \rangle/\mathrm{d}z \propto (i/2)\int(\varphi_x^* \varphi - \varphi_x \varphi^*)\mathrm{d}x$ is constant.

Similarly we have also considered 2D Airy beams. The case of an ideal 2D Airy packet was first suggested by Besieris et al. [26]. In this case a 2D SLM phase pattern (Fig. 1.6b) was imposed on the Gaussian beam and was then Fourier transformed through a spherical lens. By doing so we were able to produce finite energy Airy wave packets of the form in the right side of (1.7). The evolution diffraction dynamics of the latter 2D field configurations can be readily solved by separation of variables using the result of (1.6). The intensity distribution of such a wave is shown in Fig. 1.9a, when $w_1 = w_2$, corresponding to an x–y truncation factor of $\alpha = 0.11$. In this case, approximately 50% of the energy resides in the main intensity lobe at the corner. In general, the flexibility in separately adjusting the x–y parameters allows one to control the transverse acceleration vector of this novel 2D nondiffracting beam. In our

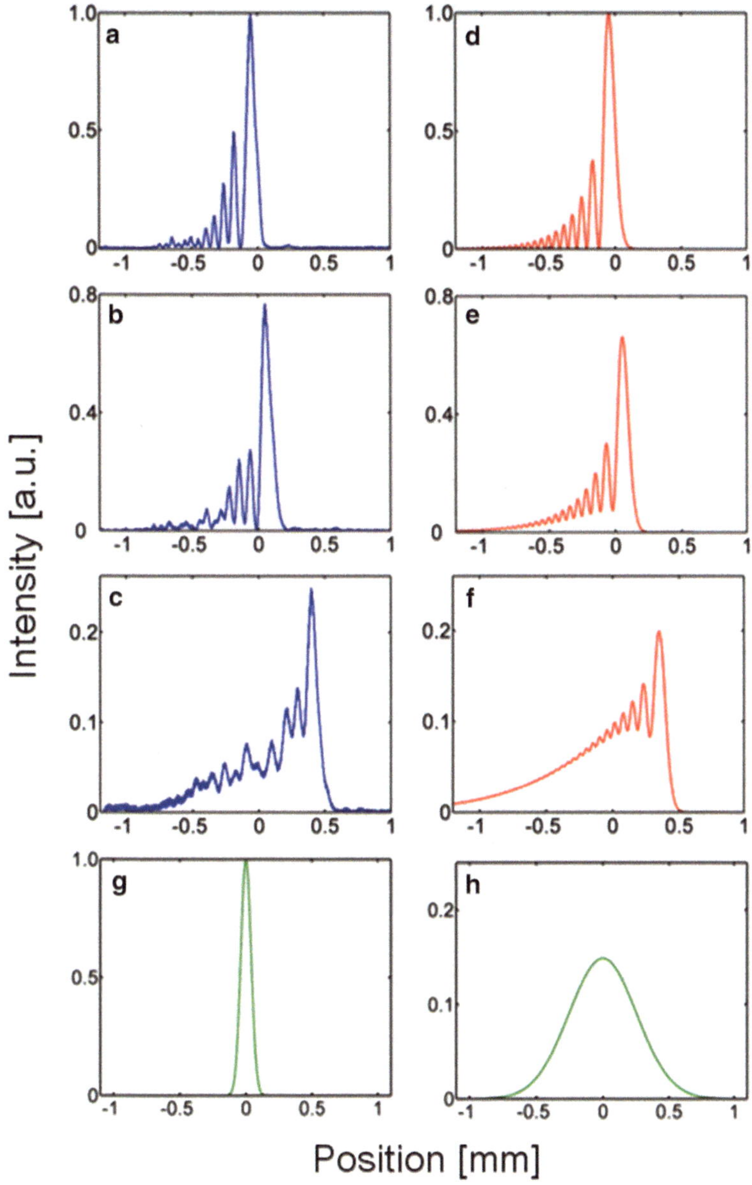

Fig. 1.7 Observed intensity cross sections of a planar Airy beam at (**a**) $z = 0$ cm, (**b**) 10 cm, and (**c**) 20 cm. Corresponding theoretical plots for these same distances (**d**), (**e**) and (**f**). (**g** and **h**) represent a comparison with a Gaussian beam having the same FWHM as the first lobe of the Airy beam

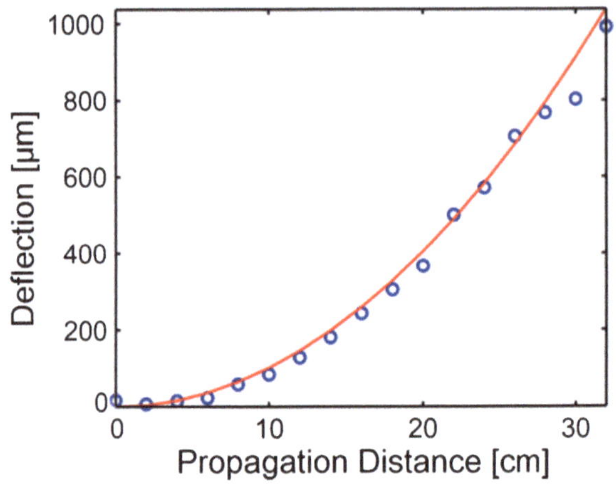

Fig. 1.8 Transverse acceleration of an Airy beam when $\alpha = 0.11$ as a function of distance. *Circles* mark experimental results while the *solid line* represents the expected theoretical deflection

experiments we considered beams with equal scales in x–y, and thus the acceleration occurred along the 45° axis. For the pattern generated, $x_0 = 53$ μm and the aperture factor is $\alpha = 0.11$. As in the 1D case, our experimental results indicate that this 2D beam propagates almost diffraction-free up to a distance of 25 cm. The main lobe keeps its spot-size (90 μm) up to a distance of ~25 cm and the beam moves on a 2D parabolic trajectory with $x_d = y_d$. The propagation and diffraction dynamics of these 2D Airy beams is shown in Fig. 1.9b–g [13].

Up to now, many different methods have been developed to generate Airy beams. As an alternative of cubic phase mask, 3/2 phase [27] or binary phase mask [28] can be explored; besides using SLM, Airy beams can also be implemented through nonlinear processing [29] or assembly of lenses [30]. In addition, Airy beams can be the direct output of a microchip laser [31].

1.2.4 Spatiotemporal Airy–Bessel Bullets

We point out that, since the Airy beams are the only type of nondiffracting wave packets found so far that exist in 1D form, they can also be synthesized in the temporal domain using dispersive elements [32]. This could lead to, for example, the observation of dispersion-free Airy pulses in optical fibers, in both the normal and anomalous dispersion regime [33]. This unique character distinguishes the Airy beams from other diffraction free beams, such as Bessel beams, Mathieu beams, etc., which have only the profile in 2D form.

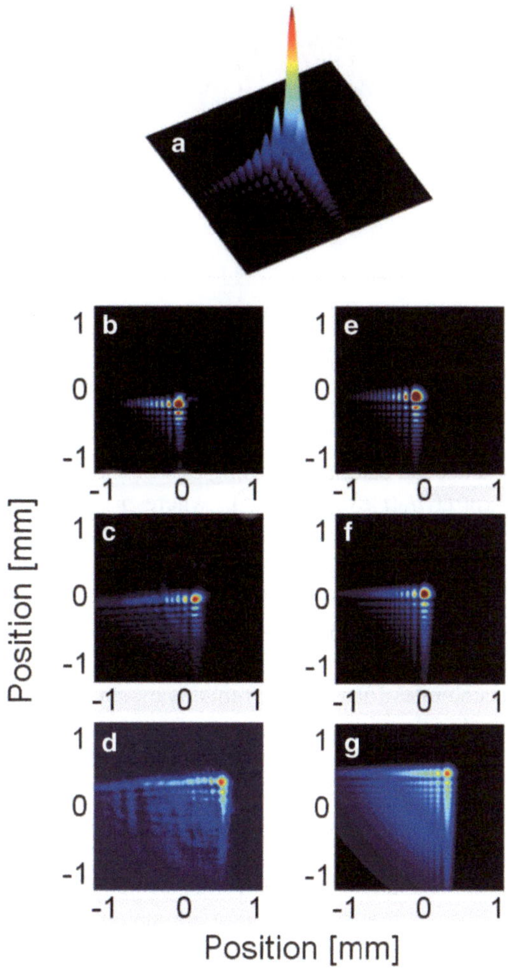

Fig. 1.9 (**a**) A schematic of a 2D Airy packet. Observed intensity distribution of a 2D Airy beam at (**b**) $z = 0$ cm, (**c**) 10 cm, and (**d**) 20 cm. Corresponding theoretical results at these same distances (**e**), (**f**), and (**g**)

Theoretically, the Airy beams in combination with other nondiffracting field configurations can also be used to describe multidimensional [(3 + 1)D] finite energy wave packets in the presence of diffraction and dispersion. In such a case, the beam envelope in the spatiotemporal domain obeys [34]:

$$i\frac{\partial \Psi}{\partial Z} + \frac{1}{2}\left(\frac{\partial^2 \Psi}{\partial X^2} + \frac{\partial^2 \Psi}{\partial Y^2} + \frac{\partial^2 \Psi}{\partial T^2}\right) = 0, \tag{1.8}$$

where in (1.8), without any loss of generality, an anomalously dispersive system was assumed. For example, a localized Airy finite energy spatiotemporal wave

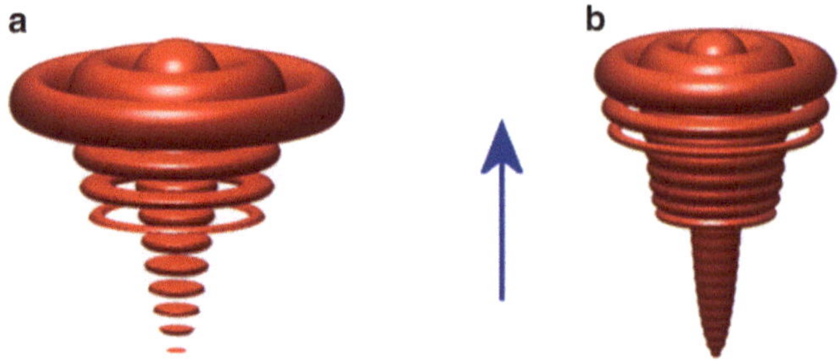

Fig. 1.10 Isosurface intensity contour plot for a spatiotemporal Airy–Gauss–Bessel wave packet (with $\alpha = 0.15$, $w_0 = 9$) (**a**) at the input $Z = 0$ and (**b**) after a normalized propagation distance of $Z = 3$. The *arrow* depicts the direction of acceleration

packet can be obtained using Bessel–Gauss beams [10], i.e., at the input $\psi = \text{Ai}(T)\exp(\alpha T)J_0(r)\exp(-r^2/w_0^2)$, where $r = (X^2 + Y^2)^{1/2}$ and w_0 is the "aperture" spot-size of the beam. Under these initial conditions, using separation of variables we find that this wave evolves according to $\psi = \varphi(Z, T)U(Z, X, Y)$, where $\varphi(Z, T)$ is given by (1.6), and $U(Z, X, Y)$ is given by the solution of Gori et al. [10]. Figure 1.10 depicts an isosurface plot of such an Airy–Bessel–Gauss wave packet at the input $Z = 0$ and after propagation to $Z = 3$. Even in this case the wave accelerates forward and remains essentially invariant.

Accelerating Airy wave packets can also be implemented in dispersive optical fibers. Equation (1.4) suggests that in the temporal domain, such an exponentially decaying Airy pulse can be produced by passing a transform limited Gaussian pulse through a system with appreciable cubic dispersion [33]. A system of this sort can be implemented using another fiber at the zero dispersion point or by employing pulse shaping techniques [35]. Acceleration pulse dynamics can then be observed in a fiber with either normal or anomalous group velocity dispersion.

Recently, spatiotemporal optical wave packets impervious to both dispersion and diffraction, referred to as *light bullets*, have been investigated by a few groups. In particular, Abdollahpour et al. [36] demonstrated the realization of intense Airy–Airy–Airy (Airy^3) light bullets by combining a spatial Airy beam with an Airy pulse in time. The Airy^3 light bullets belong to a family of linear spatiotemporal wave packets and they can withstand both diffraction and dispersion during their propagation. It was shown that the Airy^3 light bullets are robust up to the high intensity regime, since they are capable of healing the nonlinearly induced distortions of their spatiotemporal profile. Chong et al. also demonstrated Airy-based light bullets as 3D linear localized waves in free space [37]. The method employed in the latter work is independent of any particular material or nonlinearity, as the wave packets were formed by combining the Bessel beams in the transverse plane with temporal Airy pulses, which can be extended in a straight manner to explore other transversely nondiffractive beams. These versatile 3D optical bullets in free space might break through the limitations brought by other methods for generation of light bullets.

1.2.5 Radially Symmetric Autofocusing Beams

Quite recently, a new class of 2D and 3D waves that tend to autofocus in an abrupt fashion has been introduced [38]. While the maximum intensity of such a radial wave remains almost constant during propagation, it suddenly increases by orders of magnitude right before its focal point. These waves can be generated through the use of radially symmetric Airy waves or by appropriately superimposing Airy wave packets.

To analyze the properties of such waves, let us first consider the Fresnel diffraction equation in cylindrical coordinates

$$iu_z + \frac{1}{2}\left(u_{rr} + \frac{1}{r}u_r\right) = 0. \tag{1.9}$$

The propagation of an arbitrary radially symmetric initial condition $u(r, z = 0) = u_0(r)$ according to (1.9) can be computed by utilizing the Hankel transform pair

$$u(r,z) = \frac{1}{2\pi}\int_0^\infty k\tilde{u}_0(k)J_0(kr)\mathrm{e}^{-ik^2z/2}\mathrm{d}k, \quad \tilde{u}_0(k) = 2\pi\int_0^\infty ru_0(r)J_0(kr)\mathrm{d}r. \tag{1.10}$$

In particular, a radially symmetric exponentially apodized Airy beam is considered as an initial condition

$$u_0(r) = \mathrm{Ai}(r_0 - r)\,\exp[\alpha(r_0 - r)], \tag{1.11}$$

where r_0 is the radius of the main Airy ring, and α is the apodization rate. The power that the Airy ring of (1.11) carries is given by

$$P = 2\pi\int_0^\infty |u_0|^2 r\,\mathrm{d}r \approx \sqrt{\frac{\pi}{2\alpha}}\,\mathrm{e}^{2\alpha^3/3}\left[r_0 + \frac{1-4\alpha^3}{4\alpha}\right], \tag{1.12}$$

where in the computation of the above integral we extended the lower integration limit to minus infinity (alternatively for the slightly modified initial condition $\mathrm{Ai}(r_0 - r)\mathrm{e}^{\alpha(r_0-r)} - \mathrm{Ai}(r - r_0)\mathrm{e}^{\alpha(r-r_0)}$ the above formula for the power becomes exact). The propagation dynamics of such an Airy ring is depicted in Fig. 1.11 [38]. Qualitatively the dynamics can be described as following: in the early stages of propagation, r_0 is large enough and the disk $r < r_0$ is essentially dark. As a result, in the region where the amplitude is large the approximation $\nabla_\perp u \approx u_{rr}$ holds, and thus the 1D Airy solution (1.6) with $s \rightarrow r - r_0$, can approximate the propagation dynamics. From (1.6), one may expect that the maximum value of the amplitude is going to slowly decrease along z. On the other hand, as z increases, the radius of the Airy beam decreases, the power concentrates in a smaller area, and the maximum amplitude increases. In fact, the numerical simulations show

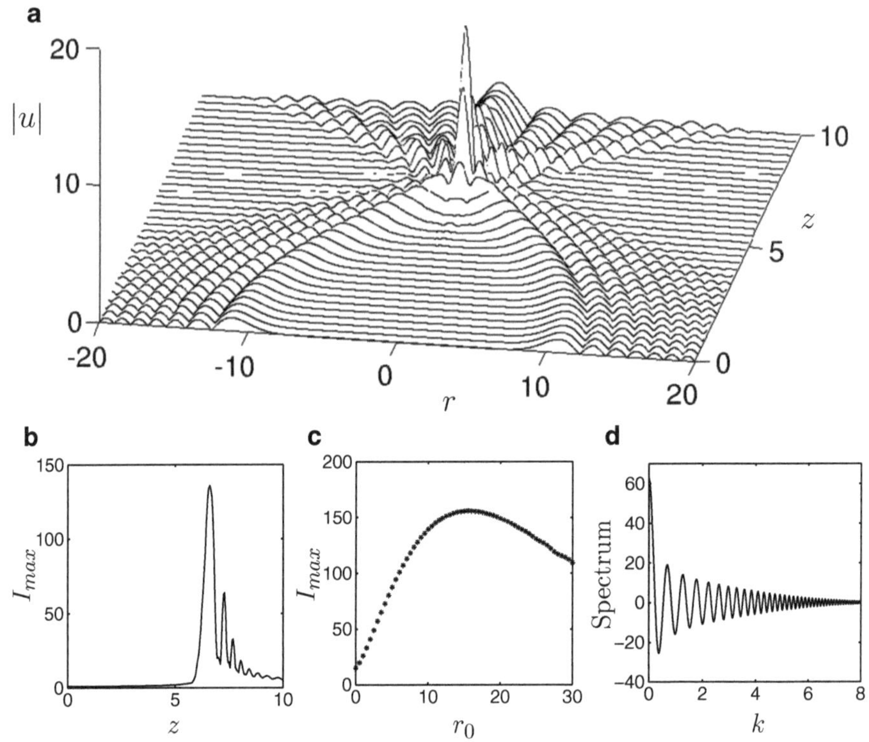

Fig. 1.11 Dynamics of radially symmetric Airy beams for $\alpha = 0.05$, $r_0 = 10$, and $I_{max}(z = 0) = 1$; (**a**) detailed plot of the central part of the propagation dynamics; (**b**) maximum intensity as a function of z; (**c**) Maximum intensity that the Airy beam reaches during propagation for different values of the initial radius r_0; (**d**) Hankel transform of the initial condition

that these two effects almost balance each other, leading to relatively small maximum amplitude changes, up to the point where the beam reaches the center (Fig. 1.11a, b). Close to the focal point, the power of the first Airy ring is concentrated in a small area around $r = 0$ and the maximum intensity at the center rapidly increases. What is behind this very abrupt increase in intensity is the lateral acceleration of the Airy beams themselves. In this case, large transverse velocities are attained and energy rushes in an accelerated fashion toward the focus. While the peak intensity remains around unity up to $z \approx 6$, it then very rapidly increases by more than 135 times at the focal point (Fig. 1.11b). For longer propagation distances, the maximum intensity starts to decrease. As can be seen in Fig. 1.11b, this decrease is not monotonic, but it exhibits oscillations, which are generated by the subsequent Airy rings. Interestingly enough, at a final stage the solution takes the form of a Bessel function with a chirped argument.

In Fig. 1.11c, the maximum intensity that the beam reaches during propagation is shown as a function of the initial radius r_0 for $\alpha = 0.05$. For small values of r_0, the Airy beam does not carry much power and thus the maximum intensity reached is also relatively small. As the value of r_0 increases, the maximum intensity also increases and for $r_0 \approx 15$ it takes its maximum value ($I_{max} \approx 156$). For even larger values of r_0, diffraction becomes significant and I_{max} starts to decrease. Note that, as shown in Fig. 1.11c, large intensity contrasts are possible for a wide range of values of r_0. Much higher values of I_{max} are possible by further suppressing diffraction (decreasing α). Figure 1.11d depicts the Hankel transform of the input field profile of this beam as a function of the radial spectral component k. The transform is a real function that oscillates between positive and negative values and its envelope is decreasing with k.

Families of abruptly autofocusing beams can also be constructed by superimposing exact Airy wave solutions. If $\phi(x,z)$ is the 1D exponentially apodized Airy wave solution as given by (1.6), then we can construct, continuous or discrete, superpositions of 2D Airy waves $\phi(x',z)\phi(y',z)$ where the coordinates (x',y') are rotated and translated with respect to (x,y) $[(x',y')^T = TR(x,y)^T$, T represents a translation, and R a rotation matrix]. A particular configuration that exhibits abruptly autofocusing dynamics consists of a continuous superposition of 2D Airy waves with centers lying on a circle and each one of them propagating towards the center of the circle [38].

The above Airy families of abruptly autofocusing waves can be generalized by considering an initial condition of the form

$$u(r, z=0) = A(r)\ \sin[q(r)], \tag{1.13}$$

where $A(r)$ is the envelope function and $q(r)$ is a sublinear chirped phase

$$q(r) = \begin{cases} C(r-r_0)^{\beta}, & r \geq r_0 \\ 0, & r < r_0 \end{cases}, \tag{1.14}$$

$C > 0$ and $1 < \beta < 2$. The term sublinear stems from the fact that the phase of a linear chirp is quadratic ($\beta = 2$). Following a ray optics approach, we find that such an initial condition generates a caustic that propagates according to

$$r = r_0 - \frac{[C\beta(\beta-1)z]^{\nu}}{\nu - 1}, \tag{1.15}$$

where $\nu = (2-\beta)^{-1}$ [39]. Note that the caustic of the Airy beam is reproduced by setting $\beta = 3/2$ resulting to the parabolic trajectory $r = r_0 - (3Cz/4)^2$. The exponent of the power law caustic can be engineered by varying the chirp coefficient. For example, for $\beta = 5/3$ the exponent of the power law caustic becomes cubic [39, 40]. In addition, different types of convex trajectories, such as exponential, are also possible [40].

Three-dimensional abruptly autofocusing waves are also possible. The corresponding anomalous dispersion paraxial equation, which is normalized such that dispersion and diffraction are equalized, is given by

$$iu_z + \frac{1}{2}(u_{xx} + u_{yy} + u_{tt}) = iu_z + \frac{1}{2}\left(u_{rr} + \frac{2}{r}u_r\right) = 0, \tag{1.16}$$

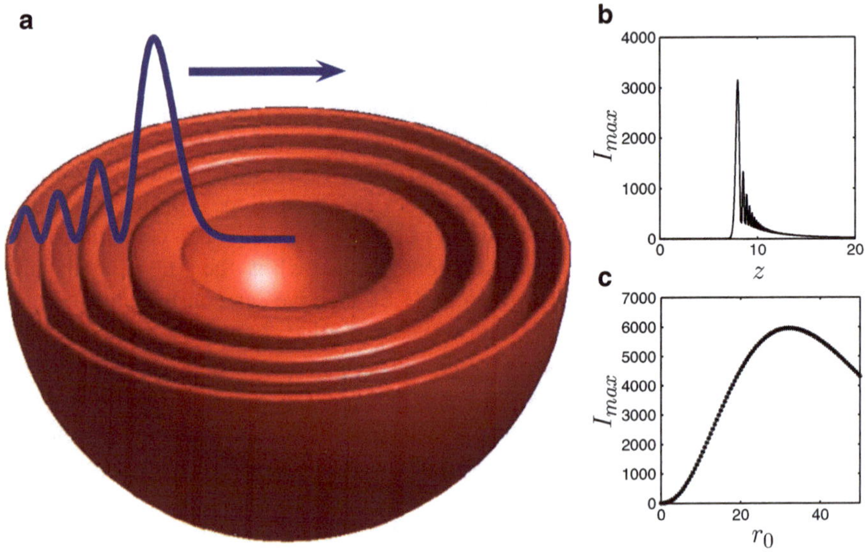

Fig. 1.12 (**a**) Isointensity hemisphere of the spherical Airy-type wave; (**b**) maximum intensity for $r_0 = 15$ as a function of z; (**c**) maximum intensity that the Airy beam reaches during propagation for different values of the radius. In (**b**), (**c**) $I_{\max}(z = 0) = 1$, $\alpha = 0.05$

where $r = \sqrt{x^2 + y^2 + t^2}$. Interestingly enough, exact expressions for the evolution of an abruptly autofocusing wave can be found in the case where the Airy beam is apodized both exponentially and with a power law r^{-1}: In this case, the solution takes the form [38]

$$u(r,z) = \frac{\phi(r_0 - r, z) - \phi(r_0 + r, z)}{r}, \tag{1.17}$$

where $\phi(x,z)$ is the 1D exponentially apodized Airy wave solution. In Fig. 1.12 dynamical properties of such a spatiotemporal wave are shown.

These radially symmetric autofocusing Airy beams have been recently generated by two research groups, along with the proposed application of such beams [15, 41, 42]. We shall discuss these applications in Sect. 1.6.4. Experimental data corresponding to the theoretical results of Fig. 1.11 are displayed in Fig. 1.13 [15]. The bottom panel shows the autofocusing beam from a side propagating view and the top panels display snapshots of transverse patterns at different propagation distances. For better visualization, the intensities in Fig. 1.13a–d have all been scaled to the same peak intensity; however, the intensity pattern without normalization would illustrate the drastic increase of peak intensity near the "focal point" of the circular Airy beam. In Fig. 1.13e, the propagation length is about 3 cm and the beam size changed from 600 to 20 μm. Our findings show a good agreement between theoretical and experimental results.

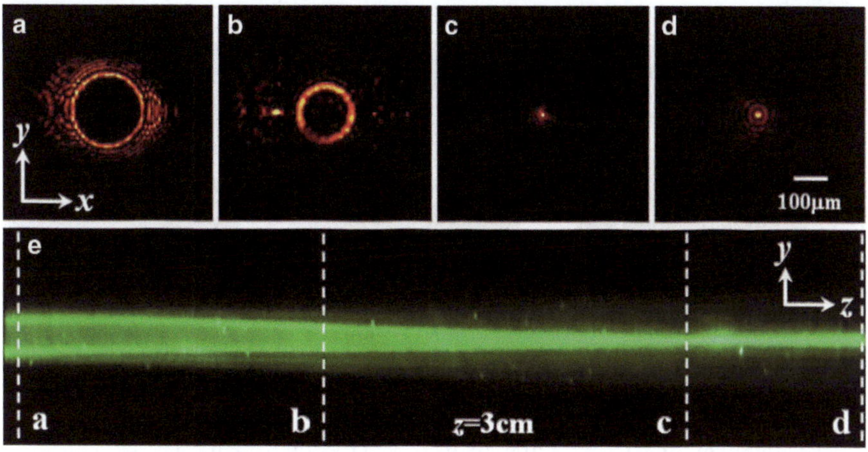

Fig. 1.13 (**a–d**) Experimental snapshots of transverse intensity patterns of the autofocusing beam (contrast enhanced) taken at different planes as marked in (**e**), the direct side-view photography of the beam taken from scattered light

1.3 Control of Ballistic Motion of Airy Beams

1.3.1 Ballistic Dynamics of 1D Airy Beams

The ballistic motion of optical Airy beams can be analyzed using the (1.1). In order to investigate the beam trajectories, we consider the input field distribution $\varphi(s, \xi = 0) = \mathrm{Ai}(s)\exp(\alpha s)\exp(ivs)$, where Ai($s$) represents the Airy function, α in the exponential truncation factor is a small positive parameter, and v is associated with the initial launch angle (or "velocity") of this beam. Under these initial conditions and from (1.1), we find that this finite-energy Airy wave evolves according to:

$$\begin{aligned}\varphi(s,\xi) = {} & \mathrm{Ai}[s-(\xi/2)^2 - v\xi + i\alpha\xi]\,\exp(\alpha s - \alpha\xi^2/2 - \alpha v\xi) \\ & \times \exp\{i[-\xi^3/12 + (\alpha^2 - v^2 + s)\xi/2 + vs - v\xi^2/2)]\}.\end{aligned} \tag{1.18}$$

From the argument of the Airy function in (1.18), one can conclude that this beam follows a ballistic trajectory in the s–ξ plane which is described by the parabola $s = v\xi + (\xi/2)^2$. In physical units, this parabolic deflection of the beam intensity features is given by $x_d = \theta z + [z^2/(4k^2 x_0^3)]$ where the actual launch angle θ in the x–z coordinates is related to the normalized v parameter through $\theta = v/(kx_0)$. The corresponding Newtonian (kinematical) equations describing this ballistics are $d^2x/dz^2 = 1/(2k^2x_0^3) = g$ and $dx/dz = gz + \theta$ where g plays here the role of "gravity". The trajectories of an ideal Airy beam (without truncation) are shown in Fig. 1.14 under different launch conditions, corresponding to three distinct regimes. More specifically, when this wave is launched upwards (when the launch angle is

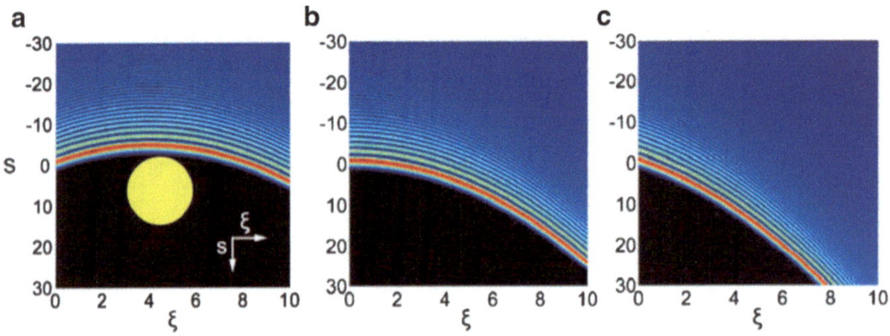

Fig. 1.14 Ballistic dynamics of an ideal Airy beam $\alpha = 0$ when (**a**) $v = -2$, (**b**) $v = 0$, and (**c**) $v = +2$. The *circle* in (**a**) represents an opaque obstacle

negative $v < 0$), the beam will initially ascend until it stalls due to downward acceleration at $\hat{z} = -\theta/g = -2k^2x_0{}^3\theta$. At this apogee point the maximum deflection is $x_{d\max} = -\theta^2k^2x_0{}^3$. From that point on, the packet will accelerate downwards as shown in Fig. 1.14a. In fact, this ballistic behavior suggests that the Airy wave packet can circumvent an opaque object O (depicted schematically in Fig. 1.14a) lying straight in its path, by following instead a curved trajectory. If on the other hand the launch angle is zero, the wave will follow a parabolic trajectory (Fig. 1.14b), similar to that predicted and demonstrated in [1, 12, 13]. The case for $v > 0$ is shown in Fig. 1.14c.

The experimental setup used to observe the ballistic dynamics of finite energy Airy wave packets is shown in Fig. 1.5. The propagation dynamics of these beams were then recorded as a function of propagation distance by translating the imaging apparatus. The origin $z = 0$ is taken at a distance f after the lens, e.g., at the point where the exponentially truncated Airy function is Fourier generated. The launch angle θ is controlled by varying the transverse displacement of the imaging lens with respect to the axis of the system. This operation is equivalent to the shifting property of Fourier transforms [43]. The ballistic dynamics of these exponentially truncated Airy beams are shown in Fig. 1.15 for various launch angles [44]. The parabolic trajectories of the intensity features of these waves were monitored up to a distance of 25 cm and the wavefront tilt angle varied from -1.33 to 0.83 mrad in order to realize the three ballistic regimes discussed above. The curves A, B, C in Fig. 1.15 were obtained for $\theta = -1.33, -1.0, -0.5$ mrad respectively. As one can clearly see, for a lunch angle of $\theta = -1.33$ mrad, the Airy beam reaches its apogee at $\hat{z} = -2k^2x_0{}^3\theta \approx 9$ cm, at which point the beam deflection is $x_{d\max} = -\theta^2k^2x_0{}^3 \approx 60$ μm, in excellent agreement with our predictions. Curve D in Fig. 1.15 corresponds approximately to a zero launch angle. On the other hand, curve E, in Fig. 1.15, is obtained for $\theta = +0.83$ mrad. For the latter scenario, the acceleration displacement is further enhanced because of downward motion. The solid lines in Fig. 1.15 correspond to the theoretical curves associated with these cases.

Following the analysis of [1, 25], the motion of the center of gravity of these finite energy Airy wave packets can be studied. As usual, the intensity centroid is

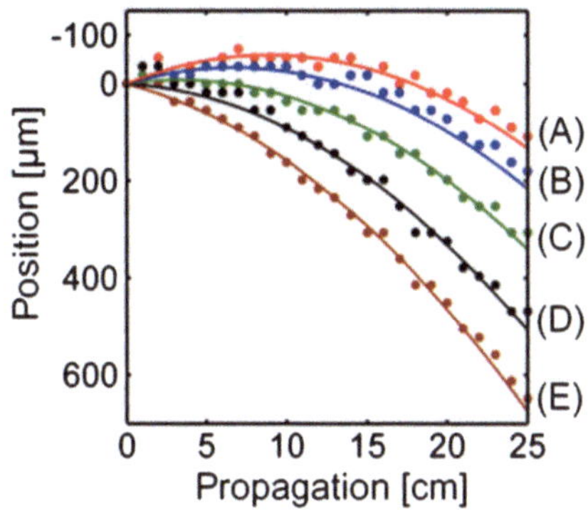

Fig. 1.15 Experimental results of Airy beam ballistics for (A) $\theta = -1.33$ mrad, (B) -1.0 mrad, (C) -0.5 mrad, (D) $+0.17$ mrad, and (E) $+0.83$ mrad

defined as $\langle s(\xi)\rangle = (1/N)\int s\left|\varphi(s,\xi)\right|^2 ds$ where in our case the constant norm is given by $N = \int\left|\varphi\right|^2 ds = (8\pi\alpha)^{-1/2}\exp(2\alpha^3/3)$. From Ehrenfest's theorem [9] it can be shown that $d\langle s\rangle/d\xi = v$, from where one can deduce that [25]:

$$\langle s\rangle = v\xi + \frac{4\alpha^3 - 1}{4\alpha}. \tag{1.19}$$

Equation (1.19) indicates that the center of gravity of these beams moves at a constant velocity (or remains invariant for $v = 0$) as a function of propagation distance. Indeed, experimental results clearly show this linear behavior for the same launch angles used in Fig. 1.15 [44]. Thus, we emphasize again that the aforementioned acceleration behavior refers to the trajectories of the local beam intensity features and is by no means in contradiction with Ehrenfest's theorem.

1.3.2 Ballistic Dynamics of 2D Airy Beams

To analyze the ballistic dynamics of 2D Airy beams, the normalized (2 + 1)D paraxial equation of diffraction is employed:

$$i\frac{\partial\varphi}{\partial z} + \frac{1}{2k}\frac{\partial^2\varphi}{\partial x^2} + \frac{1}{2k}\frac{\partial^2\varphi}{\partial y^2} = 0, \tag{1.20}$$

where φ is the electric field envelope and $k = 2\pi n/\lambda_0$ is the wavenumber of the optical wave. The evolution of a 2D finite energy accelerating Airy beam, whose field profile at the origin is given by $\varphi(x,y,z=0) = \prod_{m=x,y} \mathrm{Ai}(s_m)\exp(\alpha s_m)\exp(iv_m s_m)$, can be obtained in closed form:

$$\varphi(x,y,z) = \prod_{m=x,y} u_m(s_m,\xi_m), \tag{1.21}$$

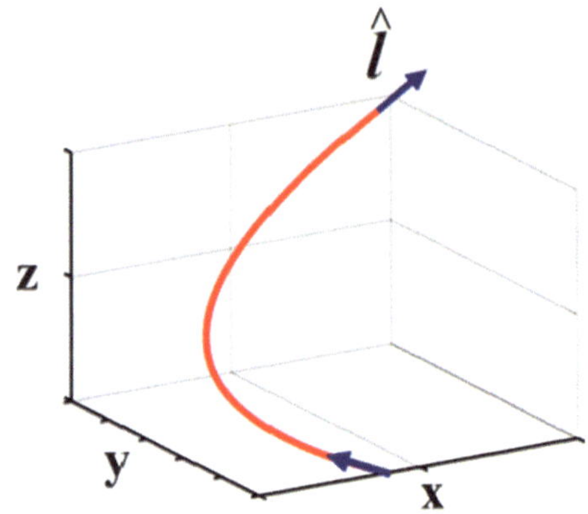

Fig. 1.16 Motion of the main lobe of a symmetric ($x_0 = y_0 = 77$ μm) 2D Airy beam when launched at $\theta_x = -2$ mrad and $\theta_y = 2$ mrad

where

$$u_m(s_m, \xi_m) = \mathrm{Ai}[s_m - (\xi_m/2)^2 - v_m\xi_m + i\alpha_m\xi_m] \exp(\alpha_m s_m - \alpha_m {\xi_m}^2/2 - \alpha_m v_m \xi_m)$$
$$\times \exp\{i[-{\xi_m}^3/12 + ({\alpha_m}^2 - {v_m}^2 + s_m)\xi_m/2 + v_m s_m - v_m{\xi_m}^2/2)]\}, \quad (1.22)$$

$\mathrm{Ai}(s_m)$ denotes the Airy function [23], $s_x = x/x_0$ and $s_y = y/y_0$ represent dimensionless transverse coordinates, with x_0, y_0 being arbitrary transverse scales, and $\xi_x = z/k{x_0}^2$ and $\xi_y = z/k{y_0}^2$ are used to normalize the propagation distance z. α_m in the exponential function is a small positive parameter associated with the effective aperture of the system, and v_m is related to the initial launch angle θ_m (or "velocity") of this beam through $\theta_m = v_m/k(x_0, y_0)$.

From (1.22), one can also directly determine the trajectory of the main ("head") lobe of the Airy beam as a function of distance. This 3D curve is given by:

$$x_d = \frac{1}{4k^2x_0^3}z^2 + \theta_x z$$
$$y_d = \frac{1}{4k^2y_0^3}z^2 + \theta_y z\,. \quad (1.23)$$

In principle this trajectory can be appropriately tailored through the magnitude and sign of the launch angles θ_m, and the scales x_0, y_0. Clearly, for zero launch angles θ_m and if $x_0 = y_0$, the main lobe of the Airy beam will move on a parabola (projected along the 45° axis in the x–y plane). On the other hand, a "boomerang-like" curve may be created if for example the "launch" angles are chosen to have opposite signs, say $\theta_x = -2$ mrad and $\theta_y = 2$ mrad (while $x_0 = y_0 = 77$ μm), as shown in Fig. 1.16. What is also very interesting is the fact that these displacements vary quadratically with the wavelength λ_0.

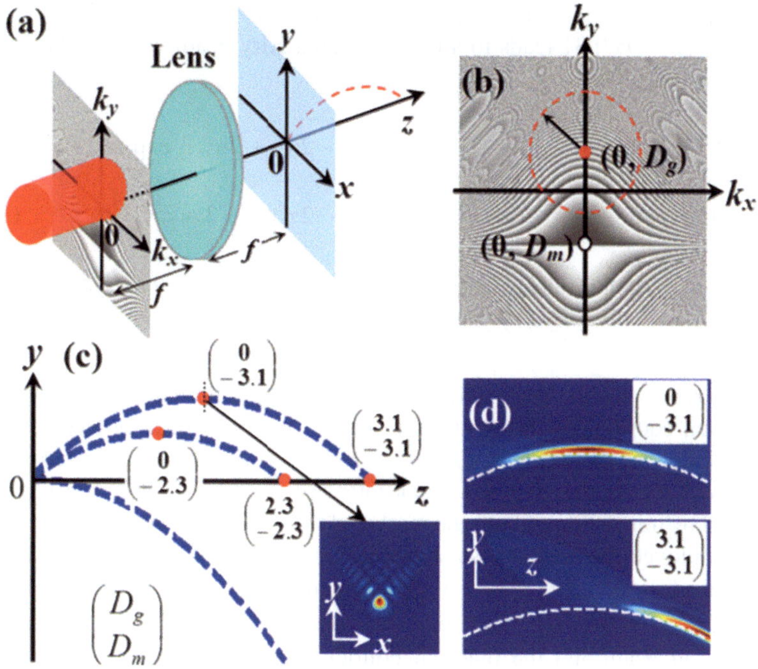

Fig. 1.17 (**a**) Schematic of input Gaussian beam, cubic phase mask, and Fourier lens used for generation of truncated Airy beam; (**b**) location of mask (center denoted by *white spot*) and input beam (marked by *red dashed circle* and center denoted by *red spot*) in Fourier plane; (**c**) illustration of different trajectories obtained at different D_g and D_m when the peak beam intensity appears at maximum heights or ranges (marked by *red spots*). The inset shows the Airy beam profile at the maximum height of upper curve. The lower curve corresponds to normal excitation at $D_g = D_m = 0$, so its peak intensity is at the starting point ($z = 0$); (**d**) Numerical simulations of beam propagation for two specific cases corresponding to the upper trajectory shown in (**c**)

1.3.3 Optimal Control of the Ballistic Motion of Airy Beams

Let us consider a typical optical system for generation of 2D Airy beams as depicted in Fig. 1.17a, where a Gaussian beam is first modulated by a cubic phase mask and then passes through a Fourier transform lens. Usually, the Gaussian beam, the mask, and the Fourier lens are set to be coaxial along z. If the lens is transversely shifted, a tilting angle will be introduced into the Airy beam. As an example, let us fix the position of the lens but allow the mask and Gaussian beam to have transverse displacements in the Fourier plane. To understand the influence of these displacements, let us first consider the 1D case. The Fourier spectrum of a truncated Airy beam can be expressed as $\exp(-\alpha w^2)\ \exp[i(w^3 - 3\alpha^2 w - i\alpha^3)/3]$, where α is a small parameter for the exponential truncation factor, and w is the normalized wave number. If the Gaussian beam and the phase mask are translated by w_g and w_m in the

Fourier plane, the resulting spectrum $\exp[-\alpha(w - w_g)^2] \exp\{i[(w - w_m)^3 - 3\alpha^2(w - w_m) - i\alpha^3]/3\}$ leads to a new truncated Airy beam with a field envelope ϕ expressed as follows:

$$\phi = Cf(s,\xi)\mathrm{Ai}[s - w_m\xi - (\xi/2)^2 + i\alpha(\xi - 2w_g + 2w_m)] \exp(iw_m s), \tag{1.24a}$$

$$C = \exp(-\alpha w_g{}^2 - \alpha w_m{}^2 + i2\alpha^2 w_m - i2\alpha^2 w_g + 2\alpha w_m w_g), \tag{1.24b}$$

$$\begin{aligned} f(s,\xi) = \exp[&\alpha s + is\xi/2 + (-iw_m{}^2/2 + i\alpha^2/2 - 2\alpha w_m + \alpha w_g)\xi \\ &+ (-\alpha/2 - iw_m/2)\xi^2 - i\xi^3/12], \end{aligned} \tag{1.24c}$$

where Ai represents the Airy function, and s and ξ are normalized transverse and longitudinal coordinates. From (1.24a), we see that the trajectory changes due to the translation of the mask as expressed by $s = w_m\xi + (\xi/2)^2$. The term $i\alpha(\xi - 2w_g + 2w_m)$ shows that the new peak-intensity position is at $\xi = 2(w_g - w_m)$, controlled by translation of both the mask and the Gaussian beam. Similar analyses can be employed for the 2D case shown in Fig. 1.17b, and the trajectory can now be expressed as $-\sqrt{2}[D_m\xi/\sqrt{2} + (\xi/2)^2]$ with a new peak-intensity position at $\xi = -\sqrt{2}D_m + \sqrt{2}D_g$ (D_g and D_m are normalized vertical displacements of the Gaussian beam and the mask in Fourier plane, respectively). Therefore, by translating the mask and Gaussian beam with respect to z-axis, the location of peak beam intensity as well as maximum height and range of the trajectory can be controlled with ease. Typical 2D numerical results are shown in Fig. 1.17c, d. For $D_g = D_m = 0$, the Airy beam propagates akin to a body projected horizontally with the peak intensity appearing at the starting point. Moving the mask to different vertical positions ($D_m \approx -2.3, -3.1$) leads to propagation of the Airy beam in a ballistic trajectory as does a batted baseball. In the case of $D_g = 0$, the peak intensity always appears at the maximum height. However, by translating also the Gaussian beam so that $D_g = -D_m$, the peak intensity appears at the maximum range ("point of fall") as demonstrated below.

Experimental results corresponding to above analyses are shown in Fig. 1.18, where a Gaussian beam is turned into a truncated Airy beam assisted with a SLM and a Fourier transform lens as shown in Fig. 1.18a. When the beam, mask, and lens are aligned coaxially, a "horizontally projected" Airy beam is generated with a decaying intensity (due to diffraction) during propagation (Fig. 1.18a). If the mask is translated slightly in vertical direction, the resulting Airy beam propagates in general ballistic trajectories with different ranges (Fig. 1.18b, d) while its peak intensity appears at the maximum heights. These different trajectories correspond to different launching angles due to the transverse displacement of the phase mask relative to the z-axis of the system. By also translating the Gaussian beam the same distance but to the opposite direction, the trajectory remains the same but the peak intensity moves to the maximum range (Fig. 1.18c, e). These experimental results agree well with our theoretical predications.

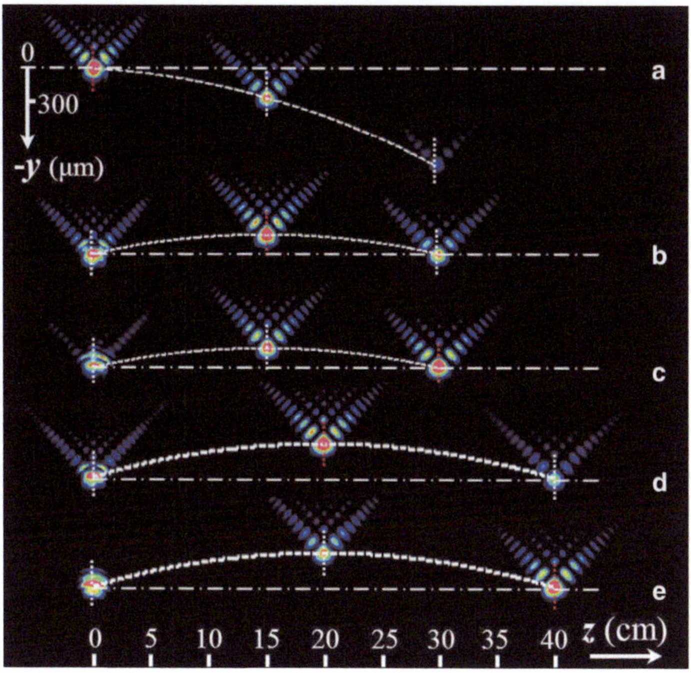

Fig. 1.18 Experimental demonstration of controlled trajectories (*white dashed curves*) of truncated Airy beams under different excitation conditions. Snapshots of transverse intensity patterns are shown at marked positions. (**a**) Normal condition when peak beam intensity is at the starting point, corresponding to lower curve in Fig. 1.17c; (**b**, **d**) peak intensity goes to the maximum height with shifting of only cubic phase mask; (**c**, **e**) peak intensity goes to the "point of fall" with additionally shifting of the Gaussian beam

If we allow both vertical and horizontal displacements of the phase mask [(D_{mx}, D_{my})] and the Gaussian beam [(D_{gx}, D_{gy})] as illustrated in Fig. 1.19a, the projectile motion of the Airy beam can be set into any arbitrary direction. Following similar theoretical analysis for Eqs. (1.24a, b, c), the (x, y) trajectory can be expressed as $\left[-D_{mx}\xi, -D_{my}\xi - \sqrt{2}(\xi/2)^2\right]$. Clearly, the Airy beam in this case undergoes uniform motion along horizontal direction while accelerating along vertical direction. As such, the Gaussian beam (even initially aiming along z-direction) can propagate to any off-axis location. The horizontal displacements of the mask and Gaussian beam will not change the location of the peak beam intensity, but they can change the Airy beam profile from symmetric (when $D_{mx} = D_{gx}$) to asymmetric (when $D_{mx} \neq D_{gx}$). An example of experimental results is shown in Fig. 1.19b, c. The peak intensity appears at the maximum height of the trajectory when the Gaussian beam is on axis ($D_{gx} = D_{gy} = 0$, shown in Fig. 1.19b) but moves to the "point of fall" when the beam is displaced vertically ($D_{gx} = 0, D_{gy} = -d$, shown in Fig. 1.19c). In this case, since $D_{mx} \neq D_{gx}$, the Airy beam starts with an asymmetric profile but evolves into a symmetric profile after restoring its peak intensity. These experimental observations are corroborated with numerical simulations [45].

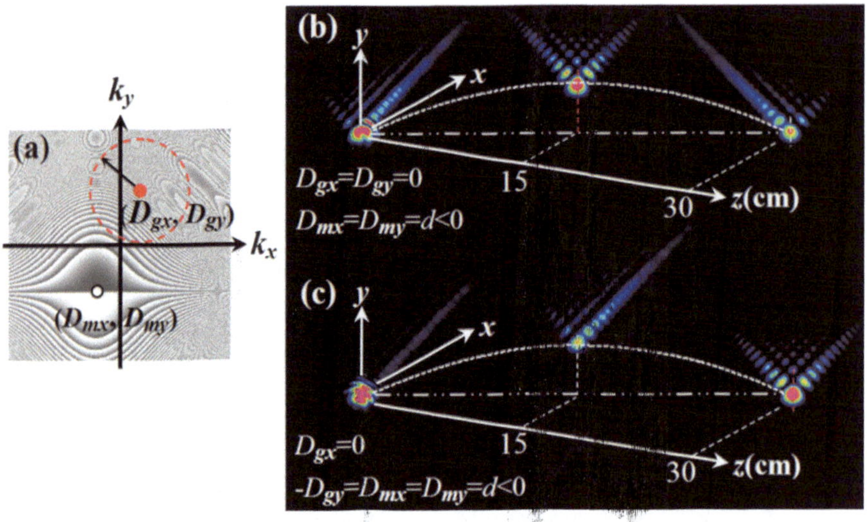

Fig. 1.19 Experimental demonstration of accelerating Airy beams with transverse uniform motion. (**a**) Relative positions of cubic phase mask and Gaussian beam in the Fourier plane; (**b**, **c**) experimental results of the trajectory and intensity pattern of the Airy beam obtained under different excitation conditions as depicted in (**a**)

1.3.4 Airy Trajectory Engineering in Dynamic Linear Index Potentials

Quite recently, we have successfully demonstrated that, with optically induced linear index potential, enhancement as well as reduction of Airy beam acceleration can be realized by changing the index gradient transversely [46]. Here we show our design of linear longitudinal (z-axis) index potential to engineer the trajectory that the Airy beam follows. In particular, let us consider the paraxial dynamics of an optical field

$$iu_z + \frac{1}{2}u_{xx} - \frac{d(z)x}{2}u = 0, \tag{1.25}$$

where $d(z)x/2$ is the transversely linear index potential with a gradient $d(z)$. In the Fourier space (1.25) becomes

$$\tilde{u}_z - \frac{d(z)}{2}\tilde{u}_k = -i\frac{k^2}{2}\tilde{u}, \tag{1.26}$$

which is equivalent to the characteristic system

$$\frac{dz}{1} = \frac{dk}{-d(z)/2} = \frac{d\tilde{u}}{-ik^2\tilde{u}/2}. \tag{1.27}$$

By integrating the above system we obtain the integral formula

$$u(x,z) = \frac{1}{2\pi}\int_{-\infty}^{\infty} \tilde{u}(k,0)\mathrm{e}^{-(i/2)\int_0^z \kappa(s)^2 \mathrm{d}s}\mathrm{e}^{i\kappa(z)x}\mathrm{d}k, \tag{1.28}$$

where $\kappa(z) = k - D(z)/2$, $D(z) = \int_0^z d(s)\mathrm{d}s$. We focus our attention on Airy type initial conditions

$$u(x,z=0) = \mathrm{Ai}(\gamma^{1/3}x), \tag{1.29}$$

where γ is the width of the Airy wave. By substituting the Fourier transform of (1.29)

$$\tilde{u}(k,z=0) = \frac{1}{\gamma^{1/3}}\exp\left[\frac{ik^3}{3\gamma}\right] \tag{1.30}$$

into (1.28) we find that

$$u(x,z) = \mathrm{e}^{\phi(x,z)}\mathrm{Ai}(\gamma^{1/3}\mu(x,z)), \tag{1.31}$$

where

$$\mu(x,z) = \frac{F_1(z)}{2} + x - \frac{\gamma z^2}{4}, \tag{1.32}$$

$$\phi = i\frac{\gamma F_1(z)z}{4} - i\frac{D(z)x}{2} - i\frac{F_2(z)}{8} + i\frac{\gamma z x}{2} - i\frac{\gamma^2 z^3}{12}, \tag{1.33}$$

and

$$F_1(z) = \int_0^z D(s)\mathrm{d}s, \quad F_2(z) = \int_0^z D^2(s)\mathrm{d}s. \tag{1.34}$$

We consider that the trajectory of the Airy beam is provided by $\mu(x,z) = 0$, i.e.,

$$x = -\frac{F_1(z)}{2} + \frac{\gamma z^2}{4}, \tag{1.35}$$

which mainly determines caustic trajectory. By generalizing the initial condition as

$$u(x,z=0) = \mathrm{Ai}(\gamma^{1/3}(x - x_0))\mathrm{e}^{i\Psi x}, \tag{1.36}$$

in which case x_0 is the initial spatial displacement and ψ is the initial tilt, the trajectory of the beam becomes

$$x = x_0 + \Psi z - \frac{F_1(z)}{2} + \frac{\gamma z^2}{4}. \tag{1.37}$$

More importantly, we can also solve the inverse problem: given a predefined Airy beam trajectory $x = g(z)$ we can determine the index gradient $d(z)$ as well as the initial condition parameters x_0, ψ that are required. Following the relevant algebra we find that

$$x_0 = g(0), \quad \Psi = g'(0), \quad d(z) = \gamma - 2g''(z). \tag{1.38}$$

In a similar fashion one can derive expressions for the dynamics of exponentially and Gaussian apodized Airy beams [47]. In particular, the trajectory of the exponentially apodized Airy beam is essentially the same as compared to the "pure" Airy beam, whereas the effective trajectory is modified in the case of a Gaussian apodization. As an example, in Fig. 1.20 we see an exponentially apodized Airy beam following different trajectories. In all cases, the potential gradient and initial condition parameters are determined by utilizing (1.38).

1.4 Self-healing of Airy Beams

1.4.1 Self-reconstructing Optical Airy Beams

Perhaps one of the most remarkable properties of any diffraction-free beam is its very ability to self-reconstruct during propagation. This characteristic is of particular importance when such waves propagate in inhomogeneous media [48]. The question naturally arises whether Airy beams can self-heal and to what extent? If so, how does this process take place and how is it affected by the beam's acceleration dynamics? For example, can an Airy beam negotiate adverse environments? In this section we review the self-healing properties of optical Airy beams. We show that this family of waves exhibits remarkable resilience against perturbations and tends to reform during propagation.

1.4.2 Babinet's Principle for a Nondiffracting Beam

The self-healing properties of a nondiffracting field configuration, when it is partially blocked by a finite opaque obstacle at $z = 0$, can be explained from Babinet's principle [49]. If the nondiffracting input field is disturbed by a finite energy perturbation $\varepsilon(x, y)$, i.e., $\varphi(x, y, z = 0) = U_{ND}(x, y, z = 0) - \varepsilon(x, y, z = 0)$, then

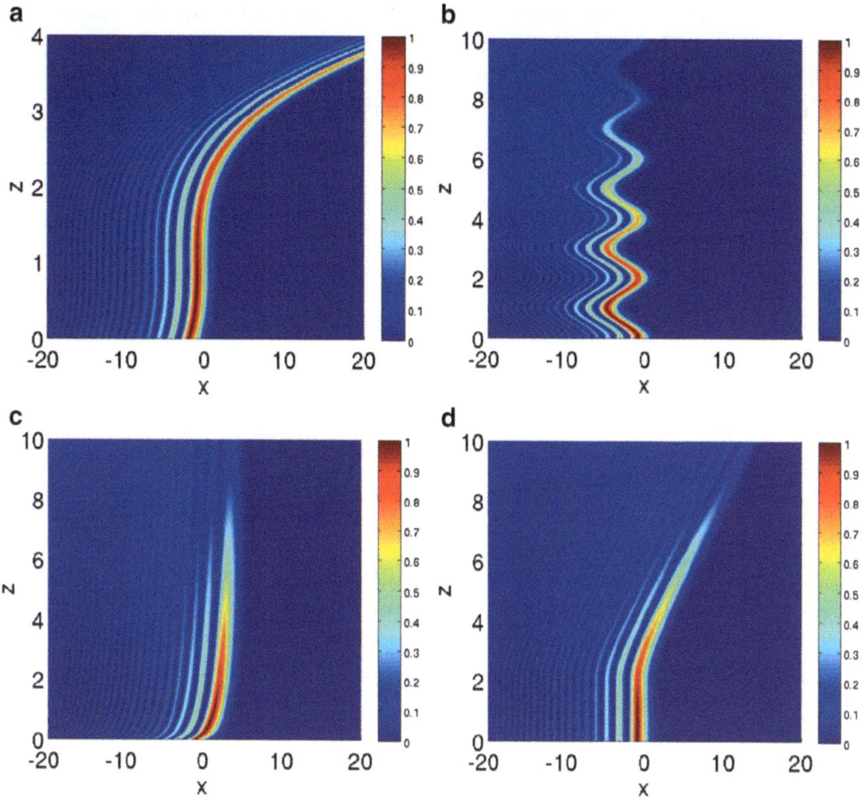

Fig. 1.20 Exponentially apodized Airy beam propagation ($\alpha = 0.05$) along predefined paths. (**a**) Power law trajectory $x = (z-1)^3$; (**b**) sinusoidal trajectory given by $x = g(z) = 2[\cos(\pi z) - 1]$; (**c**) logarithmic trajectory $x = \log(1 + 10\,z)$; (**d**) hyperbolic type trajectory given by $x = g(z) = 2[\sqrt{(z-2)^2+1} - 1]H(z-2)$ where $H(z)$ is the Heaviside step function

from (1.1) one finds that $i\varepsilon_z + (1/2k)\nabla_\perp^2\varepsilon = 0$. As a result the perturbation ε is expected to rapidly diffract as opposed to the nondiffracting beam that remains invariant during propagation. As a consequence, at large distances $|\varphi(x,y,z)|^2 = |U_{ND}(x,y,z)|^2$, and hence the nondiffracting beam reforms during propagation. This argument holds for all nondiffracting fields including the accelerating Airy beam.

1.4.3 Transverse Power Flow of an Optical Airy Beam

Of relevance to our discussion is the Poynting vector $\vec{S}$ associated with Airy optical beams. In the paraxial regime, $\vec{S}$ is given by [50]:

$$\vec{S} = \vec{S}_z + \vec{S}_\perp = \frac{1}{2\eta_0}|\varphi|^2\hat{z} + \frac{i}{4\eta_0 k}[\varphi\nabla_\perp\varphi^* - \varphi^*\nabla_\perp\varphi], \tag{1.39}$$

where $\eta_0 = \sqrt{\mu_0/\varepsilon_0}$ is the impedance of free space. $\vec{S_z}$ denotes the longitudinal component of the Poynting vector whereas $\vec{S}_\perp$ the transverse. From Eqs. (1.23) and (1.39) one can directly obtain the direction of the Poynting vector associated with an ideal 2D Airy ($\alpha_m = 0$) beam. More specifically, the angle ψ the projection of $\vec{S}$ makes with respect to x axis is given by:

$$\tan \Psi = \frac{S_y}{S_x} = \frac{\theta_y + \frac{z}{2k^2 y_0^3}}{\theta_x + \frac{z}{2k^2 x_0^3}}. \tag{1.40}$$

On the other hand, the direction of $\vec{S}$ relative to the z axis is given by:

$$\tan \delta = \frac{\sqrt{S_x^2 + S_y^2}}{S_z} = \sqrt{\left(\theta_x + \frac{z}{2k^2 x_0^3}\right)^2 + \left(\theta_y + \frac{z}{2k^2 y_0^3}\right)^2}. \tag{1.41}$$

Note that for ideal Airy beams, the Poynting vector $\vec{S}$ is at every point parallel to the unit tangent vector $\hat{l}$ of the trajectory curve of (1.23). This statement is also valid for finite energy Airy beams during the quasi-diffraction free stage of propagation. At larger distances, however, small deviations are expected to occur. In addition one can show that the polarization of the beam can evolve in a similar manner.

The reconstruction of an accelerating optical Airy beam can be monitored through the transverse component of the Poynting vector $\vec{S}_\perp$ [51].

1.4.4 Observation of Self-healing Properties of 2D Airy Beams

In order to study experimentally the self-healing properties of a 2D Airy beam, the Airy pattern is partially blocked in a controlled manner with a rectangular opaque obstacle inserted at the desired location, and then monitored for their self-reconstruction during propagation. In all cases we block a portion of its initial intensity profile. The most prominent intensity characteristic of an Airy beam happens to be its main corner lobe (as seen in Fig. 1.9a) which contains a large percentage of the beam's total power. In a first experiment, an opaque rectangular obstacle was employed to obstruct the corner main lobe of the Airy pattern (Fig. 1.21a). The FWHM of the blocked lobe feature was approximately 130 μm corresponding to $x_0 = y_0 = 77$ μm and $\alpha = 0.08$. Figure 1.21b depicts the reformation of this Airy beam after a distance of $z = 11$ cm. The self-healing of this beam is apparent. The main lobe is reborn at the corner and persists undistorted up to a distance of 30 cm (Fig. 1.21c). In our setup, the latter distance (30 cm) corresponds approximately to four diffraction lengths of the corner lobe. Our experimental observations are in excellent agreement with numerical results presented in Fig. 1.21d–f for the same propagation distances.

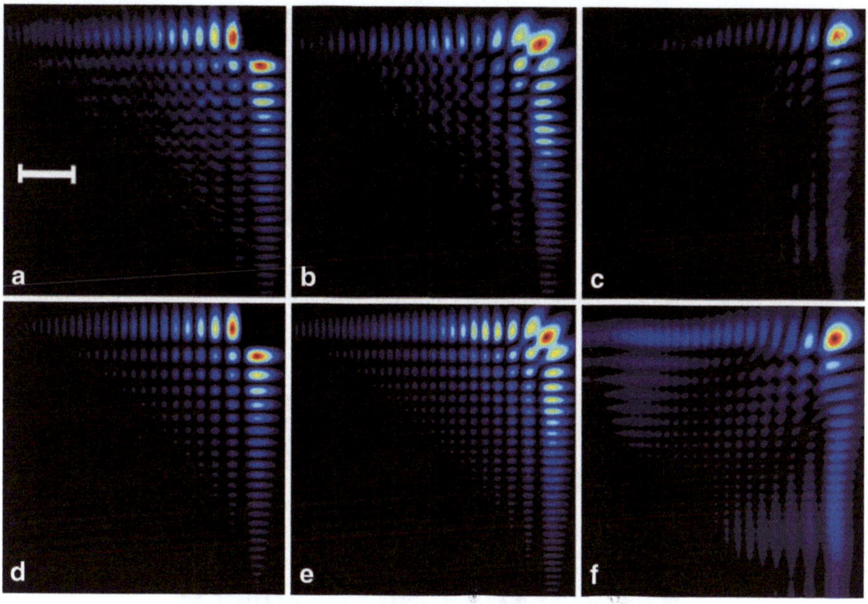

Fig. 1.21 Self-healing of an Airy beam when its main lobe is blocked. Observed intensity profile at (**a**) the input $z = 0$, (**b**) $z = 11$ cm, and (**c**) $z = 30$ cm. The corresponding numerical simulations are shown in (**d–f**)

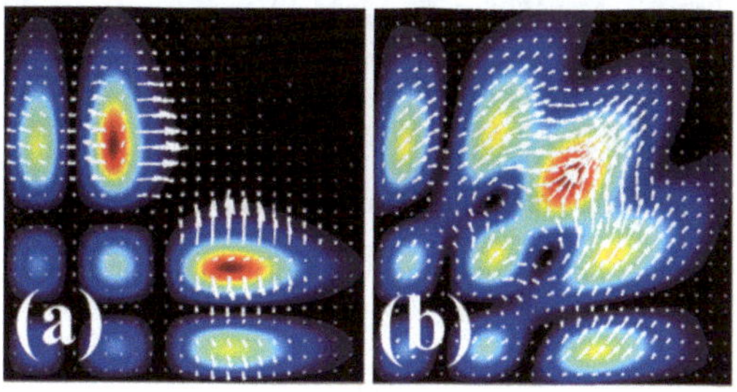

Fig. 1.22 Calculated transverse power flow $\vec{S}_{\perp}$ at (**a**) $z = 1$ cm and (**b**) $z = 11$ cm

We note that had the main lobe been launched in isolation it would have experienced a fivefold increase in the beam width over the same propagation distance, while the peak intensity would have dropped to 5% of its initial value. This is another manifestation of the nondiffracting nature of Airy beams.

In order to understand this self-healing process it is important to study the internal transverse power flow $\vec{S}_{\perp}$ within the perturbed Airy beam. To do so we use the result of (1.39). Figure 1.22a depicts the transverse flow within the Airy beam at $z = 1$ cm when the main lobe has been removed. Evidently the power flows from the side lobes towards the corner in order to facilitate self-healing. On the other hand, once

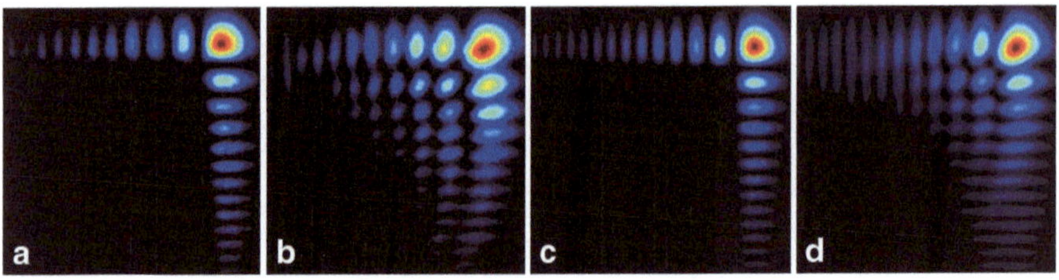

Fig. 1.23 Self-healing of an Airy beam when all the internal lobes are blocked. Observed intensity profiles at (**a**) the input $z = 0$ and (**b**) $z = 16$ cm. The corresponding numerical simulations are shown in (**c**) and (**d**)

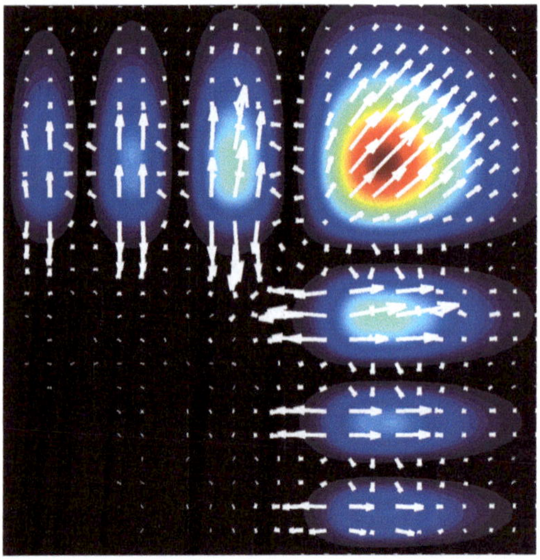

Fig. 1.24 Transverse power flow $\vec{S}_{\perp}$ revealing the self-healing mechanism at $z = 1$ cm for the incomplete Airy beam shown in Fig. 1.23

reconstruction has been reached (at $z = 11$ cm), then the internal power density around the newly formed main lobe flows along the 45° axis in the x–y plane (for $x_0 = y_0$) in order to enable the acceleration dynamics of the Airy beam (Fig. 1.22b).

So far we have experimentally demonstrated that an Airy beam can reconstruct itself when its main lobe has been blocked. It is of interest to examine whether the beam could self-heal even after more severe perturbations. In a second set of experiments we have totally blocked all the internal structure (all inner lobes) of the Airy pattern (Fig. 1.23a). Remarkably after $z = 16$ cm of propagation the beam self-heals and reconstructs in detail its fine intensity structure as depicted in Fig. 1.23b. Figure 1.23c, d shows the corresponding calculated intensity profiles for these same distances.

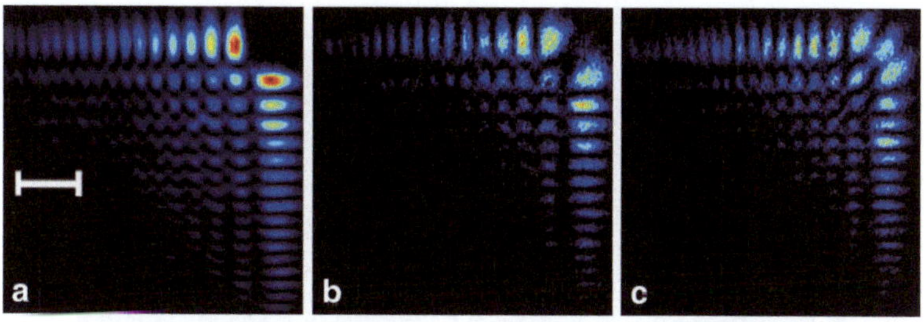

Fig. 1.25 Self-healing of an Airy beam when propagating in a suspension of 0.5 μm silica microspheres in pure water. Observed intensity profiles at (**a**) the input $z = 0$, (**b**) $z = 5$ cm, and (**c**) $z = 10$ cm

The internal power flow during the latter self-healing process is shown in Fig. 1.24. At $z = 1$ cm, the Poynting vector provides energy towards the blocked region for rebirth to occur while on the main lobe is directed along 45° in the x–y plane in order to enable the self-bending of the Airy beam.

In addition, we have also demonstrated experimentally that an Airy beam can reconstruct itself when a nonsymmetric obstruction is used. This asymmetric perturbation was carried out by blocking, for example, the first three lobes of an Airy wave packet along the y axis. Interestingly, in this physical setting, the beam not only self-heals itself but also the initially blocked part is reborn even brighter when compared to its surroundings. This is a clear manifestation of the nondiffracting character of the Airy beam [51]. In addition, this self-regeneration property can be improved by dual Airy beams [52].

1.4.5 Self-healing of Optical Airy Beams in Scattering Environments

In the previous section we have demonstrated that optical Airy beams are remarkably resilient to amplitude deformations when propagating in free space. The question is: are such self-healing Airy wave packets also robust in adverse environments? To address this question we have experimentally studied the propagation of Airy beams in scattering and turbulent media.

In order to study the self-healing dynamics of Airy beams in scattering media we have again blocked their main corner lobe (Fig. 1.25a). To do so we have prepared two different samples of silica microspheres ($n = 1.45$) suspended in pure water ($n = 1.33$). The size of the dielectric microparticles was 0.5 and 1.5 μm in diameter and thus light scattering was predominantly of the Mie type [53]. Both suspensions were 0.2% in weight concentration while the volume filling factor was 0.1%. We have ultrasonicated the prepared mixtures for 1 h, to make sure that the silica particles were monodispersedly suspended in water. The scattering cross section of the microspheres is estimated to be 0.055 and 3.76 μm^2 [54] for the small and large

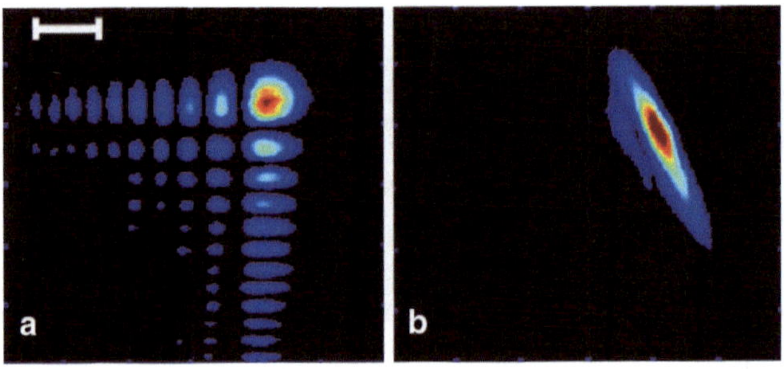

Fig. 1.26 Propagation in a turbulent medium of (**a**) an optical 2D Airy beam and (**b**) a comparable Gaussian beam

particles, respectively. These values lead to significant light scattering, enough to give a granular appearance when the beam propagates 5 cm in the water–silica mixture (diameter of 0.5 μm) (Fig. 1.25b). A longer (10 cm) cell was used to observe the complete reformation of the Airy pattern in the same scattering media. Figure 1.25c depicts the self-healing of an Airy beam after propagating 10 cm in the same environment. Besides the anticipated drop in the beam intensity due to Mie scattering, the beam still exhibits in every respect its characteristic pattern.

1.4.6 Resilient Airy Beam Propagation in a Turbulent Medium

We have also studied the effect of turbulence on an Airy beam. The turbulent environment was realized over a heated rough accordion-shaped aluminum foil above which violent heat convection air currents were generated. The turbulence was controlled by adjusting the temperature of the hotplate around 300°F. The Airy beam was then passed right above the aluminum foil up to a distance of 8 cm. In all our experiments the resilience of the Airy beam (without any initial amplitude distortions) against turbulence was remarkable (see Fig. 1.26a and the related video file in [55]). To some extent this robustness can be qualitatively understood if one considers the phase structure of the beam: alternations in phase between 0's and π's result in zero-intensity regions and these singularities can be in turn extremely stable [56, 57]. For comparison purposes we turned off the cubic phase from the SLM, thus producing a comparable Gaussian beam. This diffracting Gaussian beam was then passed through the same turbulent system. Unlike the Airy beam, the Gaussian beam was strongly deformed and it suffered massive distortions (see Fig. 1.26b and the related video file in [58]).

Another experiment we did for self-healing of an Airy beam is to reposition the peak beam intensity to a target even through disordered media using the linear control method discussed in Sect. 1.3.3. This is illustrated in Fig. 1.27a, where the peak intensity of a truncated Airy beam is supposed to land on a target located at $(x, y, z) = (0, 0, 25\text{ cm})$ along a curved trajectory (red dashed curve) after

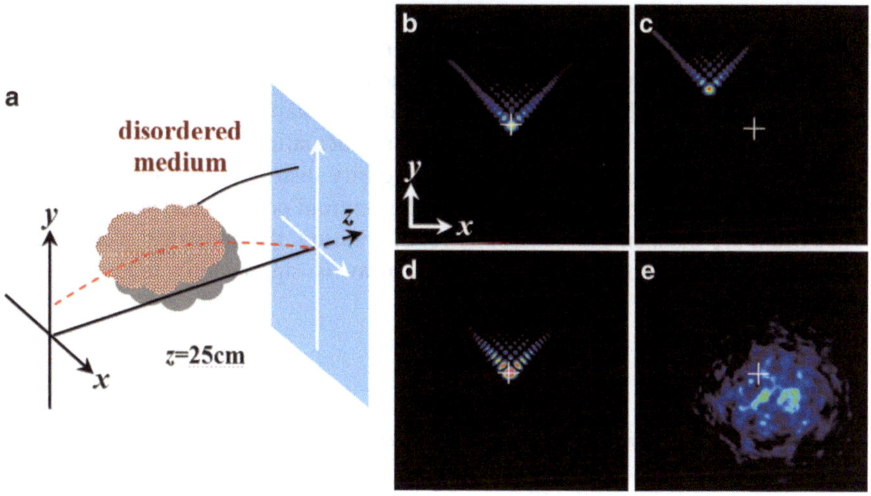

Fig. 1.27 (**a**) Schematic of Airy beam propagation through a disordered medium. The *red dashed* (*black solid*) *curve* depicts the trajectory in free-space (disordered medium); (**b**, **c**) intensity pattern of output Airy beam at $z = 25$ cm through air (**b**) and stirred salt–water mixture (**c**); (**d**) restoration of the Airy beam peak intensity at the target after translating the phase mask and input Gaussian beam; (**e**) typical output pattern of a Gaussian beam from the salt–water mixture. The *white cross* corresponds to the target point at (0, 0, 25 cm)

passing through a disordered medium. However, due to presence of the disordered media, the Airy beam path (black solid curve) is deflected off the target apart from diminishing intensity during propagation. Simply by translating the phase mask and the initial Gaussian beam, the restored peak intensity can be repositioned at the target. Corresponding experimental results obtained with a turbulent salt–water mixture are shown in Fig. 1.27b–e. First, we "aim" the Airy beam at the target after 25 cm of propagation through air. Then, salt is added and stirred in water placed in the beam path. Although the Airy beam is recovered through disordered scatters due to its self-healing property, its position in the target plane is shifted dramatically (Fig. 1.27c). The large lateral shift of the Airy beam path from Fig. 1.27b, c is caused mainly by refraction from the salt–water mixture (which has a refractive index different than that of air), while small variation of the Airy beam in its shape and location in a given output plane occurs due to turbulence of stirred mixture. By translating the mask and the Gaussian beam independently, as expected, not only the Airy beam comes back to the target, but also its peak intensity is restored (Fig. 1.27d). We emphasize that Fig. 1.27c, d were taken as a snapshot to show one example of the "fluctuating" pattern, as the shape and transverse position of the self-healing Airy beam vary slightly with time. However, the average intensity pattern is a well-defined Airy beam with its peak intensity repositioned at the target. For comparison, keeping all conditions unchanged as for Fig. 1.27d except for changing the cubic phase into uniform phase in the SLM, the Airy beam returns to a normal Gaussian beam which is severely scattered, deformed and shifted after propagating through the same salt–water mixture (Fig. 1.27e). These results suggest that Airy beams are excellent candidates for beam reposition to a given target through disordered or turbulent media, perhaps even with a feed-back system that could compensate time-dependent fluctuation.

In detail, the evolution of Airy beam intensity distribution in turbulence was examined in [59], where results show that the centroid position and skewness of an Airy beam are independent of turbulence. In addition, the numerical work in [60] found that scintillation of an Airy beam array is significantly reduced in the turbulent atmosphere.

1.4.7 Restoration and Degeneration of Deformed Airy Beams

Previous sections investigated reconstruction of either perfect or imperfect (with missing lobes) Airy beams. In this section, we discuss the self-healing of deformed Airy beams, in which none of the Airy lobes are blocked but the angle between the two "wings" differs from the regular 90° for the perfect 2D Airy beams. The wave function of a deformed Airy beam studied here can be expressed by

$$U = \mathrm{Ai}(X/x_0)\exp(\alpha X/x_0)\mathrm{Ai}(Y/x_0)\exp(\alpha Y/x_0) \tag{1.42}$$

where X and Y are respectively equivalent to $(-\sqrt{r}x - y/\sqrt{r})/\sqrt{2}$ and $(\sqrt{r}x - y/\sqrt{r})/\sqrt{2}$, x_0 is a constant governing the size of the Airy beam, and r is the parameter determining the degree of deformation of the Airy beam. $r = 1$ corresponds to the normal case for which the angle between the two "wings" of the Airy beam is equal to 90° (Fig. 1.28a). After 20 cm of linear propagation, this regular Airy beam remains nearly diffraction free. For $r < 1$, the angle between the two wings of a deformed Airy beam is obtuse as depicted in Fig. 1.28b, c. In this case, the wings shrink gradually towards the 90° position during propagation, with a slight increase of its transverse acceleration (see Fig. 1.28b, c, bottom). Inversely, for $r > 1$, the angle between the two wings of a deformed Airy beam is acute as depicted in Fig. 1.28d, e, so the wings expand gradually towards the 90° position with a slight decrease of its acceleration. Interestingly, when the angle deviation from 90° is not too big, the deformed Airy beam seems to restore well during propagation, but at large deviations, the angle can change from obtuse to acute and vice verse. In particular, it is found that a strongly deformed 2D Airy beam with a small acute angle degenerates into a quasi-1D Airy beam, as can be seen from Fig. 1.28e. This was confirmed by more detailed simulations.

To perform experimental demonstration, a cubic phase mask is utilized as before (see inset in Fig. 1.29a). After 20 cm of propagation in free space, the Airy beam shifts vertically due to its transverse acceleration. Now we modify the vertical–horizontal length *ratio* of the phase mask, which controls the angle between the two wings.

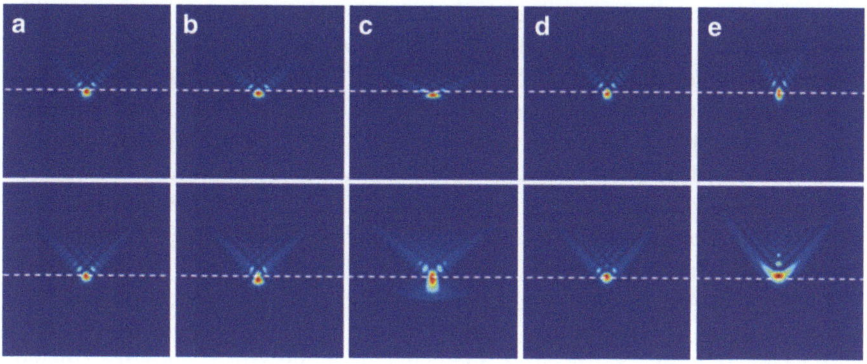

Fig. 1.28 Numerical simulations of linear propagation of deformed Airy beams. *Top* and *bottom panels* correspond to input and output after 20-cm propagation, respectively. From left to right, $r = 1, 0.89, 0.67, 1.11$, and 1.33. (**a**) A regular Airy beam at a right angle between the two wings, (**b**, **c**) deformed Airy beams at an obtuse angle, and (**d**, **e**) deformed Airy beams at an acute angle. *Dashed lines* mark the position of peak intensity of the regular Airy beam

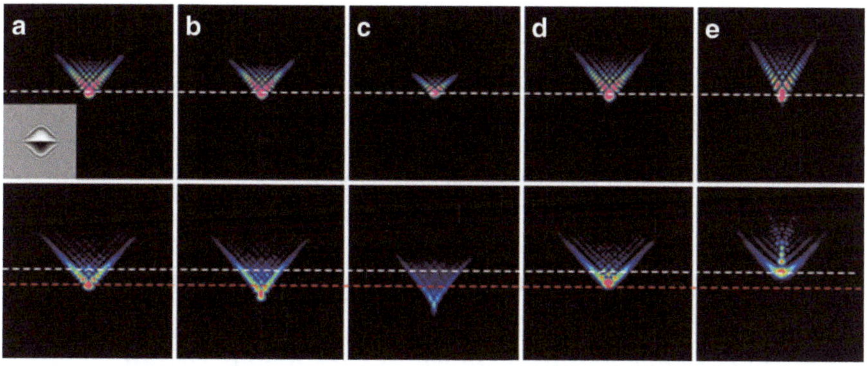

Fig. 1.29 Experimental observation of linear propagation of deformed Airy beams corresponding to Fig. 1.28. *Top* and *bottom panels* correspond to input and output after 20 cm propagation, respectively. The values of the parameter r are the same as in Fig. 1.28. The inset in (**a**) shows the unstretched cubic phase mask. The *white dashed lines* mark the position of peak intensity of the Airy beam at $z = 0$ cm, and the *red line marks* the position of peak intensity of the regular Airy beam at $z = 20$ cm (b, c) deformed Airy beams at an obtuse angle, and (d, e) deformed Airy beams at an acute angle.

We observe that the Airy beam indeed gradually opens its wings (Fig. 1.29b, c, top) when the ratio of the mask is reduced (i.e., the mask is stretched horizontally). After 20 cm propagation, the wings contract towards 90° position (Fig. 1.29b, c, bottom), in agreement with simulation. If the ratio is too small, the nondiffraction property of the Airy beam cannot persist, and the Airy beam cannot be restored as seen in Fig. 1.29c. Likewise, when the ratio of the mask is increased (i.e., the mask is stretched vertically), the wings expand and the angle of the Airy beam increases from an acute angle towards the 90° position (Fig. 1.29d, e). Apart from the wing flipping, another two interesting observations merit mentioning: firstly, as compared to the perfect Airy beam, increased acceleration is observed for "expanded" Airy beam (Fig. 1.29b, c), whereas decreased acceleration for "compressed" Airy beam (Fig. 1.29d, e). The larger change in experiment as opposed to that in simulation is due to that the area of the phase mask (hence, the size of the deformed Airy beam) is not kept constant as the mask is stretched; secondly, when the Airy beam is strongly deformed at a very small acute angle, the 2D Airy beam cannot maintain, but rather degenerates into a quasi-1D Airy beam as seen from the bottom panel of Fig. 1.29e. The experimental results agree well with our numerical simulations.

In addition, Airy beams without standard shapes are also investigated in other aspects, such as, combining Airy beams with vortex [61, 62], reducing side lobs of Airy beams [63], etc. Self-healing dominates the propagation dynamics although some other interesting phenomena associated with these nonconventional Airy beams were discussed.

In summary, we have demonstrated both theoretically and experimentally the self-healing properties of optical Airy beams. By monitoring their internal transverse power flow we have provided insight concerning the self-healing mechanism of Airy beams. We have also experimentally shown that these optical beams can be robust in adverse environments such as in scattering and turbulent media. Furthermore, we have demonstrated that an Airy beam can retain its shape and structure under turbulent conditions as opposed to a comparable Gaussian beam that suffers from massive distortion. The robust nature of Airy beams may have important implications in other areas such as atmospheric propagation and microparticle manipulation as we shall discuss in Sect. 1.6.

1.5 Nonlinear Control and Generation of Airy Beams

1.5.1 Persistence and Breakdown of Airy Beams Driven by an Initial Nonlinearity

In this section, we study the transition of Airy beams from a nonlinear to a linear medium driven initially by a self-focusing or -defocusing nonlinearity. Some unique behaviors of such nonlinearity-controlled Airy beams, including loss or persistence of acceleration, normal or anomalous diffraction were observed. In particular, an Airy beam under an initial self-defocusing nonlinearity exhibits anomalous diffraction and propagates robustly over long distance after exiting the nonlinear medium, but it breaks down in both Airy-beam pattern and acceleration when driven by a self-focusing nonlinearity [64].

The experiments were performed in a biased 1 cm-long photorefractive SBN:60 crystal (Fig. 1.30a). As before, to create a truncated Airy beam, a SLM is placed at the focal plane of the Fourier transform lens. The Airy beam ($\lambda = 532$ nm) is extraordinarily polarized, propagating first through the biased crystal under the influence of photorefractive screening nonlinearity, and then through air (free-space) for

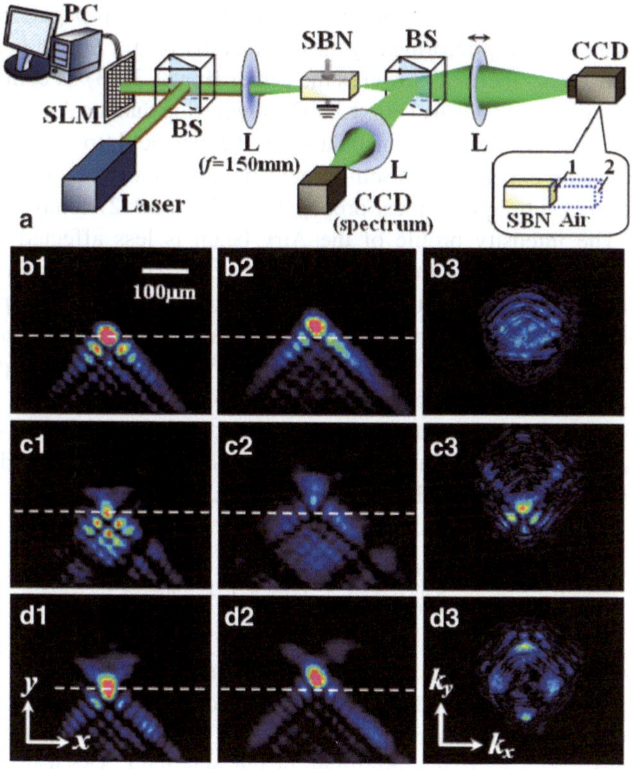

Fig. 1.30 (**a**) Schematic of experimental setup. *SLM* spatial light modulator, *SBN* strontium–barium niobate crystal. (**b–d**) Output intensity patterns of an Airy beam after 1 cm through crystal (1st column) plus another 1 cm through air (2nd column) when (**b**) no nonlinearity, (**c**) self-focusing and (**d**) self-defocusing nonlinearity is present. *White dashed line* marks the "head" position of the Airy beam at crystal output. The 3rd column shows Fourier spectra of the Airy beam corresponding to the 1st column)

another 1 cm. Solely by switching the polarity of the bias field, self-focusing and self-defocusing nonlinearity is achieved for nonlinear control of the Airy beam. The Airy beam patterns along with $\boldsymbol{k}$-space spectra are monitored by CCD cameras.

Typical experimental results are shown in Fig. 1.30b–d. When no bias field is present, the Airy beam undergoes linear propagation inside the crystal. (The photorefractive diffusion effect [65] can be neglected due to the large size of the Airy beam used here—about 50 μm for the main lobe). After another 1 cm of propagation in air, its main spot (or "head") is shifted along the vertical direction (Fig. 1.30b2) in comparison with that right at the existing face of the crystal (Fig. 1.30b1) due to the transverse acceleration. When a positive dc field of 4×10^4 V/m is applied, the Airy beam experiences a self-focusing nonlinearity and reduces its overall size with most of its energy distributed to the four spots close to the Airy head (Fig. 1.30c1). In this case, the nonlinearity seems to cause stagnation of Airy beam's acceleration, and the subsequent free-space propagation shows that the Airy beam is strongly deformed by the nonlinearity (Fig. 1.30c2). In addition, its **k**-space spectrum is "focused" towards the center (Fig. 1.30c3) as compared to the case without initial nonlinearity (Fig. 1.30b3), suggesting that the Airy beam exhibits normal diffraction. By reversing the polarity of the bias field (to -4×10^4 V/m) so that the Airy beam experiences a self-defocusing nonlinearity, its nonlinear output (Fig. 1.30d1) and subsequent linear propagation (Fig. 1.30d2) behave dramatically differently. The intensity profile of the Airy beam is less affected by the self-defocusing nonlinearity, and the peak intensity of the Airy beam after subsequent linear propagation in air is not decreased but rather increased while persistent in its acceleration. Furthermore, the Fourier spectrum reshapes into four major spots in **k**-space as shown in Fig. 1.30d3, resembling the Broullion zone (BZ) spectrum and associated anomalous diffraction behavior in photonic lattices [66–68].

These experimental observations were corroborated with numerical simulations. Propagation of an Airy beam in a biased photorefractive crystal can be described by the following nonlinear Schrödinger equation

$$\frac{\partial U}{\partial z} = \frac{i}{2k_0 n_0}\left(\frac{\partial^2 U}{\partial x^2} + \frac{\partial^2 U}{\partial y^2}\right) + ik_0 \Delta n U, \tag{1.43}$$

where U is the wave function, k_0 is the vacuum wave vector, and $n_0 = 2.3$ is the unperturbed refractive index. In the biased crystal, the nonlinearity for an e-polarized beam can be determined by $\Delta n = -0.5 n_0^3 \gamma_{33} E_0/(1 + |U|^2)$, in which $\gamma_{33} = 280$ pm/V and E_0 is the amplitude of the bias field. The wave function of an input Airy beam can be expressed as $U(x, y, z = 0) = U_0 \mathrm{Ai}(X/x_0) \exp(\alpha X/x_0)$ $\mathrm{Ai}(Y/y_0) \exp(\alpha Y/y_0)$, where U_0 is the amplitude, Ai denotes the Airy function, X and Y are respectively equivalent to $(x + y)/\sqrt{2}$ and $(-x + y)/\sqrt{2}$, x_0 and y_0 are constants governing the transverse size of the Airy beam, and again α is the decay factor for the truncated beam profile.

Numerical simulations were performed by solving (1.43) with the split-step beam propagation method (parameters U_0, x_0, and α are chosen as 7.3, 13.5×10^{-6}, and 0.11, respectively). Numerical results corresponding to Fig. 1.30(c1–c3) under the self-focusing initial nonlinearity ($E_0 = +40$ kV/m) are shown in Fig. 1.31a–c. The propagation can be better visualized from the side-view evolution as shown in Fig. 1.31d, where the dashed curve marks the path of the same Airy beam without initial nonlinearity. Clearly, the acceleration is reduced or lost as compared to the case without the nonlinear control. In Fig. 1.31e–f, we plotted the transverse energy flow of the output beam corresponding to the areas marked in Fig. 1.31a, b. Apparently, after initial nonlinear propagation, the direction of the energy flow goes towards all directions, suggesting that the phase (and thus the acceleration) of the Airy beam is destroyed by the self-focusing nonlinearity. This might be due to that nonlinear trapping and interaction among waveguides induced by the lobes of the Airy beam destructed its phase relation.

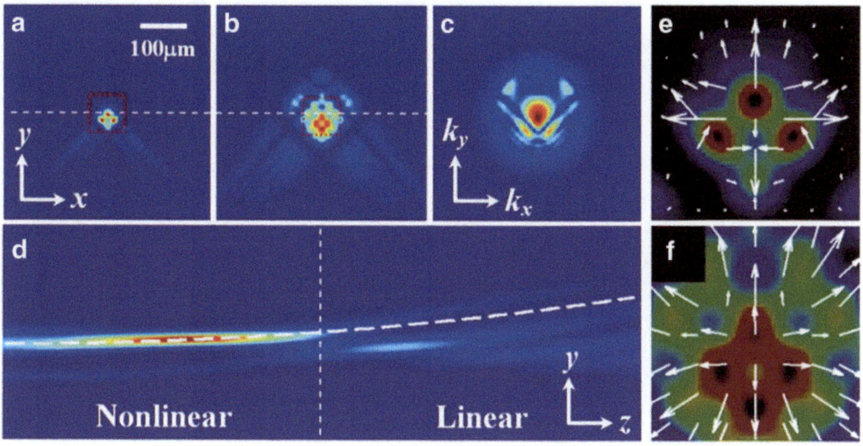

Fig. 1.31 Numerical simulation of an Airy beam propagating under an initial self-focusing nonlinearity. (**a**, **b**) Transverse intensity patterns after (**a**) 1 cm through crystal plus (**b**) another 1 cm through air. (**c**) Fourier spectrum of the output Airy beam. (**d**) Side-view of 2 cm-propagation, where the *dashed curve* represents the trajectory of the Airy beam without initial nonlinearity. (**e**, **f**) Calculated transverse energy flow around the main lobe corresponding to the *square area* shown in (**a**, **b**), respectively

Once the Airy beam is released into free-space, it behaves more like a confined Gaussian beam, showing normal diffraction without evident acceleration.

With a reversed bias field of $E_0 = -40$ kV/m, i.e., under the self-defocusing nonlinearity, our numerical results show that the Airy beam is somewhat expanded at the beginning due to the self-defocusing nonlinearity but its shape is nearly unchanged (Fig. 1.32a). In contrast to the self-focusing case, the Airy beam persists in its intensity pattern and transverse acceleration during subsequent free-space propagation (Fig. 1.32b, d). Furthermore, its power spectrum reshapes into a diamond-like pattern and is concentrated unevenly on four spots (Fig. 1.32c), resembling the first BZ [66] of an asymmetric square lattice. The energy-flow of the Airy beam is also quite different from that in the self-focusing case, since the Poynting vectors of the Airy beam line up towards the same direction around the Airy "head" (Figs. 1.32e, f). Counterintuitively, the peak intensity of the main lobe gets even stronger after subsequent linear propagation, as seen from the side-view evolution (Fig. 1.32d). This phenomenon suggests that the Airy beam might experience anomalous diffraction after initial self-defocusing nonlinearity, akin to that observed in photonic lattices [67, 68].

Indeed, if we zoom in the Airy-beam intensity pattern not far from the "head", it exhibits a square-like structure with nonuniform intensity distribution and lattice spacing. Under a self-defocusing nonlinearity, the Airy beam induces an index distribution akin to a nonuniform or chirped "backbone" lattice. This self-induced chirped lattice could exhibit properties similar to uniformed photonic lattices [69], thereby change the diffraction of the Airy beam. Using the BZ spectroscopy method [70, 71], we calculated the BZ spectrum of the induced lattice and found that the lattice

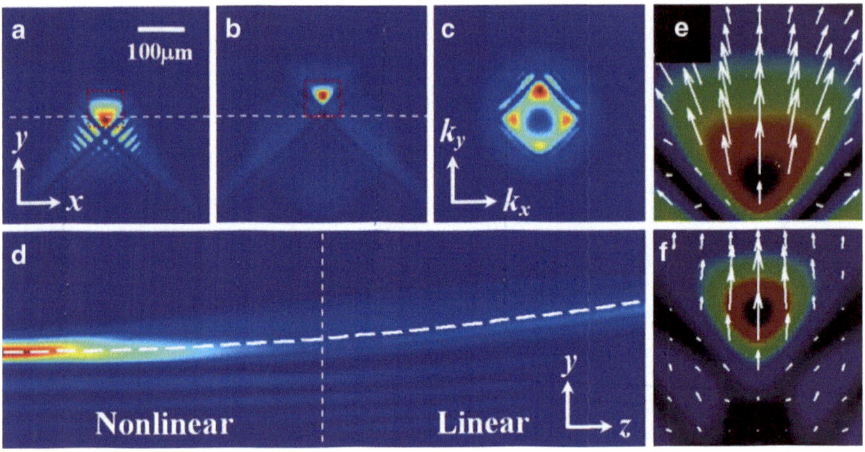

Fig. 1.32 Numerical simulation of an Airy beam propagating under an initial self-defocusing nonlinearity. Other description is the same as that for Fig. 1.31

self-induced by the Airy beam indeed shows a BZ structure [64]. Thus, the principle for anomalous diffraction observed here could be similar to that reported in [66, 68].

1.5.2 Nonlinear Generation and Control of Airy Beams

Airy beams are usually generated by using linear diffractive elements. Recently, Arie's group has proposed and demonstrated a method for nonlinear generation (i.e., through a nonlinear process) of Airy beams [29]. Specifically, it was done by a second harmonic generation (SHG) process through the design and fabrication of an asymmetric quadratic nonlinear photonic structure. The asymmetric structure induces a cubic phase front to the generated SHG output, whose Fourier transform is an accelerating Airy beam. This nonlinear generation process enables Airy beams to be obtained at new wavelengths. Indeed, frequency doubling of 1D and 2D Airy beams in a periodically poled crystal has been achieved in their experiment [72].

In addition, it was suggested theoretically that the generation of Airy beams by nonlinear processes opens several possibilities for controlling and manipulating these beams. As an example, the same group experimentally demonstrated a method to control the relative intensity along the caustic of nonlinearly generated Airy beams by controlling the phase matching conditions of the nonlinear interaction via temperature tuning. By analyzing the interactions in the Fourier space, they showed that the shaping of the beams is achieved by having noncollinear interactions. Furthermore, they studied the possibilities for all optical control by changing the pump wavelength. This includes for example nonlinear control of the beam acceleration direction, the wavelength of the output Airy beam, and the location of the Airy beam peak intensity along the same curved trajectory [73, 74].

1.5.3 Nonlinear Self-trapping of Airy-like Optical Beams

As discussed above, in practice, all nondiffracting Airy beams must be truncated, to keep the energy finite. Such truncated beams eventually diffract and lose their unique structure after long enough linear propagation. Thus, recently, research efforts have also been devoted to identify nonlinear physical mechanisms that could allow these highly localized wave packets to propagate in a true diffraction-free manner, as in the case of optical spatial solitons [75, 76].

It is well known that self-focusing can counteract the effects of dispersion and diffraction, leading to formation of optical solitons. Spatial soliton beams are usually symmetric, due to the even (second-order) nature of diffraction, and the typical nonlinear responses are local and conservative. However, more complex responses, such as the gradient-sensitive diffusion nonlinearity, can lead to a whole different class of instability and dynamics. In particular, it was suggested that highly asymmetric nonlinear action such as diffusion nonlinearity experienced by an asymmetric beam could lead to invariant propagation of the beam. The prospect of observing a diffusion-trapped exponentially contained Airy wave packet in nonlinear photorefractive media with diffusion nonlinearity was first proposed in [24]. Recently, Fleischer's group has reported the experimental observation of self-trapped Airy beams in a nonlinear medium. As opposed to screening or photovoltaic spatial solitons [75] this new class of self-localized beams owes its existence to carrier diffusion effects. The asymmetric action of two-wave mixing supports the asymmetric intensity profile of the Airy states, with a balance that is independent of the beam intensity. Furthermore, the self-trapped wave packets self-bend during propagation at an acceleration rate that is independent of the thermal energy associated with the diffusive nonlinearity. These results represent the first example of Airy solitary-like wave formation using two-wave mixing [77].

Quite recently, Segev's group has studied self-accelerating self-trapped beams in nonlinear optical media, exhibiting self-focusing and self-defocusing Kerr and saturable nonlinearities, as well as a quadratic response. In Kerr and saturable media such beams are stable under self-defocusing and weak self-focusing, whereas for strong self-focusing the beams off-shoot solitons while their main lobe continues to accelerate. These self-trapped Airy-like accelerating beams in nonlinear media propagate along parabolic trajectories and their existence curve of nonlinear generation was derived [78].

1.6 Proposed Applications

1.6.1 Optically Clearing Particles Using Airy Beams

Nondiffracting beams, such as the Bessel beam, Mathieu beam, and Airy beam, do not spread as they propagate. This property is particular useful in many applications include imaging, micromanipulation, and optical transfection.

Perhaps, one of the exciting applications of Airy beams is particle transportation and clearing along the curved paths as proposed by Dholakia's group [14]. They employed micrometer-sized colloidal particles, although their experiment can be extended directly to other chemical samples or even biological matter. As seen from Fig. 1.9a, a 2D truncated Airy beam has the form of a main spot and a number of side lobes whose intensity decays with increasing distance from the main spot, which induces optical gradient forces dragging dielectric particles towards the main spot. Then, these particles are guided along a projective trajectory as the Airy beams propagate and accelerate transversely. In their experimental setup, a finite Airy beam is used as a form of micrometer-sized "snowblower" at the bottom to blow the particles upwards and sideways. Without any motion of the light field, these samples are cleared towards the direction where the trajectory of the Airy beams point to. Due to the self-healing property of the Airy beams, such optically mediated particle clearing could be made more robust. The advanced "optical path clearing (OPC)" method mediated with Airy beams was demonstrated in the following work of Dholakia's group [79]. Rotating multiple Airy beams were used for clearing regions in a sample in a synchronized effort. The authors also suggested that the improvement for both clearing performance and efficiency might be achieved by utilizing dynamic multiple-step clearing approaches based on multiple Airy beams operated according to a time-sharing protocol.

1.6.2 Generation of Curved Plasma Channels by Airy Beams

In early studies, the plasma channels generated by femtosecond laser pulse were always along straight lines. Thus, conical radiation originating from straight filaments at different longitudinal sections of the beam overlaps in the observation plane, making the analysis of the emission more complicated. To solve this problem, femtosecond laser beams with a transverse spatial pattern in the form of a 2D Airy function were employed to write curved plasma channels by Moloney's group in collaboration with Christodoulides' group. Broadband emission from different longitudinal sections of the filament is therefore along angularly separated paths, resulting in the spatial separation of this emission in the far field. This technology has been successfully applied to different media: air and water, for which, the self-focusing collapse of the beam to a singularity is respectively arrested by plasma defocusing and the group-velocity dispersion [17, 18].

1.6.3 Generation and Control of Plasmonic Airy Beams

Another exciting application of Airy beams is their possible use in routing plasmonic energy. One-dimensional Airy beams have recently been introduced theoretically by Christodoulides' group into the field of plasmonics as the only nondiffracting solution in the planar systems of metal interfaces [21]. The analysis indicates that a new class of surface plasmons could be introduced that can freely self-bend during propagation. Due to the strong confinement of surface plasmon polaritons, the plasmonic Airy beams could be a promising candidate for subwavelength beam manipulation and on-chip signal processing in the emerging field of nanoplasmonics. Recently, Kivshar's group has proposed theoretically plasmonic Airy beam manipulation by means of linear potentials created by a wedged metal–dielectric–metal structure with one titling metal plate. They showed that the plasmonic Airy beam deflection could be enhanced, compensated, or even reversed, while still maintaining the self-healing properties [80].

Quite recently, the race for experimental generation and control of plasmonic Airy beams has accelerated, and there are at least three independent groups that reported successful demonstration of Airy surface plasmons [81–83]. In these studies, the observed Airy plasmons were excited on a metal–air interface using specially designed diffraction grating by taking advantage of the subwavelength confinement and localized propagation of tightly localized surface plasmon polaritons. The propagation of Airy Plasmon along curved parabolic trajectories was directly observed, together with their self-healing phenomenon after passing through surface defects. In particular, it was shown that the ballistic motion of the Airy plasmons could be reconfigured in real-time by either a computer addressed spatial light modulator or mechanical means [81].

1.6.4 Optical Manipulation with Morphing Autofocusing Airy Beams

Beam focusing has always been a subject of practical importance. For many applications it is desirable to have a wavefront's energy abruptly autofocused right onto a target while maintaining a low intensity profile up until that point. This feature could be useful, for example, in biomedical treatments and other nonlinear settings such as optical filamentation. Lately, a new class of autofocusing waves, as mentioned in Sect. 1.2.5, based on Airy beams has been proposed [38, 39] and observed [41, 42]. These field configurations rely on radially symmetric or circular Airy beams, and they can be established entirely on linear effects, i.e., without invoking any nonlinear self-focusing processes. In addition, a recent theoretical study has also indicated that a superposition of Airy wavelets can asymptotically "morph" into Bessel beams in their far-field [84].

Of particular interest is the potential application of these abruptly autofocusing beams. One example is processing of a thick sample by laser ablation [41]. Another example is optical guiding of nonabsorbing microparticles [15]. In the latter example demonstrated by Chen's group, an abruptly autofocusing beam was sent into an optical tweezers-like setting, where the particles were trapped and transported along the autofocusing beam. It was suggested that the circular autofocusing Airy beam might be used as a tapered channel guide for microparticles [15].

1.7 Summary

In this chapter, we provide an overview on self-accelerating Airy beams, their generation, control, and potential applications. As mentioned, driven by both fundamental interest and application potential, the number of research papers dealing with optical Airy beams soared in the last a few years, accompanied with new discoveries and surprising results. We envision that the research on self-accelerating Airy beams will keep its accelerating momentum in years to come.

This work was supported by the US National Science Foundation, the Air Force Office of Scientific Research, and by the 973 program, the National Natural Science Foundation of China (NSFC), and the Program for Changjiang Scholars and Innovative Research Team in China.

References

1. M.V. Berry, N.L. Balazs, Am. J. Phys. **47**, 264 (1979)
2. K. Unnikrishnan, A.R.P. Rau, Am. J. Phys. **64**, 1034 (1996)
3. J. Durnin, J. Opt. Soc. Am. A **4**, 651 (1987)
4. J. Durnin, J.J. Miceli, J.H. Eberly, Phys. Rev. Lett. **58**, 1499 (1987)
5. J.C. Gutiérrez-Vega, M.D. Iturbe-Castillo, S. Chávez-Cerda, Opt. Lett. **25**, 1493 (2000)
6. M.A. Bandres, J.C. Gutiérrez-Vega, S. Chávez-Cerda, Opt. Lett. **29**, 44 (2004)
7. D. McGloin, K. Dholakia, Contemp. Phys. **46**, 15 (2005)
8. R. Stützle et al., Phys. Rev. Lett. **95**, 110405 (2005)
9. L.I. Schiff, *Quantum Mechanics*, 3rd edn. (McGraw-Hill, New York, NY, 1968)
10. F. Gori, G. Guattari, C. Podovani, Opt. Commun. **64**, 491 (1987)
11. J. Durnin, J.J. Miceli, J.H. Eberly, Phys. Rev. Lett. **66**, 838 (1991)
12. G.A. Siviloglou, D.N. Christodoulides, Opt. Lett. **32**, 979 (2007)
13. G.A. Siviloglou, J. Broky, A. Dogariu, D.N. Christodoulides, Phys. Rev. Lett. **99**, 213901 (2007)
14. J. Baumgartl, M. Mazilu, K. Dholakia, Nat. Photonics **2**, 675 (2008)
15. P. Zhang, J. Prakash, Z. Zhang, M.S. Mills, N.K. Efremidis, D.N. Christodoulides, Z. Chen, Opt. Lett. **36**, 2883 (2011)
16. Z. Zheng, B. Zhang, H. Chen, J. Ding, H. Wang, Appl. Opt. **50**, 43 (2011)
17. P. Polynkin, M. Kolesik, J.V. Moloney, G.A. Siviloglou, D.N. Christodoulides, Science **324**, 229 (2009)
18. P. Polynkin, M. Kolesik, J. Moloney, Phys. Rev. Lett. **103**, 123902 (2009)
19. J. Li, W. Zang, J. Tian, Opt. Exp. **18**, 7300 (2010)
20. J. Li, X. Fan, W. Zang, J. Tian, Opt. Lett. **36**, 648 (2011)
21. A. Salandrino, D.N. Christodoulides, Opt. Lett. **35**, 2082 (2010)
22. D.M. Greenberger, Am. J. Phys. **48**, 256 (1980)

23. O. Vallée, M. Soares, *Airy Functions and Applications to Physics* (Imperial College Press, London, 2004)
24. D.N. Christodoulides, T.H. Coskun, Opt. Lett. **21**, 1460 (1996)
25. I.M. Besieris, A.M. Shaarawi, Opt. Lett. **32**, 2447 (2007)
26. I.M. Besieris, A.M. Shaarawi, R.W. Ziolkowski, Am. J. Phys. **62**, 519 (1994)
27. D.M. Cottrell, J.A. Davis, T.M. Hazard, Opt. Lett. **34**, 2634 (2009)
28. H.T. Dai, X.W. Sun, D. Luo, Y.J. Liu, Opt. Exp. **17**, 19365 (2009)
29. T. Ellenbogen, N. Voloch, A. Ganany-Padowicz, A. Arie, Nat. Photonics **3**, 395 (2009)
30. D.G. Papazoglou, S. Suntsov, D. Abdollahpour, S. Tzortzakis, Phys. Rev. A **81**, 061807(R) (2010)
31. S. Longhi, Opt. Lett. **36**, 716 (2011)
32. A. Efimov, C. Schaffer, D.H. Reitze, J. Opt. Soc. Am. B **12**, 1968 (1995)
33. M. Miyagi, S. Nishida, Appl. Opt. **18**, 678 (1979)
34. D.N. Christodoulides, N.K. Efremidis, P. Di Trapani, B.A. Malomed, Opt. Lett. **29**, 1446 (2004)
35. M. Weiner, Rev. Sci. Instrum. **71**, 1929 (2000)
36. D. Abdollahpour, S. Suntsov, D.G. Papazoglou, S. Tzortzakis, Phys. Rev. Lett. **105**, 253901 (2010)
37. A. Chong, W. Renninger, D.N. Christodoulides, F.W. Wise, Nat. Photonics **4**, 103 (2010)
38. N.K. Efremidis, D.N. Christodoulides, Opt. Lett. **35**, 4045 (2010)
39. I. Chremmos, N.K. Efremidis, D.N. Christodoulides, Opt. Lett. **36**, 1890 (2011)
40. E. Greenfield, M. Segev, W. Walasik, O. Raz, Phys. Rev. Lett. **106**, 213902 (2011)
41. D.G. Papazoglou, N.K. Efremidis, D.N. Christodoulides, S. Tzortzakis, Opt. Lett. **36**, 1842 (2011)
42. P. Zhang, J. Prakash, Z. Zhang, Y. Hu, N. K. Efremidis, V. Kajorndejnukul, D. N. Christodoulides, and Z. Chen, CLEO:QELS-Fundamental Science, paper QThS7 (2011)
43. J.W. Goodman, *Introduction to Fourier Optics*, 3rd edn. (Roberts & Company Publishers, Greenwood Village, CO, 2005)
44. G.A. Siviloglou, J. Broky, A. Dogariu, D.N. Christodoulides, Opt. Lett. **33**, 207 (2008)
45. Y. Hu, P. Zhang, C. Lou, S. Huang, J. Xu, Z. Chen, Opt. Lett. **35**, 2260 (2010)
46. Z. Ye, S. Liu, C. Lou, P. Zhang, Y. Hu, D. Song, J. Zhao, Z. Chen, Opt. Lett. **36**, 3230 (2011)
47. N.K. Efremidis, Opt. Lett. **36**, 3006 (2011)
48. V. Garces-Chavez, D. McGloin, H. Melville, W. Sibbett, K. Dholakia, Nature **419**, 145 (2002)
49. Z. Bouchal, J. Wagner, M. Chlup, Opt. Commun. **151**, 207 (1998)
50. L. Allen, M.W. Beijersbergen, R.J.C. Spreeuw, J.P. Woerdman, Phys. Rev. A **45**, 8185 (1992)
51. J. Broky, G.A. Siviloglou, A. Dogariu, D.N. Christodoulides, Opt. Exp. **16**, 12880 (2008)
52. C. Hwang, D. Choi, K. Kim, B. Lee, Opt. Exp. **18**, 23504 (2010)
53. H.C. van de Hulst, *Light Scattering by Small Particles* (Dover Publication Inc., New York, NY, 1981)
54. S. Prahl, http://omlc.ogi.edu/calc/mie_calc.html (2008)
55. http://www.opticsinfobase.org/abstract.cfm?URI=oe-16-17-12880 (Media 1)
56. J.F. Nye, M.V. Berry, Proc. R. Soc. London, Ser. A **336**, 165 (1974)
57. A. Dogariu, S. Amarande, Opt. Lett. **28**, 10 (2003)
58. http://www.opticsinfobase.org/abstract.cfm?URI=oe-16-17-12880 (Media 2)
59. X. Chu, Opt. Lett. **36**, 2701 (2011)
60. Y. Gu, G. Gbur, Opt. Lett. **35**, 3456 (2010)
61. H.T. Dai, Y.J. Liu, D. Luo, X.W. Sun, Opt. Lett. **35**, 4075 (2010)
62. H.T. Dai, Y.J. Liu, D. Luo, X.W. Sun, Opt. Lett. **36**, 1617 (2011)
63. S. Barwick, Opt. Lett. **36**, 2827 (2011)
64. Y. Hu, S. Huang, P. Zhang, C. Lou, J. Xu, Z. Chen, Opt. Lett. **35**, 3952 (2010)
65. S. Jia, J. Lee, G.A. Siviloglou, D.N. Christodoulides, J.W. Fleischer, Phys. Rev. Lett. **104**, 253904 (2010)
66. X. Wang, A. Bezryadina, Z. Chen, K.G. Makris, D.N. Christodoulides, G.I. Stegeman, Phys. Rev. Lett. **98**, 123903 (2007)
67. T. Pertsch, T. Zentgraf, U. Peschel, A. Bräuer, F. Lederer, Phys. Rev. Lett. **88**, 093901 (2002)
68. P. Zhang, C. Lou, S. Liu, J. Zhao, J. Xu, Z. Chen, Opt. Lett. **35**, 892 (2010)
69. M.I. Molina, Y.V. Kartashov, L. Torner, Y.S. Kivshar, Opt. Lett. **32**, 2668 (2007)

70. G. Bartal, O. Cohen, H. Buljan, J.W. Fleischer, O. Manela, M. Segev, Phys. Rev. Lett. **94**, 163902 (2005)
71. S. Liu, P. Zhang, X. Gan, F. Xiao, J. Zhao, Appl. Phys. B **99**, 727 (2010)
72. I. Dolev, A. Arie, Appl. Phys. Lett. **97**, 171102 (2010)
73. I. Dolev, T. Ellenbogen, A. Arie, Opt. Lett. **35**, 1581 (2010)
74. I. Dolev, T. Ellenbogen, N. Voloch-Bloch, A. Arie, Appl. Phys. Lett. **95**, 201112 (2009)
75. G.I. Stegeman, M. Segev, Science **286**, 1518 (1999)
76. D.N. Christodoulides, F. Lederer, Y. Silberberg, Nature **424**, 817 (2003)
77. S. Jia, J. Lee, J.W. Fleischer, G.A. Siviloglou, D.N. Christodoulides, Phys. Rev. Lett. **104**, 253904 (2010)
78. I. Kaminer, M. Segev, D.N. Christodoulides, Phys. Rev. Lett. **106**, 213903 (2011)
79. J. Baumgartl, T. Čižmár, M. Mazilu, V.C. Chan, A.E. Carruthers, B.A. Capron, W. McNeely, E.M. Wright, K. Dholakia, Opt. Exp. **18**, 17130 (2010)
80. W. Liu, D.N. Neshev, I.V. Shadrivov, A.E. Miroshnichenko, Y.S. Kivshar, Opt. Lett. **36**, 1164 (2011)
81. P. Zhang, S. Wang, Y. Liu, X. Yin, C. Lu, Z. Chen, X. Zhang, Opt. Lett. **36**, 3191 (2011)
82. A. Minovich, A.E. Klein, N. Janunts, T. Pertsch, D.N. Neshev, Y.S. Kivshar, Phys. Rev. Lett. **107**, 116802 (2011)
83. L. Li, T. Li, S.M. Wang, S.N. Zhu, Phys. Rev. Lett. **107**, 126804 (2011)
84. C. Hwang, K. Kim, B. Lee, Opt. Exp. **19**, 7356 (2011)

Chapter 2
Advanced Integrated Photonics in Doped Silica Glass

Marcello Ferrera, David Duchesne, L. Razzari, Marco Peccianti, Alessia Pasquazi, Yong-Woo Park, Jose Azaña, Roberto Morandotti, Brent E. Little, Sai T. Chu, and David J. Moss

2.1 Introduction

Broadband optical communication systems are rapidly becoming the key to overcome the stringent limitations imposed by standard electronic telecommunication networks. However, in order to complete the inevitable transition from electronics to photonics,

---

M. Ferrera (✉)
INRS-EMT, Varennes, QC, Canada

School of Physics and Astronomy, University of St Andrews,
Room 256, Fife, Scotland KY16 9SS, UK
e-mail: mf39@st-andrews.ac.uk

D. Duchesne
INRS-EMT, Varennes, QC, Canada

Massachusetts Institute of Technology, Cambridge, MA, USA

L. Razzari
INRS-EMT, Varennes, QC, Canada

IIT, Istituto Italiano di Tecnologia, Genoa, Italy

M. Peccianti • A. Pasquazi • Y.-W. Park • J. Azaña
INRS-EMT, Varennes, QC, Canada

R. Morandotti
INRS-EMT, Université du Québec, Varennes, J3X 1S2, Canada

B.E. Little
Infinera Corporation, Sunnyvale, CA, USA

S.T. Chu
Infinera Corporation, Sunnyvale, CA, USA

City University of Hong Kong, Hong Kong, China

D.J. Moss
IPOS/CUDOS, School of Physics, University of Sydney, Sydney, NSW, Australia

Z. Chen and R. Morandotti (eds.), *Nonlinear Photonics and Novel Optical Phenomena*, Springer Series in Optical Sciences 170, DOI 10.1007/978-1-4614-3538-9_2,

several critical requirements must be addressed, including lowering energetic demands, achieving higher efficiency, increasing bandwidth and flexibility, all within a compact form factor [1–3]. In particular, it is broadly accepted that future photonic devices must be CMOS compatible in order to exploit the existing silicon fabrication technology that has been largely developed during the last 60 years [4–7]. Following this idea, there has been a tremendous growth of hybrid optoelectronic technologies that has not only responded to the need of lowering costs, but has also enabled on-chip ultra-fast signal processing. However, these hybrid solutions are an intermediate step to achieve the ambitious goal of an all-optical technology, which would bring together the intrinsic benefit of lowering the production costs and simplifying future ultrafast communication networks.

Nowadays, the demand to implement a multifunctional all-optical chip, which could efficiently function as an optical frequency converter [8, 9], signal regenerator [10, 11], multiplexer [12–14], as well as for routing and switching optical signals [4, 12], is growing rapidly. The preference for optical networks over electronic ones originates from certain fundamental advantages. First, the optical operational bandwidth is much larger; secondly, the extremely low loss provided by modern standard fibers ensures that an optical signal (with a ~THz bit rate) can be transmitted for more than a hundred kilometers without the need of being "refreshed" or amplified. Optical systems also provide a greater resistance to electromagnetic interference and offer low-cost maintenance [15].

However, photonics is also affected by peculiar drawbacks and limitations, as is any other kind of technology. For instance, bending losses become extremely high in fibers for chip-scale size bends ($\ll$1 cm), limiting the integrability of fibers for optical networks. Moreover, their weak nonlinearity limits the practical realization (i.e., due to the additional constraints in terms of low power signals and short propagation lengths) of some fundamental operations requiring nonlinear optical phenomena, such as frequency conversion, pulse compression, etc. [16]. A typical example of a fundamental problem in modern optical communication networks is the Optical-to-Electrical-to-Optical (OEO) signal decoding. Substantially, it consists of a wavelength remapping protocol that is necessary in an optic telecommunication network whenever we must route a data stream from a channel to another with a different optical frequency carrier. This operation is becoming increasingly fundamental in today's long distance lines, and because of the absence in the market of an all-optical frequency converter, it is normally performed by means of an OEO transducer. This equipment operates the necessary I/O wavelength conversion by first converting the input optical signal into an electric one, and then encoding the obtained information into an optical output at a different frequency. Today, one of the main reasons limiting the wide spread of optical and optoelectronic technologies in modern telecommunications systems is the high cost required for converting data from the optical domain to the electric one and vice versa. Intuitively, the solution to this fundamental problem can be addressed by looking into the domain of nonlinear optics. In fact, only nonlinear processes can be responsible for phenomena such as the spectral broadening of an optical pulse or the frequency conversion. It is equally important to note however, that not all of the operations required for low-cost integrated optical telecommunications systems can be satisfied with nonlinear optics. Whereas frequency conversion seems the natural choice to eliminate OEO conversions, certain key devices such as logical memories, as well as analog and logic functionality, still need to be developed in the optical domain, where linear optics may pose as a better solution. Several alternative material platforms have been developed for photonic integrated circuits (PICs) [1, 2, 17, 18], including semiconductors such as AlGaAs, silicon, and silicon-on-insulator (SOI) [3, 5, 17, 19], as well as nonlinear glasses such as chalcogenides [20], silicon oxynitride [21], and bismuth oxides [2, 22, 23]. In addition, exotic and novel manufacturing processes have also led to new and promising structures in these materials and in regular silica fibers. Photonic crystals [24–26], optical cavities [27], and nanowires [28, 29], all make use of a tight light confinement to enhance nonlinearities, and allow for submillimeter photonic chips.

Despite the abundance of alternatives, none of the developed material platforms can simultaneously address all the requirements that a multifunctional CMOS compatible all-optical chip should possess. In fact, many nonlinear platforms require power levels that largely exceed the limit for commercial applications, whereas others have limiting factors such as saturation due to multiphoton absorption. Moreover, in certain cases the fabrication technology is simply not sufficiently advanced, resulting in a severe linear attenuation and an incomplete CMOS compatibility.

It is in response to these issues and limitations, that a new high-index doped silica glass platform was developed in 2003 by Little Optics [30]. This material, named Hydex®, combines the best qualities of single-mode fibers (SMFs), namely low propagation losses, absence of multiphoton absorption, and a robust fabrication technology, with those of semiconductor materials, such as tight bending radii and higher nonlinearities. This chapter primarily describes this new material platform, through the characterization of its linear and nonlinear properties. Our primary goal is to show important results suitable for a broad set of fundamental all-optical operations such as the generation of ultra-narrow linewidth and multiple wavelength sources, supercontinuum generation (SCG), optical pulse compression, time duration measurement of subpicosecond pulses, and ultrafast optical integration. Most of these applications make use of very important nonlinear effects and it is for this reason that this chapter is mainly focused on materials and applications for nonlinear optics. In the following sections, we present an overview of concurrent alternative material platforms and photonic structures, discussing their advantages and limitations. Subsequently, we review some fundamental theoretical concepts strictly linked with the reported experiments.

2.2 Alternative Material Platforms in Integrated Photonics

2.2.1 *Semiconductor Nonlinear Photonics*

The most basic component for optical telecommunications systems is the optical waveguide, which can serve the dual purpose of transporting and modifying the properties of the light propagating within it. The basic principle for guiding light is total internal reflection, which can occur when light propagates in a higher index core material (n_c) that is surrounded by a cladding region of lower index (n_s).

The core refractive index (n_c) and the index contrast ($\Delta n = n_c - n_s$) are normally used to describe the linear properties of the optical waveguide. In addition, these two basic parameters can also provide a rough estimate of the nonlinear properties of the structure.

To explain this, let us consider the material polarization $\mathbf{P}(\omega)$ as a function of the electric field $\mathbf{E}(\omega)$

$$\begin{aligned} \mathrm{P}(\omega) &= \varepsilon_0[\chi^{(1)}(\omega)\cdot\mathrm{E}(\omega) + \chi^{(2)}(-\omega:\omega_i,\omega_j):\mathrm{E}(\omega_i)\mathrm{E}(\omega_j) \\ &\quad + \chi^{(3)}(-\omega:\omega_i,\omega_j,\omega_k)\vdots\mathrm{E}(\omega_i)\mathrm{E}(\omega_j)\mathrm{E}(\omega_k) + \cdots], \\ &= \varepsilon_0[\chi^{(1)}(\omega)\cdot\mathrm{E}(\omega)] + \mathrm{P}_{NL}(\omega), \end{aligned} \tag{2.1}$$

where $\chi^{(i)}$ for $i = 1, 2, 3, \ldots$, indicates the ith order susceptibility tensor of rank $i + 1$, $\mathbf{P}_{\mathrm{NL}}(\omega)$ is the nonlinear polarization term, ε_0 is the free space permittivity, ω is the (angular) frequency and the notation: $\mathbf{E}(\omega_i)\mathbf{E}(\omega_j)$;:$\mathbf{E}(\omega_i)\mathbf{E}(\omega_j)\mathbf{E}(\omega_k)$; indicates tensorial products of the fields [31].

Through Miller's rule [32], it is possible to approximately express the nonlinear susceptibilities as a function of the linear susceptibility, $\chi^{(1)}$, which is in turn directly linked to the material refractive index. This in part explains why high index materials are normally associated with having a good nonlinear performance. For example, in typical semiconductors, the nonlinear Kerr coefficient n_2 takes on a series of remarkable values ranging from ~10^{-18} to 10^{-17} $\mathrm{m^2/W}$ (with a core refractive index $n_c > 3$; e.g., ~3.5 for Si and ~3.3 GaAs). For comparison, fused silica ($n_c = 1.45$) possesses an $n_2 \sim 2.6 \times 10^{-20}$ $\mathrm{m^2/W}$. On the other hand, the index contrast is the main parameter that estimates the effective area of a propagating mode in a waveguide and consequently, a large index contrast leads to a more confined mode and thus higher nonlinearities (which typically scale with the power density, or intensity, of an optical beam). III–V semiconductors, such as silicon and AlGaAs, are thus extremely interesting photonic platforms for both linear and nonlinear applications, since they possess both a large index of refraction at the telecommunication wavelength ($\lambda = 1.55$ μm) and a mature waveguide fabrication technology allowing for submicron structures with a high index contrast.

This chapter deals primarily with third order nonlinear phenomena that are related to the $\chi^{(3)}$ tensor (see (2.1)), as this is the lowest order nonlinearity in centrosymmetric media (such as glasses) where $\chi^{(2)} = 0$. The strength of these third-order nonlinear interactions can be estimated through the nonlinear parameter $\gamma = n_2\omega/cA_{\mathrm{eff}}$ [16], where n_2 is the nonlinear Kerr coefficient determined solely from the material properties, ω is the angular frequency of light, c is the speed of light in vacuum, and A_{eff} the effective area of the mode.

Taking this into account, there are three basic strategies to enhance nonlinear effects in optical devices: (1) selecting a material with a high value of n_2, (2) exploiting low-loss and long structures, and/or (3) fabricating structures that allow for tight modal confinement. In III–V semiconductors, long structures are typically

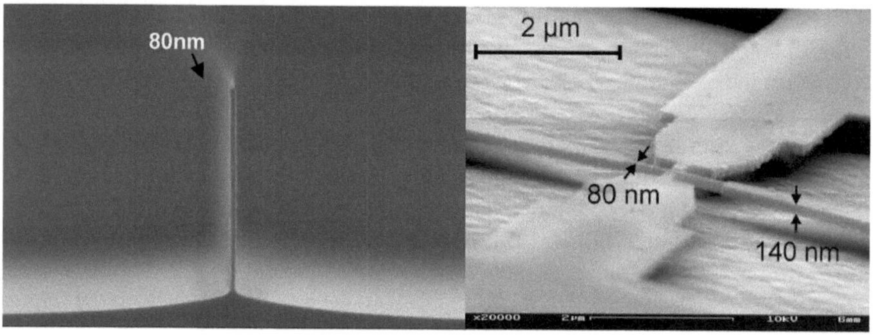

Fig. 2.1 Inverted nano-taper (80 nm in width) for an AlGaAs waveguide (*left*) and a silicon-free standing nanowire defined by high-resolution, low-energy electron-beam lithography (the pictures are taken from [37] and H. Lorenz nanophysics group report, http://www.nano.physik.uni-muenchen.de/, respectively). These images exemplify the advanced fabrication processes behind AlGaAs and silicon

not favorable as the propagation losses are typically too high. However, their well-developed fabrication processes do allow for channel or ridge waveguides open to air with a large index contrast, producing modal effective areas below 1 μm^2, see Fig. 2.1. Combined with their strong nonlinear coefficient, this can lead to extremely high values of γ exceeding 200,000 W^{-1} km^{-1} [10, 29, 33] (to be compared with SMFs which have $\gamma \sim 1$ W^{-1} km^{-1} [16]).

These optimal nonlinear features have been extensively used to demonstrate several nonlinear applications for telecommunications, including all-optical regeneration via four-wave mixing (FWM) and self-phase modulation (SPM) [10, 34], frequency conversion [8, 9, 35], and Raman amplification [36].

However, despite the high nonlinearity of III–V semiconductors, major limitations still prevent the implementation of these materials for the fabrication of future integrated devices for optical networks. Their high dispersion at telecommunication wavelengths, which is a consequence of being near the band gap of the material, drastically limits the usable length of a guiding structure. Whereas both Si and AlGaAs allow for engineering the waveguide dispersion, creating an effective quasi zero-dispersion window in the spectral region of interest, this in turn requires subwavelength waveguides in which the intensity mode profile extends over the waveguide boundaries along the etched sidewalls. As a result, a high degree of optical scattering is experienced by the light beam, thus causing high propagation losses of the order of 10 dB/cm for AlGaAs [38–40], and ~1 dB/cm for SOI [8]. In addition, although reducing the modal area enhances the nonlinear properties of the waveguide, it also drastically reduces the fiber-to-waveguide coupling. This leads to high insertion losses, which can be overcome only by using either expensive amplifiers at the output, or complicated tapers at the input. This latter solution often requires a multistep etching processes [41] (SOI waveguides make use of state-of-the-art single etch inverse tapers which limits the insertion losses to approximately 5 dB [8, 42]), in turn consistently increasing the device fabrication complexity.

Finally, another factor that limits semiconductor applicability in nonlinear optics is multiphoton absorption and specifically two-photon absorption (TPA). This effect consists in the excitation of electrons from the valence band to the conduction band via the successive absorption of two (or more) photons, mediated via a virtual state. This undesirable effect leads to a saturation of the transmitted power and consequently limits the nonlinear performance of the device. Hence, the intrinsic nonlinear figure of merit for semiconductors, defined as FOM $= n_2/\alpha_2\lambda$ (where α_2 is the TPA coefficient), is particularly low [43]. In addition, another contribution to nonlinear losses results from multiphoton-induced free carrier absorption [29, 44, 45].

2.2.2 *Nonlinear Glasses*

For applications in integrated nonlinear optics, the principal competitors to semiconductors are nonlinear glasses. Throughout the last decade, a great technical and scientific effort has been dedicated to the development and the characterization of different families of highly nonlinear glasses, including chalcogenides, silicon oxynitride, and bismuth oxides [2, 22, 23, 46]. Amongst these, chalcogenide glasses have shown optimal results mainly due to their high nonlinear refractive index. However, although these glasses have been largely studied, their fabrication process is still quite problematic [47, 48] and not compatible with standard CMOS technology. In addition, they also suffer from other issues such as photosensitivity to intense light, low hardness, and high thermal expansion coefficients, which make the realization of commercial chalcogenide optical devices problematic [49]. In addition, most of these glasses typically have a low FOM [50] due to nonlinear absorption. Whereas other high-index glasses, such as silicon oxynitride, have negligible nonlinear absorption (i.e., a virtually infinite figure of merit), they unfortunately also typically require high temperature annealing to reduce propagation losses, thereby rendering the entire fabrication process incompatible with CMOS processes.

As anticipated in Sect. 2.1, the material at the center of this chapter is a high-index doped silica glass called Hydex® [30], which can be viewed as an ideal compromise between the excellent linear properties of optical fibers (low linear losses, absence of nonlinear losses, weak dispersion) and that of semiconductors (high nonlinearity, tight field confinement, small bending radii). Films are first deposited by using standard chemical vapor deposition. Subsequently, waveguides are formed using photolithography and reactive ion etching, producing waveguide sidewalls with exceptional smoothness. The waveguides are then buried in standard fused silica glass, making the entire fabrication process CMOS compatible without any need for high temperature annealing. The typical waveguide cross section is $1.45 \times 1.5\ \mu m^2$ as shown in Fig. 2.2. The linear index at $\lambda = 1.55\ \mu m$ is 1.7, and propagation losses have been shown to be as low as 0.06 dB/cm [51, 52]. In addition, fiber pigtails have been designed for coupling to and from Hydex waveguides, with coupling losses of the order of 1.5 dB. The linear properties of this material platform have already been exploited to achieve filters with >80 dB extinction ratios [53], as well as optical sensors for biomolecules [54].

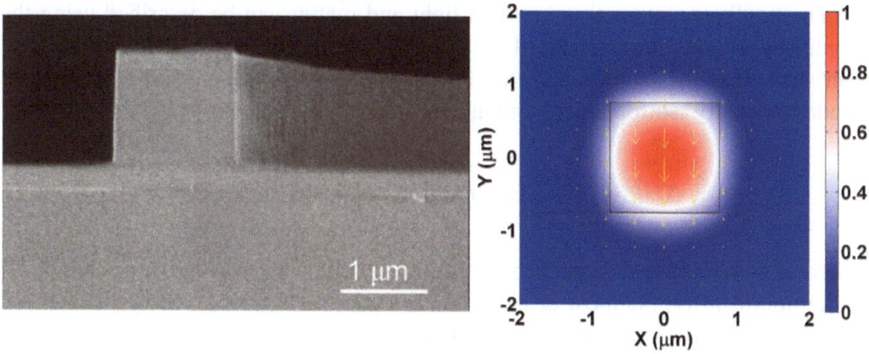

Fig. 2.2 Hydex waveguide overview. Scanning electron microscopy picture of the high-index glass waveguide prior to the upper SiO_2 deposition (*left*), and electromagnetic field distribution of the fundamental mode (*right*)

In the following section, we present the fundamental theory required to understand the subsequent experiments described in this chapter. It will be shown that this material platform can be used to fabricate state-of-the-art resonant structures capable of producing remarkable nonlinear effects at very low power levels.

2.3 Nonlinear Phenomena: Fundamentals

2.3.1 Nonlinear Light-Matter Interaction

Nonlinear optics is the study of phenomena that occurs as a consequence of an intense light-matter interaction. Whenever the electric field, associated to a propagating wave, is strong enough, the corresponding restoring potential of the electrons is an harmonic, and the material response (i.e., the specific physical quantity under investigation) depends in a nonlinear manner on the strength of the optical field.

Generally speaking, the broad family of nonlinear processes can be divided into two main categories which are the parametric and the nonparametric phenomena. A parametric process implies that the initial and the final quantum-mechanical state of the system are identical. In this way, parametric processes are mediated via virtual quantum levels where an electron can be excited for a very brief duration only (related to the uncertainty principle). In this way, the photon energy is conserved since no phonons are involved in the process. Common parametric phenomena include $\chi^{(2)}$ processes such as sum and difference frequency generation, second harmonic generation, etc., and $\chi^{(3)}$ phenomena such as third harmonic generation and FWM. Conversely, multiphoton absorption and Raman scattering are examples of nonparametric processes.

The nonlinear interaction between light and matter can be described using the nonlinear polarization introduced in (2.1). Making use of Maxwell's equation, one can show that the evolution of the electric field distribution E, in a waveguide made by locally isotropic and homogenous materials, is given by:

$$\nabla^2 E(\omega) + \frac{\omega^2}{c^2}\varepsilon(\omega)E(\omega) + \frac{\omega^2 \varepsilon_0^{-1}}{c^2}P_{\mathrm{NL}}(\omega) = 0, \tag{2.2}$$

where ε is the space-dependent permittivity defining the waveguide. For a third-order nonlinearity, the most general representation of P_{NL} is given by the interaction of four waves centered at different frequencies:

$$E = \sum_{m=1}^{4} A_m(z,t)F_m(x,y)\exp(i\omega_m t - i\beta_m z) + c.c, \tag{2.3}$$

where we have explicitly assumed four waveguide modes of (angular) frequency ω_m, propagation constant β_m, vectorial modal (spatial transverse) distribution F_m, and envelope A_m. More generally, the modal distribution can be allowed to depend on the time t and propagation distance z, although this is only necessary for ultra-short pulses and/or extreme nonlinear propagation.

The mathematical evaluation of the tensorial product of the nonlinear polarization involves 1944 complex elements, which can lead to a quite daunting analysis. Fortunately, symmetry properties, approximations, and phase matching requirements can reduce this to a simple number of terms. Specifically for the purposes of this chapter, the resultant generalized nonlinear Schrödinger (GNLS) propagation equations are given by:

$$\frac{\partial A_m}{\partial z} + \frac{\alpha_m}{2}A_m + \beta_{1,m}\frac{\partial A_m}{\partial t} + i\frac{\beta_{2,m}}{2}\frac{\partial^2 A_m}{\partial t^2} + \mathrm{HOD}$$
$$= \begin{cases} -\dfrac{\alpha_{2,m}}{2A_{\mathrm{eff}}}|A_m|^2 A_m - \mathrm{HOL} \\ -\gamma\left(\dfrac{1}{\omega_m}\dfrac{\partial}{\partial t}|A_m|^2 A_m + iT_{\mathrm{R}}A_m\dfrac{\partial}{\partial t}|A_m|^2\right) + \mathrm{MIX} \\ +i\gamma\left(|A_m|^2 + 2\sum_{p\neq m}|A_p|^2\right)A_m \\ +2i\gamma A_2^* A_3 A_4 \exp(i\Delta\beta z) \end{cases} \tag{2.4}$$

Here, the pulse envelope A_m has been normalized such that $|A_m|^2$ represents the power propagating in the waveguide of beam m, $\alpha_{1,m}$ is the linear loss, $\beta_{1,m}$ is the inverse of the pulse group velocity, $\beta_{2,m}$ is the group velocity dispersion, and the last term HOD represents higher order dispersion terms that become important for ultra-short pulses.

The terms on the right of (2.4) are a result of the nonlinear polarization. They are dominated by the nonlinear parameter given by $\gamma = n_2\omega/cA_{\text{eff}} = 3Re(\chi^{(3)})\mu_0\omega_m/(4n_m{}^2A_{\text{eff}})$, where n_m is the effective index of the mode. The modal effective area is defined by the vectorial overlap of the mode profiles:

$$A_{\text{eff}} = \frac{\left(\iint\limits_{\infty} |F_m|^2 \mathrm{d}x\,\mathrm{d}y\right)^2}{\iint\limits_{WG} |F_m|^4 \mathrm{d}x\,\mathrm{d}y}. \tag{2.5}$$

The modal profiles and the dispersion parameters are obtained by solving the linear continuous wave version of Maxwell's equations:

$$\nabla^2 F_m + \frac{\omega_m^2 \varepsilon_m}{c^2 \varepsilon_0} F_m = \beta_m^2 F_m. \tag{2.6}$$

Returning to the right-hand side of (2.4), the first term $\alpha_{2,m}$ is the TPA coefficient (related to the imaginary part of $\chi^{(3)}$), whereas HOL refers to higher order losses from higher order multiphoton absorption terms (e.g., three- and four-photon absorption). The second line on the right side is proportional to the change in time of the nonlinearity. In particular, the first term accounts for self-steepening, whereas the second is the Raman effect, caused by the noninstantaneous nature of the material nonlinearity (Raman response time T_R). The label "MIX" represents cross terms amongst the various input and generated beams arising from the Raman and self-steepening effects. We do not express these terms explicitly here as they will not be significant for any of the experiments reported further below. The third line are self-phase and cross-phase modulation terms, whereas the last line (explicitly expressed here for wave $m = 1$; the other $m = 2$, 3, and 4 components have a very similar form) is a specific FWM interaction requiring the phase matching $\Delta\beta = \beta_4 + \beta_3 - \beta_2 - \beta_1$ for $\omega_1 + \omega_2 = \omega_3 + \omega_4$. Other frequency mixing terms have been neglected on the assumption of this specific phase matching.

It is important to note that a number of approximations have been made to arrive at (2.4). Specifically, these relations hold for a glass material system where the modal profiles (F_m) of the interacting beams have a similar spatial electric field distribution (i.e., $F_1 \approx F_2 \approx F_3 \approx F_4$). We have also employed the slowly varying envelope approximation thus requiring pulses to be longer than a few cycles.

As will be shown through the various experiments described below, the GNLS takes a much simpler form in most experiments. In particular for the Hydex waveguides described in this chapter, nonlinear losses can be completely neglected, as verified experimentally in transmission experiments. This is a consequence of the very large bandgap of glass systems. The absence of multiphoton absorption (for all power levels used in the experiments reported) is extremely important, as these nonlinear losses lead to saturation of the total induced nonlinearity, which has been reported to limit several nonlinear applications in semiconductors [45].

To conclude this section, it is important to note that most nonlinear effects scale with three important parameters: the intensity, the waveguide length, and the nonlinearity (material value of $\chi^{(3)}$). In most applications we seek to minimize the energetic input and maximize the nonlinear effects, as such the only way to increase the intensity is by minimizing the area of the mode (or increasing the modal confinement). The notion to research nonlinear optics in a glass material system may at first appear counter-productive; silicon waveguides provide a much smaller mode area, a much larger material nonlinearity, and a good fabrication process which allows long waveguides to be formed. However, a combination of nonlinear losses and residual linear losses (on the order of ~1 dB/cm) prevents highly nonlinear devices. Apart from the absence of nonlinear losses, Hydex waveguides also provide exceptionally low linear losses, thereby allowing for resonant or long winding structures (recall that the intensity decays as $\exp(-\alpha z)$ and it can be shown that this leads to an effective maximum length of the order of α^{-1} for nonlinear effects).

2.4 Ultra-narrow Linewidth and Multiple-Wavelength Integrated Optical Sources

2.4.1 Enhanced Nonlinearities in Integrated Photonic Cavities

Advances in fabrication processes and technologies have allowed for the fabrication of complex integrated structures capable of locally enhancing optical nonlinearities. Amongst these we can list optical cavities such as micro-spheres, toroids, disks, and rings [55–60], and other optical components such as photonic crystals [61, 62] and spiral waveguides [51, 63, 64]. These devices exploit various strategies to maximize the nonlinear interactions. In particular, micro-cavities take advantage of constructive interference to boost the local intensity and hence the nonlinearities; on the other hand, photonic crystals can rely on an exceptionally high "engineered" refractive index permitting a tight field confinement and 90° bends, while spiral waveguides simply tend to reduce the on-chip occupancy of the device while simultaneously building up an important nonlinearity through a long optical path.

Resonators in particular have found a very broad range of applications in optics, including high-order filters [53], parametric lasers [65, 66], frequency converters [52, 55, 67], entangled photon sources [68, 69], and frequency comb generators [70]. As it will be subsequently detailed, the significant advantage of these structures is that, at resonant frequencies, a low input optical power can lead to enormous nonlinear effects due to the field enhancement provided by the cavity.

The four-port micro-ring resonator used in several experiments in this chapter is portrayed in Fig. 2.3. The waveguide cross-section (of the ring and bus) is shown in Fig. 2.2. Light is normally coupled from the input (bus) waveguide into the ring structure via vertical evanescent field coupling [71]. Light circulating inside the resonator is attenuated by the propagation losses and the ring-to-bus coupling

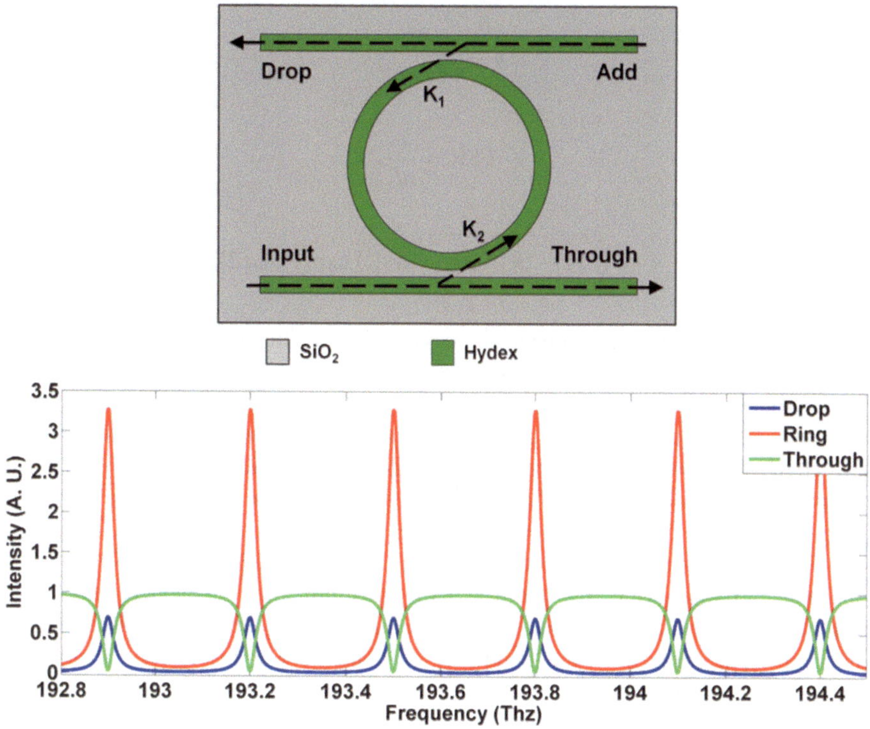

Fig. 2.3 (*Top*) Hydex micro-ring resonator scheme where the evanescent coupling coefficients (K_1, K_2) are represented. (*Bottom*) Intensity profile as a function of frequency for a wave propagating inside the ring (*red curve*), together with the theoretical transmission curves evaluated at the Drop (*blue curve*) and the Through port (*green curve*), respectively. The simulation assumes a FSR of 300 GHz, a Q of 6,863, and losses of $\alpha = 0.1/L$ (input port excited)

(two locations), whereas net gain is simultaneously obtained from an external light source that is coupled from the input bus waveguides to the ring. By using a direct analogy with a standard Fabry-Perot cavity, where the reflectivity of the mirrors has been replaced with the transmission across the bus waveguides, or with $\sqrt{1-|K_i|^2}$ where K_1 and K_2 are the field coupling coefficients, it is possible to calculate the beam intensity inside the resonator as a function of the angular frequency ω:

$$I_{\text{res}} = \frac{I_{\max}}{(1 + g\sin^2(\omega/(2\text{FSR})))}, \tag{2.7}$$

where

$$I_{\max} = I_0 \frac{|K_1|^2}{(1-r)^2}, \tag{2.8}$$

$$g = \frac{4r}{(1-r)^2}, \tag{2.9}$$

$$\mathrm{FSR} = \frac{c}{nL}, \tag{2.10}$$

$$r = \sqrt{(1-|K_1|^2)(1-|K_2|^2)}\exp(-\alpha L/2), \tag{2.11}$$

are the maximum intensity inside the cavity, the coefficient of finesse (not to be confused with the Finesse itself), the free spectral range (FSR) and the round trip amplitude attenuation factor, respectively (I_0 is the intensity of the input wave in the bus waveguide). Moreover, L is the circumference of the ring resonator, n is the effective refractive index of the structure, and α is (as previously defined) the linear propagation loss coefficient. The optical intensity spectrum inside such a resonator is presented in Fig. 2.3 (red curve), where we can identify the FSR (which is the frequency distance between adjacent resonances) and the linewidth of the resonance Δf_{FWHM} (the full width at half maximum of a resonance peak). The resonances occur at frequencies $f_{\mathrm{res}} = mc/nL$ (equally spaced if dispersion is negligible), according to the condition of constructive interference inside the cavity, where m is an integer.

In the same image we have also plotted the intensity transmission curves for the waves detected at the Drop (blue curve) and Through (green curve) output channels, respectively. The ratio between the light intensity inside the ring and that injected in the bus waveguides can be an extremely large value under resonance condition. This effect leads to a drastic increase of the nonlinear effects inside a low-loss resonator. The quality factor Q and the so-called finesse F of a resonator are two important parameters that can be used to quantify this increase, and they are respectively defined by:

$$Q = \frac{f_0}{\Delta f_{\mathrm{FWHM}}}, \tag{2.12}$$

$$F = \frac{FSR}{\Delta f_{FWHM}}. \tag{2.13}$$

The quality factor can be proved to be proportional to the ratio between the energy stored in the resonator to the energy dissipated for each round trip inside the resonator. In the next paragraph, we see that a high Q factor is extremely important for the enhancement of nonlinear effects.

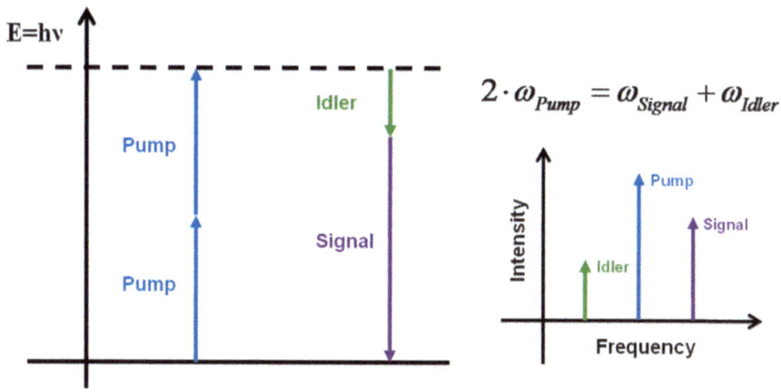

Fig. 2.4 Semi-degenerate four-wave mixing (FWM)

2.4.2 Nonlinear Wavelength Conversion in Integrated Silica Glass Resonators

The possibility of forming high-quality factor resonators primarily depends on the technological advancement of the fabrication processes. In particular, in order to have a high-quality factor micro-ring resonator, low propagation and radiative losses are essential, as well as a weak coupling coefficient between the bus waveguide and the cavity itself. The high-index doped silica glass ring resonator discussed in this chapter meets all these criteria, with propagation losses as low as 0.06 dB/cm, and negligible bending losses for radii down to 30 μm [30, 52]. High-Q resonators are ideal for applications such as narrow linewidth, multiple-wavelength sources, as shown in this section.

Here, we report on all-optical wavelength conversion via FWM by using a four-port Hydex micro-ring with a quality factor of 1,200,000, a FSR of 200 GHz, and a radius of 135 μm. Together with its negligible nonlinear absorption and CMOS compatible fabrication process, this resonator is optimal for integrated all-optical wavelength conversion.

The FWM we study involves the generation of a new “idler” wavelength by simultaneously using two different CW sources: a low power signal and a higher power pump. The quantum interaction amongst the photons is shown in Fig. 2.4 and is called semi-degenerate since two of the photons involved have the same frequency. For CW excitation, (2.4) predicts the following evolution equations:

$$\frac{\partial A_p}{\partial z} + \frac{\alpha_p}{2} A_p = i\gamma |A_p|^2 A_p + 2i\gamma A_p^* A_s A_i \exp(i\Delta\beta z), \tag{2.14a}$$

$$\frac{\partial A_s}{\partial z} + \frac{\alpha_s}{2} A_s = 2i\gamma |A_p|^2 A_s + 2i\gamma A_p^2 A_i^* \exp(-i\Delta\beta z), \tag{2.14b}$$

$$\frac{\partial A_i}{\partial z} + \frac{\alpha_i}{2} A_i = 2i\gamma \left|A_p\right|^2 A_i + 2i\gamma A_p^2 A_s^* \exp(-i\Delta\beta z), \tag{2.14c}$$

where the subscripts "p," "s," and "i" refer to the pump, signal, and idler, respectively. All temporal phenomena drop out of the equations due to the CW nature of the beams, and we neglected some cross-phase modulation terms based on the assumption that the power of the pump beam, P_p, is much greater than that of either the signal or idler, $P_p \gg P_s, P_i$. The phase matching term is given by $\Delta\beta = 2\beta_p + \beta_s - \beta_i$ and the frequencies are related through $2\omega_p = \omega_s + \omega_i$.

Considering a steady state inside the resonator (i.e., once the cavity is fully loaded), the above system of equations can be solved in the undepleted pump regime assuming a small signal gain. Specifically, given no input idler beam, we find that the wavelength conversion efficiency is given by:

$$\eta \equiv \frac{P_{i-out}}{P_{s-in}} = \gamma^2 P_{p-in}^2 L_{eff}^2 \rho^4, \tag{2.15}$$

$$L_{eff}^2 = L^2 \exp(-\alpha L) \left| \frac{1 - \exp(-\alpha L + i\Delta\beta L)}{\alpha L - i\Delta\beta L} \right|^2, \tag{2.16}$$

where we made the hypothesis that the propagation losses of all three beams were identical. There are several important factors in the above solution. First, the net wavelength conversion efficiency is quadratic with the pump power and also scales with the nonlinear parameter squared. This is as expected for third order nonlinear interactions. Moreover, the process scales with the effective length, L_{eff}, and with the resonator enhancement factor ρ to the fourth power. This enhancement factor (ρ^4) for the Hydex high-Q resonator has been theoretically calculated to be as high as 10^{10}. This extremely impressive factor results from a combination of low propagation losses and a careful cavity design with very small coupling coefficients. If we assume a low loss propagation regime, the resonator enhancement factor ρ can be rewritten in terms of the cavity finesse and/or the Q factor as [52]:

$$\rho \approx \frac{F}{\pi} = \frac{FSR}{f_0} \cdot \frac{Q}{\pi}. \tag{2.17}$$

This relation unequivocally explains why high finesse and high quality factor cavities are so widely used in nonlinear optics in order to enhance the efficiency of wavelength conversion processes.

From (2.16), we see that another limiting factor is the effective length. Indeed, if we assume perfect phase matching, the effective length reduces to $L_{eff} = \alpha^{-1} \exp(-\alpha L/2)(1 - \exp(-\alpha L))$, which for high losses decays exponentially as function of L, thereby severely affecting the wavelength conversion process.

Wavelength conversion in the high quality factor Hydex micro-ring resonator was examined experimentally. The set-up used for the semi-degenerate FWM

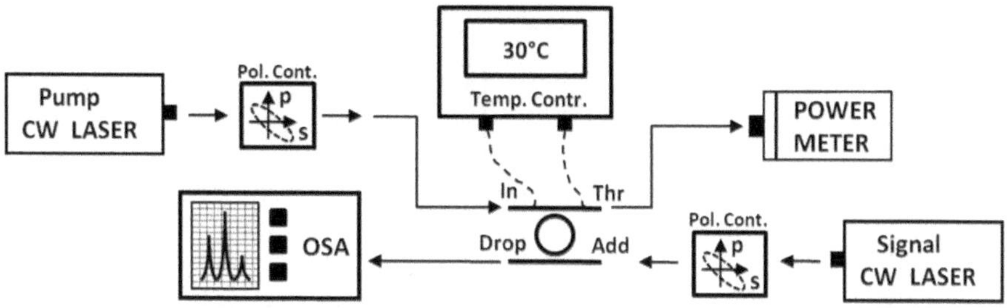

Fig. 2.5 Experimental set-up used to characterize the ring resonator and to measure the generated idler via FWM. Two tunable fiber CW lasers are used, one at the Input port and another at the Add port, whose polarizations are both set with inline fiber polarization controllers to match the ring resonances. The output spectrum and power are collected at the Drop and Through ports. A temperature controller is used to regulate the temperature of the device

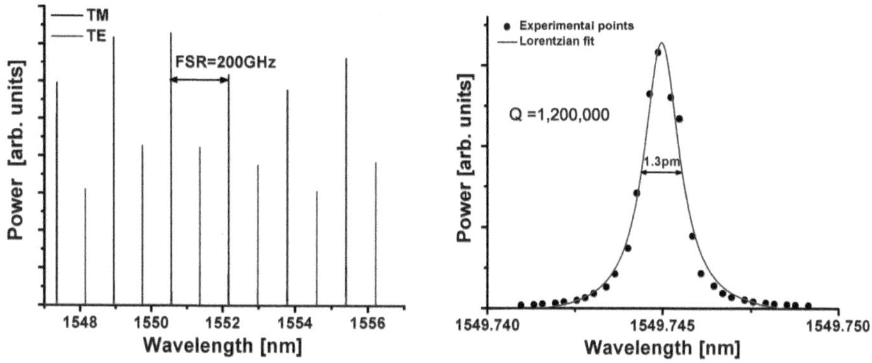

Fig. 2.6 Input-Drop response of the micro-ring resonator (*left*) and zoom of a representative transverse electric (TE) resonance around 1,549 nm (*right*)

experiments is depicted in Fig. 2.5, and consists of two CW lasers, two polarizers, a power meter, and a spectrometer. A Peltier cell is also used in order to stabilize the temperature of the device from thermal fluctuations in the lab. A linear characterization, performed by simply recording the output power at the Drop port (input port excited), allowed us to identify the resonances and the ring parameters (Q factor, FSR) as shown in Fig. 2.6.

For the nonlinear frequency mixing experiment, the wavelength of the pump laser was tuned to a ring resonance while a second laser, tuned to a different resonance, stimulated the conversion process. Figure 2.7 summarizes the results of two different FWM experiments where the pump and signal lasers are tuned to adjacent resonances, and to two resonances spaced six FSRs apart from each other. In both cases, the internal conversion efficiency was estimated to be −26 dB with only 8.8 mW of input power and 1.25 mW of signal power (according to 2.15, and quoting the wave powers inside the bus waveguide). The onset of cascaded FWM

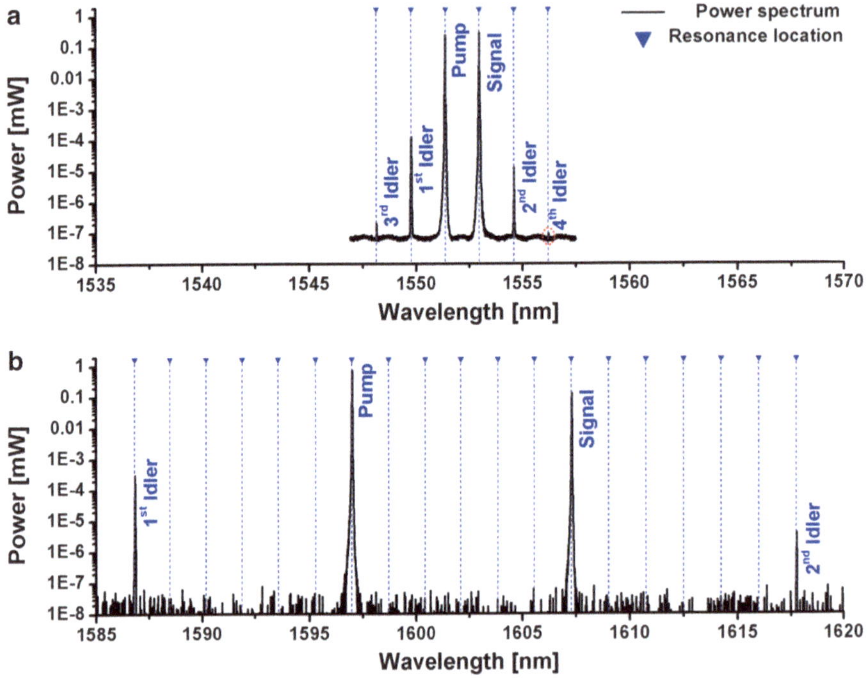

Fig. 2.7 FWM experimental results for pump and signal wavelengths tuned to adjacent resonances for the TE mode (200 GHz) (*upper plot*) and six resonances apart (~1.2 THz) for the TM mode (*lower plot*). The third and fourth idlers in the *upper plot* demonstrate the onset of cascaded FWM

can also be observed whereby the pump and first idler interact to generate the third idler or where the second idler mixes with the signal wave to generate the fourth idler (red dashed circle in Fig. 2.7a).

This results are in part due to the relatively large (experimentally calculated) γ factor of 220 W^{-1} km^{-1} [51], as compared to SMFs ~1 W^{-1} km^{-1}, and, more importantly, due to the low losses, resulting in a large intensity enhancement factor, which is orders of magnitude higher than in semiconductors, where losses are typically the limiting factor [8, 38]. Recent results in SOI have also shown impressive, and slightly higher, conversion efficiencies using CW power levels. However, as can be seen in [8], saturation due to free carrier absorption via TPA limits the overall process, whereas in silica-doped Hydex glass it has been shown [51] that no saturation effects occur for up to more than 25 GW/cm^2 of intensity (note that the pump intensity in the ring is only ~0.08 GW/cm^2 at resonance for 5 mW of input power).

Another critical parameter that was required to achieve these results was phase matching. Whereas all resonator modes automatically satisfy the phase matching relation, there is no guarantee that the generated idler wave is at a resonant

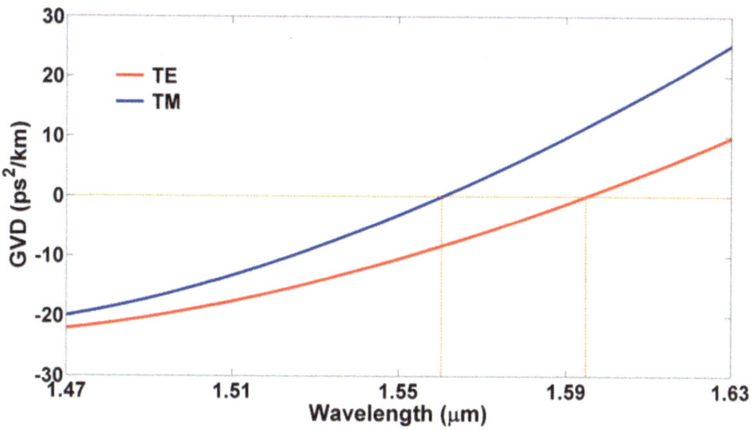

Fig. 2.8 Group velocity dispersion (i.e., β_2) of the resonator obtained by fitting the experimentally measured resonance frequencies. The fit includes dispersion terms up to fourth order and it is performed for both the TE (*red curve*) and TM (*blue curve*) modes

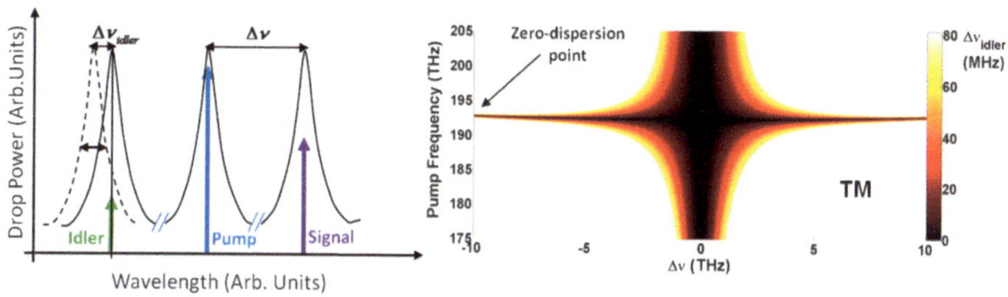

Fig. 2.9 Phase matching diagram associated to FWM in a micro-ring resonator. Explicative scheme of the idler detuning resonance due to dispersion (*left*). Plot of the idler detuning $\Delta\nu_{\text{idler}}$ as a function of both the signal-to-pump frequency distance $\Delta\nu$ and the pump frequency (*right*). The regions in *white* are areas where FWM is not possible, whereas the *colored* regions denote possible FWM with the color indicating the degree of frequency mismatch (*black* implies perfect phase matching)

frequency. Indeed, energy conservation dictates the output frequency of the FWM process to be $\omega_{\text{i}} = 2\omega_{\text{p}} - \omega_{\text{s}}$. This frequency will only be aligned to a cavity resonance provided that the dispersion in the overall resonator is negligible (see Fig. 2.8—left). The dispersion in this system was evaluated by measuring the frequency dependence of the FSR, from which the propagation constant was extracted and fit to a fourth order polynomial (i.e., dispersion parameters up to β_4) as can be seen in Fig. 2.9 [67]. The dispersion is extremely close to 0 in most of the C-band, and thus allowed for phase matching in our experiments.

Moreover, this dispersion data can be used to predict the maximum bandwidth over which FWM is observable. This is obtained by considering that the generated idler frequency must be within the line width of the closest resonance:

$$\Delta\omega_i = |\omega_i - \omega_{res}| < \Delta\omega_{FWHM}/2, \tag{2.18}$$

where ω_i, ω_{res}, and $\Delta\omega_i$ are the idler frequency obtained from energy conservation, the closest ring resonance to the idler, and the correspondent idler detuning, respectively. Whenever this condition is satisfied, the generated idler is within the bandwidth of the resonance, phase matching is fulfilled, and FWM is experimentally achievable [67]. In Fig. 2.8 (right), we plot the above defined frequency shift (note $\Delta\nu_{idler} = \Delta\omega/2\pi$) as a function of the signal to pump frequency distance $\Delta\nu = (\omega_p - \omega_s)/2\pi$ at a given pump frequency. The region in white represents lack of phase matching; whereas the colored region represents possible FWM (black implies perfect phase matching). The curvature in the plot is a result of high order dispersive terms. It can be seen that FWM can be accomplished in the vicinity of the zero dispersion point (see Fig. 2.8—right) up to ±10 THz (80 nm) away from the pump. This extraordinary result comes from the low dispersion of the resonator, which permits appreciable phase matching over a broad bandwidth at low power. At higher input powers the net phase mismatch also contains a term proportional to the pump power; a result of cross-phase modulation. It can be shown that the gain bandwidth increases with pump power for an anomalous dispersion regime [16].

2.4.3 *Optical Hyper-parametric Oscillator*

The previous section demonstrated the effectiveness of Hydex resonators for nonlinear optical applications through an experiment of semi-degenerate FWM. Frequency conversion was obtained using a pump and a signal wave, the latter being used to stimulate (or seed) the process. However, use of this signal laser is not mandatory, but allows for an increased efficiency, while simultaneously allowing the possibility to tune the idler wave in a broad frequency range.

In this subchapter, we exploit hyper-parametric gain to achieve a fully integrated CMOS compatible multiple wavelength source. By resonantly pumping a single CW light source in our high-*Q* (1.2 million) micro-ring resonator, we achieve parametric oscillation of multiple lines over a very broad (>200 nm) spectral range.

When only the pump photons are present, we call the FWM process spontaneous. Effectively, the photons necessary to complete the transition displayed in Fig. 2.4 are supplied by the quantum vacuum. The vast difference between seeded and spontaneous FWM is that the vacuum is broadband, leading to broadband wavelength conversion. However, as (2.4) shows, the idler power depends on the signal power (vacuum here) which is inherently low.

Notwithstanding this, in a resonator this small gain, proportional to the pump power, can be large enough to overcome the losses within the cavity. In this case, oscillation can occur whereby a pair of idlers (to be identified with the signal and idler of the previous section) can grow out of vacuum via self-seeded parametric gain. A device capable of achieving this is called an optical hyper-parametric oscillator (OHPO).

A parametric oscillator is quite similar to a laser in which we have substituted the active medium with a nonlinear material [72, 73]. In a laser the stimulated emission stems from the passage of electrons from a higher to a lower energy level, whereas in a parametric oscillator the process is mediated by virtual levels. Similarly to the lasing process, the parametric oscillation is a threshold phenomenon that occurs when the total roundtrip gain equals the roundtrip losses. Once we surpass the threshold in the cavity, the oscillation wavelength and pump power in the resonator become clamped, and an increase in pump power further amplifies the oscillating mode, eventually initiating a process of cascaded FWM that generates a series of wavelengths at multiples of the initial spacing between the pump frequency and the oscillating frequency. The majority of OPOs exploit $\chi^{(2)}$ phenomena and are subject to very stringent phase matching conditions.

Here, the parametric gain is provided by the next highest nonlinearity $\chi^{(3)}$, and the process is said to be "hyper-parametric" [65]. Below the oscillation threshold, the gain of the vacuum is approximately exponential:

$$I_{\text{noise}}(z) = I_{\text{noise}}\big|_{z=0} \exp(gz), \tag{2.19}$$

where g is the parametric gain expressed by:

$$g = 2|\Omega|\sqrt{-\left(\frac{\beta_4}{24}\Omega^2 + \frac{\beta_2}{2}\right)\left(\frac{\beta_4}{24}\Omega^4 + \frac{\beta_2}{2}\Omega^2 + 2\gamma P_{\text{p}}\right)}, \tag{2.20}$$

and $\Omega = (\omega - \omega_{\text{p}})$ is the frequency detuning from the pump. Equation (2.20) assumes dispersion up to fourth order (as characterized experimentally). The factor g represents a gain (i.e., $g > 0$) only if we are under the condition of anomalous dispersion (requiring β_2 or/and $\beta_4 < 0$).

By equating the roundtrip losses to the parametric gain in the resonator, we deduce that the pump power threshold should be ~54 mW. A plot of (2.20), for a pump wavelength of 1,544.15 nm and a power of 54 mW, produces the graph presented in Fig. 2.10, from which we determine the maximum parametric gain to be at 1,590 nm.

Figure 2.11 shows the setup used for the optical parametric oscillation experiment. A CW pump laser (Anritsu Tunics Plus, 150 kHz linewidth), amplified by an EDFA (Amonics HEDFA), was passed through a wavelength tunable band-pass filter (1 nm, 3 dB bandwidth) to eliminate amplified spontaneous emission (ASE) from the amplifier. The pump laser was then properly polarized, coupled into the ring resonator, and tuned to one of its resonances while the power was monitored by a power meter placed at one of the arm of a 99%-1% splitter.

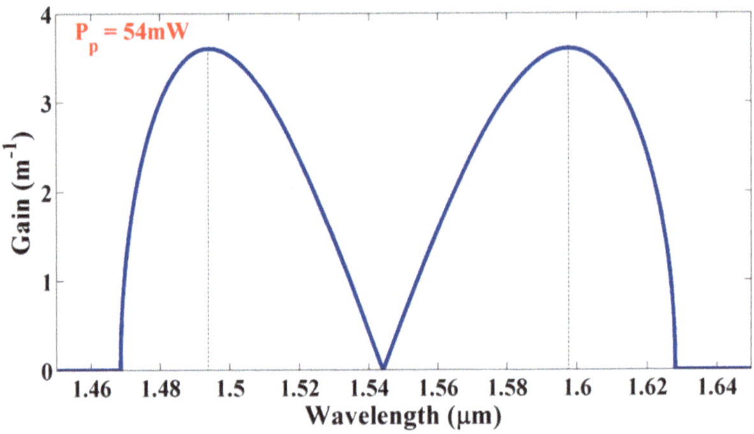

Fig. 2.10 Plot of the parametric gain for a pump wavelength of 1,544.15 nm and a power of 54 mW

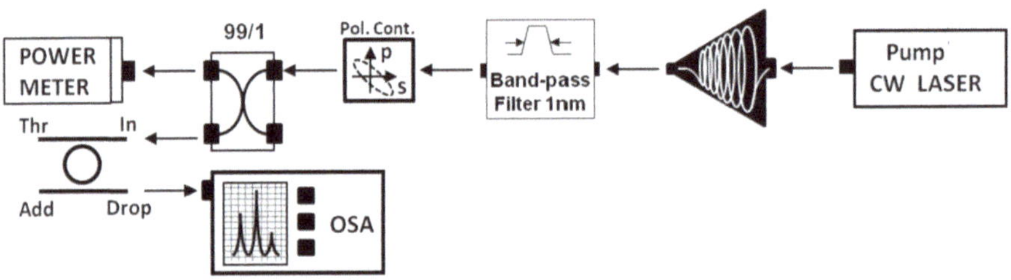

Fig. 2.11 Schematic of the optical hyper-parametric oscillation experimental set-up

During the experiment a "soft thermal lock" condition was established in order to keep the optical power inside the cavity constant (the temperature increase led to a modification of the refractive index and hence a shift of the resonance frequencies). The thermal lock procedure consists in tuning the pump wavelength to one resonance of the ring, then slowly following its drift due to cavity heating, until a thermal equilibrium is reached. The optical output spectrum for different values of the input power was then recorded by means of an optical spectrum analyzer (OSA) (ANDO AQ6317B).

Our pump laser linewidth was fully coupled inside the FWHM of the selected resonance while no overlap between the laser tail and the ring resonances was observed (the band-pass filter used reduced the noise to below −70 dB). This allowed the system to evolve according to parametric oscillation rather than stimulated FWM. Figure 2.12 shows the output spectra for a TM polarized pump, tuned to a resonance at 1,544.15 nm, for different values of the input power.

From this set of plots, we can make some important conclusions. The lasing condition occurs for a resonance located at 1,596.98 nm, which is close to the location for the peak of the parametric gain predicted by calculations. The threshold

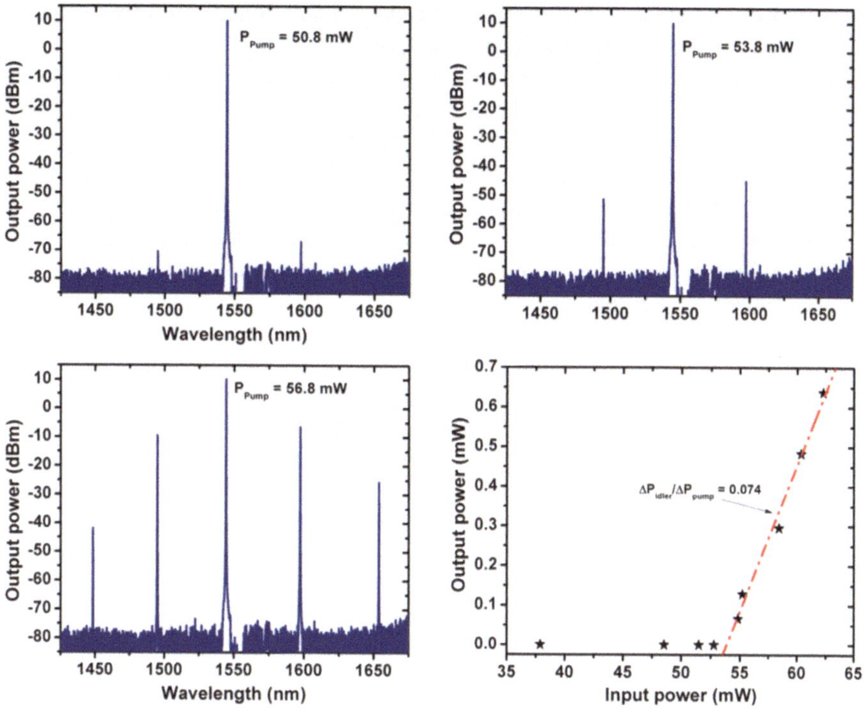

Fig. 2.12 Output spectra of the hyper-parametric oscillator and oscillation threshold diagram. The spectra show the evolution of the spontaneous FWM process as a function of the input pump power at 50.8, 53.8, and 56.8 mW (pump on a resonance at 1,544.15 nm). The last graph (*bottom-right*) represents the output power at the drop port for the single oscillating mode at 1,596.98 nm versus pump power. A differential slope efficiency of 7.4% is extracted from a linear fit (*red dashed line*) above threshold

power was determined to be 54 mW (all the powers are quoted inside the bus waveguide unless stated otherwise) while the single-line slope efficiency (above threshold at 1,596.98 nm) was estimated to be 7.4%. For an input power of 101 mW (at 1,544.15 nm [65]), we obtained a remarkable conversion efficiency of 9% (9 mW amongst all the oscillating modes). This overall efficiency is an outstanding result, especially when taking into account that the efficiency of FWM in ring resonators [8, 67] is typically below −20 dB at these pump powers. Figure 2.12 also shows that as the pump power keeps increasing the cascade, FWM takes over with the same spacing imposed by the parametric oscillation.

The experiment was repeated by tuning the pump to other cavity resonances located into a region of normal dispersion. In these cases, we did not observe oscillation even for pump powers much larger than the threshold power we used at 1,544.15 nm. This result is consequently consistent with the predicted behavior of the parametric gain, requiring anomalous dispersion. It is worth mentioning that the maximum frequency spacing we have achieved in our optical parametric oscillator (52.83 nm or >6 THz) is unprecedented and potentially very useful for applications where high repetition rates are required.

In fact, if we assume a phase relation amongst the different components of a broadband frequency comb, the latter, in the time domain, corresponds to a high repetition rate train of optic pulses. However, the most immediate advantage of having an integrated multiple wavelength source would be the possibility to develop a compact and cost-effective wavelength-division multiplexing (WDM) system. Such an integrated device would also drastically reduce the complexity of today's WDM devices, where an array of laser sources could be replaced by one single source with a resonator.

Here, we have shown how we can exploit integrated optical cavities to generate coherent narrow linewidth light in a broad spectral range. However, one can think to use other approaches to design this kind of optical sources. For example, this could be achieved using a very broadband "white light" source, in conjunction with an ultra-narrow band pass filter. In this case, there is an added advantage of not being restricted to cavity resonances and allowing for continuous tuning, which can be an important tool for spectroscopic applications [74]. The following paragraph elucidates the possibility to employ Hydex integrated structures for this alternative configuration.

2.4.4 *Broad Band Light Generation in an Integrated Spiral Waveguide*

In this paragraph, low loss Hydex glass waveguides will be exploited to study SCG using a 45-cm long waveguide that is tightly confined on chip via a spiral geometry. Here, we seek to exploit the length dependence of the nonlinearity. Compared with the resonators introduced above, this spiral allows a continuous generation of new frequencies (and not discrete resonances). SCG typically refers to the generation of broadband light from spectrally narrow pulses through a mixture of nonlinear effects [75]. The SCG process can be used for a wide range of applications such as the realization of multiwavelength and tunable sources, spectroscopy, bio-imaging, pulse compression, and optical coherence tomography.

The waveguide used in our experiments has the same cross-sectional core area as that used for the ring resonator, namely $1.45 \times 1.5\ \mu m^2$. It is contained on a footprint area as small as $2.25 \times 2.25\ mm^2$, and is schematically shown in Fig. 2.13. Due to the significant index contrast (17% between clad and core), spiral structures can be fabricated with negligible bending loss, thereby allowing the possibility to exploit the very long interaction length for nonlinear effects.

As a result of the broad transparency of these waveguides (from the UV to well into the mid-IR) and the very low dispersion, they naturally offer the best conditions for very broad SCG. As previous studies have shown [76], SCG is the result of a cascade of nonlinearities, and has been shown to be very efficient when a pump wavelength close to a zero-dispersion point is used. Here, we present SCG in the spiral waveguide by pumping light in proximity of two different zero-dispersion wavelengths, namely at

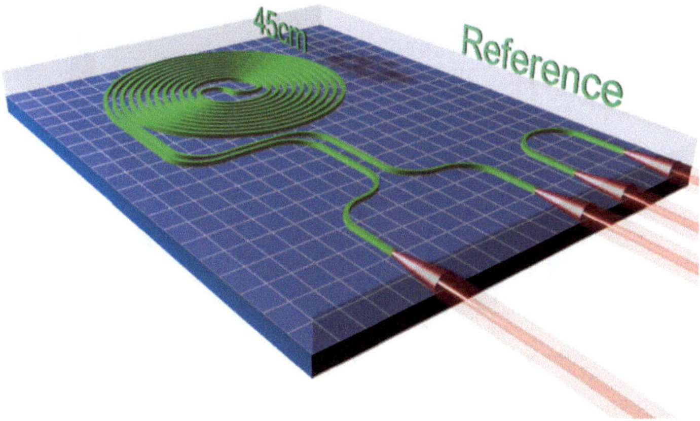

Fig. 2.13 Top-down schematic view of the 45-cm long spiral waveguide

1,288 nm where the normal dispersion regime lies at shorter wavelengths, and also at 1,550 nm where the normal dispersion regime is at longer wavelengths (see ref. [63] for more details on the various dispersion regimes).

SCG can be numerically modeled assuming a single-beam propagation with the dominant nonlinear effects being: SPM, cross-phase modulation, intra-pulse four-wave mixing, soliton fission, Raman self-frequency shift (SFS), and the generation of Cherenkov radiation. This complex array of nonlinear interactions can be modeled by the generalized nonlinear equation (2.4):

$$\begin{aligned}&\frac{\partial A}{\partial z}+\frac{\alpha}{2}A+\beta_1\frac{\partial A}{\partial t}+i\frac{\beta_2}{2}\frac{\partial^2 A}{\partial t^2}-\frac{\beta_3}{6}\frac{\partial^3 A}{\partial t^3}-i\frac{\beta_4}{24}\frac{\partial^4 A}{\partial t^4}\\&=i\gamma|A|^2A-\gamma\left(\frac{1}{\omega}\frac{\partial}{\partial t}|A|^2A+iT_{\mathrm{R}}A\frac{\partial}{\partial t}|A|^2\right).\end{aligned} \tag{2.21}$$

This model takes into account dispersion up to fourth order (as measured experimentally) and takes into account ultra-short pulse phenomena. Since a precise estimation of the Raman effect in these waveguides is still under investigation, the Raman response of fused silica was used as a first order approximation for the simulations below. The input waveform was taken to be a Gaussian pulse with a small chirp ($C=-0.1$) in order to match the experimentally measured pulse spectrum and autocorrelation. In addition, we also consider that the input pulse is propagating in a quasi-TE mode with the experimentally determined dispersion coefficients: $\beta_2=10.5$ ps^2/km, $\beta_3=0.26$ ps^3/km, and $\beta_4=0.0035$ ps^4/km at 1,550 nm.

The experiment was performed by using a Spectra Physics OPO generating 100 fs pulses (200 fs at 1,300 nm), determined via autocorrelation, with a repetition rate of 80 MHz. The spectral bandwidth at -20 dB for the input pulse was 110 nm at 1,550 nm and 55 nm at 1,288 nm [63]. The pulses produced by the OPO were first attenuated and then sent to a 99/1 fiber coupler. While 99% of the signal was sent to

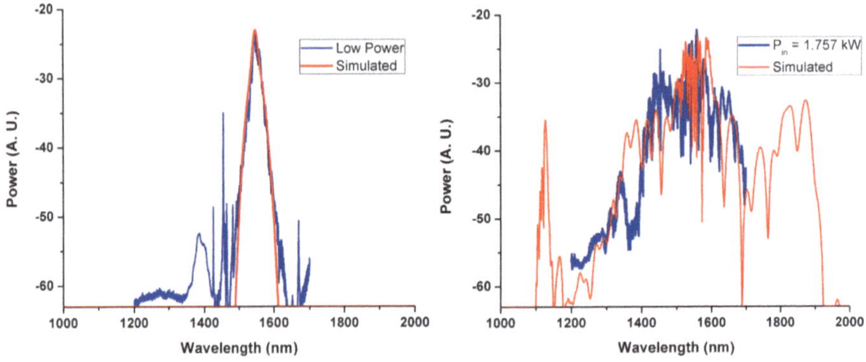

Fig. 2.14 (*Left*) Simulated (in *red*) and experimentally measured (in *blue*) input pulse spectra for the SCG experiment. (*Right*) SC spectrum at the output of the 45-cm long spiral waveguide when the input peak power is 1.76 kW. Note that the simulated pulse spectral broadening predicts new frequency components even beyond the experimental measurements which were limited to the range of 1,200–1,700 nm by the OSA

the input of the spiral waveguide, the remainder was directed to a power meter to continuously monitor the input power. The output spectrum was finally recorded by using an OSA for various input power levels; a detailed experimental set-up can be found in [63]. The output and input spectrum for the experiments at 1,550 and 1,288 nm are presented in Figs. 2.14 and 2.15, respectively.

At 1,550 nm the input bandwidth is shown to broaden by more than a factor of 3, achieving more than 300 nm of spectral width using an input peak power of 1.76 kW. Note that the bandwidth is theoretically predicted to be much broader on the long wavelength side (>1,700 nm), but this was not detectable experimentally due to the limited range of our spectrometer. The results are even more impressive at 1,288 nm, where the broadening is significantly wider (the −30-dB bandwidth is more than 600 nm) and at a lower peak power of 1 kW. As determined from the simulations, soliton fission, dispersive wave generation, Raman self-frequency shifting coupled with FWM are the dominant effects responsible for the broadening at 1,550 nm. The significant differences between the 1,288- and 1,550-nm results are attributed primarily to the different dispersion regimes, and we anticipate that the higher order dispersive terms near 1,288 nm are significantly larger than those at 1,550 nm. Moreover, soliton instability is strictly governed by higher order dispersion terms, and thus could initiate the onset of SCG earlier at 1,300 nm.

These results are extremely promising, and demonstrate the exploitation of long structures, possible due to the exceptionally low propagation losses, to generate significant nonlinear effects, while requiring an extremely small device footprint. Moreover, the broad transparency of the Hydex glass could be utilized to extend the spectrum to the visible and into the UV. Together with a narrow band tunable filter, this device shows much promise as a tunable wavelength source.

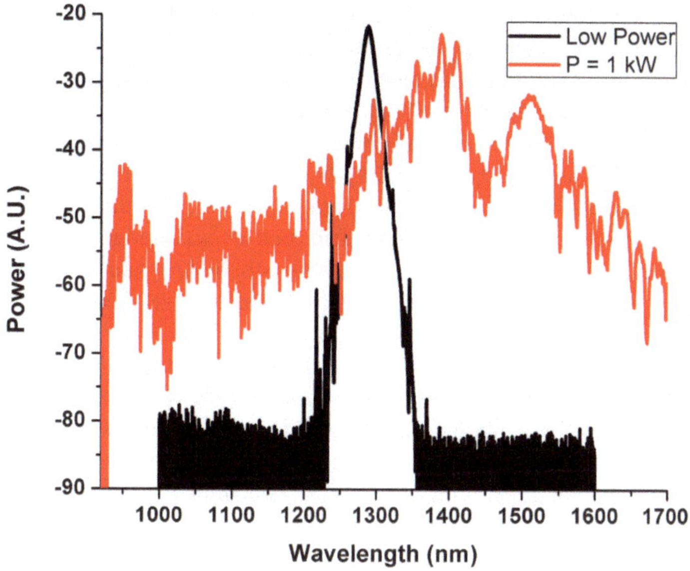

Fig. 2.15 SCG using a pumping wavelength of 1,288 nm. The experimental input (in *black*) and output (in *red*) pulse spectra are shown. The output spectrum, obtained at a peak power of 1 kW, has a bandwidth of more than 300 nm at −20 dB

2.5 Ultrafast All-Optical Processing

2.5.1 *Ultra-short Pulses Frequency Conversion via FWM in Integrated Glass Waveguides*

The final part of this chapter is focused on ultrafast optical processing, with particular attention to four very important signal processing operations: (1) ultra-short pulse frequency conversion, (2) optical compression, (3) time width estimation of subpicosecond pulses, and (4) optical temporal integration. While the first three of these operations have all been carried out by using the 45-cm Hydex spiral waveguide presented in the previous section, the fourth makes use of the same micro-ring cavity used for the FWM and OPO experiments previously described. Moreover, it is important to emphasize that the optical integration presented here is a linear phenomenon and demonstrates the versatility of this integrated platform is not limited uniquely to the nonlinear regime.

Whereas the previous section dealt with the generation of new frequencies via SCG, a more selective frequency conversion with a specific output frequency target can be more efficient and may find diverse applications. In particular, all-optical approaches based on ultrafast optical nonlinearities, such as pulse FWM, have been successfully employed for optical signal processing. When implemented in a waveguide geometry, they provide useful instruments able to overcome the bandwidth limitations associated with electronic approaches. FWM in waveguides has been used to demonstrate several fundamental operations on pulses for optical signal processing, ranging from frequency conversion to signal reshaping, regeneration, and amplification [20, 23, 29, 34, 77, 78], and more recently for temporal signal measurement, which will be discussed in Sect. 5.2. For this reason, there is a strong interest in demonstrating parametric gain in waveguides. Gain in highly nonlinear waveguides has been demonstrated in silicon [79] and in some nonlinear glasses like heavy metal oxides [23] or chalcogenide [78]. In particular, a net parametric gain of 1.8 dB was first reported in dispersion engineered silicon nanowires [79] and since then, a net overall gain (including all losses) of 16 dB was obtained in chalcogenide glass waveguides [78]. However, as mentioned above, for both of these platforms, linear and nonlinear losses can limit performance—particularly for silicon where saturation due to nonlinear absorption of the pump and subsequent generation of free carriers clamps the maximum achievable gain at about 10 W of peak pump power [79].

In this section, the remarkable nonlinear properties of the Hydex spiral waveguide are explored for wavelength conversion of ultra-short pulses by exploiting FWM. We report net parametric gain [80] obtained with subpicosecond pump-probe pulses. We achieve +16.5 dB conversion efficiency, with 40 W of pump (peak) power, and observe no saturation. The low dispersion of the waveguide allows us to observe FWM on a remarkably large bandwidth of over 100 nm.

As in the previous section, we modeled pulse propagation in the waveguide with the GNLS equation (2.21), where we neglected the Raman effect. Equation (2.21) was integrated via a standard pseudo-spectral approach, and Gaussian pulses were assumed for both the input pump and the signal envelopes.

The measurements were carried out with an OPO system that generates 180 fs (bandwidth = 30 nm) long pulses at a repetition rate of 80 MHz. The broadband pulse source was split and filtered by way of two tunable Gaussian bandpass filters operating in transmission with a −3 dB bandwidth of 5 nm (equivalent to a pulsewidth of ~700 fs), in order to obtain synchronized and coherent pump and signal pulses at two different wavelengths. The pump and signal pulses were combined into a standard SMF using a 90/10 beam splitter and then coupled into the spiral waveguide. Pulse synchronization was adjusted by means of an optical delay line, while power and polarization were controlled with a polarizer and a $\lambda/2$ plate. Both pump and probe polarizations were aligned to the quasi-TE mode of the device.

Figure 2.16 shows the experimental results for three different signal excitations at $\lambda_s = 1{,}480$, 1,490, and 1,500 nm with 3 mW of power, respectively. The pump wavelength was set to $\lambda_p = 1{,}525$ nm in all the three cases and the pump peak power coupled inside the waveguides was varied from 3 to 40 W. Cascaded FWM was observed for pump powers larger than 30 W. The asymmetry visible in the interaction (the cascaded FWM induced by the idler is stronger than the one induced by the signal) indicates a non-negligible contribution due to the β_3 term, consistent with a low absolute value of β_2 [16], as confirmed by our fitting (see Fig. 2.9).

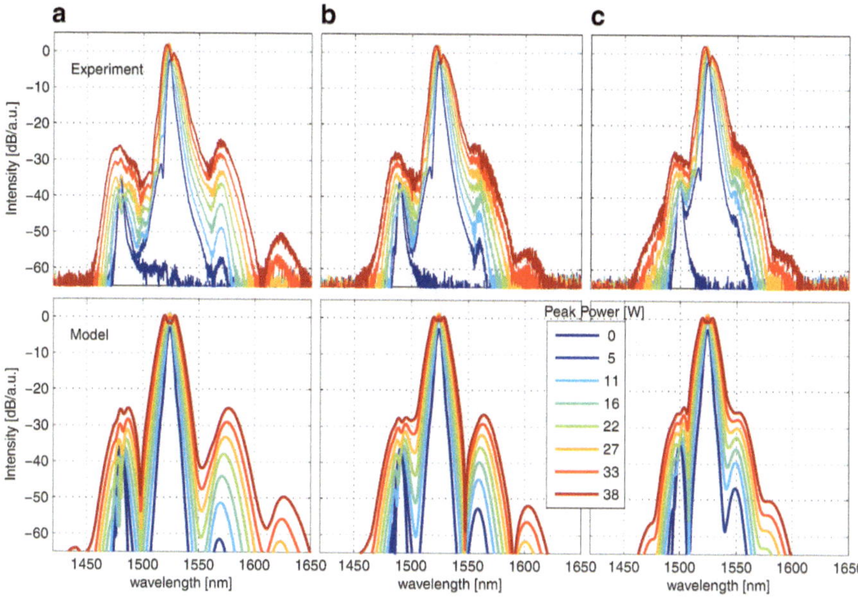

Fig. 2.16 Experimental (*top*) and theoretical (*bottom*) signal intensity spectrum for a 1,525-nm pump and a 1,480- (**a**), 1,490- (**b**), and 1,500-nm (**c**) signal. The legend lists the pump peak powers, while the signal peak power is 3 mW

The experiments show a larger gain for wavelengths closer to the pump using powers that are lower than 13 W, while the gain becomes almost flat (in a range of over 100 nm) for powers larger than 30 W, as can be seen in Fig. 2.16b. We define the "on/off" conversion efficiency η_{i}, as the ratio of the idler wave's $i(t)$ transmitted energy to the power of the signal wave $s(t)$ [79]:

$$\eta_i = \lim_{T\to\infty} \frac{\int_{-T/2}^{T/2} |i(t)|^2 \mathrm{d}t}{\int_{-T/2}^{T/2} |s(t)|^2 \mathrm{d}t}. \tag{2.22}$$

This allows us to account for various pulse shapes and for the spectral broadening due to cross-phase modulation, which lowers the spectral intensity. The net "on-chip" gain is then the on/off gain minus the propagation loss. The experimental on/off efficiency versus pump peak power is shown in Fig. 2.17c, d along with the theoretical calculations for both a CW and a pulsed pump. The modeling and experiment agree quite well. The CW case represents the maximum achievable gain for a given pump peak power, as it maximizes relation (2.22), and it is insensitive to detrimental effects, such as spectral broadening and temporal walk-off, that can limit the efficiency in the pulsed regime. For a 38-W pump power,

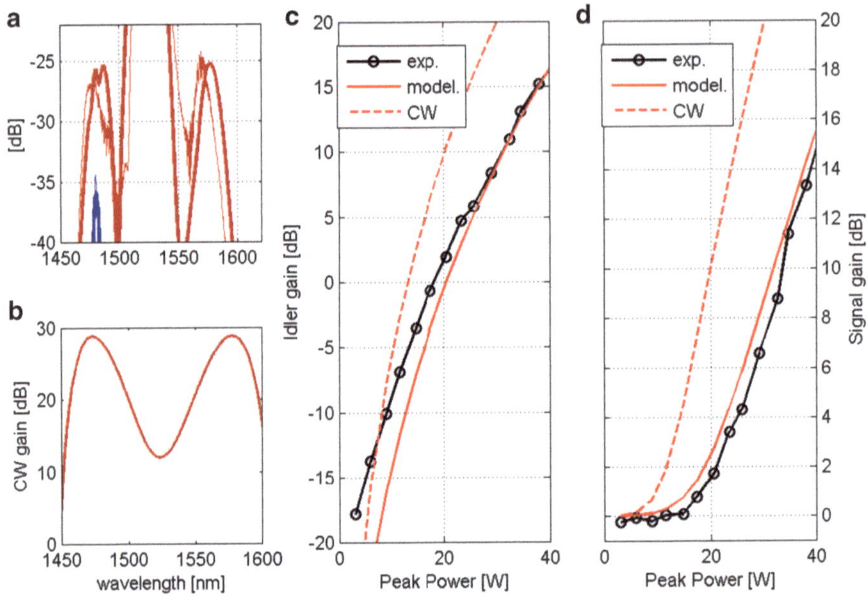

Fig. 2.17 Gain for a 1,480-nm signal: (**a**) Spectra for a 3-mW peak power signal alone (*blue*) and with a 40-W pump (*red*). (**b**) CW gain. (**c**, **d**) FWM gain for idler and signal, respectively: measurement (*black dots*), simulated pulsed model (*red continuous line*), and for a CW regime (*red dashed line*)

we measured a maximum on/off FWM conversion efficiency of +16.5 dB from signal to idler. This translates into a net on-chip conversion efficiency of +13.7 dB and a gain of +12.3 dB, when the overall propagation loss of 2.7 dB is included. We note that even at the highest pump powers used in these experiments, we do not observe any signs of saturation and this is largely due to the negligible nonlinear absorption in this platform.

This demonstration of parametric gain via pulsed FWM in a CMOS compatible doped silica glass waveguide is promising for all-optical ultrafast signal processing applications such as frequency conversion, optical regeneration, and ultrafast pulses generation. As an interesting application of pulsed FWM signal processing, we show in the next paragraph pulse measurements via time-to-frequency conversion based on FWM interaction.

2.5.2 *Time Lens Measurement of Subpicosecond Optical Pulses*

This section is dedicated to the ultrafast optical pulse characterization technique based on time lenses (TL) [81]. Ultrafast nonlinearities can be conveniently exploited for temporal signal imaging through the so-called space–time duality [81–91]. In simple terms, refractive optical elements such as lenses and prisms have temporal analogs that perform equivalent functions in the time domain, as they are represented by the same mathematical equations. The importance of this equivalence is associated to the fact that a vast category of optical instruments for diffractive beams rely on a suitable combination of lenses and free-space beam propagation. Thus, equivalent concepts can be employed for processing temporal signals by properly combining TLs, and dispersive elements.

More specifically, we define a TL as a physical mechanism that induces a quadratic phase profile on an input temporal optical signal, in analogy to a spatial lens, which adds a quadratic curvature to the phase front of a transmitted beam. The best results to date for a TL process have been obtained by using a parametric process, such as three-wave mixing (TWM) or degenerate FWM [88–92]. In particular, the FWM approach [90–92] has the advantage of being suitable for centrosymmetric materials, such as silicon or glass, and thus for fundamental platforms in integrated optics.

TLs can also be used for time-to-frequency domain conversion. This mechanism has proven particularly useful for measuring the time-domain intensity profiles of ultrafast optical waveforms. In the spatial case, it is well known that the optical field transmitted by a lens at its focal plane is the Fourier Transform (FT) of the spatial input field distribution one focal length in front of the lens. This concept, when translated into its temporal counterpart, allows one to capture the time-domain intensity profile through a simple optical spectrum measurement.

In practice, the pulse under test (PUT) is initially stretched with a dispersive element of length L and group velocity dispersion β_2 (i.e., a spool of SMF), resulting in a temporal phase curvature $\phi_F = \beta_2 L$, and is then coupled into the waveguide as the signal for the parametric interaction. This signal (i.e., the PUT with the appropriate chirp applied) plays the role of the input waveform to the TL. A highly chirped Gaussian pulse, much longer in time than the signal duration, acts as the pump of the parametric interaction and it plays the role of the lens itself. Only the pump chirp ϕ_P is relevant, while its temporal amplitude can be neglected. The TL effective phase curvature resulting from a FWM process is $\phi_F = \phi_P/2$, due to the quadratic dependence on the pump profile of the idler for a FWM interaction [90–92]. We note that for a TWM-based TL, the condition would be $\phi_F = \phi_P$ [88, 89].

The idler behaves as the (output) waveform transmitted by the TL, since it acquires the signal temporal shape and the TL phase curvature ϕ_F that exactly compensates the signal chirp. In this configuration, the idler energy spectrum maps the temporal intensity shape of the PUT according to the following time-to-frequency scaling law:

$$t = \phi_F \omega. \tag{2.23}$$

A temporal imaging system is then realized by measuring the idler with a spectrometer. Figure 2.18 summarizes the principle of time-to-frequency conversion using a TL based on FWM.

In our experiments, subpicosecond pulses for the pump and the signal were prepared from a 17-MHz repetition rate mode-locked fiber source, providing pulses at $\lambda = 1{,}550$ nm. The pulses were spectrally broadened after propagation in a nonlinear

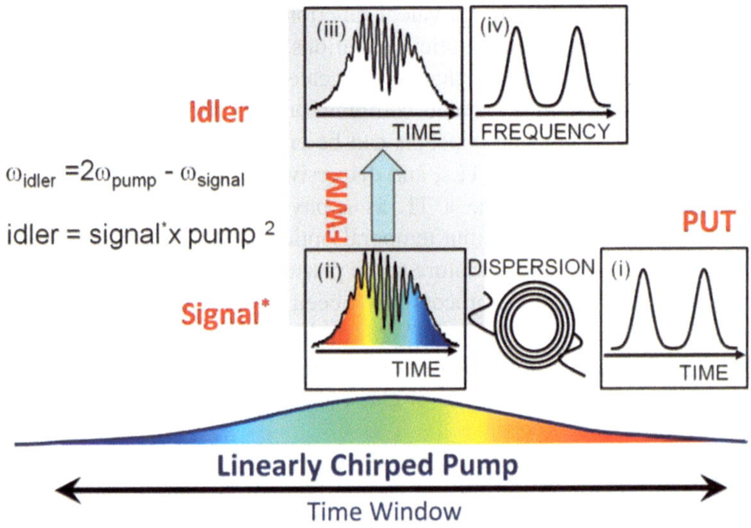

Fig. 2.18 Principle of time-to-frequency conversion using a TL based on the FWM experienced by the PUT when interacting with a linearly chirped pump pulse. In the case illustrated here, the PUT first undergoes a sufficiently large dispersion to enter the so-called temporal Fraunhofer regime, where the time-domain optical waveform at the input of the TL (signal) is proportional to the spectral shape (amplitude variation) of the PUT and is phase chirped due to the large dispersion (*i*). When the signal interacts with the pump (*ii*) the nonlinear frequency conversion produces an idler with the temporal shape of the signal but with a chirp exactly compensated by the pump chirp (*iii*). The output of the TL (idler) is transform limited and its temporal shape is proportional to the spectrum of the PUT. As a result, the spectrum at the TL output is proportional to the temporal amplitude profile of the PUT (*iv*)

fiber. As in the previous FWM experiment (described in Sect. 5.1), the PUT and the pump pulses were coherently synchronized by way of two tunable Gaussian band pass filters operating in transmission with a −3-dB bandwidth of 5 nm (equivalent to a pulse width of ~700 fs) and 8 nm (equivalent to a pulse width of ~570 fs), centered at 1,560 and 1,530 nm, respectively. An interferometer was used to shape the PUT in a double pulse waveform, and a movable mirror controlled their relative delay.

The dispersion of the pulses was controlled using spools of SMF. The pump was stretched to approximately 100 ps with 2 km of SMF. The signal dispersion was carefully adjusted to obtain half the dispersion of the pump. Pulses were first amplified with a standard erbium-doped fiber amplifier (EDFA) and were then coupled into the 45-cm long waveguide. The pump peak power at the input waveguide was 1 W, and an OSA provided the output spectrum.

Figure 2.19 shows the resulting idler spectrum generated by the device as a function of the delay between the two peaks of the double pulse waveform used as PUT. By measuring the spectral separation of the two idler peaks, a calibration curve was obtained (see Fig. 2.20). The measurement shows the linear trend expected by relation (2.23). We obtained a calibration factor of 14.2 ps/nm, used

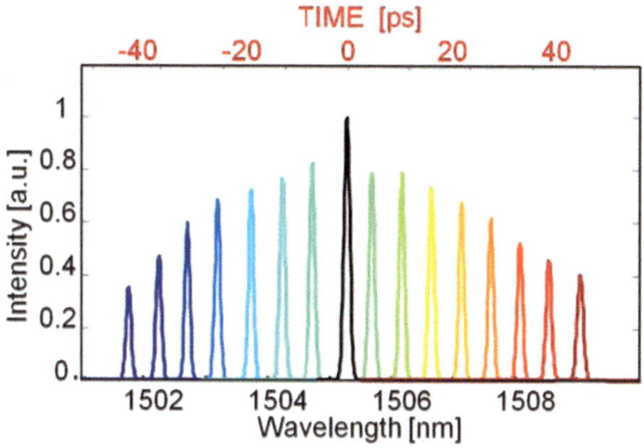

Fig. 2.19 Idler spectrum for a PUT consisting of two delayed Gaussian pulses, for different temporal delays. Here, the pump dispersion is equivalent to 2 km of SFM. The first replica (black Gaussian pulse, at the center of the picture) does not change in time, while the second replica is recorded for different temporal delays and is represented with different colors

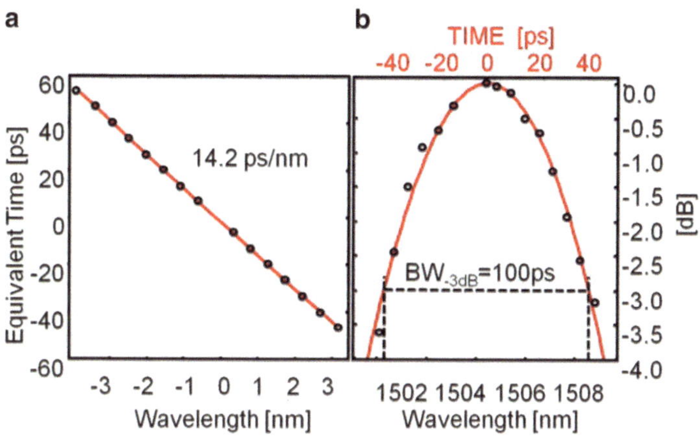

Fig. 2.20 Calibration curve for the delay between the two Gaussian pulses composing the PUT (extracted from in Fig. 2.19): experimental data extracted from Fig. 2.19 (*black dots*) and related linear interpolation (*red line*). (**b**) Maxima of the Gaussian pulses in Fig. 2.19 for different time delays (*black dots*) and related parabolic interpolation of the experimental maxima (*red curve*). We note that the intensity of the imaged pulses varies because the approximation of a constant pump intensity does not hold for the large delays used in this experiment

to calculate the corresponding temporal axis of the spectral measurement, shown at the top of Fig. 2.20, in red. The peak value of the delayed pulse replica in the measured idler spectrum versus wavelength is shown in Fig. 2.20b: this measurement allowed us to precisely estimate the temporal recording length of our instrument, obtaining a −3-dB time window of 100 ps.

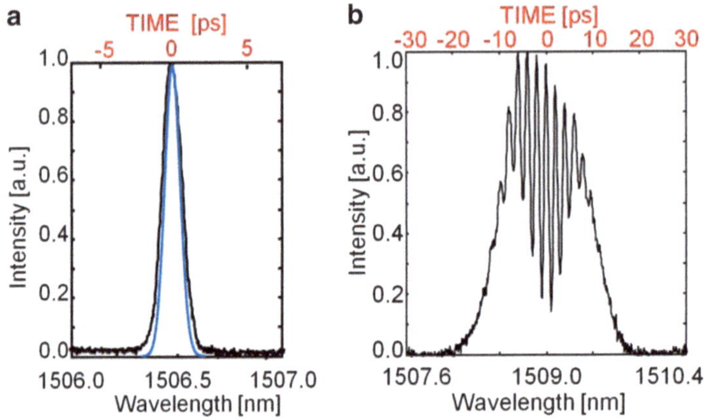

Fig. 2.21 (**a**) Idler spectrum (*black curve*) for a PUT consisting of a transform limited Gaussian pulse with 5-nm bandwidth. The converted timescale (*red axis at top*) is obtained using the calibration factor of 14.2 ps/nm extrapolated from Fig. 2.20. The *blue curve* is the FT of the signal (i.e., the PUT) spectrum, representing the transform limited PUT temporal waveform. (**b**) Idler spectrum of a temporal shape with a large time bandwidth product, revealing 800-fs long oscillations over a temporal duration of 40 ps

The image of a single, transform limited Gaussian pulse is shown in Fig. 2.21a, together with the result of the TL experiment (with the transform limited temporal shape obtained from the FT of the spectral measurement of the signal), yielding a Gaussian pulse width of ~700 fs. This was also confirmed by autocorrelation measurements. In comparison, the TL measurement yields a pulse width of ~900 fs. This is a result of two effects. First, the resolution of our TL-FWM experiment is ~400 fs: a $\tau_0 = 70$ fs temporal pump pulse duration in the transform limited section corresponds to a resolution of $\tau_0/\sqrt{2}$ ~400 fs [89–91]. This broadens the pulsewidth to ~800 fs. The additional broadening to ~900 fs arises from our TL system being slightly out of focus; equivalent to propagation through ~5 m of SMF. Indeed, we could control the signal dispersion with an accuracy of approximately 10 m of SMF, due to synchronization constraints between the signal and the pump. This is due to the fact that, when the length of the SMF spool used to disperse the PUT is modified, the relative delay between pump and signal is also affected.

An instrument with a temporal resolution of 400 fs over a time window >100 ps possesses a time bandwidth product >250. This is comparable with previous results obtained in silicon [89–91]. Better performances in terms of both output idler conversion efficiency and time-bandwidth product can be reached for larger pump excitations, limited by the amplifier used in our set-up but not by our device. As our platform is not affected by nonlinear absorption, higher power can be employed without distortion inducing aberrations in the TL. This temporal imaging instrument is well suited for the measurement of pulses with complex temporal features over large scales. As is visible from Fig. 2.21b, a temporal shape consisting of an oscillation with a time scale of 800 fs over a temporal window larger than 40 ps can be successfully imaged.

2.5.3 All-Optical Pulse Compressor

Our study of pulse compression is primarily based on recent advances in PICs, and on how they have become a necessity for the future of ultrafast telecommunication networks. Their development will simultaneously be able to meet the exponentially growing demand for bandwidth, lower costs, use a smaller footprint, and reduce energetic consumption compared to today's electronic technologies. In particular, the design and fabrication of an all-optical temporal compressor would bring great benefits to many applications such as optical metrology and imaging, where the compression of pulses down to subpicoseconds generates a broad bandwidth pulse particularly suited for optical coherent tomography. However, these kind of ultra-short pulses are very difficult to be directly obtained from commercially available laser sources. A standard strategy to circumvent this problem is to exploit devices that make use of ultrafast nonlinearities. In this case, pulse compression is obtained by first spectrally broadening a transform limited optical pulse, via nonlinear propagation in a normally dispersive optical fiber, and then re-phasing it via linear anomalous dispersion (the order of these two steps can be interchanged) [64, 81, 93].

SPM-based spectral broadening and dispersion for ps and sub-ps pulses can be effectively modeled according to (2.4) by the following expression:

$$\frac{\partial A}{\partial z} + \frac{\alpha}{2}A + \beta_1 \frac{\partial A}{\partial t} + i\frac{\beta_2}{2}\frac{\partial^2 A}{\partial t^2} = i\gamma |A_m|^2 A_m. \tag{2.24}$$

The other nonlinear terms of (2.4) were dropped on the assumption of (1) lack of phase matching for frequency conversion, (2) pulse durations around 1 ps, and (3) moderate power levels as to not excite higher order effects. The equation is governed by two main effects: dispersion and SPM. Under suitable conditions, these two effects can act at different length scales and can thus be treated independently. To gain some insight on the effect of the nonlinear contribution to (2.24), it is useful to look at the no-dispersion limit, which can be readily solved to obtain:

$$A' = A_0 \exp(-\alpha_1 z/2) \exp(i\gamma |A_0|^2 [1 - \exp(-\alpha_1 z)]/\alpha_1), \tag{2.25}$$

where $A' = A(z,t - \beta_1 z)$. The nonlinear term introduces a nonlinear chirp in the temporal phase, which in the frequency domain corresponds to spectral broadening (the generation of new frequencies), i.e., self-phase modulation (SPM). Ultra-short pulses can be generated by first broadening the spectrum of a propagating optical pulse in a normally dispersive nonlinear medium, thereby creating new spectral components via SPM and then by propagating in an anomalous dispersion regime, thereby rephasing the pulse with an opposite chirp. Technically, this approach can lead to the shortest possible pulse having a certain frequency content, which is defined as a transform limited pulse.

Figure 2.22 shows the experimental set-up used to demonstrate pulse compression in the 45-cm long Hydex spiral waveguide. This can be achieved in Hydex due

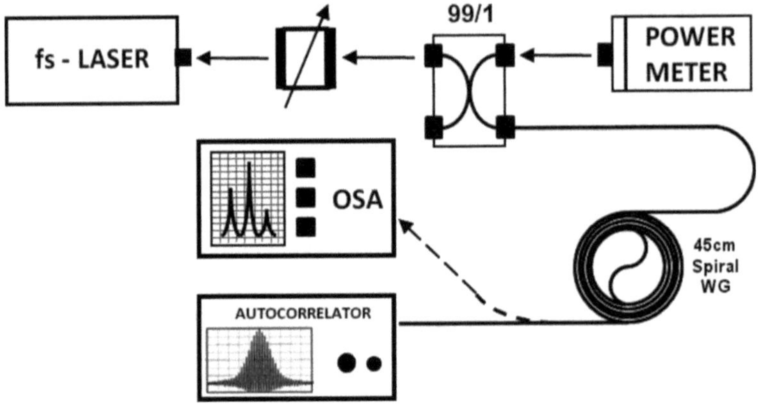

Fig. 2.22 Sketch of the experimental setup for the optical pulse compression measurements

to a weak dispersion regime (for pulses of ps-scale durations or longer) near 1,550 nm. A standard fiber laser (Pritel FFL) was used as an input source, with pulses having a temporal waist of $T_0 = 0.94$ ps (1.10 ps FWHM), a total dispersion quadratic chirp of 0.17 ps/nm (which can be compensated by approximately 10 m of SMF), a repetition rate of 16.9 MHz, and a central wavelength $\lambda = 1{,}550$ nm.

The experimental setup is portrayed in Fig. 2.22. By means of a polarization controller and an attenuator, we were able to arbitrarily set both the polarization and power. The latter was estimated by using a power meter coupled at the output of a 1% channel from a 99/1 fiber coupler. For the data analysis, both an autocorrelator and an OSA were used to characterize the input and output pulse. The total length of the SMF used was 8 m. More specifically, the distance from the laser to the chip input was 7.33 m, while an additional 0.66 m of fiber was added to the chip output. Most of the initial source chirp was compensated during the propagation along the initial 7.33 m of input fiber, and the remaining input pulse residual chirp was estimated to be equivalent to the value induced by propagating approximately 2 m in an SMF.

Figure 2.23 shows both the pulse autocorrelations (intensity: normalized unit) and spectra (power: normalized unit) recorded at the chip output right after the 0.66-m of SMF. These plots indicate that the pulse is temporally compressed as we increase the pulse peak power. The pulse time width estimation was obtained by fitting the experimental intensity autocorrelation to a Gaussian field defined by $A(T) = A_0 \exp(-(T/T_0)^2)$, where T is a time renormalized to the group velocity, i.e., $T = t - \beta_1 z$. The final results are characterized by pulses with a time waist T_0 spanning from 0.65 to 0.45 ps and energies (peak powers) from 15 pJ (19 W) to 71.2 pJ (98 W).

The pulse spectrum in Fig. 2.23b clearly shows the double peak feature of an ongoing SPM arising at high powers. We also performed the experiments using a reference 1 mm long waveguide, and verified that the compression was due to the nonlinear chirp acquired by the 45-cm waveguide and did not take place in the input or output fiber pigtails used to couple from SMFs to the optical chip.

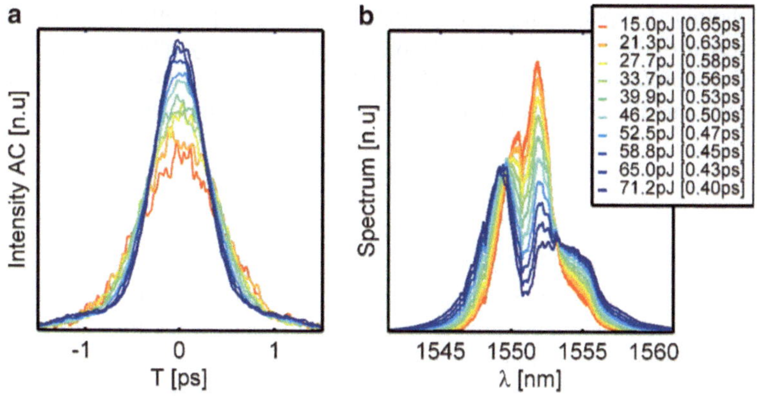

Fig. 2.23 Pulse compression at different pulse energies: autocorrelation (**a**) and spectrum (**b**) obtained by collecting the output of the spiral waveguide after 0.66 m of SMF. The Gaussian best-fitted pulse waist is indicated between *brackets*

We stress that the role played by the output fiber is to compensate for the nonlinear chirp induced by the integrated device. Although in this work the compensation is not performed on-chip, the total required dispersion compensation is quite low, at around 10 fs/nm. This level of dispersion can be easily achieved on-chip by engineering the waveguide parameters. For instance, our platform could easily allow for modification of the GVD by simply changing the waveguide width and height [64].

2.5.4 Optical Integrator

In this section, we demonstrate the possibility of exploiting the linear properties of Hydex for performing ultrafast all-optical processing via the use of novel integrated photonic devices. Specifically, we report on the possibility of using micro-ring resonators to perform temporal-integration of complex waveforms with features down to ≈8 ps.

The design and fabrication of basic photonic blocks (equivalent to basic electronic devices) are steps of paramount importance to facilitate the passage from electronics to photonics [92]. In particular, integrators and differentiators, which were largely used as fundamental modules for the realization of ALUs (Arithmetic Logic Units) during the dawn of integrated electronics, have yet to be implemented all-optically in today's photonic systems. Here, we focus our attention on optical integrators [94]. Photonic temporal integrators can be used to fabricate ultrafast computational units devoted to solve ordinary differential equations (ODEs), which in turn play a fundamental role in many fields of science and engineering [95].

A photonic temporal integrator is a device capable of performing the cumulative time integral of the complex temporal envelope of an arbitrary optical input [96]. From basic signal processing theory [97–100], we find that an integrator is characterized by a temporal impulse response $h(t)$ proportional to the step function $u(t)$:

$$\begin{cases} u(t) = 0 & \text{for} \quad t<0, \\ u(t) = 1 & \text{for} \quad t>0, \end{cases} \tag{2.26}$$

where t is the time variable. In order to realize such a physical system, we require a structure capable of storing the intensity of an incoming optical radiation and subsequently releasing a continuous signal proportional to the total stored field at each instant of time. This functionality, which in electronics is realized by a simple capacitor, is challenging in optics due to the intrinsic nature of photons—which typically cannot be spatially localized over long intervals. An optical integrator is potentially very useful for a specific set of fundamental applications such as data processing, pulse shaping, data storage, and optical computing.

In the spectral domain, the transfer function $H(\omega)$ of an ideal integrator is expressed by:

$$H(\omega) \propto \frac{1}{j(\omega_0 - \omega)}, \tag{2.27}$$

where ω is the optical frequency variable, and ω_0 is the carrier frequency of the signal to be processed. From this function, it is immediately clear that an optical integrator near ω_0 should in principle rely on a transmission >1 and diverge to infinity at ω_0.

One possible approach to emulate the integrator transfer function is based on the use of resonant cavities. Considering a Fabry-Perot interferometer, for a certain fraction of its FSR, its temporal impulse response can be described by [101]:

$$h(t) \propto \exp(-kt) \cdot u(t), \tag{2.28}$$

where $k = (-1/T) \cdot \ln(r^2\gamma)$, T is the round trip propagation time in the cavity ($T = 1/\text{FSR}$), r is the mirror reflectivity, and γ represents the round-trip field amplitude gain, which can be either >1 or <1 depending on whether light is propagating in an active or an absorbing medium, respectively. Figure 2.24 shows that around a specific resonance (whose shape is well approximated by a Lorentzian function), the transfer function of an ideal integrator (black curve) resembles that of an optical resonator (red curves). In the same image, the comparison is made for two different resonators (case "a" and "b," respectively) that only differ in terms of FSRs. The dashed boxes represent the ranges in which the resonator characteristic resembles that of the ideal integrator.

From Fig. 2.24, it is also clear how a larger FSR corresponds to a broader integration bandwidth. Since the FSR is inversely proportional to the physical dimensions of a cavity, the device size reduction can bring the double advantage of widening the integration bandwidth while simultaneously reducing the component footprint.

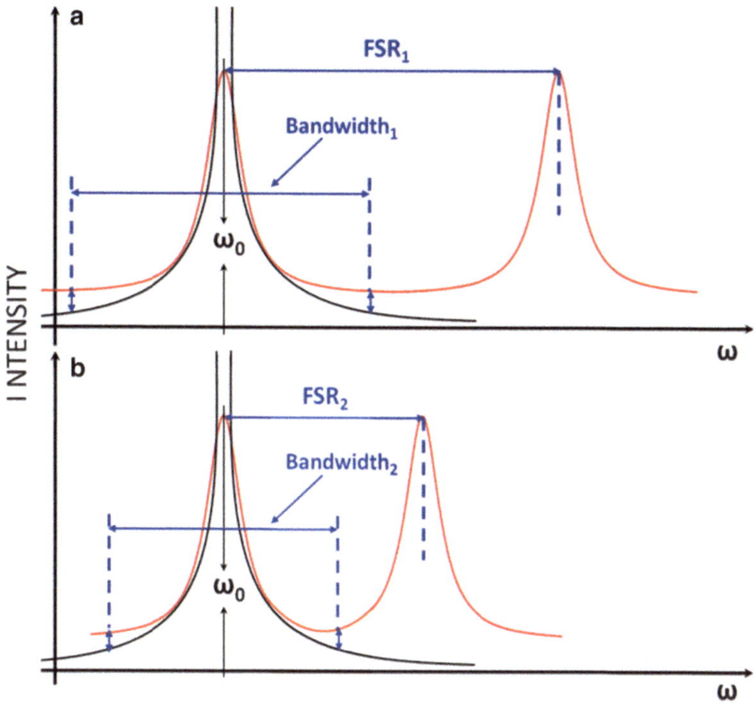

Fig. 2.24 Typical fit between the spectral intensity transfer functions of a general resonator (*red curve*) and that of an ideal integrator (*black curve*). The plot makes a comparison in terms of integration bandwidth between two resonators (case "a" and "b") with two different FSRs. A larger FSR (FSR 1 > FSR 2) corresponds to a broader integration bandwidth (bandwidth 1 > bandwidth 2)

The performance near ω_0 can be improved by using an active device, in which losses can be compensated by gain [102]. However, the active integrator approach is also affected by major disadvantages such as a high noise level (produced from spontaneous emission), a high energy consumption, and remarkable challenges encountered throughout the fabrication. These limitations are often so severe that the passive approach is often preferable to its active counterpart, despite significant drawbacks in terms of throughput (defined as the input–output power efficiency) and the time-bandwidth product.

Realization of photonic integrators has also been explored through other techniques, including the use of feedback-based photonic filters [103] and fiber Bragg gratings (FBG) [99]. These solutions, however, despite representing a fundamental improvement, suffer from important limitations. Specifically, the photonic filter strategy is very limited in terms of processing speed, whereas the performance of FBG integrators are fundamentally related with the reflectivity, which cannot be lower than 99.99% for acceptable operation. Here, we show that the ring resonator described earlier in this chapter presents an ideal alternative for on-chip all-optical temporal integration.

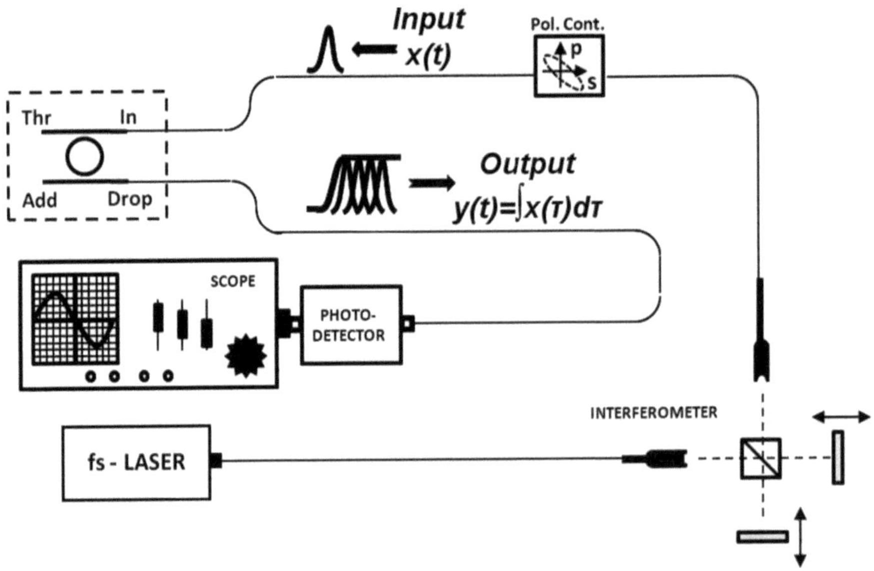

Fig. 2.25 Experimental set-up for all-optical integration

Figure 2.25 shows the experimental set-up used for optical integration. In our system, a fiber laser (PriTel FFL) emits transform limited pulses at a repetition rate of 16.9 MHz. The pulses have a time duration of approximately 7.5 ps (measured by using a Fourier transform spectral interferometry technique). The light beam is then sent to a pulse shaper (Michelson interferometer) which is used to set three different kinds of waveforms: (1) the optical wave directly generated by the laser source (obtained by simply shielding one arm of the interferometer—see Fig. 2.26a), (2) a train of in-phase pulses (inset in Fig. 2.26b), and (3) a pair of π-phase-shifted pulses (inset in Fig. 2.26c). By coarsely varying the optical path difference of the interferometer, the temporal distance between the pulses was set at 275 ps for both the inputs (2) and (3).

The mutual phase-shift between pulses was set by using a piezo controller mounted on one of the two mirrors of the pulse shaper. The waveforms were then sent to a polarization controller in order to select a TE ring resonance at 1,559.46 nm. Light coupled at the Input port is collected at the drop port at resonance, and is recorded and visualized by a fast oscilloscope (Tektronix CSA8200 signal analyzer) operating in sampling mode.

The output normalized intensities for the (1), (2), and (3) waveforms are reported in Fig. 2.26a–c, respectively. The correspondent theoretical cumulative integrals are represented by the solid red curves. In order to record the fast features of the system response to the input (1) (see inset (a) in Fig. 2.26a), we also used an amplified optical detector with a time constant ~8 ps. This optical component was placed between the ring output and the oscilloscope input (see Fig. 2.25). Since the

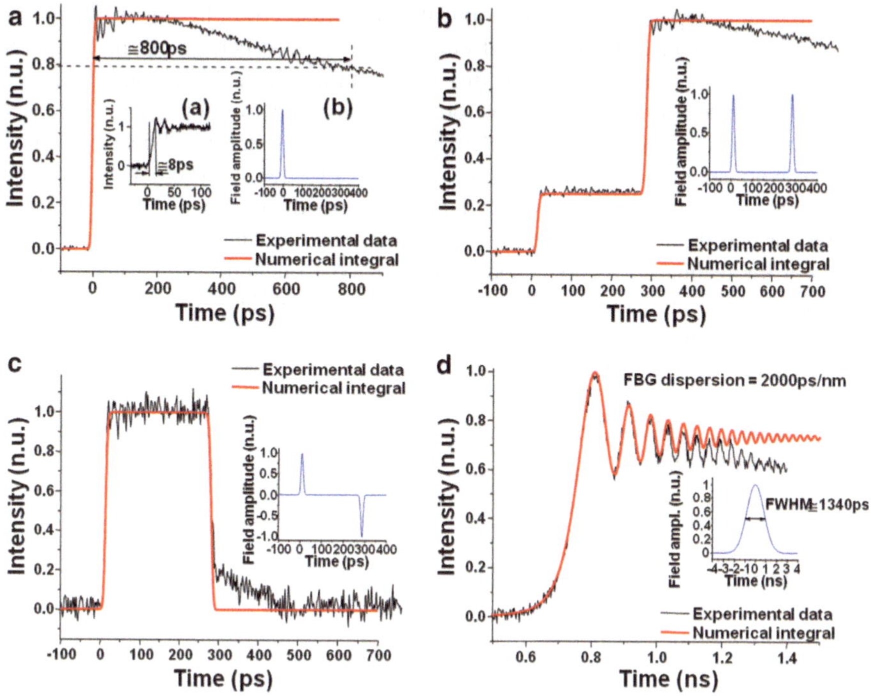

Fig. 2.26 Analysis of the integrator temporal response. The main plots represent both the experimental (*black curve*) and the theoretical (*red curve*) time integrals of the various intensity optical waveforms under analysis. These input waveforms are reported in the corresponding insets (*blue curves*). The measurement is performed for different cases: an ultra-short optical pulse directly generated by the laser source (**a**); in-phase pulses with a relative temporal delay of 275 ps (**b**); π-phase-shifted pulses with a relative temporal delay of 275 ps (**c**); and a linearly chirped (quadratic phase) broadband optical pulse, with a field amplitude FWHM time duration of ~1,340 ps (**d**). The impulse response (**a**, inset (**a**)) was obtained by using a fast (~8 ps) amplified photo-detector. The dispersed pulse was calculated by assuming an ideal quadratic phase variation on the measured temporal amplitude pulse profile according to the linear dispersion value (~2,000 ps/nm) of the fiber-dispersive element used in our experiments

input pulse and the integrator bandwidths almost coincide, Fig. 2.26a can be considered the impulse time response of our system.

Figure 2.26a also shows that the integration time window of the ring is as long as 800 ps (defined as the time required to decay to 80% of the maximum). The resulting time-bandwidth product of our integrator is consequently evaluated to be 100, obtained by taking the ratio between the integration time window and the rising time. This value is one order of magnitude better than what is achievable by using state-of-the-art electronic integrators [104] and in our case it is accomplished by using a fully CMOS compatible and passive device [105].

It is worth noticing that the integration of the waveforms (2) and (3) differs only because our integrator is phase sensitive. This important sensitivity allows for certain unique applications, such as optical data storage [106]. In order to underline this very important characteristic of our device, we also performed the temporal integral of a waveform obtained by dispersing in time the laser output.

The dispersed input optical waveform was obtained by propagating the laser pulses through a fiber-dispersive element, resulting in strongly linearly chirped pulses with an intensity FWHM time duration of ~950 ps (field FWHM time duration of ~1,340 ps). The chirped pulses were then launched into the ring resonator and the temporal intensity waveform was measured at the output (drop port) of the resonator. Figure 2.26d shows the experimental results (black curve) together with the theoretical time integral (red curve) of the chirped optical pulse. Despite the complexity of the output waveform, the experimental and theoretical curves agree well over timescales longer than the resolution of the device (~8 ps). This last result suggests that it should be possible to recover the temporal phase information of a given arbitrary optical waveform from the temporal intensity profile of its integral.

2.6 Outlooks and Conclusions

In this book chapter, we have presented a recently developed high-index doped silica glass material platform that is potentially useful for a very broad range of linear and nonlinear applications in integrated photonics. This material combines the optimal linear properties of SMFs such as very low propagation losses, and a robust fabrication process, together with those typical of semiconductors and other nonlinear glasses, such as significant nonlinearities. This material allows for the fabrication of micrometric waveguides having very low linear losses of the order of 0.06 dB/cm and negligible nonlinear losses for peak intensities tested up to 25 GW/cm^2. The 17% index contrast achievable with respect to a silica glass cladding allows for a very tight mode confinement down to 1.5 μm^2—in turn leading to a nonlinear γ parameter as large as 220 W^{-1} km^{-1}. The processes behind the Hydex waveguides fabrication are fully CMOS compatible with no need for high temperature postannealing and they also allow for very tight bend radii down to 30 μm. The latter being the key for integrating complex spiral guiding structures onto chip size areas (<1 cm^2).

We have shown that although semiconductors possess a much larger nonlinearity, this parameter cannot often be fully exploited because of the detrimental (linear and) nonlinear absorption. In particular, we have presented and described how the Hydex platform can be employed for the realization of efficient ultra-narrow linewidths and multiple-wavelength integrated optical sources, as well as for fabricating fundamental optical blocks for ultrafast optical processing. This was demonstrated through experiments on semi-degenerate FWM, on the generation of an HOPO, and by means of a SCG source. In the second part of the chapter, which mostly focused on the ultrafast signal processing, Hydex devices were used to realize an integrated pulse frequency converter, optical time lens, compressor, and integrator.

Only two fundamental structures were explored: a 135-μm radius ring cavity, and a 45-cm long spiral waveguide. Nonetheless, using these two devices, we proved: (a) a narrow linewidth (160 MHz FWHM) and efficient (−26 dB) frequency conversion with a predicted tunability range of 160 nm (20 THz); (b) an optical CW parametric oscillator with a single line efficiency of 7.4% and a threshold power of 54 mW; (c) a spectral bandwidth broadening of more than 300 nm at both 1,290 and 1,550 nm for an input pulse spectrum of 110 nm (at −20 dB); (d) parametric gain via pulsed FWM with a maximum on/off conversion efficiency of +16.5 dB from signal to idler (for a 38-W pump power); (e) a time lens with a temporal resolution of 400 fs over a time window >100 ps (time bandwidth product >250); (f) pulse compression spanning from 0.65 to 0.40 ps for pulse energies (peak power) varying from 15 pJ (19 W) to 71.2 pJ (98 W); and (g) all-optical phase-sensitive temporal integration over a bandwidth of 200 GHz. All these results undoubtedly place Hydex technology amongst the most promising for future ultrafast all-optical telecommunication networks.

References

1. A. Alduino, M. Paniccia, Interconnects: wiring electronics with light. Nat. Photon. **1**, 153–155 (2007)
2. B.J. Eggleton, S. Radic, D.J. Moss, Nonlinear Optics in Communications: From Crippling Impairment to Ultrafast Tools, in *Optical Fiber Telecommunications V: Components and Sub-systems*, ed. by I.P. Kaminow, T. Li, A.E. Willner, 5th edn. (Academic, Oxford, 2008). Chap. 20. ISBN ISBN 978-0123741714
3. G. Lifante, *Integrated Photonics* (Wiley, England, 2003). ISBN ISBN 978-0470848685
4. N. Izhaky, M.T. Morse, S. Koehl, O. Cohen, D. Rubin, A. Barkai, G. Sarid, R. Cohen, M.J. Paniccia, Development of CMOS-compatible integrated silicon photonics devices. IEEE J. Sel. Top. Quant. Electron. **12**, 1688–1698 (2006)
5. L. Tsybeskov, D.J. Lockwood, M. Ichikawa, Silicon photonics: CMOS going optical. Proc. IEEE **97**, 1161–1165 (2009)
6. R. Won, M. Paniccia, Integrating silicon photonics. Nat. Photon. **4**, 498–499 (2010)
7. A. Liu, L. Liao, Y. Chetrit, J. Basak, H. Nguyen, D. Rubin, M. Paniccia, Wavelength division multiplexing based photonic integrated circuits on silicon-on-insulator platform. IEEE J. Sel. Top. Quant. Electron. **16**, 23–32 (2010)
8. A.C. Turner, M.A. Foster, A.L. Gaeta, M. Lipson, Ultra-low power parametric frequency conversion in a silicon microring resonator. Opt. Express **16**, 4881–4887 (2008)
9. S. Venugopal Rao, K. Moutzouris, M. Ebrahimzadeh, Nonlinear frequency conversion in semiconductor optical waveguides using birefringent, modal and quasi-phase-matching techniques. J. Opt. A: Pure Appl. Opt. **6**, 569–584 (2004)
10. R. Salem, M.A. Foster, A.C. Turner, D.F. Geraghty, M. Lipson, A.L. Gaeta, Signal regeneration using low-power four-wave mixing on silicon chip. Nat. Photon. **2**, 35–38 (2008)
11. V.G. Ta'eed, M. Shokooh-Saremi, L. Fu, D.J. Moss, M. Rochette, I.C.M. Littler, B.J. Eggleton, Y. Ruan, B. Luther-Davies, Integrated all-optical pulse regeneration in chalcogenide waveguides. Opt. Lett. **30**, 2900–2902 (2005)
12. B.G. Lee, X. Chen, A. Biberman, X. Liu, I.-W. Hsieh, C.-Y. Chou, J.I. Dadap, F. Xia, W.M.J. Green, L. Sekaric, Y.A. Vlasov, R.M. Osgood, K. Bergman, Ultrahigh-bandwidth silicon photonic nanowire waveguides for on-chip networks. IEEE Photon. Tech. Lett. **20**, 398–400 (2008)

13. N.S. Bergano, Wavelength division multiplexing in long-haul transoceanic transmission systems. J. Lightwave Tech. **23**, 4125–4139 (2005)
14. T.A. Ibrahim, V. Van, P.-T. Ho, All-optical time-division demultiplexing and spatial pulse routing with a GaAs/AlGaAs microring resonator. Opt. Lett. **27**, 803–805 (2002)
15. L.G. Kazovsky, N. Cheng, W.-T. Shaw, D. Gutierrez, *Broadband Optical Access Networks* (Wiley, England, 2011). ISBN ISBN 978-0-470-18235-2
16. G.P. Agrawal, *Nonlinear Fiber Optics* (Academic, San Diego, CA, 2006). ISBN ISBN 978-0123695161
17. T.L. Koch, U. Koren, Semiconductor photonic integrated circuits. IEEE J. Quant. Electron. **27**, 641–653 (1991)
18. B.E. Little, S.T. Chu, Towards very large-scale integrated photonics. Optic. Photon. News **11**, 24–29 (2000)
19. B. Jalali, S. Fathpour, Silicon photonics. J. Lightwave Tech. **24**, 4600–4615 (2006)
20. M.D. Pelusi, V.G. Ta'Eed, M.R.E. Lamont, S. Madden, D.-Y. Choi, B. Luther-Davies, B.J. Eggleton, Ultra-high nonlinear As2S3 planar waveguide for 160 Gb/s optical time-division demultiplexing by four-wave mixing. IEEE Photon. Tech. Lett. **19**, 1496–1498 (2007)
21. K. Wörhoff, L.T.H. Hilderink, A. Driessen, P.V. Lambeck, Silicon oxynitride. A versatile material for integrated optics applications. J. Electrochem. Soc. **F149**, F85–F91 (2002)
22. V.G. Ta'eed, N.J. Baker, L. Fu, K. Finsterbusch, M.R.E. Lamont, D.J. Moss, H.C. Nguyen, B. J. Eggleton, D.-Y. Choi, S. Madden, B. Luther-Davies, Ultrafast all-optical chalcogenide glass photonic circuits. Opt. Express **15**, 9205–9221 (2007)
23. J.H. Lee, K. Kikuchi, T. Nagashima, T. Hasegawa, S. Ohara, N. Sugimoto, All-fiber 80-Gbit/s wavelength converter using 1-m-long bismuth oxide-based nonlinear optical fiber with a nonlinearity gamma of 1100 W^{-1} km^{-1}. Opt. Express **13**, 3144–3149 (2005)
24. A. Arie, N. Voloch, Periodic, quasi-periodic, and random quadratic nonlinear photonic crystals. Laser Photon. Rev. **4**, 355–373 (2010)
25. P. Russell, Photonic crystal fibers. Science **299**, 358–362 (2003)
26. E. Yablonovitch, T.J. Gmitter, K.M. Leung, Photonic band structure: the facecentered-cubic case employing nonspherical atoms. Phys. Rev. Lett. **67**, 2295–2298 (1991)
27. A. Gondarenko, J.S. Levy, M. Lipson, High confinement micron-scale silicon nitride high Q ring resonator. Opt. Express **17**, 11366–11370 (2009)
28. L.M. Tong, R.R. Gattass, J.B. Ashcom, S.L. He, J.Y. Lou, M.Y. Shen, I. Maxwell, E. Mazur, Subwavelength-diameter silica wires for low-loss optical wave guiding. Nature **426**, 816–819 (2003)
29. M.A. Foster, A.C. Turner, M. Lipson, A.L. Gaeta, Nonlinear optics in photonic nanowires. Opt. Express **16**, 1300–1320 (2008)
30. B. E. Little, *A VLSI Photonics Platform*, Optical Fiber Communication Conference, (Atlanta, Georgia, Optical Society of America, 2003), 444–445
31. E.A.S. Bahaa, M.C. Teich, *Fundamentals of Photonics* (Wiley, England, 1991). ISBN ISBN 978-0471839651
32. R.W. Boyd, *Nonlinear Optics* (Academic, San Diego, CA, 2008). ISBN ISBN 978-0123694706
33. H. Rong, Y.-H. Kuo, A. Liu, M. Paniccia, O. Cohen, High efficiency wavelength conversion of 10 Gb/s data in silicon waveguides. Opt. Express **14**, 1182–1188 (2006)
34. R. Salem, M.A. Foster, A.C. Turner, D.F. Geraghty, M. Lipson, A.L. Gaeta, All-optical regeneration on a silicon chip. Opt. Express **15**, 7802–7809 (2007)
35. P.P. Absil, J.V. Hryniewicz, B.E. Little, P.S. Cho, R.A. Wilson, L.G. Joneckis, P.-T. Ho, Wavelength conversion in GaAs micro-ring resonators. Opt. Lett. **25**, 554–556 (2000)
36. H. Rong, S. Xu, O. Cohen, O. Raday, M. Lee, V. Sih, M. Paniccia, A cascaded silicon Raman laser. Nat. Photon. **2**, 170–174 (2008)
37. M. Volatier, D. Duchesne, R. Morandotti, R. Arès, V. Aimez, Extremely high aspect ratio GaAs and GaAs/AlGaAs nanowaveguides fabricated using chlorine ICP plasma etching with N2 promoted passivation. Nanotechnology **21**, 134014 (2010)

38. G.A. Siviloglou, S. Suntsov, R. El-Ganainy, R. Iwanow, G.I. Stegeman, D.N. Christodoulides, R. Morandotti, D. Modotto, A. Locatelli, C. De Angelis, F. Pozzi, C.R. Stanley, M. Sorel, Enhanced third-order nonlinear effects in optical AlGaAs nanowires. Opt. Express **14**, 9377–9384 (2006)
39. M. Borselli, T.J. Johnson, O. Painter, Measuring the role of surface chemistry in silicon microphotonics. Appl. Phys. Lett. **88**, 131114 (2006)
40. A. Jouad, V. Aimez, Passivation of air-exposed AlGaAs using low frequency plasma-enhanced chemical vapor deposition of silicon nitride. Appl. Phys. Lett. **89**, 092125 (2006)
41. I. Moerman, P.P. Van Daele, P.M. Demeester, A review on fabrication technologies for the monolithic integration of tapers with III–V semiconductor devices. IEEE J. Sel. Top. Quant. Electron. **3**, 1308–1320 (1997)
42. V.R. Almeida, R.R. Panepucci, M. Lipson, Nanotaper for compact mode conversion. Opt. Lett. **28**, 1302–1304 (2003)
43. H.K. Tsang, Y. Liu, Nonlinear optical properties of silicon waveguides. Semicond. Sci. Tech. **23**, 064007 (2008)
44. E. Dulkeith, Y.A. Vlasov, X. Chen, N.C. Panoiu, R.M. Osgood Jr., Self-phase-modulation in submicron silicon-on-insulator photonic wires. Opt. Express **14**, 5524–5534 (2006)
45. T.K. Liang, H.K. Tsang, Role of free carriers from two-photon absorption in Raman amplification in silicon-on-insulator waveguides. Appl. Phys. Lett. **84**, 2745–2747 (2004)
46. R.M. de Ridder, K. Worhoff, A. Driessen, P.V. Lambeck, H. Albers, Silicon oxynitride planar waveguiding structures for application in optical communication. IEEE J. Sel. Top. Quant. Electron. **4**, 930–937 (1998)
47. W.T. Li, Y.L. Ruan, B. Luther-Davies, A. Rode, R. Boswell, Dry-etch of As2S3 thin films for optical waveguide fabrication. J. Vac. Sci. Tech. **23**, 1626–1632 (2005)
48. Y.L. Ruan, W.T. Li, R. Jarvis, N. Madsen, A. Rode, B. Luther-Davies, Fabrication and characterization of low loss rib chalcogenide waveguides made by dry etching. Opt. Express **12**, 5140–5145 (2004)
49. M. Shokooh-Saremi, V.G. Ta'eed, N.J. Baker, I.C.M. Littler, D.J. Moss, B.J. Eggleton, Y. Ruan, B. Luther-Davies, High-performance Bragg gratings in chalcogenide rib waveguides written with a modified Sagnac interferometer. J. Opt. Soc. Am. B **23**, 1323–1331 (2006)
50. H.C. Nguyen, K. Finsterbusch, D.J. Moss, B.J. Eggleton, Dispersion in nonlinear figure of merit of As2Se3 chalcogenide fibre. Electron. Lett. **42**, 571–572 (2006)
51. D. Duchesne, M. Ferrera, L. Razzari, R. Morandotti, B.E. Little, S.T. Chu, D.J. Moss, Efficient self-phase modulation in low loss, high index doped silica glass integrated waveguides. Opt. Express **17**, 1865–1870 (2009)
52. M. Ferrera, L. Razzari, D. Duchesne, R. Morandotti, Z. Yang, M. Liscidini, J.E. Sipe, S. Chu, B. Little, D.J. Moss, Low power continuous-wave nonlinear optics in doped silica glass integrated waveguide structures. Nat. Photon. **2**, 737–740 (2008)
53. B.E. Little, S.T. Chu, P.P. Absil, J.V. Hryniewicz, F.G. Johnson, F. Seiferth, D. Gill, V. Van, O. King, M. Trakalo, Very high-order microring resonator filters for WDM applications. IEEE Photon. Tech. Lett. **16**, 2263–2265 (2004)
54. A. Yalcın, K.C. Popat, J.C. Aldridge, T.A. Desai, J. Hryniewicz, N. Chbouki, B.E. Little, O. King, V. Van, S. Chu, D. Gill, M. Anthes-Washburn, M.S. Unlu, B.B. Goldberg, Optical Sensing of Biomolecules Using Microring Resonators. IEEE J. Sel. Top. Quant. Electron. **12**, 148–155 (2006)
55. I.H. Agha, Y. Okawachi, M.A. Foster, J.E. Sharping, A.L. Gaeta, Four-wave-mixing parametric oscillations in dispersion-compensated high-Q silica microspheres. Phys. Rev. A **76**, 043837 (2007)
56. D.H. Broaddus, M.A. Foster, I.H. Agha, J.T. Robinson, M. Lipson, A.L. Gaeta, Silicon-waveguide-coupled high-Q chalcogenide microspheres. Opt. Express **17**, 5998–6003 (2009)
57. T.J. Kippenberg, S.M. Spillane, K.J. Vahala, Kerr-nonlinearity optical parametric oscillation in an ultrahigh-Q toroid microcavity. Phys. Rev. Lett. **93**, 083904 (2004)
58. M. Soltani, Q. Li, S. Yegnanarayanan, A. Adibi, Improvement of thermal properties of ultra-high Q silicon microdisk resonators. Opt. Express **15**, 17305–17312 (2007)

59. S. Yegnanaryanan, M. Soltani, Q. Li, E.S. Hosseini, A.A. Eftekhar, A. Adibi, Microresonator in CMOS compatible substrate. J. Nanosci. Nanotechnol. **10**, 1508–1524 (2010)
60. K.J. Vahala, Optical microcavities. Nature **424**, 839–846 (2003)
61. A. Tandaechanurat, S. Ishida, D. Guimard, M. Nomura, S. Iwamoto, Y. Arakawa, Lasing oscillation in a three-dimensional photonic crystal nanocavity with a complete bandgap. Nat. Photon. **5**, 91–94 (2011)
62. J. Bravo-Abad, A. Rodriguez, P. Bermel, S.G. Johnson, J.D. Joannopoulos, M. Soljacic, Enhanced nonlinear optics in photonic-crystal micro-cavities. Opt. Express **15**, 16161–16176 (2007)
63. D. Duchesne, M. Peccianti, M.R.E. Lamont, M. Ferrera, L. Razzari, F. Légaré, R. Morandotti, S. Chu, B.E. Little, D.J. Moss, Supercontinuum generation in a high index doped silica glass spiral waveguide. Opt. Express **18**, 923–930 (2010)
64. M. Peccianti, M. Ferrera, L. Razzari, R. Morandotti, B.E. Little, S.T. Chu, D.J. Moss, Subpicosecond optical pulse compression via an integrated nonlinear chirper. Opt. Express **18**, 7625–7633 (2010)
65. L. Razzari, D. Duchesne, M. Ferrera, R. Morandotti, S. Chu, B.E. Little, D.J. Moss, CMOS compatible integrated optical hyper-parametric oscillator. Nat. Photon. **4**, 41 (2010)
66. J.S. Levy, A. Gondarenko, M.A. Foster, A.C. Turner-Foster, A.L. Gaeta, M. Lipson, CMOS-compatible multiple-wavelength oscillator for on-chip optical interconnects. Nat. Photon. **4**, 37–40 (2009)
67. M. Ferrera, D. Duchesne, L. Razzari, M. Peccianti, R. Morandotti, P. Cheben, S. Janz, D.-X. Xu, B.E. Little, S. Chu, D.J. Moss, Low power four wave mixing in an integrated, micro-ring resonator with Q = 1.2 million. Opt. Express **17**, 14098–14103 (2009)
68. S. Clemmen, K.P. Huy, W. Bogaerts, R.G. Baets, P. Emplit, S. Massar, Continuous wave photon pair generation in silicon-on-insulator waveguides and ring resonators. Opt. Express **17**, 16558–16570 (2009)
69. M. Pelton, C. Santori, G.S. Solomon, O. Benson, Y. Yamamoto, Triggered single photons and entangled photons from a quantum dot microcavity. Eur. Phys. J. D **18**, 179–190 (2002)
70. J. Chen, Z.H. Levine, J.Y. Fan, A.L. Migdall, Frequency-bin entangled comb of photon pairs from a Silicon-on-Insulator micro-resonator. Opt. Express **19**, 1470–1483 (2011)
71. D. Marcuse, *Theory of Dielectric Optical Waveguides* (Academic, Boston, MA, 1991). ISBN ISBN 978-0124709515
72. Q. Lin, T.J. Johnson, R. Perahia, C.P. Michael, O.J. Painter, A proposal for highly tunable optical parametric oscillation in silicon micro-resonators. Opt. Express **16**, 10596–10610 (2008)
73. J.A. Giordmaine, R.C. Miller, Tunable coherent parametric oscillation in LiNbO3 at optical frequencies. Phys. Rev. Lett. **14**, 973–976 (1965)
74. W. Denzer, G. Hancock, M. Islam, C.E. Langley, R. Peverall, G.A.D. Ritchie, D. Taylor, Trace species detection in the near infrared using Fourier transform broadband cavity enhanced absorption spectroscopy: initial studies on potential breath analytes. Analyst **136**, 801–806 (2011)
75. D.-I. Yeom, E.C. Mägi, M.R.E. Lamont, M.A.F. Roelens, L. Fu, B.J. Eggleton, Low-threshold supercontinuum generation in highly nonlinear chalcogenide nanowires. Opt. Lett. **33**, 660–662 (2008)
76. J.M. Dudley, G. Genty, S. Coen, Supercontinuum generation in photonic crystal fiber. Rev. Mod. Phys. **78**, 1135–1184 (2006)
77. S. Radic, C.J. McKinstrie, A.R. Chraplyvy, G. Raybon, J.C. Centanni, C.G. Jorgensen, K. Brar, C. Headley, Continuous-wave parametric gain synthesis using nondegenerate pump four-wave mixing. IEEE Photon. Tech. Lett. **14**, 1406–1408 (2002)
78. M.R. Lamont, B. Luther-Davies, D.-Y. Choi, S. Madden, X. Gai, B.J. Eggleton, Net-gain from aparametric amplifier on a chalcogenide optical chip. Opt. Express **16**, 20374–20381 (2008)

79. M.A. Foster, A.C. Turner, J.E. Sharping, B.S. Schmidt, M. Lipson, A.L. Gaeta, Broad-band optical parametric gain on a silicon photonic chip. Nature **441**, 960–963 (2006)
80. A. Pasquazi, Y. Park, J. Azaña, F. Légaré, R. Morandotti, B.E. Little, S.T. Chu, D.J. Moss, Efficient wavelength conversion and net parametric gain via Four Wave Mixing in a high index doped silica waveguide. Opt. Express **18**, 7634–7641 (2010)
81. A. Pasquazi, Y. Park, S.T. Chu, B.E. Little, F. Légaré, R. Morandotti, J. Azaña, D.J. Moss, Time lens measurement of subpicosecond optical pulses in CMOS compatible high index glass waveguides. IEEE J. Sel. Top. Quant. Electron. **PP**(99), 1–8 (2011)
82. B.H. Kolner, M. Nazarathy, Temporal imaging with a time lens. Opt. Lett. **14**, 630–632 (1989)
83. B.H. Kolner, Generalization of the concepts of focal length and f-number to space and time. J. Opt. Soc. Am. A **11**, 3229–3234 (1994)
84. J. Azana, N.K. Berger, B. Levit, B. Fischer, Spectral Fraunhofer regime: time-to-frequency conversion by the action of a single time lens on an optical pulse. Appl. Opt. **43**, 483–490 (2004)
85. L.Kh. Mouradian, F. Louradour, V. Messager, A. Barthelemy, C. Froehly, Spectro-temporal imaging of femtosecond events. IEEE J. Quant. Electron. **36**, 795–801 (2000)
86. M.T. Kauffman, W.C. Banyai, A.A. Godil, D.M. Bloom, Time-to-frequency converter for measuring picosecond optical pulses. Appl. Phys. Lett. **64**, 270–272 (1994)
87. C.V. Bennett, R.P. Scott, B.H. Kolner, Temporal magnification and reversal of 100 Gb/s optical-data with an up-conversion time microscope. Appl. Phys. Lett. **65**, 2513–2515 (1994)
88. C.V. Bennett, B.H. Kolner, Principles of parametric temporal imaging. I. System configurations. IEEE J. Quant. Electron. **36**, 430–437 (2000)
89. M.A. Foster, R. Salem, D.F. Geraghty, A.C. Turner-Foster, M. Lipson, A.L. Gaeta, Silicon-chip-based ultrafast optical oscilloscope. Nature **456**, 81–84 (2008)
90. M.A. Foster, R. Salem, Y. Okawachi, A.C. Turner-Foster, M. Lipson, A.L. Gaeta, Ultrafast waveform compression using a time-domain telescope. Nat. Photon. **3**, 581–585 (2009)
91. R. Salem, M.A. Foster, A.C. Turner-Foster, D.F. Geraghty, M. Lipson, A.L. Gaeta, Optical time lens based on four-wave mixing on a silicon chip. Opt. Lett. **33**, 1047–1049 (2008)
92. H. Nishihara, M. Haruna, T. Suhara, *Optical Integrated Circuits* (McGraw Hill, New York, NY, 1989). ISBN ISBN 978-0070460928
93. W.J. Tomlinson, R.H. Stolen, C.V. Shank, Compression of optical pulses chirped by self-phase modulation in fibers. J. Opt. Soc. Am. B **1**, 139 (1984)
94. K. Ogata, *Modern Control Engineering*, 5th edn. (Prentice Hall, New Jersey, 2001). ISBN ISBN 978-0136156734
95. G.F. Simmons, *Differential Equations with Applications and Historical Notes*, 2nd edn. (McGraw-Hill, New York, NY, 1991). ISBN ISBN 978-0070575400
96. J. Azaña, Proposal of a uniform fiber Bragg grating as an ultrafast all-optical integrator. Opt. Lett. **33**, 4–6 (2008)
97. R. Slavik, Y. Park, M. Kulishov, R. Morandotti, J. Azaña, Ultrafast all-optical differentiators. Opt. Express **14**, 10699–10707 (2006)
98. Y. Park, J. Azaña, R. Slavik, Ultra-fast all-optical first and higher-order differentiators based on interferometers. Opt. Lett. **32**, 710–712 (2007)
99. Y. Park, T.-J. Ahn, Y. Dai, J. Yao, J. Azaña, All-optical temporal integration of ultra-fast pulse waveforms. Opt. Express **16**, 17817–17825 (2008)
100. N.Q. Ngo, Design for an optical temporal integrator based on a phase shifted fiber Bragg grating in transmission. Opt. Lett. **32**, 3020–3022 (2007)
101. C.K. Madsen, J.H. Zhao, *Optical Filter Design and Analysis: A Signal Processing Approach* (Wiley, New York, NY, 1999). ISBN ISBN 978-0471183730
102. M.T. Hill, H.J.S. Dorren, T. de Vries, X.J.M. Leijtens, J.H. den Besten, B. Smalbrugge, Y.-S. Oei, H. Binsma, G.-D. Khoe, M.K. Smit, A fast low-power optical memory based on coupled micro-ring lasers. Nature **432**, 206–209 (2004)

103. N.Q. Ngo, L.N. Binh, Optical realization of Newton-Cotes-based integrators for dark soliton generation. J. Lightwave Tech. **24**, 563–572 (2006)
104. C.-W. Hsue, L.-C. Tsai, Y.-H. Tsai, Time-constant control of microwave integrators using transmission lines. IEEE Trans. Microw. Theor. Tech. **54**, 1043–1047 (2006)
105. X. Ding, X. Zhang, X. Zhang, D. Huang, Active micro-ring optical integrator associated with electro-absorption modulators for high speed low light power loadable and erasable optical memory unit. Opt. Express **17**, 12835–12848 (2009)
106. M. Ferrera, Y.-W. Park, L. Razzari, B.E. Little, S.T. Chu, R. Morandotti, D.J. Moss, J. Azaña, On-chip CMOS-compatible all-optical integrator. Nat. Commun. **1**, 29 (2010)

Chapter 3
Linear and Nonlinear Wave Dynamics in Amorphous Photonic Lattices

Mikael Rechtsman, Alexander Szameit, and Mordechai Segev

Conventional intuition in solid-state physics holds that in order for a solid to have an electronic band-gap, it must be periodic, allowing the use of Bloch's theorem. Indeed, the free-electron approximation seems to imply that Bragg scattering in periodic potentials is a necessary condition for the formation of a band-gap. But this is obviously untrue: looking through a window reveals that glassy silica (SiO_2), although possessing no order at all, still displays a band-gap spanning the entire photon energy range of visible light, without absorption. Several experimental studies have probed the properties of the band-gap in such "amorphous" electronic systems using spectroscopic techniques [1], time-of-flight measurements [2], and others. With the major progress in photonic crystals [3, 4], it is natural to explore amorphous photonic structures with band-gaps, where the actual wavefunction can be observed directly, and hence, many physical issues can be studied at an unprecedented level. Indeed, amorphous photonic media have been studied theoretically in several pioneering papers (e.g., [5, 6]), and experiments in the microwave and optical regimes have demonstrated the existence of a band-gap [5]. However, amorphous band-gap media have never been studied experimentally in the optical regime. Particularly in optics, the full beauty of disorder can be revealed: optics offers the possibility to precisely engineer the potential strength and period, as well as the unique opportunity to employ nonlinearity under controlled conditions, which could unravel unknown features that are much harder to access experimentally in other systems. In this chapter, we review recent developments [7] on amorphous photonic lattices: a two-dimensional array of randomly organized evanescently coupled waveguides. We demonstrate that the bands in this medium, comprising inherently localized Anderson states, are separated by gaps, despite the total lack of Bragg scattering. We find that amorphous photonic lattices support the existence of strongly localized defect states, whose widths is much narrower than the Anderson localization length. We show the existence of a region of

M. Rechtsman (✉) • A. Szameit • M. Segev
Physics Department, Solid State Institute, Technion, Haifa 32000, Israel
e-mail: mcr@technion.ac.il

Z. Chen and R. Morandotti (eds.), *Nonlinear Photonics and Novel Optical Phenomena*, Springer Series in Optical Sciences 170, DOI 10.1007/978-1-4614-3538-9_3,

negative effective mass (anomalous diffraction), which could be demonstrated experimentally by superimposing a weak spatial modulation on the random potential (refractive index), and observing transport. In this setting, a wavepacket with a negative effective mass moves opposite to the direction it would have moved had it had positive effective mass. Finally, we numerically demonstrate the existence of discrete solitons in amorphous photonic lattices and discuss their similarities and differences from discrete solitons in a periodic setting.

3.1 The Linear Regime

The presence of a band-gap in the electronic energy spectrum of amorphous solids, materials whose atoms possess no long-range order, was first put on firm theoretical grounds by Weaire and Thorpe, who proved rigorously that a gap existed in a model of amorphous silicon (a-Si) [8]. This called for a reevaluation of the intuition held by solid-state physicists (which is still common today) that Bragg scattering in periodic potentials is responsible for the opening of band-gaps at the boundary of the Brillouin zone. A wide range of experimental and theoretical techniques have been used to study these fascinating materials, including X-ray photoemission spectroscopy [9], time-of-flight methods [2], and others on the experimental side, as well as numerical [10] and analytic [11] methods on the theoretical side. Recently, interest has been increasing in the topic of band-gaps in amorphous photonic materials. For example, two-dimensional dielectric composites with a random geometry have been shown to possess a photonic band-gap [5], a large photonic band-gap has been found numerically in a tetrahedral network connected by dielectric cylinders [11], and new classes of two-dimensional noncrystalline photonic band-gap materials have been introduced [6]. In these, the band-gap is a range of frequencies for which propagation of electromagnetic waves is forbidden. A prime example of a technological application of an amorphous photonic system is the random laser [12], for which strong suppression of electromagnetic transport is the essential requirement.

In this chapter, review our recent theoretical and experimental results [7] on amorphous photonic lattices: a new type of optical structure, which provide a test-bed for the properties of general amorphous systems, and at the same time offer applications unique to photonics. The lattices are composed of individual waveguides with no long-range order whatsoever; that is, the structure is neither periodic nor quasi-periodic. Rather, such amorphous structures lack Bragg diffraction peaks altogether (to distinguish from quasicrystals which do exhibit pronounced Bragg diffraction [13, 14]). To this end, we distribute the waveguides similar to the atoms in a liquid, which is explained in more detail in Methods Section. Such structures result in random refractive index distributions $n(x,y)$, with one particular realization shown in Fig. 3.1a. In Fig. 3.1b, we show the microscope image of a sample corresponding to the design of Fig. 3.1a, fabricated using the "direct laser writing" method [15]. This system indeed exhibits no periodicity, as proven by the square of the Fourier transform of the structure function (so-called spatial power

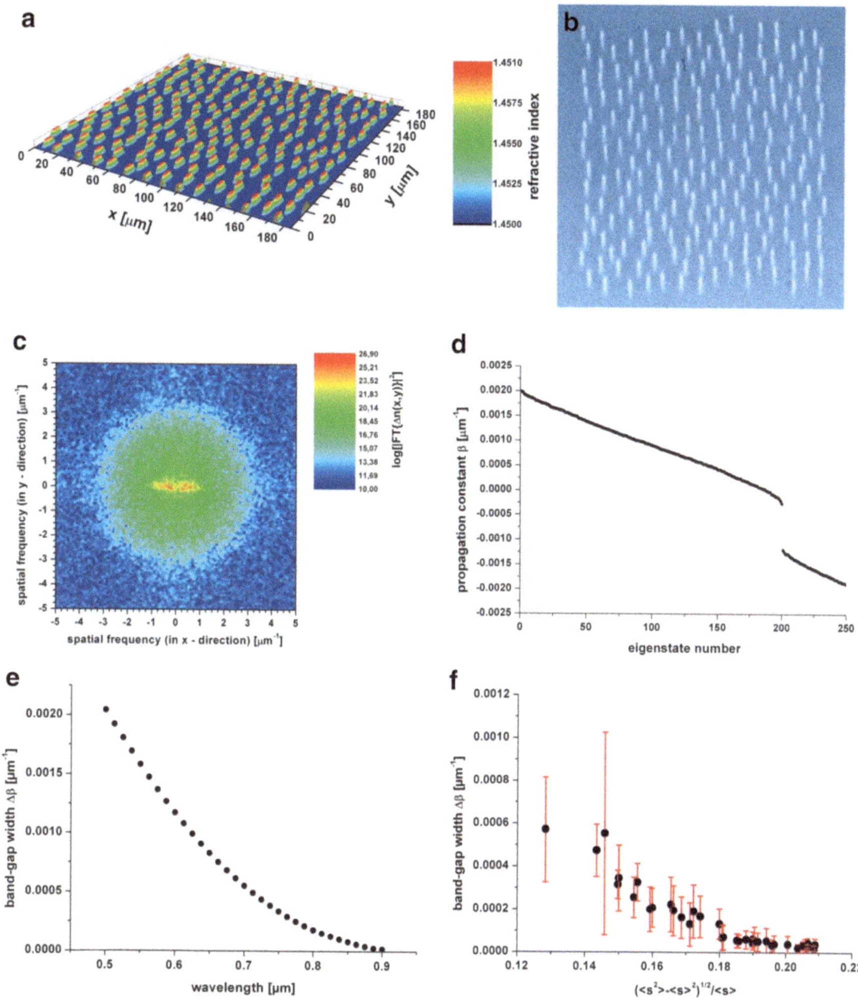

Fig. 3.1 Amorphous photonic lattice and its band structure (theory and experiments). (**a**) Refractive index profile, $n(x,y)$, of an amorphous waveguide structure, as a function the coordinates (x,y), in the plane transverse to the propagation direction of the beam (z). The index varies from 1.45 to $1.45 + 9 \times 10^{-4}$. (**b**) Optical microscope image of the input facet of the amorphous waveguide structure, as described mathematically in (3.2). (**c**) Fourier transform of $n(x,y)$, the refractive index profile given in (**a**). There are clearly no Bragg peaks here, indicating the complete lack of periodicity of this system. (**d**) Band structure (eigenvalue spectrum) of the optical analog to the Schrödinger operator, given in (3.1) for the amorphous waveguide structure shown in (**a**), at incident wavelength $\lambda = 633$ nm. A large band-gap is clearly present, despite the lack of any periodicity. (**e**) The width of the gap as a function of the optical wavelength. (**f**) The width of the gap as a function of the variance of the waveguide positions. Clearly, the gap survives a considerable amount of disorder before it closes

spectrum) displayed in Fig. 3.1c, indicating the lack of any Bragg peaks. The positions of the waveguides as shown in Fig. 3.1a, b are generated using a Metropolis Monte Carlo simulation of a liquid with periodic boundary conditions, where the component atoms interact via an isotropic pair potential of the Yukawa form, $v(r) = v_0 e^{-(r-r_0)/l}/r$, where $r_0 = 14$ μm and $l = 2.8$ μm. The area of the box is 184.2 μm^2. For our system, we take $k_B T/v_0 = 1.0$ and the particle number density is $\rho = 2/\sqrt{3}r_0^2$, where T is the temperature, l is the screening length, and k_B is Boltzmann's constant. At this temperature and density, the N-particle liquid is well above its freezing point.

The evolution of the optical wave $\Psi(x,y,z)$ in our system is described by the Schrödinger-type equation

$$i\lambdabar \frac{\partial \Psi(x,y,z)}{\partial z} = -\left[\frac{\lambdabar^2}{2n_0}\nabla_\perp^2 + \Delta n(x,y)\right]\Psi(x,y,z), \tag{3.1}$$

where $\lambdabar = \lambda/2\pi$ is the reduced wavelength, n_0 is the ambient refractive index of the bulk (fused silica, in our experiments), and $\Delta n(x,y)$ is the refractive index distribution caused by the random distribution of identical waveguides. There is an evident similarity between (3.1) and the quantum mechanical Schrödinger equation, when one replaces $\lambdabar$ with the reduced Planck constant $\hbar$, the ambient refractive index n_0 with the particle mass m, the optical potential $\Delta n(x,y)$ by the quantum potential $-V(x,y)$, and the spatial coordinate z with time t. Hence, in optics the quantum-mechanical evolution in time of a two-dimensional wavefunction is mapped onto the propagation of an optical wavepacket along the spatial z-direction [16, 17]. The z-independence of our setting is analogous to the time-independent Hamiltonian in quantum mechanics. This feature has been used in many recent experiments, demonstrating concepts from solid-state physics using paraxial optical settings [16, 17]. One important example is Anderson localization [18], which has recently been realized in photonic lattices [19, 20] using the transverse localization scheme [21]. To obtain the band structure of our system, one needs to solve the optical analog of the time-independent Schrödinger equation, which is derived by a separation of variables and substituting $\Psi(x,y,z) = \varphi(x,y)e^{i\beta z}$ with the propagation constant $\beta = -E/\lambdabar$ into (3.1). The eigenvalue E represents the energy in the corresponding quantum mechanical setting. However, the resulting eigenvalue equation cannot be solved by applying Bloch's theorem, since in our setting $\Delta n(x,y)$ is not a periodic function. Rather, as known from the theory of Anderson localization in two-dimensions, the eigenmodes are fully localized functions ("Anderson states") [22]. Consequently, one must solve using the full refractive index profile in order to find the eigenvalue spectrum. We do that for the structure shown in Fig. 3.1a, b using the plane-wave expansion method [4], and generate Fig. 3.1d, depicting the value of the propagation constant β for an optical wavelength of 633 nm. In these calculations, we employ periodic boundary conditions on a system composed of 200 waveguides. All waveguides have identical structure (slightly elliptic, due to fabrication constraints), and the refractive index step defining them is $\Delta n = 9 \times 10^{-4}$. As Fig. 3.1d clearly shows, there is a sizeable gap in the spatial spectrum, despite a total lack of periodicity (and lack of Bragg scattering). This defies a common argument for the presence of band-gaps [23], which states that gaps open at the boundary of the Brillouin zone because the degeneracy of states there is broken by the periodicity of the potential. This argument fails here because it is based on a perturbation theory where the potential is assumed to be weak, which is inapplicable for this system. This theory implies that the presence of Bragg peaks is a necessary condition for the formation of band-gaps, which clearly contradicts the physical findings in amorphous systems [5, 6, 8–11].

One of the nice features offered by optical systems described by (3.1) is the ability to test the features of the system through tuning parameters independently, with the most notable one being the optical wavelength λ, which can be tuned continuously. Figure 3.1e shows the dependence of the bad-gap width on the optical wavelength, revealing that the gap is exponentially decreasing with λ (when all other parameters are fixed), until it closes at 820 nm. This can be understood directly from (3.1): an increasing wavelength leads to a larger "kinetic energy" [transverse Laplacian in (3.1)], and thus a relative weakening of the potential that induces the gap in the first place. As shown below, this wavelength dependence of the gap provides an efficient tool for exploring the properties of amorphous photonic media.

Figure 3.1 calls for some intuition for the existence of a band-gap in amorphous photonic systems, as well as for its wavelength dependence, in spite of the complete absence of Bragg diffraction peaks. The Bragg peaks are absent due to the random distribution of waveguides, whereas the existence of a gap has been traditionally associated with the presence of order [23]. In the amorphous photonic medium displayed in Fig. 3.1, all waveguides have the same structure, but their spacings are randomly distributed, having a particular variance around a mean value. As such, it is instructive to plot the size of the gap as a function of the (normalized) variance of the interwaveguide spacing (which is directly related to the variance in the coupling coefficient between adjacent waveguides, and equivalent to the variance in the hopping parameter in the tight-binding model). We plot that in Fig. 3.1f, for $\lambda = 633$ nm. As this figure clearly shows, there is a sizeable band-gap as long as the normalized variance does not exceed 18%. When the system has a crystalline structure and thus the interwaveguide spacing has zero variance, it exhibits Bragg peaks and has a large band-gap. In a disordered pattern, Bragg resonance is absent. However, the gap survives even for a rather large variance, until it closes around ~18%, where the bands merge. This explanation holds not only when the bands arise from guided modes of the individual waveguide (bound states of a single potential well), but also when the bands arise from unbound states, as is the case for the experimental structure of Fig. 3.1 (where each waveguide has only one guided mode).

In order to visualize the gap experimentally, we introduce a defect waveguide in the structure: a single waveguide with a refractive index maximum that is lower by $\Delta n_d = 4.5 \times 10^{-4}$ than the maxima of all other waveguides. [In tight-binding calculations, this would correspond to a negative defect]. Band structure calculations show that this procedure results in a single defect state that resides

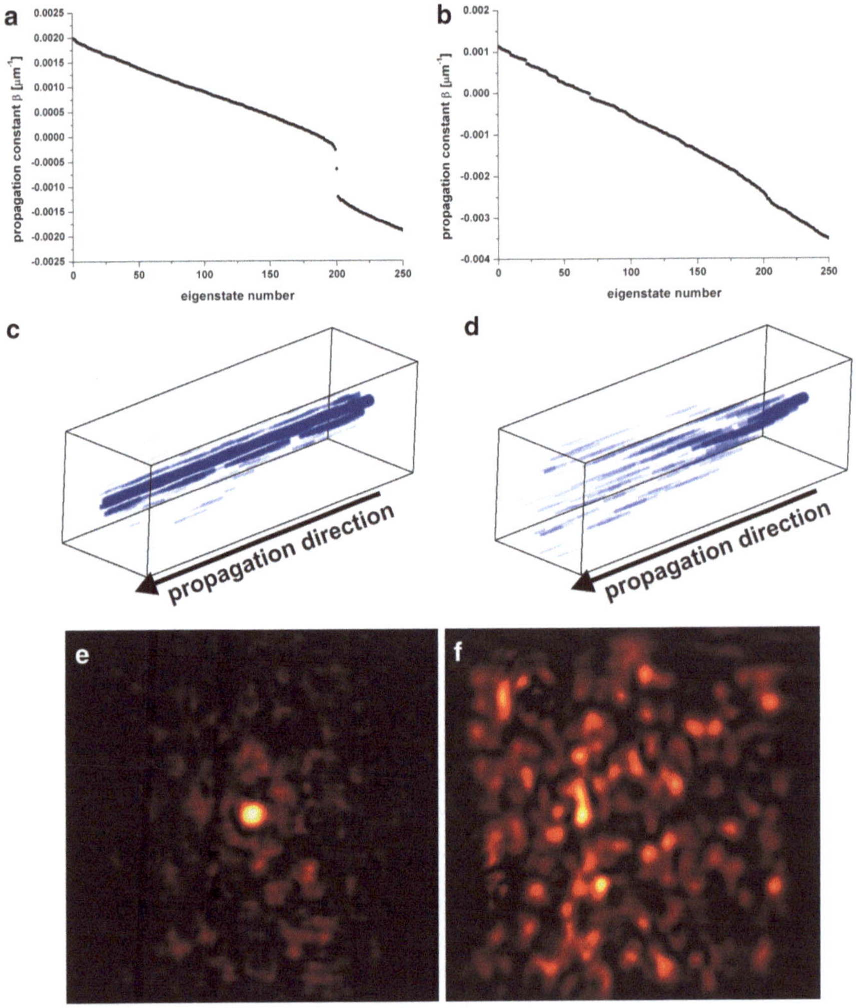

Fig. 3.2 Propagation of a light beam launched into a defect state, at an optical wavelength where the defect resides in the gap (*left column*) and at a wavelength where there is no gap (*right column*) (theory and experiments). (**a**) Band structure (eigenvalue spectrum) of the waveguide array given in Fig. 3.1a, including the presence of a defect waveguide with index of refraction 1.45 to 1.45 + 4.5 $\times$ 10^{-4}, at incident wavelength $\lambda = 633$ nm. The eigenvalue in the center of the gap corresponds to a strongly localized mode centered on the defect waveguide. (**b**) Is analogous to (**a**), but at incident wavelength $\lambda = 875$ nm, where the background potential is too weak to yield a band-gap or a localized defect mode. (**c** and **d**) Are simulated profiles of the propagation of an initial wavepacket with a Gaussian shape with standard deviation 5 μm, centered on the defect waveguide for incident wavelength of $\lambda = 633$ nm and $\lambda = 875$ nm, respectively. At $\lambda = 633$ nm, the large majority of the wave's power stays localized within the defect mode, whereas for $\lambda = 875$ nm, when the defect mode is no longer present, it diffracts away. (**e** and **f**) Are experimental results; they display the beam output from the waveguide array. In (**e**), 633 nm light was incident directly on the defect guide; the light excited the defect mode and thus stayed localized within that particular waveguide. In (**f**), 875 nm light was incident on the defect guide; it did not remain localized due to the absence of a gap. Rather, the light emerging from the amorphous structure is now distributed over hundreds of waveguides

directly in the band-gap (Fig. 3.2a), for an optical wavelength of $\lambda = 633$ nm. By contrast, at $\lambda = 875$ nm, the gap is extremely small; hence, the defect state occurs where the bands merge (Fig. 3.2b). Consequently, when we launch a $\lambda = 633$ nm beam directly into the defect waveguide, the beam stays strongly confined in both transverse directions throughout propagation, because it excites a highly localized defect state, as highlighted by the simulation shown in Fig. 3.2c (carried out with a standard split-step beam-propagation code). That is, the coupling to all nearby waveguides is greatly suppressed, in spite of their close proximity, because light is guided in a defect state residing in a sizeable band-gap. By contrast, at $\lambda = 875$ nm there is no gap; hence, a 875-nm beam launched into the same waveguide does not stay confined but couples (tunnels) to other waveguides, resulting in major expansion of the beam, as the beam is propagating along z (Fig. 3.2d). Figure 3.2e and f shows our experimental results, depicting the intensity structure of the beam at the output facet of the waveguide arrays, for $\lambda = 633$ nm and $\lambda = 875$ nm, respectively. Clearly, the tight confinement of the light in the defect state in Fig. 3.2e and its lack thereof in Fig. 3.2f echoes the simulation results of Fig. 3.2c and d, respectively. Thus, by demonstrating the presence of the defect state, we have experimentally proved the existence of a band-gap in this amorphous optical system.

The amorphous photonic lattice employed here is fabricated by the laser direct-writing method in a fused silica sample [15]. We used a Ti–Sapphire laser system operating at a wavelength of 800 nm, a repetition rate of 100 kHz and a pulse length of 170 fs. A permanent change in the molecular structure of the material can be realized by tightly focusing ultrashort laser pulses into a transparent bulk material, causing nonlinear absorption. In fused silica, this induces a permanent increase in the refractive index with approximately the dimensions of the focus of the microscope objective focusing the writing beam. By moving the sample transversely with respect to the beam, a continuous modification of the refractive index is obtained, which creates a waveguide in the volume of the bulk silica. For the fabrication of our 2 cm long samples, the average power was adjusted to 32 mW and the writing velocity was set to 90 mm/min. The waveguides form a transverse refractive index profile of the form

$$n(x,y) = n_0 + \sum_{j=1}^{N} \Delta n_j e^{-\left((x-x_j)^2/\sigma_x^2 + (y-y_j)^2/\sigma_y^2\right)^3} \equiv n_0 + \Delta n(x,y), \tag{3.2}$$

which is invariant in the propagation direction z. The ambient refractive index of the fused silica is $n_0 = 1.45$, $N = 200$ is the number of waveguides, Δn_j and (x_j, y_j) are the refractive index increase and position of the jth waveguide, respectively. The parameters $\sigma_x = 1.5$ μm and $\sigma_y = 1.5$ μm describe the transverse length and width of the waveguides, and $\Delta n(x,y)$, the deviation from the ambient refractive index, represents the potential in (3.1).

In many amorphous systems, it is interesting to observe how the band-gap closes and what happens to defect states residing in the gap as the bands merge.

We explored this issue in our system in detail, simply by launching a beam into the defect waveguide, and varying the wavelength. At $\lambda = 750$ nm, the gap is still present (Fig. 3.3a); hence, the defect state is decoupled from the band, and light remains confined to the "defect waveguide" where it was launched (Fig. 3.3b). At $\lambda = 800$ nm, the gap is almost closed (Fig. 3.3c); hence, the defect state becomes part of the *Urbach tail* [24] of the second band, i.e., states that leak into the gap due to the presence of disorder. Increasing the wavelength even further, the defect state itself becomes more and more delocalized. Eventually, at $\lambda = 850$ nm, the gap is closed and the defect state is fully absorbed into the band (Fig. 3.3e, f).

There is a sharp distinction between localization within a defect state and the phenomenon of Anderson localization. Here, we demonstrate that a defect-state in an amorphous system is in fact much more localized than Anderson states. Our system is two-dimensional (x and y, while z plays the role of time, in the quantum problem), where it is known that all states are inherently localized with any amount of disorder [22], i.e., the bands are composed of localized Anderson states. It is therefore interesting to use our amorphous photonic system to compare transverse (Anderson) localization of light [19–21] with the transverse localization of light in a defect state residing within the gap. Figure 3.4 shows the results, exemplifying the fact that a defect state (residing in the gap) is much more localized than the Anderson localization length. To this end, the experiments on Anderson localization were carried out through ensemble averaging as explained in [19]. This is because the corresponding quantum model is an expectation value problem; hence, one has to average over multiple realizations of the disorder to obtain a meaningful results (the quantum system is self-averaging given long enough evolution time, but the propagation distance in our photonic system is not large enough to experience self-averaging, and hence the ensemble average). Figure 3.4a, b depicts the ensemble-averaged experimental intensity profile of light trapped in a defect mode (Fig. 3.4a), averaged over ten samples, and light that is Anderson localized (Fig. 3.4b), averaged over 30 samples. Comparing Figs. 3.3b and 3.4a reveals that the defect state is clearly invariant under averaging. By contrast, examining a wave-packet composed of Anderson modes (which are part of the band) in a single realization (Fig. 3.4c) shows no confinement whatsoever. This emphasizes the importance of averaging over multiple realizations: when the evolution is short-ranged, as in the transverse localization scheme, only ensemble-averaging reveals Anderson localization, as nicely shown by Fig. 3.4b. Figure 3.4d shows the simulated results of the experiment of Fig. 3.4b, with an ensemble-average taken over 100 realizations of disorder. Comparing the Anderson localization length and the width of a defect mode that resides in the gap, shows clearly that the width of the defect states is much smaller that the localization length. This is evident from the experiment (Fig. 3.4e; cross-sections taken through Fig. 3.4a, b), and from the simulation (Fig. 3.4f). The intuition behind this is that the defect state exhibits an isolated eigenvalue; hence, there are no other states of similar energy with which it may hybridize and thus delocalize. On the other hand, although the other eigenstates are themselves localized, they may be thought of as being composed of a set of eigenstates in different

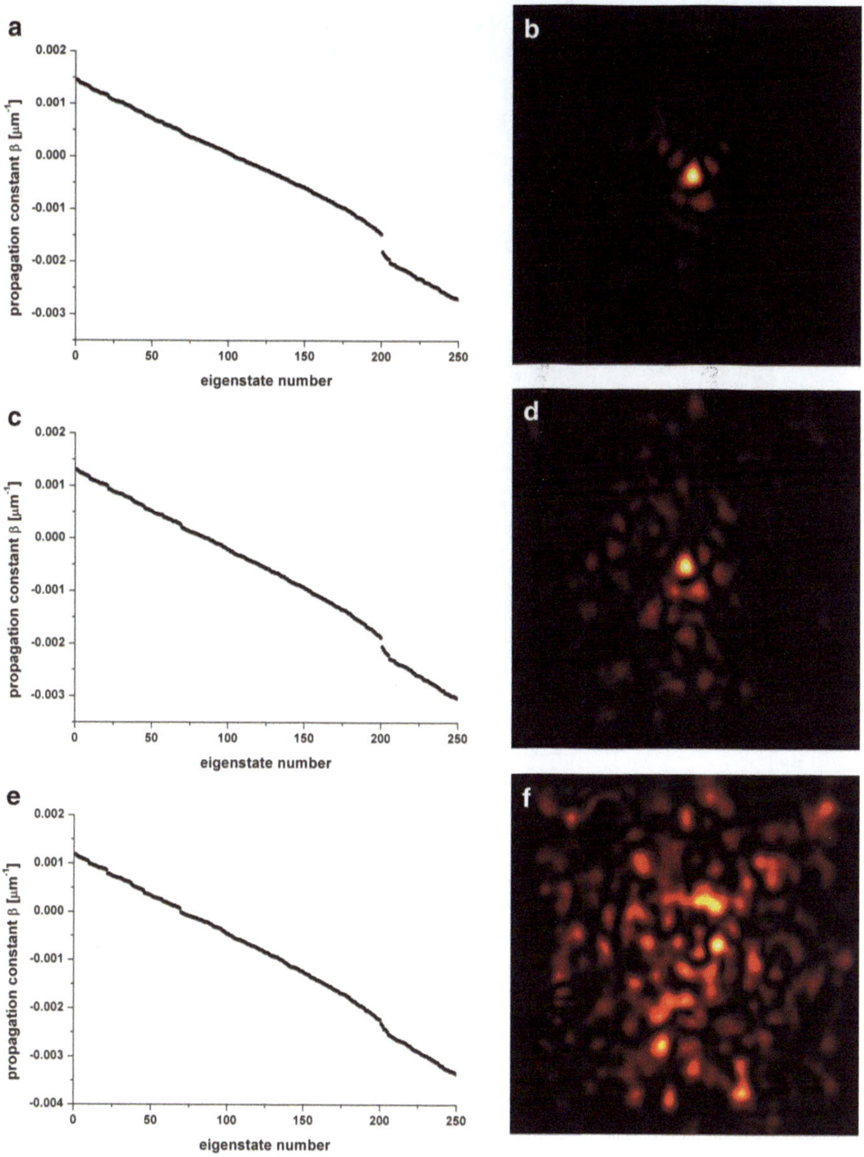

Fig. 3.3 Transformation of a defect state residing in the gap into states residing in the band, as the optical wavelength is increased (theory and experiments). (**a**) Band structure for the disordered waveguide array with a defect waveguide of index of refraction $1.45 + 4.5 \times 10^{-4}$, and at wavelength $\lambda = 750$ nm. In (**b**), we show experimental results of the output pattern (the absolute square of the wavefunction) where the input beam is launched into the defect waveguide. (**c**) And (**d**) are analogous to (**a**) and (**b**) respectively, but for $\lambda = 800$ nm; (**e** and **f**) are for $\lambda = 850$ nm. At this wavelength, the defect state may be considered to be part of the Urbach tail; as seen here, the state delocalizes as the band-gap gets smaller and it is drawn into the band

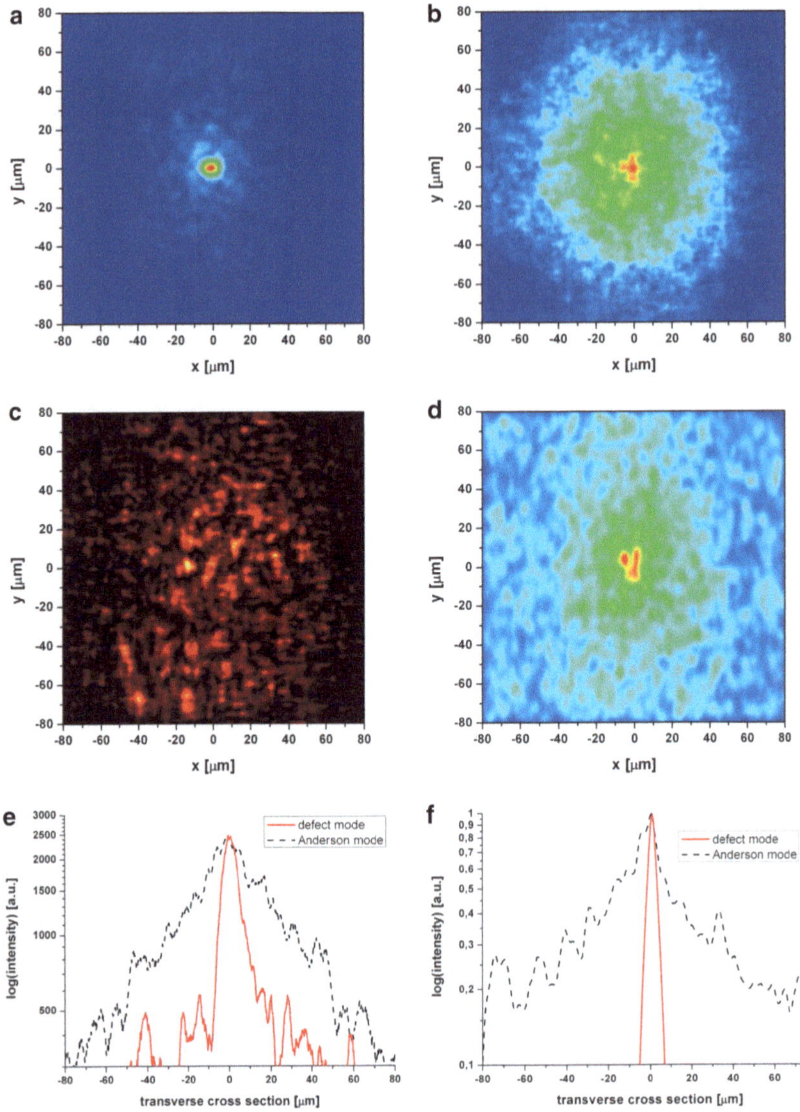

Fig. 3.4 Defect state in an amorphous photonic lattice compared to the localization length arising from Anderson states, highlighting that the defect state is much more localized, in spite of the fact that all states in an amorphous structure are localized (theory and experiments). (**a**) Experimental ensemble-average output beam for light launched into the defect waveguide (average taken over ten samples with different random realizations of the distribution of waveguides, all with the same statistics). Clearly, ensemble-averaging does not affect the shape of the defect mode. (**b**) Experimental ensemble-average output beam for 633 nm light input on 30 different nondefect waveguides in different local environments of the disordered pattern. The ensemble-average wavepacket exhibits Anderson localization. (**c**) A single experimental realization of the output beam when light is incident on a nondefect waveguide. By itself this beam does not display the exponential decay of Anderson localization; indeed Anderson localization only emerges upon ensemble-averaging. (**d**) Numerical simulation of Anderson localization at 633 nm over 100 realizations of the disordered pattern. (**e**) Semi-log plot of the light intensity vs. transverse direction for the ensemble-averaged experiments of (**a** and **b**). The linear behavior of the broader curve indicates exponential decay of the wavefunction, the hallmark of Anderson localization. The defect state is clearly not Anderson localized. (**f**) Numerical results for curves corresponding to those in (**e**) derived from beam-propagation simulations

environments hybridizing with one another, hence "spreading out". Eigenvalue differences in different local environments account for the ultimate localization of these wavefunctions, but these differences are extremely small compared to the relative isolation of the defect state.

The amorphous waveguide array is an excellent test-bed for studying general properties of waves in amorphous systems. We show here, by use of simulations, that the concept of *effective mass* at the band edges carries over faithfully from the periodic to the amorphous case. In photonic lattices, it is convenient to quantify the effective mass through the variation of the transverse velocity as a function of transverse momentum, which can be readily varied as one launches the input optical beam into the photonic lattice [25]. There, the effective mass is defined as the inverse of the second derivative of the propagation constant with respect to the transverse momentum [16, 17]. However, in our amorphous system, the dispersion relation is discontinuous; hence, one cannot simply vary the transverse momentum of the launch beam and observe the variation of transverse velocity, as is done in photonic lattices—relying on having Bloch modes that are inherently extended states. Here, the Anderson states are all localized, and they have vanishing transverse velocity (the potential is z-invariant); hence, one cannot use this method. Instead, we quantify the concept of effective mass directly through Newton's second law: we introduce a known variation of the potential (which must vary much slower than the spacing between waveguides), launch a wavepacket, and observe its trajectory. An example is shown in Fig. 3.5, where we add a weak, slowly varying, sinusoidal function to the refractive index variation, namely $\Delta n(x,y) \rightarrow \Delta n(x,y) + \alpha\sin(2\pi x/L)$, where L is the width of the sample, and the coordinate (0,0) is taken to be at the center [26]. Then, we construct a beam (wavepacket) from a superposition of eigenstates at close vicinity (such that effect mass can be defined, as in a periodic system), and launch it near the center coordinate at the input facet, gradually cutting off the wavepacket such that it lies in the center of the amorphous photonic lattice. With $\alpha > 0$ the force acts in the $+x$ direction. Hence, for a positive effective mass, the beam would be deflected towards $+x$, whereas for a negative effective mass it would propagate in the $-x$ direction, opposite to the force direction. The amount of deflection is proportional to effective mass (in the lowest order approximation). We demonstrate this concept through beam-propagation simulations, and display the results in Fig. 3.5. In Fig. 3.5a and g the beam is composed of modes taken from the center of the first band and the low-momentum edge of the second band, respectively, whereas in Fig. 3.5d the beam is composed of modes from the edge of the first band. Consequently, the beam, which was always launched at the center coordinate of the input facet (Fig. 3.5b, e, h), is propagating toward $+x$ in the former cases (Fig. 3.5c, i), and in the $-x$ direction in the latter case (Fig. 3.5f). Hence, the effective mass is positive at the center of the first band, then negative at the edge of the first band, positive again at the low-momentum edge of the second band, etc. That is, the concept of effective mass, which has been extensively studied in crystalline (perfectly periodic) systems, carries over to amorphous systems. Undoubtedly, this result suggests interesting experiments, especially in the

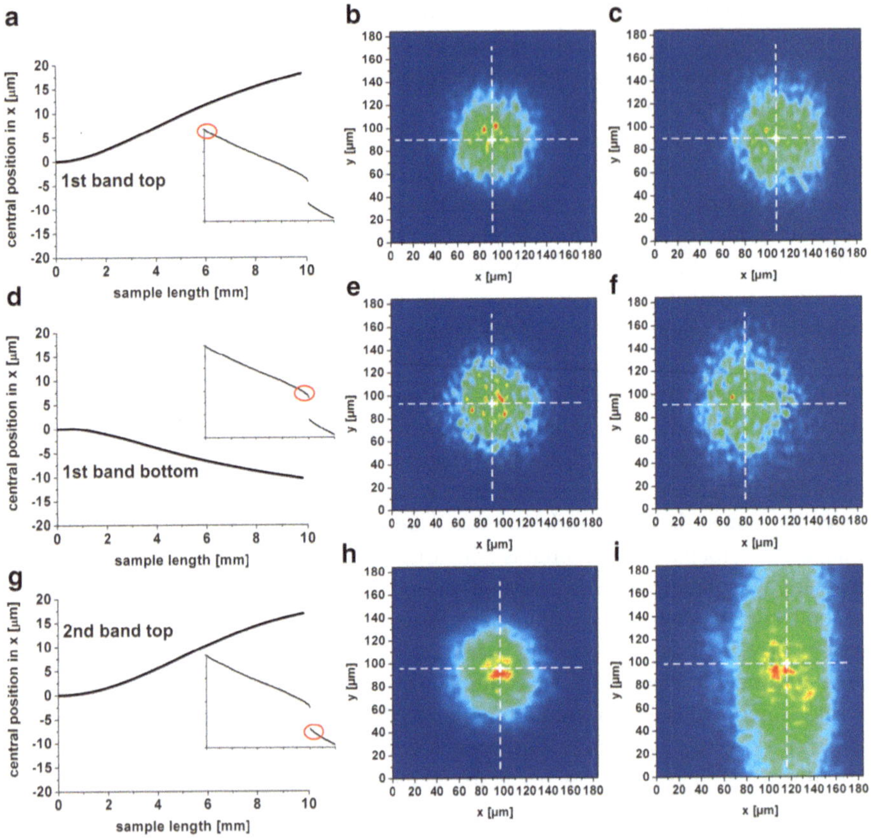

Fig. 3.5 The concept of effective mass and its impact on propagation of wavepackets in amorphous photonic lattices (simulations). Simulated propagation of wavepackets composed of five states near the edges of bands. A slowly varying force is applied by adding a sinusoidal function to the refractive index $n(x,y)$. The wavepacket is accelerated by the force, and its movement is proportional to the effective mass. The rows display results with wavepackets taken from the center of the first band (*top*), edge of the first band (*center*), and edge of the second band (*bottom*). The columns show, *from left to right*, the transverse displacement (x) as a function of propagation distance (z), the input wavepacket launched at the center of the lattice, and the output wavepacket emerging after 2 cm of propagation, displaced in correspondence with the effective mass. Clearly, the output wavepacket is displaced to the right (*top*), to the left (*center*) and to the right (*bottom*), proving that the effective mass at the band edges is positive, negative and positive, respectively, which is consistent with periodic lattices

nonlinear regime—where an attractive (self-focusing) nonlinearity would act exactly the opposite in the regimes of positive and negative effective mass [27]. Unfortunately, utilizing these ideas to demonstrate the concept of effective mass through transport in our amorphous photonic structure requires propagation distances that are beyond our current experimental reach. However, it is likely that such experiments would become practical within a short time.

3.2 The Nonlinear Regime: Solitons

In the previous section, we discussed amorphous photonic lattices and presented experimental results in the linear regime. In the present section, we will introduce nonlinearity and study its effects in these systems. To do so, we will use the discrete nonlinear Schrodinger equation (rather than the continuous one used in the previous section) because it is (arguably) simpler, certainly less computationally costly, and captures the physics related to the first band almost as accurately as the continuous model. The discrete nonlinear Schrodinger equation for the Kerr nonlinearity (in reduced units) is written as:

$$i\partial_z \psi_i + \sum_j t(r_{ij})\psi_j + \gamma|\psi_i|^2\psi_i = 0, \tag{3.3}$$

where z is the longitudinal propagation coordinate; γ is the nonlinear Kerr coefficient; Ψ_i is the value of the wavefunction in waveguide i; $t(r) = \mathrm{e}^{-3(r/a-1)}$ is a function of distance that gives the coupling between the waveguides; a is the average spacing between the waveguides; and the summation is taken over all waveguides. In the case where the waveguides are arranged periodically, it is well established that there are stationary solutions to this nonlinear equation—solitons—that were predicted by Christodoulides and Joseph [28], and were later observed experimentally by Eisenberg et al. [16] in one dimension and Fleischer et al. [27] in two dimensions. In the amorphous (or simply disordered) case, the existence of soliton solutions has been predicted [29, 30], but they have not been observed at the time of writing. Interestingly, these disordered discrete solitons (as we will show below) can be either very similar to standard periodic lattice solitons or quite different from them. Besides being of fundamental interest, the existence of these solitons are very important to the currently ongoing discussion of Anderson localization in nonlinear systems [19, 20, 31, 32], and whether in fact nonlinearity can act to reinforce localization or to destroy it.

In the present work, we focus solely on the focusing Kerr nonlinearity ($\gamma > 0$) since calculations show that the results for the defocusing ($\gamma > 0$) case are very similar. The amorphous pattern used is depicted in Fig. 3.1a, and is derived (as described in the previous section) from a snapshot of the simulation of a two-dimensional liquid. Since we are solving for stationary solutions to (3.3), we seek solutions that are z-harmonic ($\psi \rightarrow \psi \mathrm{e}^{i\beta z}$ where β is called the propagation constant), meaning that (3.3) reduces to

$$-\beta\psi_i + \sum_j t(r_{ij})\psi_j + \gamma|\psi_i|^2\psi_i = 0. \tag{3.4}$$

Solutions to this equation are solved for using Newton's method [33]. These soliton solutions can be parameterized by β, and their so-called "existence curve" [34] is a plot of their power (square of their L^2-norm, given by $P = \sum_i |\psi_i|^2$) vs. β, the propagation constant.

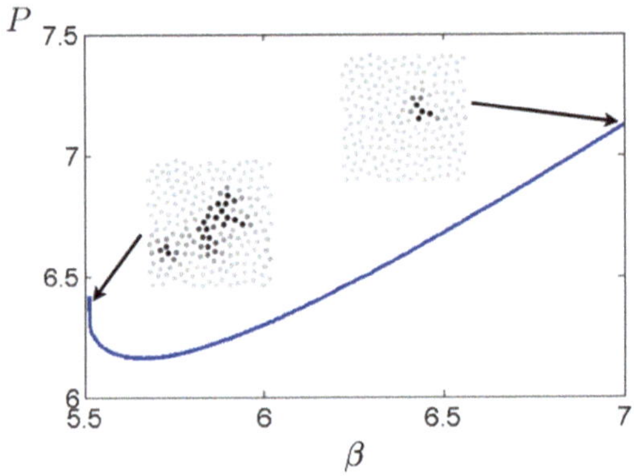

Fig. 3.6 Existence curves for soliton solutions of (3.4). This is one member of a family of solutions that monotonically decreases at lower values of β, then reaches a minimum P (meaning it has a power threshold), and then increases monotonically as β tends towards infinity. The effective width of the soliton wavefunction gets smaller with increasing β. This solution is strongly reminiscent of the behavior of discrete periodic lattice solitons

There are a number of classes of solutions, as described in ref. [30]. One class of solutions to (3.4) is shown in Fig. 3.6. At the edge of the band containing the linear eigenstates (below $\beta = 5.5$) the power of the soliton decreases as a function of β, then reaches a minimum and increases indefinitely. In this case, the soliton wavefunction steadily decreases in width as a function of increasing β. This is strongly reminiscent of the behavior of discrete solitons in periodic lattices [33]. This type of behavior is seen for a large number of solutions, not all of them approaching the band edge as the one shown in Fig. 3.6 does.

Another class of solitons is shown in Fig. 3.7. In this class, solitons bifurcate directly from the linear modes of the system at zero power, meaning that they are without a power threshold. This behavior is fundamentally different from that in periodic lattice solitons, as well as the class of solutions described in the previous paragraph. As shown in the figure, solitons bifurcate from linear modes at the band edge (the first mode) as well as those within the band. The latter are observed to undergo abrupt changes in slope for values of β within, or close to the band, most likely due to resonant interaction with other modes of the system (avoided crossings). Note that the solitons bifurcating from within the band are not necessarily more localized than the linear modes from which they originate. This is shown clearly in Fig. 3.7. The reason for this is that as the solitons resonantly interact with others, they take on some of the character of the other (perhaps more delocalized) modes. However, for sufficiently large β, these solutions will get smaller and smaller in extent until they become strongly localized. In this case, the focusing nonlinearity has induced a potential that is so deep that light cannot escape it.

The physical intuition for the zero power threshold observed for these solutions may be stated as follows. Disordered systems are characterized by the phenomenon of Anderson localization [18]. In both one and two dimensions, this means that all

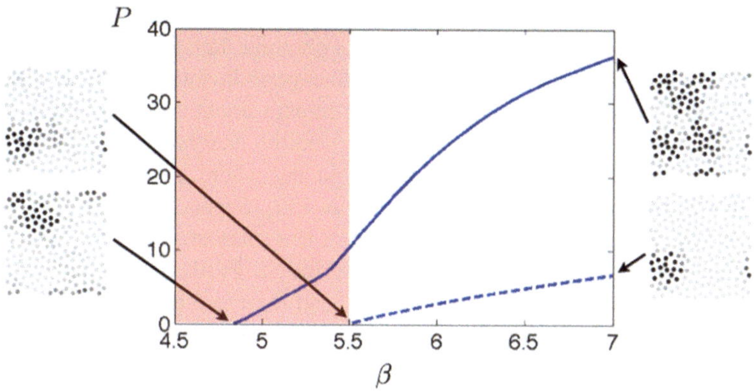

Fig. 3.7 Existence curves for soliton solutions of (3.4). These existence curves are for solitons of a different class than that depicted in Fig. 3.6: they bifurcate directly from linear modes at zero power. The *dashed curve* bifurcates from the first linear mode (highest propagation constant, β). Within increasing propagation constant, the effective width of the mode decreases. The *solid curve* is the existence curve of a soliton that bifurcates from a linear mode in the band. The effective width does not increase decrease monotonically with β, but it does converge to a single waveguide as $\beta \to \infty$. The *shaded region* denotes the band containing the propagation constants of the linear modes. Note that periodic boundary conditions are used, therefore the left and right sides of the arrays may be considered in contact with one another, as well as the top and bottom

of the linear eigenstates of the system become exponentially localized, in contrast to periodic systems wherein all eigenstates are extended. These localized eigenmodes may be thought of as defect states of the system, much like a band-gap mode associated with a vacancy in the lattice or an interstitial site (or any other local defect). Since the mode is already present, a high "amount" of nonlinearity is not necessary to localize the mode, and therefore the nonlinear mode may be formed for arbitrarily small power. This reasoning accounts for the lack of power threshold in this class of solutions.

3.3 Conclusions

In this chapter, we reviewed experimental and theoretical results [7] on amorphous photonic structures. We touched here only a few of the intriguing issues related to this topic. However, it is already evident that amorphous photonic systems provide an elegant tool to explore the universal features of strong disorder in general. In this work, we have shown that the concept of a band-gap has much wider importance and implications than traditionally conceived—in the context of crystalline structures. We have found disorder-induced defect states, which are fundamentally different from the Anderson states comprising the bands in amorphous media. We have shown that the concept of effective mass still holds in amorphous systems, and suggested a scheme to measure it experimentally. Finally, we discussed the issue of solitons in amorphous photonic lattices and described how they exhibit qualitatively different behavior than periodic lattice solitons. There are a number of important open questions that remain about solitons in disordered/ amorphous systems. First of all, a full linear stability analysis will be important in determining which of these solitons are stable and where, as well as whether the Vakhitov–Kolokolov stability criterion applies [35]. Secondly, a full enumeration of all soliton solutions is beyond the scope of this chapter, but is necessary for a completely comprehensive quantitative description. Furthermore, we have only handled the focusing Kerr nonlinearity here, and therefore other types of nonlinearity (i.e., higher order, saturable or nonlocal [36]) may exhibit different behavior. Preliminary calculations on the defocusing Kerr nonlinearity show very similar results to the focusing case, which is similar to periodic systems. In sum, the combination of disorder and nonlinearity provides a formidable challenge to optics and photonics research, but a very important and fascinating one.

These ideas and experiments are only the beginning of a new direction [7]. Many intriguing questions arise, and new ideas come up. Would nonlinear phenomena such as spontaneous pattern formation and modulation instability exist in amorphous systems? Can this system support shock waves? How do solitons move through the random potential, like waves or like particles? The experimental setting described in this chapter will help to explore these and many other questions, and to eventually understand the true universal spirit of general amorphous systems.

References

1. K. Morigaki, *Physics of Amorphous Semiconductors* (Imperial College Press, London, 1999)
2. T. Tiedje, J.M. Cebulka, D.L. Morel, B. Abeles, Evidence for exponential band tails in amorphous silicon hydride. Phys. Rev. Lett. **46**(21), 1425–1428 (1981)
3. E. Yablonovitch, Inhibited spontaneous emission in solid-state physics and electronics. Phys. Rev. Lett. **58**, 2059 (1987)
4. J.D. Joannopolous, S.G. Johnason, J.N. Winn, R.D. Meade, *Photonic Crystals: Molding the Flow of Light* (Princeton University Press, Princeton, NJ, 2008)
5. C. Jin, X. Meng, B. Cheng, Z. Li, D. Zhang, Photonic gap in amorphous photonic materials. Phys. Rev. B **63**(19), 195107 (2001)
6. M. Florescu, S. Torquato, P.J. Steinhardt, Designer disordered materials with large, complete photonic band gaps. PNAS **106**, 20658–20663 (2009)
7. M.C. Rechtsman, A. Szameit, F. Dreisow, M. Heinrich, R. Keil, S. Nolte, M. Segev, Amorphous photonic lattices: band gaps, effective mass, and suppressed transport. Phys. Rev. Lett. **106**, 193904 (2011)
8. D. Weaire, M.F. Thorpe, Electronic properties of an amorphous solid. I. A simple tight-binding theory. Phys. Rev. B **4**(8), 2508–2520 (1971)
9. L. Ley, S. Kowalczyk, R. Pollak, D.A. Shirley, X-ray photoemission spectra of crystalline and amorphous Si and Ge valence bands. Phys. Rev. Lett. **29**(16), 1088–1092 (1972)
10. J.D. Joannopoulos, M.L. Cohen, Electronic properties of complex crystalline and amorphous phases of Ge and Si. I. Density of states and band structures. Phys. Rev. B **7**(6), 2644–2657 (1973)

11. K. Edagawa, S. Kanoko, M. Notomi, Photonic amorphous diamond structure with a 3(d) photonic band gap. Phys. Rev. Lett. **100**(1), 13901 (2008)
12. H. Cao, Y.G. Zhao, S.T. Ho, E.W. Seelig, Q.H. Wang, R.P.H. Chang, Random laser action in semiconductor powder. Phys. Rev. Lett. **82**(11), 2278–2281 (1999)
13. D. Shechtman, I. Blech, D. Gratias, J.W. Cahn, Metallic phase with long-range orientational order and no translational symmetry. Phys. Rev. Lett. **53**, 1951 (1984)
14. D. Levine, P.J. Steinhardt, Quasicrystals: a new class of ordered structures. Phys. Rev. Lett. **53**, 2477 (1984)
15. A. Szameit, S. Nolte, Discrete optics in femtosecond laser-written photonic structures. J. Phys. B **43**(16), 1–25 (2010)
16. H.S. Eisenberg, Y. Silberberg, R. Morandotti, A.R. Boyd, J.S. Aitchison, Discrete Spatial Optical Solitons in Waveguide Arrays, Phys. Rev. Lett. **81**, 3383–3386 (1998)
17. F. Lederer, G.I. Stegeman, D.N. Christodoulides, G. Assanto, M. Segev, Y. Silberberg, Discrete solitons in optics. Phys. Rep. **463**, 1–126 (2008)
18. P.W. Anderson, Absence of diffusion in certain random lattices. Phys. Rev. **109**, 1492 (1958)
19. T. Schwartz, G. Bartal, S. Fishman, M. Segev, Transport and Anderson localization in disordered two-dimensional photonic lattices. Nature **446**(7131), 52–55 (2007)
20. Y. Lahini, A. Avidan, F. Pozzi, M. Sorel, R. Morandotti, D.N. Christodoulides, Y. Silberberg, Anderson localization and nonlinearity in one-dimensional disordered photonic lattices. Phys. Rev. Lett. **100**, 013906 (2008)
21. H. De Raedt, A. Lagendijk, P. de Vries, Transverse localization of light. Phys. Rev. Lett. **62**, 47–50 (1988)
22. E. Abrahams, P.W. Anderson, D.C. Licciardello, T.V. Ramakrishnan, Scaling theory of localization: absence of quantum diffusion in two dimensions. Phys. Rev. Lett. **42**(10), 673–676 (1979)
23. N. Ashcroft, D. Mermin, *Solid State Physics* (Brooks Cole, United States, 1976)
24. F. Urbach, The long-wavelength edge of photographic sensitivity and of the electronic absorption of solids. Phys. Rev. **92**(5), 1324–1324 (1953)
25. H.S. Eisenberg, Y. Silberberg, R. Morandotti, J.S. Aitchison, Diffraction management. Phys. Rev. Lett. **85**, 1863–1866 (2000)
26. D. Weaire, D. Hobbs, Electrons and holes in amorphous silicon. Phil. Mag. Lett. **68**(4), 265–272 (1993)
27. J.W. Fleischer, M. Segev, N.K. Efremidis, D.N. Christodoulides, Observation of two-dimensional discrete solitons in optically-induced nonlinear photonic lattices. Nature **422**, 147–150 (2003)
28. D.N. Christodoulides, R.I. Joseph, Discrete self-focusing in nonlinear arrays of coupled waveguides. Opt. Lett. **13**, 794–796 (1988)
29. N.K. Efremidis, K. Hizanidis, Disordered lattice solitons. Phys. Rev. Lett. **101**, 143903 (2008)
30. N.K. Efremidis, Two-dimensional disordered lattice solitons. Opt. Lett. **34**, 596–598 (2009)
31. H. Veksler, Y. Krivolapov, S. Fishman, Spreading for the generalized nonlinear Schrödinger equation with disorder. Phys. Rev. E **80**, 037201 (2009)
32. S. Flach, D.O. Krimer, Ch. Skokos, Universal Spreading of Wave Packets in Disordered Nonlinear Systems, Phys. Rev. Lett. **102**, 024101 (2009)
33. P.G. Kevrekidis, *The Discrete Nonlinear Schrödinger Equation: Mathematical Analysis, Numerical Computations and Physical Perspectives* (Springer-Verlag 2009)
34. M. Segev, M. Shih, GC Valley, J. Opt. Soc. Am. B **13**, 706 (1996)
35. N.G. Vakhitov, A.A. Kolokolov, Stationary solutions of the wave equation in a medium with nonlinearity saturation. Radiophys. Quant. Electron. **16**, 783–789 (1973)
36. Carmel Rotschild, Oren Cohen, Ofer Manela, Mordechai Segev, Tal Carmon, Solitons in Nonlinear Media with an Infinite Range of Nonlocality: First Observation of Coherent Elliptic Solitons and of Vortex-Ring Solitons, Phys. Rev. Lett. **95**, 213904 (2005).

Chapter 4
Nonlinear Control of Multicolor Beams in Coupled Optical Waveguides

Dragomir N. Neshev, Andrey A. Sukhorukov, and Yuri S. Kivshar

4.1 Introduction

Photonic structures with a periodic modulation of the optical refractive index play an important role in the studies of the fundamental aspects of wave dynamics [1, 2]. In particular, photonic crystals, layered media, or closely spaced optical waveguides enable manipulation of the key phenomena governing optical beam propagation: spatial refraction and diffraction. Arrays of coupled optical waveguides are particularly attractive as an experimental testbed due to their easier fabrication and characterization, as well as because of the opportunities they offer for enhanced nonlinear effects as a result of the large propagation distances in such structures. The physics of beam propagation in optical waveguide arrays is governed by the coupling of light between neighboring waveguides and the subsequent interference of the coupled light. Since both the coupling and the interference processes are sensitive to the light wavelength, the output intensity profiles can be drastically different for each spectral component of the input beam. This is a particular concern in many practical cases, including ultra-broad bandwidth optical communications, manipulation of ultra-short pulses or supercontinuum radiation, where the bandwidth of the optical signals can span over a wide frequency range.

Initially, most of the research on light propagation in waveguide arrays has been focused on monochromatic light propagations; however, recently strong research attention has attracted the investigation of the rich wave dynamics of multiple color beams, including studies on the propagation of broad-bandwidth and multicolor optical beams in periodic photonic structures.

D.N. Neshev (✉) • A.A. Sukhorukov • Y.S. Kivshar
Nonlinear Physics Centre, Research School of Physics and Engineering,
Australian National University, Canberra, 0200 ACT, Australia
e-mail: Dragomir.Neshev@anu.edu.au

Z. Chen and R. Morandotti (eds.), *Nonlinear Photonics and Novel Optical Phenomena*, Springer Series in Optical Sciences 170, DOI 10.1007/978-1-4614-3538-9_4,

The most significant focus in such studies is the nonlinear propagation of multicolor beams because the nonlinearity provides interaction and coupling between the different spectral components. These interactions lead to a range of new phenomena, such as phase locking of the spectral components, as well as to novel types of phase transitions of the nonlinear output beam profiles. This behavior is analogous to the synchronization of oscillations of nonlinearly coupled oscillators in nonlinear dynamics [3], but can have rather unique features in case of coupled optical waveguides. The unique features come from the fact that the nonlinear interaction of spectral components also leads to the well-known effect of localization of the beam inside the array.

Synchronization of oscillations and wave localization are two most fundamental nonlinear phenomena that have been driving the field of nonlinear dynamics for decades. Synchronization and phase locking [4] are known to exist in systems of different physical origin due to external driving and coupling between elements. Examples include the synchronous flashing of fireflies and the pulsation of laser arrays [5, 6]. On the other hand, the localization of waves in nonlinear lattices, such as waveguide arrays, is also determined by coupling between the individual lattice sites [7]. As inter-site coupling and nonlinearity govern the two phenomena, a natural question is if they can be linked together and what kind of novel fundamental effects can arise due to their interplay.

In this chapter, we describe the two representations of such interplay associated with different types of nonlinear interactions: (1) coherent interactions—when there is energy exchange between the spectral components and (2) incoherent interactions—when the different spectral components interact through cross-phase modulation, but there is no energy exchange between them. We note that in both cases, the localization of the multicolor beam is closely linked to phase locking of the spectral components at the neighboring waveguides; however, in the case of parametric coherent interactions this phase locking can be also accompanied by a phase transition from in-phase to out-of-phase state for one of the spectral components.

The chapter is organized as follows. In Sect. 4.2, we provide the general description of multicolor light propagation in waveguide arrays, and describe the distinction of the two cases of nonlinear interactions between the spectral components. In Sect. 4.3, we present the studies on incoherent interaction between the spectral components, while in Sect. 4.4, we show how the parametric driving between the spectral components influences the beam dynamics and leads to phase transition of the phase profile for one of the components. Finally in Sect. 4.5 we provide some further discussions and outlook to the open problems in the field of polychromatic light propagation.

4.2 Multicolor Light Propagation

In this section, we discuss the general features of polychromatic beam diffraction in planar photonic structures with a modulation of the refractive index along the transverse spatial dimension (Fig. 4.1a), including optically induced lattices and periodic waveguide arrays [8–19]. The physical mechanism of beam diffraction in

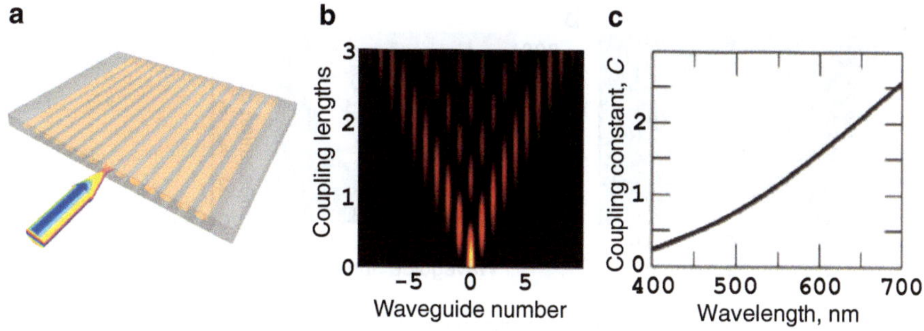

Fig. 4.1 Light propagation in waveguide arrays. (**a**) Schematic of the waveguide array structure excited by a narrow input beam. (**b**) Characteristic linear propagation of the beam inside the array, demonstrating discrete diffraction formed due to coupling between the waveguides. (**c**) Dependence of the coupling constant C on the wavelength

such structures is based on the coupling between the modes of neighboring waveguides [12, 20, 21]. When the beam is coupled into a single waveguide at the input, it experiences *discrete diffraction* where most of the light is directed into the wings of the beam (Fig. 4.1b). This is in sharp contrast to the diffraction of Gaussian beams in bulk materials where the maximum intensity is in the beam center for any propagation distance.

The light in the array couples from one waveguide to another due to the spatial overlaps of the waveguide modes. This coupling is characterized by the coupling constant C. Since the mode profile and confinement depend on the wavelength, the coupling constant (Fig. 4.1c) and correspondingly the discrete diffraction pattern exhibit strong spectral dispersion. The mode overlap at neighboring waveguides is usually much stronger for the red-shifted spectral components [22], which therefore diffract faster than their blue counterparts. This leads to spatial redistribution of the colors of the polychromatic beam, which increases along the propagation direction, see Fig. 4.2a. As a result, at the output the red components dominate in the beam wings, while the blue components are dominant in the central region, see Fig. 4.2b.

A more accurate mathematical description of the polychromatic beam propagation for optical sources with a high degree of spatial coherence, such as supercontinuum light generated in photonic-crystal fibers, can be based on a set of propagation equations for the polychromatic beam envelope $A(x,z) = \sum_{m=1}^{M} A_m(x,z)$, where $A_m(x, z)$ are the amplitudes of the different frequency components at vacuum wavelengths λ_m. Since the refractive index contrast in photonic-lattice structures is usually of the order of 10^{-4} to 10^{-2}, we can consider the beam propagation under the paraxial approximation [23–25]. For a one-dimensional (1D) array of coupled optical waveguides, such as the one depicted in Fig. 4.1a, the beam evolution is described by a system of coupled nonlinear Schödinger equations

$$i\frac{\partial A_m}{\partial z} + \frac{\lambda_m}{4\pi n_0(\lambda_m)}\frac{\partial^2 A_m}{\partial x^2} + \frac{2\pi}{\lambda_m}\Delta n(x;\lambda_m)A_m + G_m(\mathbf{A})A_m = 0, \tag{4.1}$$

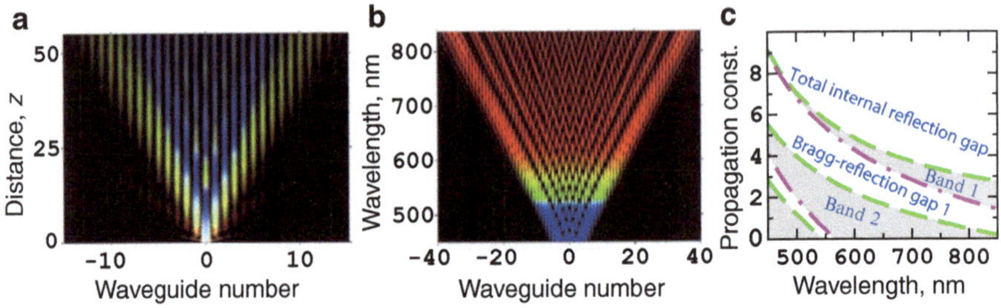

Fig. 4.2 Propagation of multicolor (polychromatic) light beams. (**a**) Numerical simulation of polychromatic beam diffraction: real-color image of beam evolution inside the array. (**b**) Spectrally resolved output intensity profile. (**c**) Dependence of photonic bandgap structure on wavelength. The simulations correspond to the parameters of $LiNbO_3$ waveguide arrays [23, 24]

where $\mathbf{A} = \{A_1, A_2, \ldots\}$, x and z are the transverse and longitudinal coordinates, respectively, and $n_0(\lambda_m)$ is the background refractive index. The function $G_m(\mathbf{A})$ describes the nonlinear interaction between the frequency components. $\Delta n(x; \lambda_m)$ describes the effective refractive index modulation, which depends on the vertical mode confinement in the planar guiding structure. Since the vertical mode profile changes with wavelength, the dispersion of the effective index modulation is strongly affected by the geometry of the photonic structure.

For an array of optical waveguides in $LiNbO_3$ (as in our experiments shown below), the modulation can be accurately described as $\Delta n(x;\lambda) = \Delta n_{max}(\lambda)\cos^2(\pi x/d)$, where d is the lattice period, and the wavelength dependence of the effective modulation depth $\Delta n_{\max}(\lambda)$ can be calculated numerically or determined by matching the experimentally measured waveguide coupling [23, 24]. Even if the material and geometrical dispersion effects are weak, the beam propagation would still strongly depend on its frequency spectrum [25] since the values of λ_m appear explicitly in (4.1).

The linear propagation $[G(\mathbf{A}) = 0]$ of optical beams through a periodic lattice can be fully characterized by decomposing the input profile in a set of spatially extended eigenmodes, called Bloch waves [26, 27]. The Bloch-wave profiles can be found as solutions of the linearized equation (4.1) in the form $A_m(x,z) = \psi_j(x;\lambda_m)$ $\exp[i\beta_j(K_b;\lambda_m)z + iK_b x/d]$, where $\psi_j(x; \lambda_m)$ has the periodicity of the underlying lattice, $\beta_j(K_b; \lambda_m)$ are the propagation constants, K_b are the normalized Bloch wavenumbers, and j is the band number. At each wavelength, the dependencies of the longitudinal propagation constant (along z) on the transverse Bloch wavenumber (along x) are periodic, $\beta_j(K_b; \lambda_m) = \beta_j(K_b \pm 2\pi; \lambda_m)$, and are fully characterized by their values in the first Brillouin zone, $-\pi \le K_b \le \pi$. These dependencies have a universal character [1, 26, 27], where the spectrum consists of nonoverlapping bands separated by photonic bandgaps. Consistent with our physical interpretation of mode coupling, the beam propagation will exhibit strong frequency dispersion. Indeed, the position and the width of bands and gaps are strongly sensitive to the wavelength of the light. This is shown in Fig. 4.2c which plots the propagation constants of the top and the bottom of each Bloch band as a function of the wavelength. Clearly, with increasing of the wavelength, the bands are getting wider and the gaps narrower.

Importantly, the rate of beam diffraction is determined by the curvature of the dependencies $\beta_j(K_b; \lambda_m)$. For an input beam coupled to a single waveguide, the first band is primarily excited, and the beam diffraction rate is determined by $max_{K_b}|\partial^2\beta_1/\partial K_b^2|$. The rate of diffraction increases at longer wavelengths where the band is wider and the gap narrower. Again, this conclusion is in full agreement with the physical interpretation presented above using the concept of coupling between waveguide modes.

The nonlinearity can dramatically affect the modes of the system leading to a number of fundamental effects, such as beam localization, soliton formation, and phase locking of the spectral components. In general, one can distinguish between two cases of nonlinear interactions between the spectral components: incoherent and coherent. For each case, the function $G(\mathbf{A})$ takes a different form that depends on the specific physical realization of the nonlinearity. The physical effects, however, can be intuitively explained: In the case of *coherent interactions*, there is mixing and exchange of energy between the spectral components, e.g., four-wave mixing processes with third order nonlinearity [28] or three-wave mixing interactions due to quadratic nonlinearity [15, 29]. In the case of *incoherent interactions*, there is no energy exchange between the frequency components and they interact through the effect of cross-phase modulation. Such interactions usually occur in materials with slow nonlinear response, such as thermal [30] or photorefractive [23, 31, 24].

Here we discuss how coherent and incoherent interactions affect the spatial beam profile during its propagation inside the array. In particular, we are interested in the changes of the phase structure of the different spectral components due to the interplay between linear coupling and nonlinear beam localization. We note that such interplay can lead to dramatic phase transformations in the beam structure when the nonlinear localization is inside the Bragg reflection gap of the periodic structure (Fig. 4.2c). Such phase transitions are attributed to the finite width of this gap and do not occur in lattices with positive (self-focusing) nonlinearity when the localization occurs inside the total internal reflection gap [32] (see also Fig. 4.2c). It is therefore an intriguing problem to understand how the phase relations of the spectral components in the polychromatic beam are established in the localization process, given the strong dependence of the bandgap spectrum on wavelength.

Below, we discuss two scenarios of phase transitions that occur for polychromatic beams with incoherent and coherent type of nonlinear interactions.

4.3 Incoherent Interaction of Spectral Components

In this section, we present the effects of phase locking of the different frequency components due to the interplay of incoherent nonlinear interactions of the spectral components and linear coupling between the waveguides.

4.3.1 Theoretical Approach

At high laser powers, the spectral components of the polychromatic beam interact incoherently with each other (no new frequency components are generated) due to the intensity-dependent change of the optical refractive index [33] (cross-phase modulation). In this case, one can write the nonlinear term in (4.1) as

$$G_m(\mathbf{A}) = \frac{\gamma_m}{M} \sum_{j=1}^{M} \sigma(\lambda_j)|A_j|^2, \tag{4.2}$$

where γ is the nonlinear coefficient for the mth spectral component, M is the total number of components, and $\sigma(\lambda)$ is a weight coefficient describing the sensitivity of the nonlinearity to wavelength. Importantly, this model of nonlinearity also describes the photorefractive interaction of polychromatic beams in Lithium Niobate ($LiNbO_3$) waveguide arrays in the regime of weak saturation, as in our experiments presented below.

Physically, the nonlinearity in $LiNbO_3$ is photorefractive and arises due to charge excitations by light absorption and corresponding separation of these charges due to diffusion [34]. A characteristic property of this photovoltaic photorefractivity is that an increase of the beam intensity leads to a decrease of the material refractive index [35, 36]. Whereas the nonlocal effects of the photorefractive nonlinearity may also affect the nonlinear beam propagation [37], this effect is weak under our experimental conditions.

The spectral response of the nonlinearity depends on the crystal doping and stoichiometry, and it may vary from crystal to crystal. In general, however, light sensitivity appears in a wide spectral range with a maximum for the blue spectral components [38], but the sensitivity extends well in the near infrared region [39]. In our analysis, we approximate the photosensitivity dependence by a Gaussian function $\sigma(\lambda) = \exp[-log(2)(\lambda - \lambda_b)^2/\lambda_w^2]$ with $\lambda > \lambda_b = 400$ nm and $\lambda_w = 150$ nm. Note that by making a transformation $\tilde{A}_m = A_m\sqrt{\sigma(\lambda_m)}$, the sensitivity function can be rescaled to unity, $\tilde{\sigma}(\lambda_m) = 1$, and therefore the presented results are also directly applicable for other shapes of the sign-definite photosensitivity functions.

In our numerical calculations, we choose a large number of frequency components ($M = 100$) in order to accurately model the polychromatic beam power spectrum generated in experiment. The simulations reveal that the input beam experiences self-trapping above a critical power level. Figure 4.3a shows the combined output of the array as a function of total power, and Fig. 4.3b shows the power in the central waveguide for different spectral components. It is clear from Fig. 4.3 that there is a sharp transition between the regimes of diffraction and soliton formation, associated with collective localization of spectral components from blue to red and infrared. The lower degree of localization of red components is due to the stronger diffraction and effectively weaker nonlinearly induced potential which is inversely proportional to the wavelength. The sharp self-trapping occurs because the length of waveguide array is several times larger than the diffraction lengths for *all spectral components*.

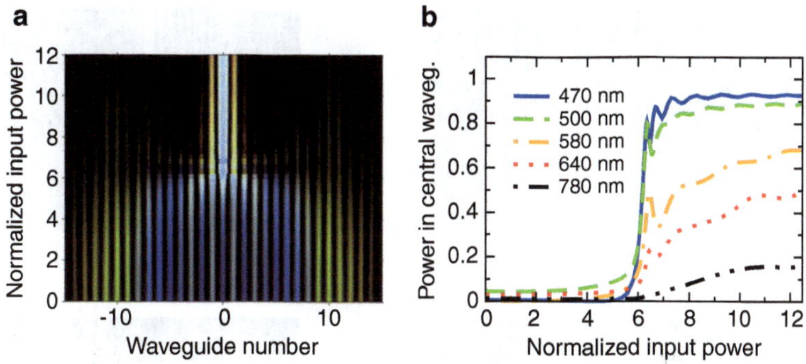

Fig. 4.3 Numerically calculated dependence of the output beam characteristics on the input power: (**a**) transformation of the output beam profile and (**b**) fraction of output power in the central waveguide for different spectral components

Accordingly, all components either diffract or experience collective self-trapping. This effect differs drastically from the beam reshaping under the conditions when diffraction of short-wavelength components is very weak [23] as in that case the power increase provides a gradual onset of localization at the output without a collective power threshold.

If we analyze in details the propagation of the polychromatic beam, we find that just above the power threshold for the soliton formation, one can still observe a strong diffraction at the initial propagation stage, especially for the red-shifted spectral components (Fig. 4.4a). With propagation, however, a well-localized polychromatic beam is formed. For higher input powers (Fig. 4.4e), the effect of diffraction is further reduced and the soliton preserves its input "white" color.

The physical mechanism of beam localization can be understood from the plots of the density of states vs. the propagation constants. Therefore, we perform a Fourier transform in z and then integrate the power spectrum over the central region of six waveguides to exclude the contribution of diffracted waves. The obtained spectra are presented as density plots (white color correspond to larger amplitudes) in Fig. 4.4b, f. We observe that at the threshold power for beam self-trapping, the propagation constants are shifted inside the Bragg-reflection gap through the nonlinear self-action (Fig. 4.4b). Note that the spectrum for blue components is shifted deeper inside the gap, whereas the red components have spectra very close to the gap edge. This explains the weaker localization of red components as shown in Fig. 4.4c. As the power increases, the spectrum shifts further inside the gap for all components (Fig. 4.4f), and accordingly the localization becomes stronger (Fig. 4.4g). Because in both cases the beam shape remains constant with propagation inside the array and the propagation constant of all spectral components is inside the Bragg-reflection gap we can conclude that the polychromatic beam forms a *polychromatic gap soliton*.

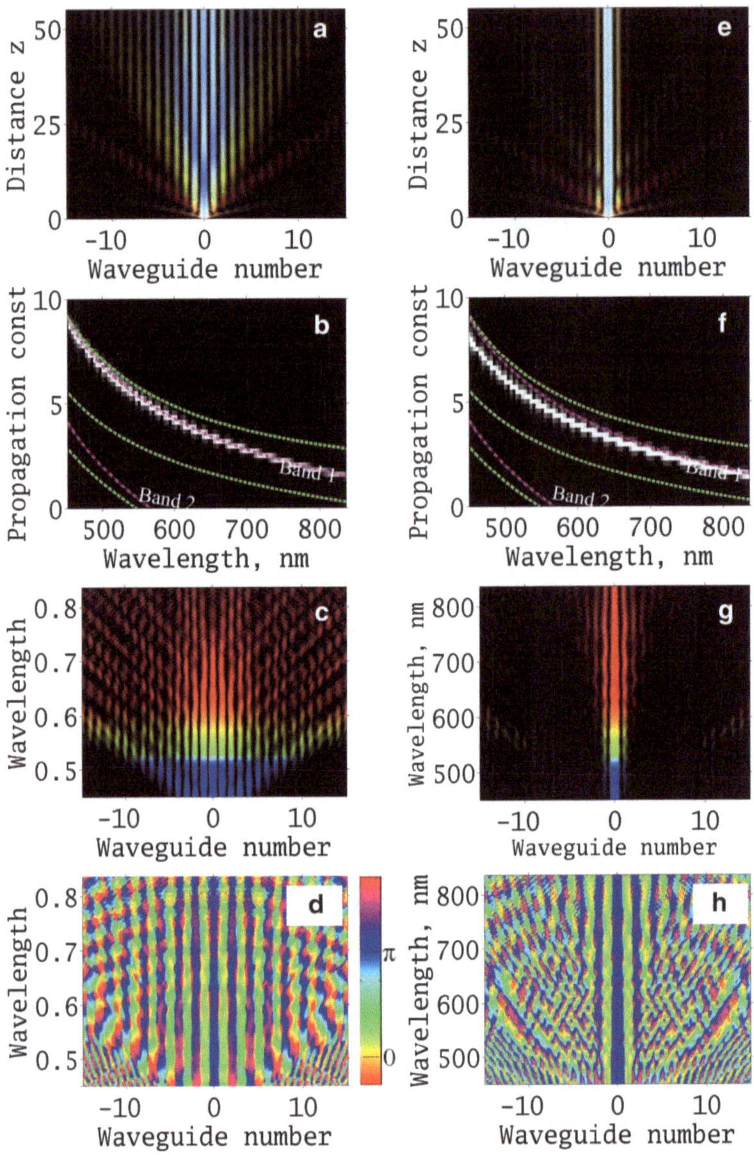

Fig. 4.4 Numerically calculated beam evolution for two different powers: (**a–d**) $P_0 \simeq 6$ and (**e–h**) $P_0 \simeq 12$. Shown are (**a, d**) the beam propagation dynamics inside the arrays; (**b, f**) density of states for the localized state, superimposed on the linear bandgap diagram; (**c, g**) spectrally resolved output intensity profiles, and (**d, h**) output phase profiles of the individual spectral components

Most importantly, the sharp power transition from diffraction to localization is linked to a phase transition in the beam spatial profile. Above the critical power level for localization, all spectral components lock their phase structure, such that they all have a staggered phase profile. To clarify this phase transition, we plot the phase profiles of all spectral components in Fig. 4.4d, h. These plots clearly show the simultaneous appearance of the staggered phase structure for all individual spectral components, when the beam experiences nonlinear self-trapping. Hence, such localization represents a uniquely different physical picture compared to the theoretically studied spatially incoherent white-light solitons in lattices supported by a focusing nonlinearity [32] where the defined phase relation is not directly associated with the soliton formation.

4.3.2 Experimental Observation of Multicolor Beam Self-trapping and SpectralPhase Locking

The key for experimental realization of the predicted phenomena of self-trapping and phase locking is the combination of periodic structure with a broadband nonlinear response and high-spatial coherence, high optical intensity polychromatic light with a broad frequency spectrum. The natural choice of such a light source is provided by the effect of supercontinuum generation [40]. In the process of supercontinuum generation, spectrally narrow laser pulses are converted into the broad supercontinuum spectrum through several processes [41, 42], including self-phase modulation, soliton formation, soliton break-up due to higher order dispersion, and Raman shifting of the solitons, leading to non-solitonic radiation in the short-wavelength range. Supercontinuum radiation has proven to be an excellent tool for characterization of bandgap materials [43], it possesses high spatial coherence [40], as well as high brightness and intensity required for nonlinear experiments [44].

In our experiments, we used a supercontinuum light beam generated by femtosecond laser pulses (140 fs at 800 nm from a Ti:Sapphire oscillator) coupled into 1.5 m of highly nonlinear photonic crystal fiber (Crystal Fiber NL-2.0–740 with engineered zero dispersion at 740 nm) [23]. The spectrum of the generated supercontinuum spans over a wide frequency range (typically more than an optical octave). After re-collimation and attenuation, the supercontinuum beam is refocused by a microscope objective (×20) to a single channel of the waveguide array (see Figs. 4.1a and 4.5a).

The optical waveguides are fabricated by in-diffusion of a thin (100Å) layer of titanium in a X-cut, 50 mm long monocrystal lithium niobate wafer [18]. The waveguides are single mode for all spectral components of the supercontinuum. Arrays with different periodicity and index contrast were tested in our experiments.

After coupling to the array, its output is imaged by a microscope objective (×5) onto a color CCD camera, where a dispersive 60° (glass SF-11) prism could be inserted between the imaging objective and the camera (Fig. 4.5a) in order to resolve spectrally all components of the supercontinuum. Additionally, a reference supercontinuum beam is used for interferometric measurement of the phase structure

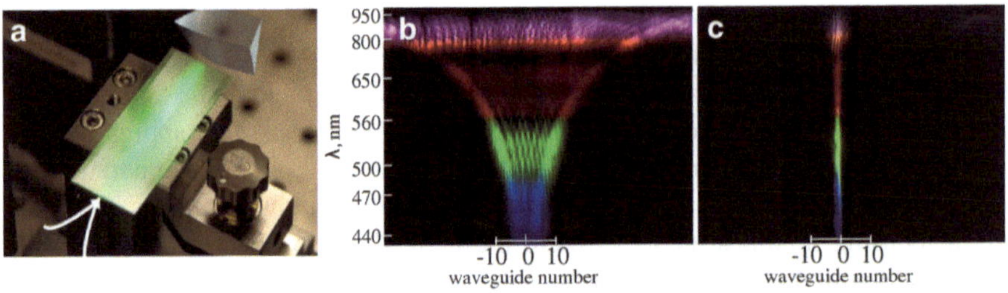

Fig. 4.5 (**a**) Experimental setup: The array is excited with a white light focused into a single waveguide. At the output the light is dispersed vertically by a prism and imaged onto a CCD camera. Spectrally resolved measurements of the beam profiles at the output face of waveguide array ($d = 10$ μm). (**b**) Polychromatic discrete diffraction at low laser power (0.01 mW). (**c**) Nonlinear localization and formation of polychromatic gap soliton at a higher (6 mW) power [24]

of the output beam [24]. To compensate for the pulse delay and pulse spreading inside the $LiNbO_3$ waveguides, this reference beam is sent through a variable delay line, implemented in a dispersion compensated interferometer, including an additional 5 cm long bulk $LiNbO_3$ crystal (to equalize the material dispersion). In this way, interferometric measurements are possible for ultra-wide spectral range.

To obtain a detailed insight into the spectral distribution at the array output, we resolve in the vertical direction the individual spectral components by the prism and acquire a single shot two-dimensional image providing spatial resolution in horizontal and spectral resolution in the vertical (orthogonal) direction. This technique enables precise determination of the spectral distribution at the array output. The image in Fig. 4.5b depicts the spectrally resolved discrete diffraction of the supercontinuum beam in an array of optical waveguides with a period $d = 10$ μm when the input beam is focused to a single waveguide. The diffraction of the beam is weakest for the blue spectral components, which experience weak coupling, while the diffraction is strongest for the infrared components. We note that the spectral scale in Fig. 4.5b is not linear due to the nonlinear dispersion of the prism. The spectrally resolved discrete diffraction provides a visual illustration of the separation of colors in the waveguide array. This separation occurs as the light is concentrated predominantly in the beam wings rather than in the center, a typical property of the discrete diffraction. The obtained diffraction pattern also allows for exact characterization of the linear dispersion parameters of the periodic structure and, in particular, waveguide coupling. In our sample we measured that the discrete diffraction length varies from 1 cm, for the blue (480 nm), to less than 0.2 cm, for the red (800 nm) spectral components. These values correspond to a total propagation distance of 5.5 and 27.5 discrete diffraction lengths for the blue and red spectral components, respectively. The propagation of few diffraction lengths for all spectral components is advantageous for nonlinear experiments, e.g., formation of solitons [45, 46], and facilitates strong phase transformations in the nonlinear regime.

In agreement with the theoretical predictions, we observe strong spatial localization of the supercontinuum light as its input power is increased

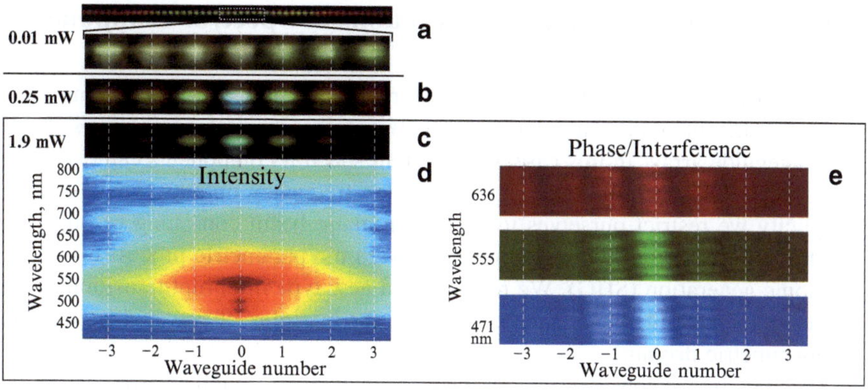

Fig. 4.6 Experimental observation of polychromatic gap soliton: (**a–c**) Real-color CCD camera images of the output beam intensity profile: (**a**) Diffraction profile at low power. (**b**, **c**) Nonlinear localization and formation of polychromatic gap soliton with increasing supercontinuum power. (**d**) Spectrally resolved measurements of the profile (**c**). (**e**) Interferograms of the output beam profile (**c**) with a tilted reference supercontinuum beam, imaged at three different wavelengths as indicated by labels

(Figs. 4.5c and 4.6a–c). In a narrow range of input powers (150–250 μW), the beam profile narrows from over 50 waveguides (Fig. 4.6a) down to three central waveguides (Fig. 4.6b). This transition, indicating the formation of a polychromatic soliton, happens over a range of 100 μW only and appears extremely sharp comparing with the fact that further localization down to a single waveguide of the array requires additional increase of 1.5 mW (15 times larger) (Fig. 4.6c). The important characteristic of this localization is the fact that it combines all wavelength components (from blue to red) of the supercontinuum spectrum (Fig. 4.6d). Localization around 700–750 nm wavelengths is not visible due to the lower spectral intensity of these components in the input supercontinuum spectrum, whereas localization over the six central waveguides is observed at 800 nm wavelength.

Taking advantage of the high spatial coherence of the supercontinuum light, we also perform interferometric measurement of the localized output profile. In our white-light interferometer, the reference beam is slightly tilted in the vertical plane in comparison to the probe beam, thus resulting in interference fringes. The interference patterns have different periods depending on wavelength, and therefore had to be imaged separately using a tunable linear filter (LVF, Ocean Optics) mounted in front of the CCD camera. The interference patterns recorded at high power (1.9 mW) are presented in Fig. 4.6e. They show that the interference fringes between neighboring waveguides are shifted by half a period hence, the probe beam phase changes by π between the neighboring waveguides. Most remarkably, such staggered phase structure appears simultaneously in an ultra-broad spectral range from blue (470 nm) to red (above 630 nm), providing direct evidence that all spectral components are simultaneously phase locked to a staggered phase structure in the process of beam localization and formation of a polychromatic gap soliton.

4.4 Coherent Parametric Interactions of Spectral Components

Next we describe the effects of phase transformation of polychromatic beams in the case of coherent nonlinear interactions between the spectral components inside the waveguide array. In this case, the different spectral components can exchange energy between each other leading to complex frequency outputs. Therefore, for simplicity we restrict ourselves to a polychromatic beam consisting of only two spectral components which are coupled through the parametric process of second harmonic generation (SHG). We reveal that a new type of abrupt transition in the phase structure of the beam can occur due to the interplay of waveguide coupling and parametric driving.

Specifically, we consider the laser beam propagation through an array of closely spaced optical waveguides in media with quadratic nonlinearity that facilitates frequency conversion and energy exchange between a fundamental wave (FW) and a second-harmonic (SH) beam. In such structures, the SH dynamics is governed by two mechanisms of energy exchange (Fig. 4.7a): (1) an effective driving force by the FW [47] at each lattice site, and (2) direct coupling of SH waves between the lattice sites due to the overlap of the neighboring waveguide modes. We show that each of these mechanisms could lead to synchronization of SH dynamics and formation of different phase patterns. Mechanism (1) is dominant when the FW (and the corresponding effective driving force) extends over many lattice sites, whereas mechanism (2) dominates when the FW exhibits nonlinear self-trapping to a single lattice site. We specially design our experimental conditions to observe this interplay, overcoming for the first time the constraint of all previous experiments where the second mechanism was suppressed due to the inhibition of linear coupling for the SH modes [7].

The key observation of our studies is that as the optical power is increased and the beam becomes more localized, the output beam profile at the second harmonic exhibits a sharp transformation from in-phase (unstaggered) to out-of-phase (staggered) pattern between the neighboring waveguides [29]. This type of behavior is in sharp contrast to the incoherent case (Sect. 4.3), where all frequency components were either de-phased or locked to the same phase structure.

4.4.1 Theoretical Approach

We develop the theoretical description of this coherent nonlinear interaction for the particular case of an array of coupled waveguides in periodically poled lithium niobate (PPLN) (Fig. 4.7) that features strong quadratic nonlinear response. To demonstrate the existence of a phase transition, we explore the nonlinear interaction between the FW_{00} and SH_{02} modes, whose intensity profiles are shown in Fig. 4.7a. Both of these modes show linear coupling of comparable strength. This is in contrast to all previous experiments on discrete quadratic solitons [7] that

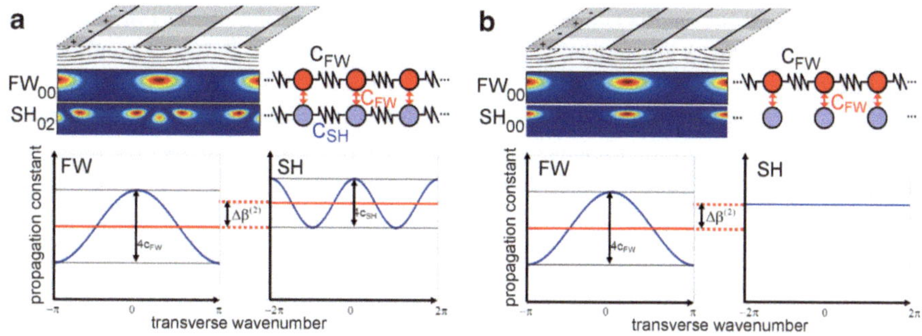

Fig. 4.7 Scheme and dispersion diagrams of the nonlinear coherent system. (**a**) Strong coupling between the SH modes. (**b**) No coupling between the SH modes where the only single component is only responsible for the beam dynamics inside the array. (*top*) Sketch of the PPLN sample with the corresponding refractive index profile. (*middle*) Intensity profiles of the FW and SH modes. (*bottom*) Dispersion relations of the FW (*left*) and SH (*right*) modes

utilized the nonlinear interaction between FW_{00} and SH_{00} (Fig. 4.7b) where the SH_{00} modes experience negligible linear coupling due to their strong localization.

The spatial beam evolution in quadratic nonlinear waveguide arrays can be modeled by a system of normalized discrete equations for the mode amplitudes at each waveguide in the array [7],

$$i\frac{dA_n}{dz} + c_{FW}(A_{n+1} + A_{n-1}) + A_n^* B_n = 0,$$

$$i\frac{dB_n}{dz} + c_{SH}(B_{n+1} + B_{n-1}) - \Delta\beta B_n + A_n^2 = 0. \tag{4.3}$$

The last terms in these equations account for the effective nonlinearity function [$G(\mathbf{A})$, as defined in (4.1)], z is the propagation distance normalized to z_s, and A_n, B_n are the normalized FW and SH mode amplitudes at the nth waveguide, respectively. Note that the total power in the array $P = P_{FW} + P_{SH}$ is conserved, where $P_{FW} = \sum_n |A_n|^2$ and $P_{SH} = \sum_n |B_n|^2$. The real coefficients $c_{FW,SH} = z_s\pi/(2L^c_{FW,SH})$ determine the coupling strength between the neighboring waveguides, where $L_{FW,\,SH}{}^c$ are the physical coupling lengths. The phase mismatch between the FW and SH waves (accounting for the periodic poling) is characterized by the value of $\Delta\beta$ that depends strongly on the FW frequency.

In the linear regime, the beam dynamics is governed by the dispersion relations for the Bloch modes of the lattice [48]: $A_n(z) = A_0(z=0)\exp(ik_{FW}n + i\beta_{FW}z)$ for the FW, and $B_n(z) = B_0(z=0)\exp(ik_{SH}n + i\beta_{SH}z)$, for the SH. Here $\beta_{FW} = 2\,c_{FW}\cos(k_{FW})$ and $\beta_{SH} = 2c_{SH}\cos(k_{SH}) - \Delta\beta$. The characteristic dispersion relations are schematically shown in Fig. 4.7(graphs). For propagation constants outside the linear bands, the waves exhibit evanescent decay due to the photonic bandgap [49], $A_n = \kappa_1^{-|n|}$ and $B_n = \kappa_2^{-|n|}$, where $\kappa_j = [1 + (1-\eta_j^2)^{1/2}]/\eta_j$ with $\eta_1 = 2c_{FW}/\beta_{FW}$ and $\eta_2 = 2c_{SH}/(\beta_{SH} + \Delta\beta)$. We note that for propagation constants below the bands ($\beta_{\mathrm{FW}} < -2c_{\mathrm{FW}}$ and $\beta_{SH} < -2c_{SH} - \Delta\beta$), $\kappa_j < 0$, i.e., the evanescent waves are staggered, with π phase oscillations between neighboring waveguides.

Next, we demonstrate that nonlinear parametric coupling of FW and SH waves can dramatically modify the phase pattern of the propagating waves. We note that most efficient SH generation occurs when the waves are spatially localized, due to enhanced local field intensities. Stronger localization occurs at higher optical powers due to self-focusing as there is a nonlinear phase shift for both components arising due to the energy exchange between the FW and the SH [50].

In order to reveal the generic relation between the nonlinear localization and phase locking, we first analyze the stationary localized states or fixed points of the system. Their power dependence provides insight into the bifurcation properties of the system, revealing possible phase transitions. These solutions have the form $A_n(z) = A_n(z=0)\exp(i\beta z)$ and $B_n(z) = B_n(z=0)\exp(2i\beta z)$ [51, 7]. Here β is a real parameter, which simultaneously defines the FW ($\beta_{\mathrm{FW}} = \beta$) and SH ($\beta_{\mathrm{SH}} = 2\beta$) propagation constants due to nonlinear synchronization. By substituting these expressions into (4.3), we obtain a set of nonlinear equations for the real amplitudes of FW and SH.

Whereas the solutions of these equations can be found only numerically, we identify the phase transition effect analytically by analyzing the tails of localized solitons, where $|A_n| \to 0$ and $|B_n| \to 0$ for $|n| \gg 0$. The solution for the FW tail is the same as for linear evanescent waves, $A_n \approx \kappa_1^{|-n|}$. For the SH wave, the nonlinear term representing the effective FW driving force cannot be neglected even in the small-amplitude limit. We perform asymptotic analysis and derive asymptotic expressions for the SH beam tails: $B_n \approx \kappa_2^{-|n|}$ if $|\kappa_2| < \kappa_1^2$ and $B_n \approx (\kappa_1^2)^{-|n|}$ if $|\kappa_2| > \kappa_1^2$. In the first case, the SH tail profile corresponds to a linear evanescent wave solution, whereas in the second case the SH tail is fully determined by the FW. A nontrivial phase transformation occurs when the propagation constant is below the bands, $\beta < \beta_l = min(-2c_{FW}, -c_{SH} - \Delta\beta/2)$, since in this case $\kappa_j < 0$. While the FW tails are always staggered, the SH tail exhibits a *phase transition* at

$$\beta_{s\pm} = -c_{FW}^2 c_{SH}^{-1} \pm \sqrt{c_{FW}^4 c_{SH}^{-2} + 2c_{FW}^2 - \Delta\beta c_{FW}^2 c_{SH}^{-1}}. \tag{4.4}$$

The SH tails are staggered for $\beta < \beta_{s-}$ and $\beta > \beta_{s+}$ and unstaggered for $\beta_{s-} < \beta < \beta_{s+}$. Importantly, for weak or zero coupling of the SH mode ($c_{\mathrm{SH}} \simeq 0$), $\beta_{s\pm}$ diverges and no phase transition is possible. Therefore in all previous experiments with quadratic waveguide arrays, no such phase transition could be observed. Under our experimental conditions, the coupling lengths are practically constant in the frequency range around $\Delta\beta = 0$, with values $L_{\mathrm{SH}}{}^c \simeq L_{\mathrm{FW}}{}^c \simeq 20$ mm. By choosing the scaling coefficient $z_s = 2L_{FW}^c/\pi$, the corresponding normalized coupling constants are $c_{\mathrm{FW,\,SH}} \simeq 1$.

The SH structure in the center and in the tails of the solutions can be different. Figure 4.8 presents the numerically calculated parameter regions for odd-type

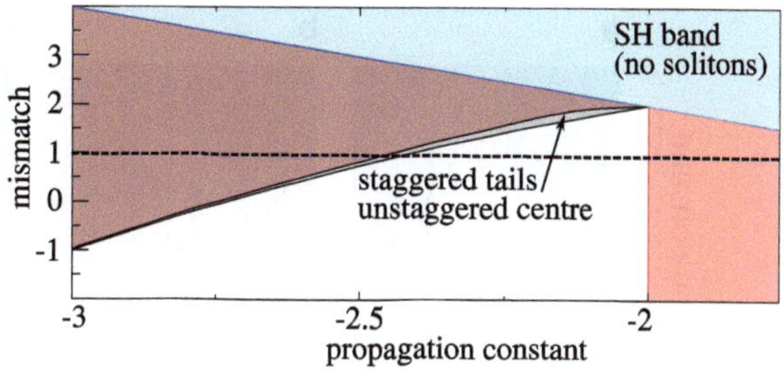

Fig. 4.8 Regions of existence and different topologies of the odd soliton solutions with $c_{\mathrm{FW}} = c_{\mathrm{SH}} = 1$

solutions (centered on a lattice site [7, 51]), where the boundaries for phase transitions in the tails and for the entire soliton are indicated. Figure 4.9a, b shows the intensity and phase profiles of the odd-type solutions vs. β for $\Delta\beta = 1$. For low absolute values of β, the FW component of the solution is staggered and the SH component is unstaggered. If the value of β is decreased below β_{s-}, the FW component becomes very narrow. Hence, the SH is driven only at a few waveguides and becomes independent of the FW component. Here the SH component undergoes a *phase transition* to the staggered state. Such transition has never been shown before, although soliton solutions with staggered/staggered and staggered/unstaggered FW/SH components have been reported [49, 51, 52]. Unambiguous signatures of the phase transition are observed in the spatial Fourier spectrum (Fig. 4.9c). While the FW spectrum is always confined around the edge of the Brillouin zone ($k_{\mathrm{FW}} \simeq \pm\,\pi$), the SH Fourier spectrum switches between the center and the edge of the Brillouin zone, corresponding to a transition from unstaggered to staggered profile.

It is relevant to note that such phase transition is uniquely different to the phase locking presented in Sect. 4.3. While in both cases, the phases of the spectral components are well defined, due to the coupling of the spectral components here we have sharp switching of the phase state of only one component. In contrast, no phase switching was present in the case of incoherent coupling of the spectral components. Thus, the phase transition predicted here is a unique feature appearing due to the parametric process and the energy exchange between spectral components.

Figure 4.9d shows the soliton power corresponding to Fig. 4.9a–c. The monotonic dependence of the power on the propagation constant is generic for all soliton families exhibiting the phase transition since they all bifurcate from the FW band edge. As such, all odd-type solitons are stable [51]. For values of the propagation constant below the phase transition threshold, the SH power of the solution is much smaller than the FW power (Fig. 4.9e), and indeed in this regime the SH is fully driven by the FW.

We also analyze even-type solutions of (4.3), where the FW profile is centered between neighboring lattice sites [7, 51]. We find that similar to the odd solutions, a phase transition occurs for the SH tails, see Fig. 4.10. However, the SH amplitudes at the two central sites are forced to have the same phase due to the even symmetry of the solution. Accordingly, the energy is always concentrated in the center of the Brillouin zone (Fig. 4.10c). This demonstrates the possibility to partially suppress

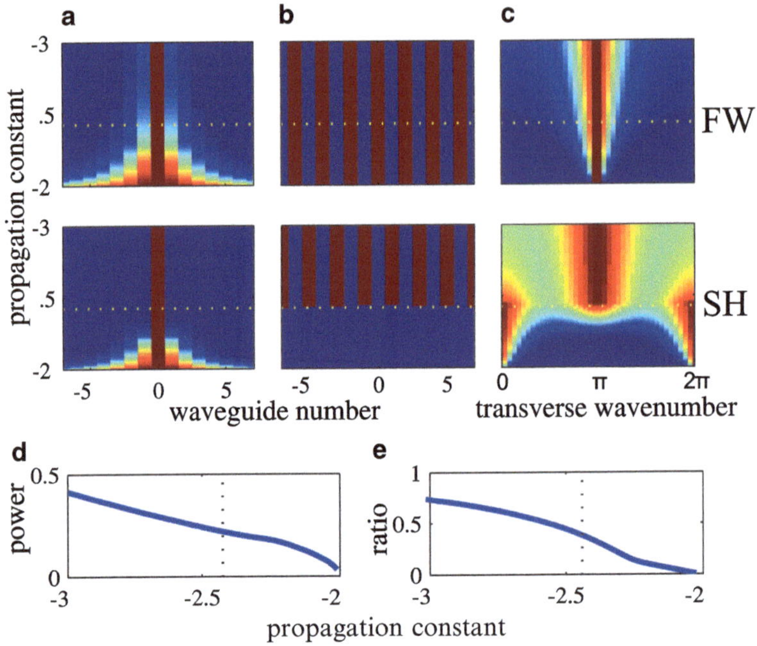

Fig. 4.9 (**a–c**) Soliton families for $\Delta\beta = 1$ (indicated by the *dashed line* in Fig. 4.8): (**a**) absolute values of the mode amplitudes, (**b**) phases where *blue* corresponds to 0 and *red* to π, (**c**) absolute values of the spatial Fourier spectra, (**d**) total power, and (**e**) power ratio P_{SH}/P_{FW}. The *dotted lines* in (**b-f**) mark $\beta_{s\pm}$ where the SH tail profiles exhibit phase transition

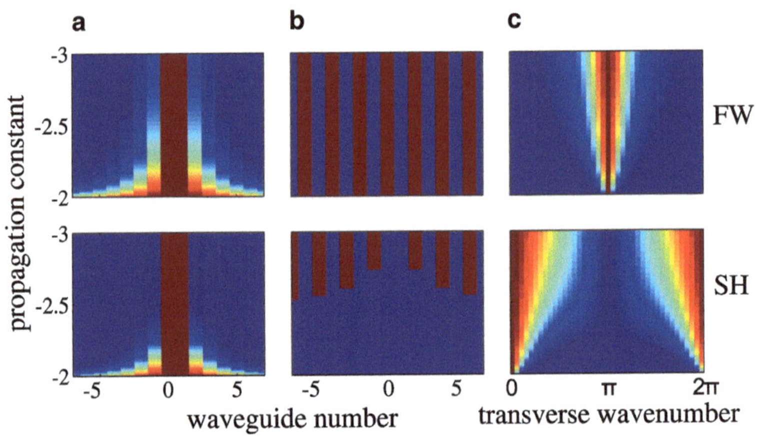

Fig. 4.10 Soliton families with even symmetry: (**a**) mode amplitudes, (**b**) phases, and (**c**) Fourier spectra. Notations and parameters correspond to Fig. 4.9b–d

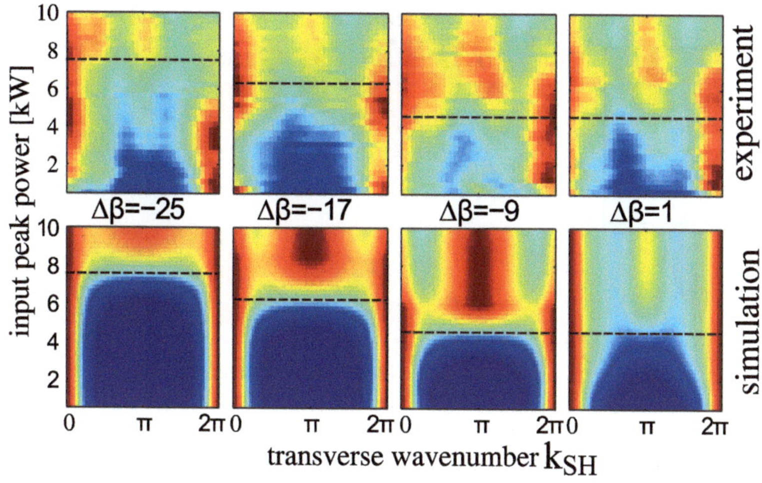

Fig. 4.11 Experimental (*upper row*) and numerical (*bottom row*) power dependence of the SH spatial output spectrum for different normalized mismatches $\Delta\beta$. The *dashed lines* mark the transition powers

the phase transition. Although the even solitons exhibit symmetry-breaking instability and tend to transform to odd solitons [51], we confirm below that such instability develops relatively slowly, so that the even symmetry states can be observed in the experiment.

4.4.2 Experimental Results

Next, we study the predicted phase transition experimentally. In our experiments, only the FW beam is launched into the $LiNbO_3$ waveguide array, leading to dynamical reshaping involving generation of SH and focusing. Nevertheless, the key predictions based on the analysis of stationary solutions are fully confirmed. We excite the array with 5.2 ps pulses generated by a tunable optical parametric amplifier at FW wavelengths around 1500 nm. The beam is shaped into an elliptic input beam with a horizontal/vertical FWHM of 63 μm/2.8 μm. To obtain the staggered FW profile the beam is tilted at a Bragg angle. The input power, controlled with a half-wave plate and a polarizer, is monitored before coupling to the sample. The array consists of 101 parallel waveguides with a pitch of 15 μm, made by titanium indiffusion in a 71 mm long PPLN crystal [53]. The sample is contained in an oven and heated to 220 °C to prevent photorefractive effects. After the sample, the powers of the transmitted FW and the generated SH components are measured and their intensity distributions are recorded by an InGaAs and CCD camera, respectively. To obtain the spatial Fourier spectrum of the SH, we employ a lens and an additional CCD camera.

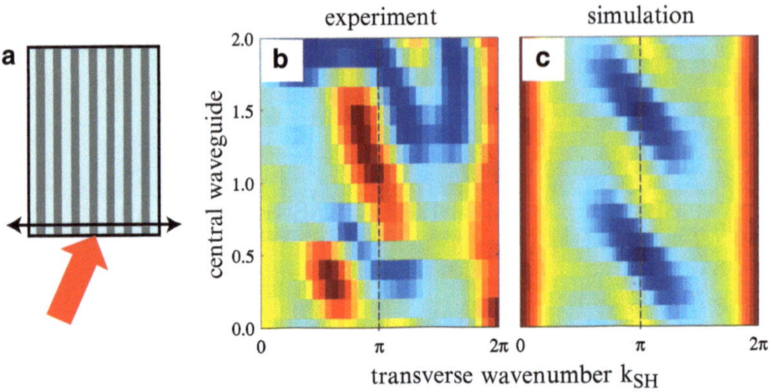

Fig. 4.12 (**a**) Excitation scheme of the array showing the transverse translation of the sample. (**b**) Experimental and (**c**) numerical dependencies of the SH spatial spectrum on the center of the excitation for $\Delta\beta = -9$ and an input peak power of 10kW. The *dashed line* marks $k_{SH} = \pi$

The upper row of Fig. 4.11 shows results of the power-dependent measurement of the SH Fourier spectrum for different phase mismatches, as determined by the input wavelength. For low input powers, the SH Fourier spectrum is concentrated around $k_{SH} = 0, 2\pi$, corresponding to unstaggered SH. For FW input powers above a mismatch-dependent threshold, staggered SH components are generated at $k_{SH} = \pi$. This is an unambiguous signature of the localization-controlled phase transition, as found for the stationary states.

To validate the interpretation of the experimental results and to explain possible differences to the stationary case, we carry out simulations of the time-dependent coupled mode equations including group velocity mismatch and pulse dispersion [15]. The simulation results are plotted in the bottom row of Fig. 4.11 and agree well with the measured data. Measured and simulated FW peak power thresholds for phase transition of the SH show a decrease from 7.5 kW for $\Delta\beta = -25$ to 5.5 kW for $\Delta\beta = 1$. In contrast to the predictions for stationary solutions, the SH transformation is not complete since the wings of the pulse remain in their initial state. Thus, we always measure nonzero SH powers at $k_{SH} = 0, 2\pi$.

Another remarkable feature found in the stationary solutions is the absence of the complete phase transition for even symmetry (see Fig. 4.10). Figure 4.12 shows a set of experimental results for different transverse shifts of the sample with respect to the broad input beam (see Fig. 4.12a). The measured SH spectral power at $k_{SH} = \pi$ (Fig. 4.12b) depends strongly on the position of the input beam. When the excitation is centered on a waveguide (odd), staggered SH is generated according to Fig. 4.11. For even excitation, centered between two waveguides, the SH power at $k_{SH} = \pi$ vanishes. This shows that the symmetry dependence of the phase transition found in the stationary solution is a robust generic property. The comparison with the time-dependent simulations (Fig. 4.12c) again shows good agreement. The deviations between measurements and simulations (compare Fig. 4.12b, c) are due to inhomogeneities of sample and input coupling, as well as to the general restrictions of the coupled mode equations.

4.5 Discussions and Outlook

In this chapter, we have presented our recent theoretical and experimental results on all-optically controlled spatial reshaping and localization of multi-color light beams in nonlinear waveguide arrays. In the case of incoherent interactions supported by defocusing photorefractive nonlinearity, all spectral components develop a staggered phase structure at larger optical powers leading to the formation of polychromatic lattice solitons. In contrast, coherent interactions between the fundamental and the SH spectral components in quadratic nonlinear waveguide arrays lead to a new type of phase transition for the SH phase structure which abruptly switched between staggered and unstaggered states, whereas the FW preserves the staggered phase. The observed localization induced phase transition is a generic phenomenon present in other nonlinear discrete systems and we anticipate that it can also occur in systems such as Bose-Einstein condensates on optical lattices and gene networks in living cells [54].

Here, we also outline selected recent developments on polychromatic light control in photonic lattices and waveguide arrays. This is a vibrant and rapidly developing research area, where many new approaches for beam shaping in the linear and nonlinear regimes have been suggested and experimentally demonstrated in recent years. We have mentioned in Sect. 4.2 that the waveguide mode coupling tends to be stronger at longer wavelengths. It was suggested in [55] that the intrinsic wavelength dependence of diffraction strength in waveguide arrays can be compensated in the linear regime by geometrically induced dispersion resulting from periodic waveguide bending. This approach can underpin broadband diffraction management, where diffraction can be made wavelength independent in a frequency range of up to 50% of the central frequency. It becomes possible to realize propagation regimes which are not possible for closely spaced straight waveguides, in particular diffraction of light can be suppressed in the regime of polychromatic dynamic localization [55, 56], an effect that generalizes the concept of dynamic localization originally introduced for electrons in crystals subjected to alternating electric field [57]. Polychromatic dynamic localization can also be realized in zigzag lattices [58] with long-range coupling between waveguide modes [59]. Another fundamental effect known as Bloch oscillations, predicted as oscillatory motion of electrons in crystals subjected to constant electric field [60], can also be generalized for polychromatic light beams in waveguide arrays containing sections with different constant curvatures [61].

Furthermore, curved waveguide arrays provide a flexible platform for manipulating polychromatic light patterns based on Talbot effect [62], and polychromatic beam splitting. These effects have been demonstrated in a three-waveguide structure realizing light propagation regime analogous to stimulated Raman adiabatic passage [63]. Finally, in two-dimensional arrays, waveguide bending allows one to control effective geometry and dimensionality of the photonic lattice [64, 56].

In the nonlinear regime, when spectral components interact incoherently, the longitudinal modulation of the waveguide coupling can facilitate all-optical switching of polychromatic light between two coupled waveguides [65], whereas in curved waveguide arrays nonlinearity leads to symmetry breaking and suppression of dynamic localization at low powers and formation of polychromatic diffraction-managed solitons only at higher powers [66]. Curved waveguide arrays support new types of defect-free surface waves [67, 68] which nonlinear switching was demonstrated with monochromatic light [69]; however, switching of polychromatic light has only been considered at the edges of straight waveguide arrays [70]. We can anticipate that polychromatic light switching at edges of curved waveguides may offer enhanced flexibility for tailoring all-optical spatial–spectral beam reshaping.

Results on phase transitions mediated by coherent nonlinear interactions presented in Sect. 4.4 call for further investigations of this fundamental phenomenon in quadratic waveguide arrays of different configurations. In particular, we expect that phase transition properties may depend nontrivially on the geometry of two-dimensional photonic lattices and on the type of excited localized solutions such as spatial discrete vortex solitons [71, 7] or spatial–spectral vortex solitons [72]. We also note that the effect of coherent nonlinear interactions on beam dynamics in longitudinally modulated waveguide arrays has not yet been considered, and presents an interesting open area for future research.

Acknowledgment The authors acknowledge the substantial and valuable contribution of a large number of research students and collaborators to the results summarized in this chapter. Especially we would like to thank A. Mitchell, F. Setzpfandt, I. L. Garanovich, A. Dreischuh, and T. Pertsch. The work has been supported by the Australian Research Council.

References

1. J.D. Joannopoulos, R.D. Meade, J.N. Winn, *Photonic Crystals: Molding the Flow of Light* (Princeton University Press, Princeton, 1995)
2. C. Denz, S. Flach, Yu. S. Kivshar (eds.), *Nonlinearities in Periodic Structures and Metamaterials* (Springer-Verlag, Berlin, 2009)
3. S.H. Strogatz, *Nonlinear Dynamics and Chaos: With Applications to Physics, Biology, Chemistry, and Engineering* (Westview Press, Cambridge, 1994)
4. A. Pikovsky, M. Rosenblum, J. Kurths, *Synchronization: A Universal Concept in Nonlinear Sciences* (Cambridge University Press, Cambridge, 2003)
5. A.T. Winfree, Science **298**(5602), 2336 (2002)
6. L. Larger, J.M. Dudley, Nature **465**(7294), 41 (2010)
7. F. Lederer, G.I. Stegeman, D.N. Christodoulides, G. Assanto, M. Segev, Y. Silberberg, Phys. Rep. **463**(1–3), 1 (2008)
8. D.N. Christodoulides, R.I. Joseph, Opt. Lett. **13**(9), 794 (1988)
9. H.S. Eisenberg, Y. Silberberg, R. Morandotti, A.R. Boyd, J.S. Aitchison, Phys. Rev. Lett. **81**(16), 3383 (1998)

10. J.W. Fleischer, T. Carmon, M. Segev, N.K. Efremidis, D.N. Christodoulides, Phys. Rev. Lett. **90**(2), 023902 (2003)
11. D. Neshev, E. Ostrovskaya, Y. Kivshar, W. Krolikowski, Opt. Lett. **28**(9), 710 (2003)
12. D.N. Christodoulides, F. Lederer, Y. Silberberg, Nature **424**(6950), 817 (2003)
13. A.A. Sukhorukov, Y.S. Kivshar, H.S. Eisenberg, Y. Silberberg, IEEE J. Quantum Electron. **39**(1), 31 (2003)
14. A. Fratalocchi, G. Assanto, K.A. Brzdakiewicz, M.A. Karpierz, Opt. Lett. **29**(13), 1530 (2004)
15. R. Iwanow, R. Schiek, G.I. Stegeman, T. Pertsch, F. Lederer, Y. Min, W. Sohler, Phys. Rev. Lett. **93**(11), 113902 (2004)
16. F. Chen, M. Stepic, C.E. Ruter, D. Runde, D. Kip, V. Shandarov, O. Manela, M. Segev, Opt. Express **13**(11), 4314 (2005)
17. J.W. Fleischer, G. Bartal, O. Cohen, T. Schwartz, O. Manela, B. Freedman, M. Segev, H. Buljan, N.K. Efremidis, Opt. Express **13**(6), 1780 (2005)
18. M. Matuszewski, C.R. Rosberg, D.N. Neshev, A.A. Sukhorukov, A. Mitchell, M. Trippenbach, M.W. Austin, W. Krolikowski, Y.S. Kivshar, Opt. Express **14**(1), 254 (2006)
19. D.N. Neshev, A.A. Sukhorukov, W. Krolikowski, Y.S. Kivshar, J. Nonlinear Opt. Phys. Mater. **16**(1), 1 (2007)
20. A.L. Jones, J. Opt. Soc. Am. **55**(3), 261 (1965)
21. S. Somekh, E. Garmire, A. Yariv, H.L. Garvin, R.G. Hunsperger, Appl. Phys. Lett. **22**(1), 46 (1973)
22. R. Iwanow, D.A. May-Arrioja, D.N. Christodoulides, G.I. Stegeman, Y. Min, W. Sohler, Phys. Rev. Lett. **95**(5), 053902 (2005)
23. D.N. Neshev, A.A. Sukhorukov, A. Dreischuh, R. Fischer, S. Ha, J. Bolger, L. Bui, W. Krolikowski, B.J. Eggleton, A. Mitchell, M.W. Austin, Y.S. Kivshar, Phys. Rev. Lett. **99**(12), 123901 (2007)
24. A.A. Sukhorukov, D.N. Neshev, A. Dreischuh, W. Krolikowski, J. Bolger, B.J. Eggleton, L. Bui, A. Mitchell, Y.S. Kivshar, Opt. Express **16**(9), 5991 (2008)
25. K. Motzek, A.A. Sukhorukov, Y.S. Kivshar, Opt. Express **14**(21), 9873 (2006)
26. P. Yeh, *Optical Waves in Layered Media* (John Wiley & Sons, New York, 1988)
27. P.S.J. Russell, T.A. Birks, F.D. Lloyd-Lucas, in *Confined Electrons and Photons*, ed. by E. Burstein, C. Weisbuch (Plenum, New York, 1995), pp. 585–633
28. I. Babushkin, A. Husakou, J. Herrmann, Y.S. Kivshar, Opt. Express **15**(19), 11978 (2007)
29. F. Setzpfandt, A.A. Sukhorukov, D.N. Neshev, R. Schiek, Y.S. Kivshar, T. Pertsch, Phys. Rev. Lett. **105**, 233905 (2010)
30. P.D. Rasmussen, F.H. Bennet, D.N. Neshev, A.A. Sukhorukov, C.R. Rosberg, W. Krolikowski, O. Bang, Y.S. Kivshar, Opt. Lett. **34**(3), 295 (2009)
31. A.A. Sukhorukov, D.N. Neshev, Y.S. Kivshar, Opt. Express **15**(20), 13058 (2007)
32. R. Pezer, H. Buljan, G. Bartal, M. Segev, J.W. Fleischer, Phys. Rev. E **73**(5), 056608 (2006)
33. M. Mitchell, M. Segev, Nature **387**(6636), 880 (1997)
34. R.W. Boyd, *Nonlinear Optics*, 3rd edn. (Academic Press, San Diego, 2008)
35. G.C. Valley, M. Segev, B. Crosignani, A. Yariv, M.M. Fejer, M.C. Bashaw, Phys. Rev. A **50**(6), R4457 (1994)
36. H. Buljan, T. Schwartz, M. Segev, M. Soljacic, D.N. Christodoulides, J. Opt. Soc. Am. B **21** (2), 397 (2004)
37. Z.Y. Xu, Y.V. Kartashov, L. Torner, Opt. Lett. **31**(13), 2027 (2006)
38. R.R. Shah, D.M. Kim, T.A. Rabson, F.K. Tittel, J. Appl. Phys. **47**(12), 5421 (1976)
39. F. Jermann, M. Simon, E. Kratzig, J. Opt. Soc. Am. B **12**(11), 2066 (1995)
40. J.M. Dudley, G. Genty, S. Coen, Rev. Mod. Phys. **78**(4), 1135 (2006)
41. J.K. Ranka, R.S. Windeler, A.J. Stentz, Opt. Lett. **25**(1), 25 (2000)
42. P.S.J. Russell, Science **299**(5605), 358 (2003)
43. M.H. Qi, E. Lidorikis, P.T. Rakich, S.G. Johnson, J.D. Joannopoulos, E.P. Ippen, H.I. Smith, Nature **429**(6991), 538 (2004)
44. M. Balu, J. Hales, D.J. Hagan, E.W. Van Stryland, Opt. Express **13**(10), 3594 (2005)

45. G.I. Stegeman, M. Segev, Science **286**(5444), 1518 (1999)
46. Y.S. Kivshar, G.P. Agrawal, *Optical Solitons: From Fibers to Photonic Crystals* (Academic Press, San Diego, 2003)
47. S.M. Saltiel, A.A. Sukhorukov, Y.S. Kivshar, Prog. Optics **47**, 1 (2005)
48. T. Pertsch, T. Zentgraf, U. Peschel, A. Brauer, F. Lederer, Phys. Rev. Lett. **88**(9), 093901 (2002)
49. A.A. Sukhorukov, Y.S. Kivshar, O. Bang, C.M. Soukoulis, Phys. Rev. E **63**(2), 016615 (2000)
50. A. Kobyakov, F. Lederer, Phys. Rev. A **54**(4), 3455 (1996)
51. T. Peschel, U. Peschel, F. Lederer, Phys. Rev. E **57**(1), 1127 (1998)
52. M.I. Molina, R.A. Vicencio, Y.S. Kivshar, Phys. Rev. E **72**(3), 036622 (2005)
53. R. Iwanow, R. Schiek, G. Stegeman, T. Pertsch, F. Lederer, Y. Min, W. Sohler, Opto-Electron. Rev. **13**(2), 113 (2005)
54. D. McMillen, N. Kopell, J. Hasty, J.J. Collins, Proc. Natl. Acad. Sci. U.S.A. **99**(2), 679 (2002)
55. I.L. Garanovich, A.A. Sukhorukov, Y.S. Kivshar, Phys. Rev. E **74**(6), 066609 (2006)
56. A. Szameit, I.L. Garanovich, M. Heinrich, A.A. Sukhorukov, F. Dreisow, T. Pertsch, S. Nolte, A. Tuennermann, Y.S. Kivshar, Nat. Phys. **5**(4), 271 (2009)
57. D.H. Dunlap, V.M. Kenkre, Phys. Rev. B **34**(6), 3625 (1986)
58. N.K. Efremidis, D.N. Christodoulides, Phys. Rev. E **65**(5), 056607 (2002)
59. S. Longhi, F. Dreisow, M. Heinrich, T. Pertsch, A. Tunnermann, S. Nolte, A. Szameit, Phys. Rev. A **82**(5), 053813 (2010)
60. F. Bloch, Z. Phys. **52**, 555 (1928)
61. S. Longhi, Opt. Lett. **34**(14), 2174 (2009)
62. I.L. Garanovich, Phys. Lett. A **372**(21), 3922 (2008)
63. F. Dreisow, M. Ornigotti, A. Szameit, M. Heinrich, R. Keil, S. Nolte, A. Tunnermann, S. Longhi, Appl. Phys. Lett. **95**(26), 261102 (2009)
64. I.L. Garanovich, A. Szameit, A.A. Sukhorukov, T. Pertsch, W. Krolikowski, S. Nolte, D. Neshev, A. Tuennermann, Y.S. Kivshar, Opt. Express **15**(15), 9737 (2007)
65. I.L. Garanovich, A.A. Sukhorukov, Opt. Lett. **32**(5), 475 (2007)
66. X.Y. Qi, I.L. Garanovich, A.A. Sukhorukov, W. Krolikowski, A. Mitchell, G.Q. Zhang, D.N. Neshev, Y.S. Kivshar, Opt. Lett. **35**(9), 1371 (2010)
67. I.L. Garanovich, A.A. Sukhorukov, Y.S. Kivshar, Phys. Rev. Lett. **100**(20), 203904 (2008)
68. A. Szameit, I.L. Garanovich, M. Heinrich, A.A. Sukhorukov, F. Dreisow, T. Pertsch, S. Nolte, A. Tunnermann, Y.S. Kivshar, Phys. Rev. Lett. **101**(20), 203902 (2008)
69. X.Y. Qi, I.L. Garanovich, Z.Y. Xu, A.A. Sukhorukov, W. Krolikowski, A. Mitchell, G.Q. Zhang, D.N. Neshev, Y.S. Kivshar, Opt. Lett. **34**(18), 2751 (2009)
70. A.A. Sukhorukov, D.N. Neshev, A. Dreischuh, R. Fischer, S. Ha, W. Krolikowski, J. Bolger, A. Mitchell, B.J. Eggleton, Y.S. Kivshar, Opt. Express **14**(23), 11265 (2006)
71. D.N. Neshev, T.J. Alexander, E.A. Ostrovskaya, Yu.S. Kivshar, H. Martin, I. Makasyuk, Z. Chen, Phys. Rev. Lett. **92**, 123903 (2004)
72. Z.Y. Xu, A.A. Sukhorukov, Opt. Lett. **34**(8), 1168 (2009)

Chapter 5
Spatial Beam Dynamics Mediated by Hybrid Nonlinearity

Peng Zhang, Cibo Lou, Yi Hu, Sheng Liu, Jianlin Zhao, Jingjun Xu, and Zhigang Chen

5.1 Introduction

Nonlinear wave dynamics has fascinated scientists for over two centuries because of its fundamental and technological applications in a variety of research areas including mathematics, physics, aerodynamics, oceanography, chemistry, and

P. Zhang (✉)
The Key Laboratory of Space Applied Physics and Chemistry, Ministry of Education and Shaanxi Key Laboratory of Optical Information Technology, School of Science, Northwestern Polytechnical University, Xi'an 710072, China

Department of Physics and Astronomy, San Francisco State University, San Francisco 94132, CA, USA
e-mail: pengzhang.npu@gmail.com

C. Lou • J. Xu
The Key Laboratory of Weak-Light Nonlinear Photonics, Ministry of Education and TEDA Applied Physics School, Nankai University, Tianjin 300457, China

Y. Hu
Department of Physics and Astronomy, San Francisco State University, San Francisco 94132, CA, USA

The Key Laboratory of Weak-Light Nonlinear Photonics, Ministry of Education and TEDA Applied Physics School, Nankai University, Tianjin 300457, China

INRS-EMT, University of Quebec, 1650 Blvd. Lionel Boulet, Varennes, QC, CanadaJ3X 1S2,

S. Liu • J. Zhao
The Key Laboratory of Space Applied Physics and Chemistry, Ministry of Education and Shaanxi Key Laboratory of Optical Information Technology, School of Science, Northwestern Polytechnical University, Xi'an 710072, China

Z. Chen
Department of Physics and Astronomy, San Francisco State University, San Francisco 94132, CA, USA

The Key Laboratory of Weak-Light Nonlinear Photonics, Ministry of Education and TEDA Applied Physics School, Nankai University, Tianjin 300457, China

Z. Chen and R. Morandotti (eds.), *Nonlinear Photonics and Novel Optical Phenomena*, Springer Series in Optical Sciences 170, DOI 10.1007/978-1-4614-3538-9_5,

biology, just to mention a few [1–5]. In optics, nonlinear optical media have served as an ideal platform for exploring various fundamental issues in nonlinear systems [6, 7]. The nonlinear material response results in complex changes in the spatio-temporal structure of light, leading to a host of intriguing nonlinear phenomena such as light-induced scattering [8], wave mixing [9], phase conjugation [10], and self-trapping [11–14]. Interestingly, nearly all materials including crystals [15], liquids [16], and even gases [17] can exhibit perceptible nonlinearity at sufficiently high light intensities.

In the past two decades, photorefractive media became one of the most popular materials for exploring nonlinear beam dynamics in both continuous and discrete regimes. In contradistinction with other standard nonlinear optical materials such as Kerr [16], saturable [17], and nonlocal [18, 19] materials, photorefractive crystals exhibit a relatively large nonlinear response at as low as microwatt power levels with large flexibility [20–25]; nearly all parameters influencing nonlinear beam propagation can be easily controlled. For example, the amplitude of the nonlinearity can be adjusted by varying the bias field, and the degree of saturation can be adjusted by a homogeneous illumination with an additional background beam. More interestingly, by a simple reversal of the polarity of the bias field, either self-focusing or self-defocusing nonlinearity can be achieved in the same crystal [24, 25]. Finally, the nonlinear refractive index changes can be either erased by uniform light illumination or fixed by electrical means [26]. Thus far, a variety of nonlinear self-trapped states, better known as *optical spatial solitons*, including families of fundamental bright solitons, dark and vortex solitons, vector solitons, and incoherent solitons, along with their particle-like coherent and incoherent interactions have been extensively explored in photorefractive media [27–39].

In recent years, there has been a growing research interest in nonlinear beam dynamics in photonic lattices (also known as closely spaced waveguide arrays) [40, 41]. Even in the linear regime, light propagation in photonic lattices can exhibit many intriguing phenomena mediated from the photonic band-gap structures, including anomalous diffraction and refraction, and Bloch and Rabi oscillations [42–46] which has no counterpart in continuum systems. In the presence of self-focusing or self-defocusing nonlinearity, normal or anomalous diffraction of light can be suppressed, leading to discrete or gap solitons [47, 48]. In photorefractive media, photonic lattice structures in 1-, 2-, or 3-dimensions with various symmetries can be created solely by illuminating them with periodic optical fields created either from multi-beam interference [49] or optical amplitude masks [50]. Such optically induced photonic lattices exhibit large reconfigurability and provide ideal settings for studying the basic properties of wave propagations in discrete systems. Up to now, many discrete phenomena have been demonstrated in these optically induced photonic lattices [51–59], including linear tunable negative refraction, band-gap guidance by defects, Anderson localization in disordered lattices, and nonlinear spectrum reshaping, Bessel ring lattices and rotary solitons, discrete vortex solitons, embedded solitons, and random phase lattice solitons, following the observation of fundamental discrete and gap solitons [60–62].

Generally speaking, the nonlinearities can be divided into two categories: self-focusing and self-defocusing nonlinearity. Since self-focusing and self-defocusing nonlinearity can be established by changing the polarity of the bias field in the same photorefractive crystal [24, 25], a natural question arises: is it possible for a nonlinear material to support both nonlinearities simultaneously under an identical experimental setting? If so, how would that affect the nonlinear beam dynamics? To answer these questions, in this chapter, we introduce a new type of nonlinearity, the hybrid nonlinearity. This is a type that occurs when self-focusing and self-defocusing nonlinearity co-exist in the same material under identical conditions. We show how such hybrid nonlinearity can be created in a biased photorefractive crystal by formulizing the problem based on Kukhtarev's band transport model. Then we present a brief overview of our recent work on the unusual nonlinear beam dynamics supported by the enhanced anisotropy and nonlocality of the hybrid nonlinearity in both continuous and discrete regimes. Specifically, elliptical optical solitons, stabilization of nonlinear optical vortices, and orientation-induced transition between bright and dark solitons in *homogeneous media* will be discussed. Then, in *discrete media* (photonic lattices), we show our recent work on band-gap engineering and light manipulation based on ionic-type photonic lattices, optical "saddle" solitons unique to the hybrid nonlinearity, along with earlier work on elliptical discrete solitons and orientation-induced transitions of soliton-trains between different band-gaps.

5.2 Theoretical Formulations

The mechanism dominating the nonlinear process in a biased photorefractive crystal is the so-called screening effect, where the light excited charge carriers drift to screen the external electric field, leading to a nonuniform field distribution and refractive index modulation via the electro-optic effect inside the crystal [24, 25]. In a conventional bias setting, the direction of the external field is always set to be parallel (for self-focusing) [27] or antiparallel (for self-defocusing) [29] to the crystalline c-axis of the photorefractive crystal. However, to obtain hybrid nonlinearity, the direction of the biased electric field is set to be along an arbitrary direction. Such a bias scheme is termed a nonconventionally biased (NCB) condition [63–65].

To establish a theoretical model governing the hybrid nonlinearity under NCB conditions, we consider a coordinate system in biased photorefractive crystals as illustrated in Fig. 5.1, where an elliptical light beam propagating perpendicular to the c-axis is shown as an example. The angles of the c-axis and the external bias field E_0 with respect to the x-axis are denoted by θ_c and θ_e, respectively. To determine the light-induced space charge field under the NCB condition, we start from Kukhtarev's band transport model [66]. By neglecting diffusion field and

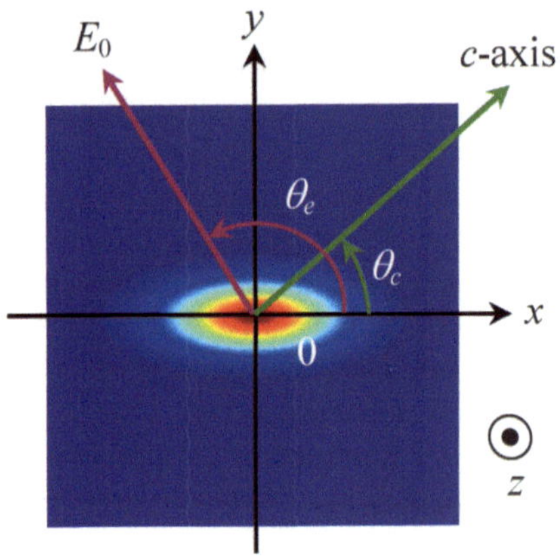

Fig. 5.1 Geometry of the coordinate system for nonconventionally biased photorefractive crystals (after ref. [65])

photovoltaic effect, the material response of the NCB photorefractive crystal under static condition can be written as [67]

$$(sI + \beta)(N_D - N_D^+) = \gamma_R n N_D^+, \tag{5.1a}$$

$$\nabla \cdot \mathbf{J} = 0, \tag{5.1b}$$

$$\mathbf{J} = e\mu n \mathbf{E}, \tag{5.1c}$$

$$\rho = e(N_D^+ - N_A - n), \tag{5.1d}$$

$$\nabla \cdot (\varepsilon \mathbf{E}) = \rho, \tag{5.1e}$$

where N_D, N_D^+, N_A, and n are the density of donors, ionized donors, acceptors, and conducting electrons, respectively, β and s are the thermal and photoexcitation coefficients, I is the intensity of light beam, γ_R is the recombination constant, $\mathbf{E}$ is the amplitude of the static electric field, e is the elementary charge, ε is the static dielectric tensor, ρ and $\mathbf{J}$ are the charge and the electric current densities, respectively, and μ is the electron mobility.

Taking the approximation $N_D^+ \approx N_A$ [20, 66] and from (5.1a), one can derive a simple expression for the electron density:

$$n = \frac{(\beta + sI)(N_D - N_A)}{\gamma_R N_A}. \tag{5.2}$$

Inserting (5.1c) and (5.2) into (5.1b) yields

$$\nabla \cdot [(\beta + sI)\mathbf{E}] = 0. \tag{5.3}$$

By denoting the light-induced electrostatic potential as Φ, the total electric field inside the crystal takes the form of $\mathbf{E} = \hat{x}E_0 \cos\theta_e + \hat{y}E_0 \sin\theta_e - \nabla\Phi$ with $\hat{x}$ and $\hat{y}$ being unit vectors. Furthermore, by introducing a dimensionless potential $\varphi = \Phi/E_0$ and a dimensionless light intensity $I = I/I_d$, where $I_d = \beta/s$, (5.3) can be rewritten as

$$\nabla^2\varphi + \nabla\varphi \cdot \nabla \ln(1+I) = \frac{\partial \ln(1+I)}{\partial x} \cos\theta_e + \frac{\partial \ln(1+I)}{\partial y} \sin\theta_e. \tag{5.4}$$

By solving (5.4), we can readily obtain the light-induced space charge field: $\mathbf{E}_s = -E_0\nabla\varphi$, which results in modulation of the refractive index of the crystal according to linear electro-optic effect.

To quantify the optically induced nonlinear refractive index changes, it is necessary to first analyze the deformation of the index ellipsoid of the crystal under a general electric field. Assuming the electric field has components along all three principal axes, according to the linear electro-optic effect, it can be easily seen that the directions of the new major axes and the magnitudes of the respective indices of the ellipsoid will be changed [68]; thereby an initial uni-axial crystal will be transformed into a biaxial one. However, after quantitative analyses with the typical experimental parameters, it is found that only the *c*-axis component of the electric field can introduce perceptible index modulation in a photorefractive crystal, i.e., the rotations of the *c*-axis and the changes of the refractive indices introduced by the components perpendicular to the *c*-axis can be neglected [68, 69]. Based on above analyses, the light-induced refractive index changes due to $\mathbf{E}_s$ can be determined by

$$\Delta n_e = \frac{1}{2} n_e^3 \gamma_{33} E_0 \left(\frac{\partial\varphi}{\partial x} \cos\theta_c + \frac{\partial\varphi}{\partial y} \sin\theta_c \right), \tag{5.5}$$

where n_e is the extraordinary refractive index and γ_{33} is the electro-optic coefficient. Therefore, the nonlinear beam dynamics in NCB photorefractive crystals obeys the following dimensionless nonlinear Schrödinger equation:

$$\left(\frac{\partial}{\partial z} - \frac{i}{2}\nabla^2 \right) B(\mathbf{r}) = i \left(\frac{\partial\varphi}{\partial x} \cos\theta_c + \frac{\partial\varphi}{\partial y} \sin\theta_c \right) B(\mathbf{r}), \tag{5.6}$$

where $B(\mathbf{r})$ is the amplitude of the optical field, the dimensionless coordinates (x, y, z) are related to the physical coordinates (x', y', z') by the expressions $(x, y) = (kl)^{1/2}(x', y')$ and $z = lz'$, where $l = 0.5kn_e{}^2\gamma_{33}E_0$, and k is the wave number of light in the crystal.

5.3 Hybrid Nonlinearity

5.3.1 *One-Dimensional Case*

When the long axis of the elliptical beam depicted in Fig. 5.1 becomes infinite, Eq. (5.1a–e) will degenerate into a one-dimensional (1D) problem with $\partial \ln(1 + I)/\partial x = 0$ and $\partial\varphi/\partial x = 0$. For a bright input beam, we can find the 1D analytical solution for (5.4): $\partial\varphi/\partial y = I \sin\theta_e/(1 + I)$. Therefore, the equation governing the nonlinear propagation of the 1D input beams becomes

$$\left(\frac{\partial}{\partial z} - \frac{i}{2}\frac{\partial^2}{\partial y^2}\right)B(y,z) = i \sin\theta_c \sin\theta_e \frac{I}{1+I}B(y,z). \tag{5.7}$$

Thus, the normalized light-induced refractive index changes can be described by $\Delta n = |E_0|I \sin\theta_c \sin\theta_e/(1+I)$. Obviously, the type of the nonlinearity experienced by 1D input beams in NCB crystals depends on the values of θ_c and θ_e, which defines the beam orientation relative to the c-axis and bias field. Now, we consider the nonlinearity versus the beam orientation under a fixed bias condition. Figure 5.2 depicts Δn versus θ_c at different $(\theta_e - \theta_c)$. It is obvious that, under the conventional bias condition $E_0 \parallel c$, the crystal can exhibit either a self-focusing nonlinearity ($\Delta n > 0$) at $(\theta_e - \theta_c) = 0$ or a self-defocusing nonlinearity ($\Delta n < 0$) at $(\theta_e - \theta_c) = \pi$, but not both for a given $(\theta_e - \theta_c)$. However, for the NCB case at $(\theta_e - \theta_c) = \pi/2$, for which the bias field is perpendicular to the c-axis ($E_0 \perp c$), the crystal can exhibit self-focusing or self-defocusing nonlinearity depending on the beam orientation. That is to say, under such a NCB condition, if two stripe beams are launched into the crystal oriented, respectively, at $\theta_c = \pi/4$ and $-\pi/4$ at same time, they will exhibit self-focusing and -defocusing nonlinearity simultaneously in the same crystal under identical bias conditions (see Fig. 5.2), and therefore, *hybrid nonlinearity*.

The nonlinear beam propagation of a stripe Gaussian beam under the hybrid nonlinearity has been both experimentally and numerically demonstrated in a $5 \times 5 \times 6.7(c)\text{mm}^3$ SBN:60 crystal. The direction of the bias field (about 1.2 kV) is perpendicular to the c-axis, and is kept constant during our experiments. The polarization direction of the input beam is always parallel to the c-axis.

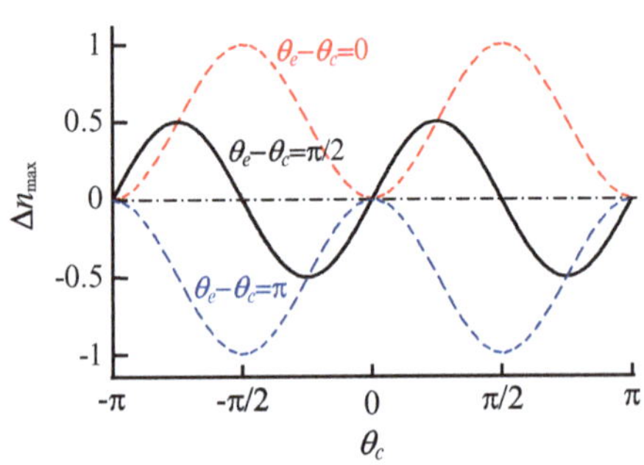

Fig. 5.2 Calculated one-dimensional light-induced refractive index changes (Δn) versus beam orientation (θ_c) at different fixed bias conditions ($\theta_e - \theta_c$) (after ref. [65])

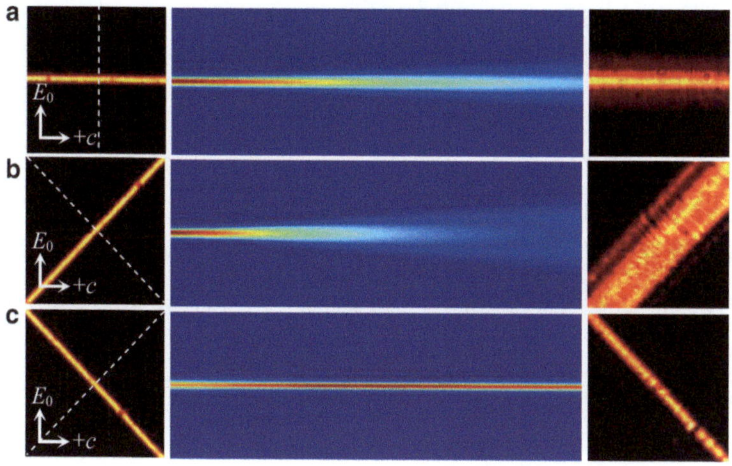

Fig. 5.3 Experimental (*left* and *right*) and numerical (*middle*) results of propagations of a Gaussian beam at $\theta_e - \theta_c = \pi/2$, but $\theta_c = 0$ (**a**), $-\pi/4$ (**b**), and $\pi/4$ (**c**). *Left*: Input beam; *Middle*: Evolution of the intensity profile along the *dashed line* in the *left panel*; *Right*: Output beam (after ref. [65])

Figure 5.3 displays the typical experimental and simulation results. It is clear that the input beams indeed experience linear diffraction, self-defocusing, and self-focusing at different orientations under identical bias conditions.

5.3.2 Two-Dimensional Case

To visualize the hybrid nonlinearity in two-dimensional (2D) cases, numerical procedures have to be employed to solve (5.4–5.6). The light-induced refractive index changes under different bias conditions induced by an input Gaussian beam as shown in Fig. 5.4a are depicted in Fig. 5.4b–d, where Fig. 5.4b, c corresponds to the conventional case $E_0 \parallel +c\,(\theta_e = \theta_c = 0)$ and $E_0 \parallel -c\,(\theta_c = \pi,\ \theta_e = 0)$, respectively, and Fig. 5.4d corresponds to the typical NCB condition $E_0 \perp c$ $(\theta_c = 0,\ \theta_e = \pi/2)$. Figure 5.4e describes the linearly diffracted output beam pattern, while the nonlinear output beam patterns under different bias conditions, corresponding to Fig. 5.4b–d, are depicted in Fig. 5.4f–h, respectively. Figure 5.4i shows the FWHMs of the beam profiles along the dashed lines in Fig. 5.4e, h versus the propagation lengths z. From Fig. 5.4, we can see that different bias conditions will cause different index changes as well as various nonlinear beam propagations. Under conventional bias conditions, although the peripheral regions of the index changes possess opposite sign with respect to the central part, the nonlinearity experienced by the input beam is mainly determined by the index changes of the central part, resulting in self-focusing in Fig. 5.4f and self-defocusing in Fig. 5.4g. While under the NCB condition at $E_0 \perp c$, the distribution of the index changes

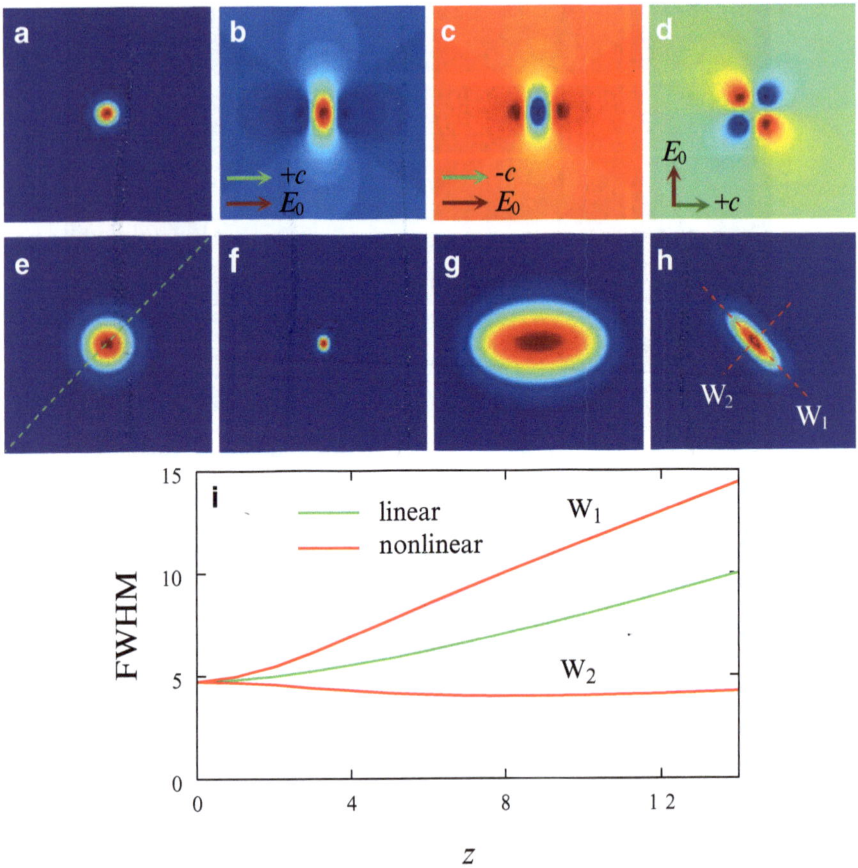

Fig. 5.4 Numerical simulations on light-induced refractive index changes and nonlinear beam propagation of a two-dimensional Gaussian beam, where (**a**) is the input beam, (**b–d**) are corresponding to the refractive index changes induced by (**a**) under different bias conditions indicated by the *arrows*, (**e**) is the output beam pattern after linear propagation, (**f–h**) depict the nonlinear output beam patterns under the same bias conditions with that in (**b–d**), and (**i**) depict the FWHMs of the beams versus propagation lengths z along the *dashed lines* shown in (**e**) and (**h**) (after ref. [65])

becomes very distinct in comparison with the conventional ones. From Fig. 5.4d, it can be seen that the index change at the center of the input beam is zero, and the maxima of the index changes occur at the positions far away from the beam center, showing a typical nonlocality. More interestingly, along different diagonal directions across the beam center, the index changes always possess identical sign, but the sign and the maximum index changes can be dramatically different in different directions, representing an enhanced anisotropy. From Fig. 5.4h, i, it is clear that the input beam experiences self-focusing and self-defocusing at the same time, i.e., hybrid nonlinearity, leading to an elliptical output beam pattern.

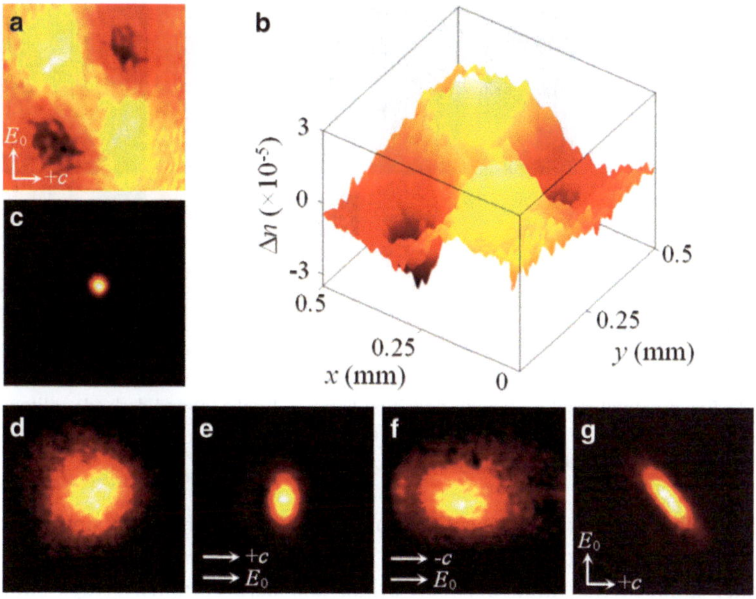

Fig. 5.5 Experimental results on light-induced refractive index changes and nonlinear beam propagations of a 2D Gaussian beam. (**a**) and (**b**) depict the 2D and 3D displays of the measured refractive index changes at $E_0 \perp c$, corresponding to the center part of Fig. 5.4d. (**c–g**) are the experimental results for beam propagations, where (**c**) is the input beam, and (**d–g**) are the linear and nonlinear output beam patterns corresponding to Fig. 5.4e–h, respectively (after ref. [65])

The experimental demonstrations are depicted in Fig. 5.5, where the light-induced refractive index changes at $E_0 \perp c$ (see Fig. 5.5a, b) are directly visualized by employing digital holography with similar setup as that used in refs. [64, 65, 70]. Figures 5.5d–g display outputs after linear and nonlinear propagation of an input 2D Gaussian beam (see Fig. 5.5c) corresponding to Fig. 5.4e–h, respectively. By comparing Fig. 5.5 with Fig. 5.4, it is clear that the experimental results are in good agreement with the theoretical predictions.

5.4 Nonlinear Beam Dynamics in Homogenous Media

5.4.1 *Transition Between Bright and Dark Solitons*

As shown in Sect. 5.3.1, in the presence of hybrid nonlinearity, one can switch the type of the nonlinearity solely by changing the optical beam orientation. This enables an optically induced transition from bright to dark solitons without having to reverse the bias field [71]. From (5.7), one can readily conclude that, in the 1D case, although hybrid nonlinearity offers the opportunity to create bright and dark

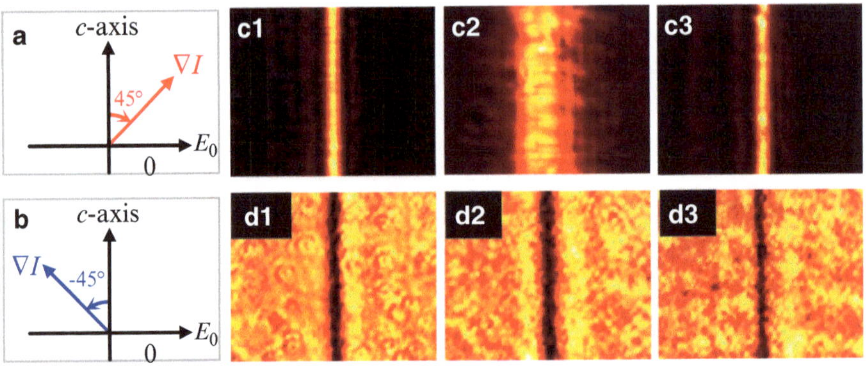

Fig. 5.6 Experimental results showing transition between bright (**c**) and dark (**d**) solitons under hybrid nonlinearity. (**a–b**) Geometry of beam orientation. In (**c–d**), (**1–3**) are for input, linear diffracted, and self-trapped beams, respectively (after ref. [71])

solitons in the same setting, the soliton solutions would share forms similar to those under conventional nonlinearities [24, 25]. To experimentally demonstrate the optically induced transition between bright and dark solitons, setups similar to those used in refs. [27, 29] are utilized. The beam orientations at the NCB condition $E_0 \perp c$ are illustrated in Fig. 5.6a, b. Typical experimental results in an SBN crystal are depicted in Fig. 5.6c–d. As expected, without nonlinearity (no bias field is applied in this case), both the bright Gaussian beam and dark notch undergo linear diffraction. By switching on the bias field, in the same crystal and with identical conditions, the Gaussian beam will experience self-focusing to balance its normal linear diffraction, resulting in a bright soliton. Meanwhile, the dark notch experiences self-defocusing resulting in a dark soliton. These experimental results are in good agreement with the numerical results presented in ref. [71].

5.4.2 *Elliptical Optical Solitons*

In the presence of enhanced anisotropy and nonlocality, the 2D soliton solutions under hybrid nonlinearity would take a more complicated form than is seen in conventional case [64]. From (5.6), the solitonary solutions can be found in the standard form $B(x,y,z) = b(x,y)\exp(i\beta z)$, where β is the propagation constant, and the real envelope $b(x, y)$ satisfies the following equation:

$$\left(\beta - \frac{1}{2}\nabla^2\right) b(x,y) = \left(\frac{\partial \varphi}{\partial x}\cos\theta_c + \frac{\partial \varphi}{\partial y}\sin\theta_c\right) b(x,y), \tag{5.8}$$

where φ is determined by (5.4) with $I = |b(x,y)|^2$. We numerically solve the eigenproblem (5.8) by employing the renormalized iterative procedure [72]. By setting $\theta_e = \theta_c = 0$, the conventional condition, we derive the same results as

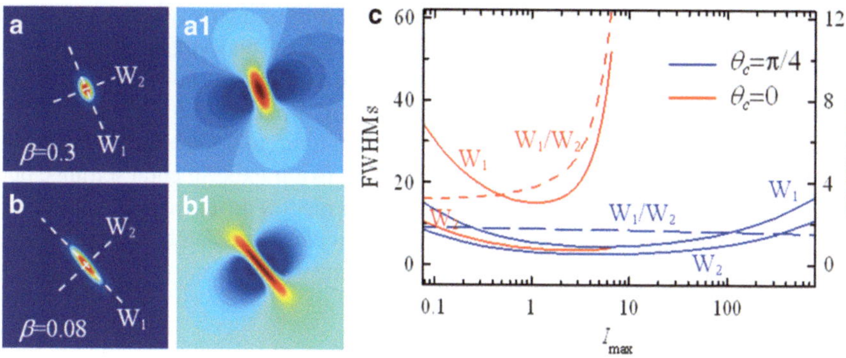

Fig. 5.7 Numerical results for elliptical optical solitons under hybrid nonlinearity. Soliton solutions (*left*) and their induced refractive index changes (*right*) at $\theta_e = \pi/2$, but $\theta_c = \pi/4$ (**a**) and $\theta_c = 0$ (**b**); (**c**) FWHMs of the solitons and their ratio versus $I_{\max}$ (after ref. [64])

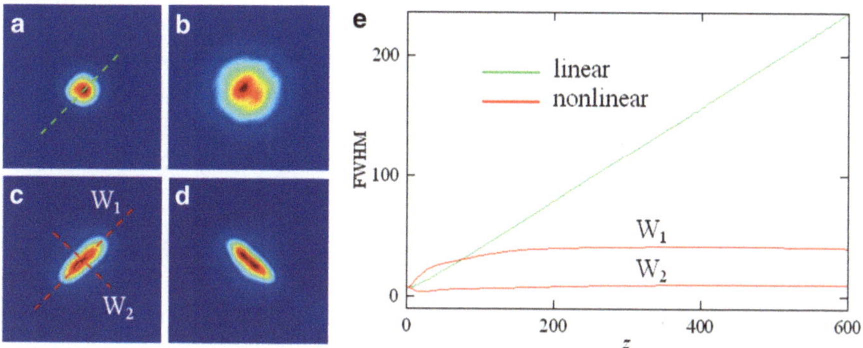

Fig. 5.8 Experimental results for observing elliptical soliton formations under hybrid nonlinearity (**a–d**) and BPM simulation of soliton evolution process (**e**). (**a**) Input Gaussian beam. (**b–d**) Output beam profiles for a linearly diffracted beam, and the solitons formed at $\theta_c = 0$ but $\theta_e = -\pi/2$ and $\pi/2$, respectively. (**e**) FWHMs of the beams versus propagation lengths z (after ref. [64])

those presented in refs. [73, 74]. Figure 5.7 shows some typical simulation results under NCB conditions, from which it can be seen that the hybrid nonlinearity indeed can support bright solitons. However, due to the enhanced anisotropy, the solitons are severely distorted, exhibiting nontrivial elliptical profiles. Moreover, the soliton-induced refractive index changes are dominated by the positive part, indicating that the solitons mainly experience self-focusing nonlinearity, even though the crystal has hybrid nonlinearity. This coincides with the previous discussions on the orientation-dependent nonlinearity in the 1D case.

To experimentally observe such elliptical solitons under hybrid nonlinearity, a circular Gaussian beam as shown in Fig. 5.8a is launched into an SBN:60 crystal with the bias field perpendicular to the c-axis. Without nonlinearity, the input beam would experience linear diffraction, leading to an output intensity pattern with an increased beam size, as shown in Fig. 5.8b. With a 3.6 kV bias voltage (over 5 mm) but $\theta_e = \pm\, \pi/2$ and $\theta_c = 0$, the circular input beam evolves into an elliptical soliton at a particular orientation with a typical beam size ratio $W_1/W_2 \approx 3.5$ (see Fig. 5.8c, d). To confirm such a soliton evolution behavior from a circular Gaussian beam to an elliptical soliton, a BPM simulation corresponding to our experimental settings, but with a much longer propagation length is performed. The results are shown in Fig. 5.8e, from which we can see clearly that a circular Gaussian beam indeed can evolve into an elliptical soliton under hybrid nonlinearity.

5.4.3 *Stabilization and Breakup of Optical Vortices*

Vortices are often found in nature and share many common properties in different physical systems [75]. In optics, vortices are termed as "phase singularities" in light waves, and carry nontrivial orbital angular momentums (OAMs) [76]. In the past few decades, the interplay between natural diffraction of helical-phase structures of optical vortices (OVs) and nonlinear effects has been investigated extensively in various nonlinear media [30, 31, 77–83]. The intrinsic anisotropy and nonlocality mediated by photorefractive nonlinear materials lead to unusual features of dynamical vortex propagation [81–83]. Specifically, it was found that the anisotropy of the nonlinearity might be responsible for the nonlinear decay of higher-order vortices [82], whereas the nonlocality could dramatically suppress the azimuthal modulation instability of the vortex beam for formation of bright ring vortex solitons [80, 83]. Due to the enhanced anisotropy and nonlocality, one would expect hybrid nonlinearity to play a nontrivial role in the nonlinear beam dynamics of OVs. In the following, we will show two examples of suppression of the breakup of a single-charged vortex and aggravation of the breakup of double-charged higher-order vortices [84].

To analyze the nonlinear propagation of OVs, we first calculate the nonlinear refractive index change with an input vortex beam taken to be

$$B(\mathbf{r}) = \sqrt{I_0}\mathbf{r}\exp\left(-\frac{|\mathbf{r}|^2}{\sigma^2}\right)\exp(im\theta), \tag{5.9}$$

where $r = (x^2 + y^2)^{1/2}$ and θ is the azimuth angle, σ is the diameter of the Gaussian beam in which the vortex embeds, m is a signed integer called topological charge, and I_0 is the characteristic intensity of the vortex. By numerically solving (5.4) and (5.5), the calculated refractive index changes Δn induced by the vortex beam as depicted in Fig. 5.9a under different conditions are shown in Fig. 5.9b–d. From Fig. 5.9b, c, it can be seen that the vortex circular ring induces Δn possessing inhomogeneous pseudoelliptical rings with two lobes of either high or low index change around the center, which implies that the crystal exhibits mainly self-focusing or -defocusing nonlinearity with some anisotropy. In addition, the two

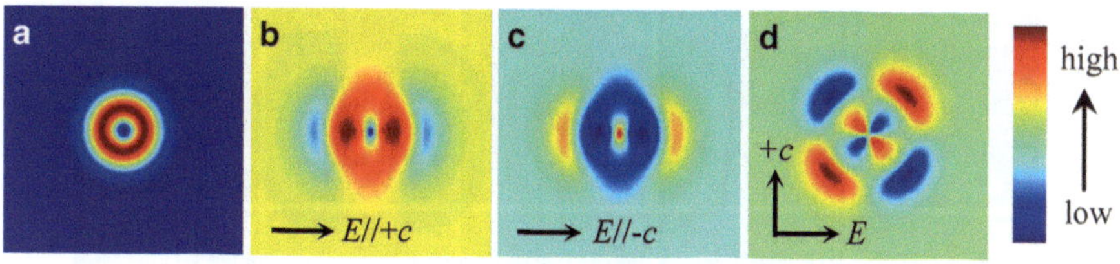

Fig. 5.9 Numerical simulations of the refractive index change in a biased photorefractive crystal induced by a vortex beam. (**a**) Input vortex beam. (**b–d**) Index changes at $E_0 \parallel +c$ ($\theta_c = \theta_e = 0$), $E_0 \parallel -c$ ($\theta_c = \pi$, $\theta_e = 0$), and $E_0 \perp c$ ($\theta_c = 0$, $\theta_e = \pi/2$), respectively (after ref. [84])

peaks of the Δn coincide with the intensity maximum of the vortex beam, reflecting relatively weak nonlocality. However, from Fig. 5.9d, we can see that an interesting structure of Δn arises under hybrid nonlinearity; it has a symmetric profile, but with opposite signs along the two diagonal directions and each diagonal direction involves higher and lower lobes simultaneously, representing a more complex anisotropic nonlinearity. Moreover, Δn is almost zero at points where the intensity of the vortex beam is maximum, exemplifying an enhanced nonlocality.

By solving (5.4), (5.6), and (5.9) with BPM, the nonlinear evolutions of the single-charged ($m = 1$) vortex beam depicted in Fig. 5.9a are numerically simulated under different bias conditions. The results are shown in Fig. 5.10. Similar to the anisotropic boundary condition as mentioned in ref. [79], the original anisotropic index change depicted in Fig. 5.9b expedites the breakup of the vortex beam at $E_0 \parallel +c$ with the self-focusing nonlinearity, as shown in Fig. 5.10a. The nontrivial interplay between the inherent angular momentum of the vortex beam and the anisotropic nonlinearity makes the two collapsed fragments undergo both clockwise and anticlockwise damped rotations in turn during the propagation. While in the case of $E_0 \parallel -c$, as shown in Fig. 5.10b, the self-defocusing nonlinearity stretches the vortex beam into an ellipse with its major axis along the c-axis and leads to the formation of two lobes. Meanwhile, the two lobes gradually flow clockwise along the elliptical ring, driven by the helical-phase structure. However, the energy flow will be retarded by the anisotropic nonlinearities, as demonstrated in ref. [81]. Figure 5.10c presents the results of the beam's evolution in the presence of the hybrid nonlinearity at $E_0 \perp c$. By comparing with the above two conventional cases, it can be found that the vortex beam experiences a more stable nonlinear evolutionary process under hybrid nonlinearity. Notwithstanding the coexistence of self-focusing and -defocusing nonlinearities, it neither collapses into two separated lobes as happens at $E_0 \parallel +c$, nor does it expand or stretch as much as it does at $E_0 \parallel -c$. This characteristic nonlinear dynamic manifests the suppression of the azimuthal modulation instability exerted by the enhanced nonlocality in the presence of hybrid nonlinearity. Such behaviors are further confirmed by the dynamics of OAMs during the vortex evolutions as shown in Fig. 5.10d. It can be seen that, in comparison with the conventional cases, the loss of OAM during nonlinear propagation is dramatically reduced in the NCB case, indicating the suppression of azimuthal modulation instability.

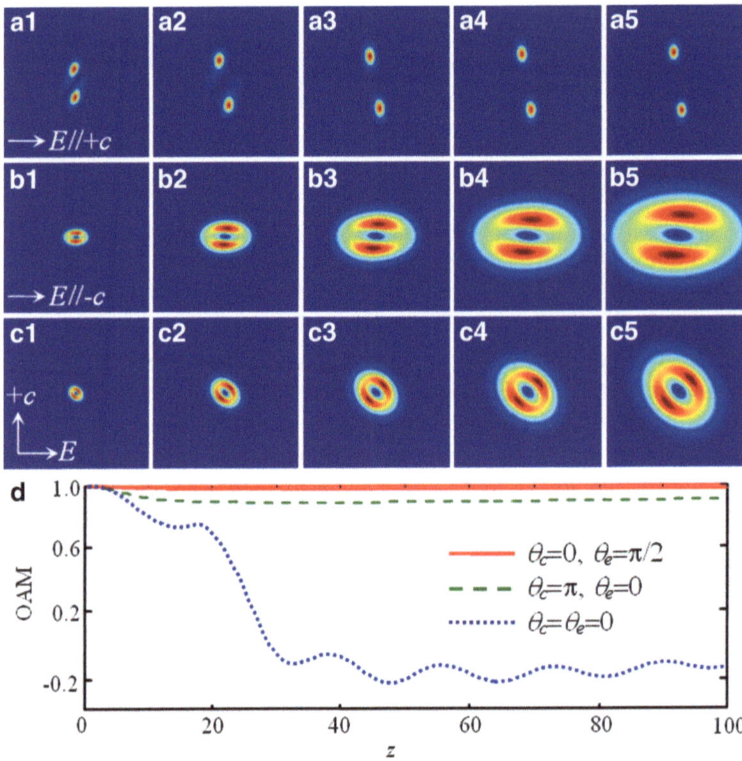

Fig. 5.10 Nonlinear evolution of the single-charged vortex beam at $E_0 \parallel +c$ ($\theta_c = \theta_e = 0$) (**a**), $E_0 \parallel -c$ ($\theta_c = \pi$, $\theta_e = 0$) (**b**), and $E_0 \perp c$ ($\theta_c = 0$, $\theta_e = \pi/2$) (**c**). From *left* to *right* are the output intensity profiles at different normalized propagation distances $z = 20$, 40, 60, 80, and 100, respectively; (**d**) OAM versus z (after ref. [84])

The experimental demonstrations of the nonlinear evolution of a single-charged vortex under different bias conditions are performed in an SBN:60 crystal. Typical results are shown in Fig. 5.11. The vortex beam is created with a computer-generated holographic mask and a spatial filter [85]. According to Fig. 5.11d, under self-focusing nonlinearity ($E_0 \parallel +c$), the single-charged vortex beam, as expected, will break up into two filaments, which gradually rotate counterclockwise. In Fig. 5.11e, we find that the anisotropic self-defocusing nonlinearity ($E_0 \parallel -c$) causes the vortex beam to stretch along the c-axis, and the ellipticity of the beam profile to increase gradually with the enhancement of the external bias field. Figure 5.11f displays the nonlinear output of NCB case with the external field perpendicular to the c-axis. Although the nonlinear evolution is similar with that in Fig. 5.11e, the orientation of the induced elliptical output possesses an angle of about 45° with respect to the c-axis, and the output intensity patterns are much smaller. These experimental observations agree with the numerical expectations as shown in Fig. 5.10.

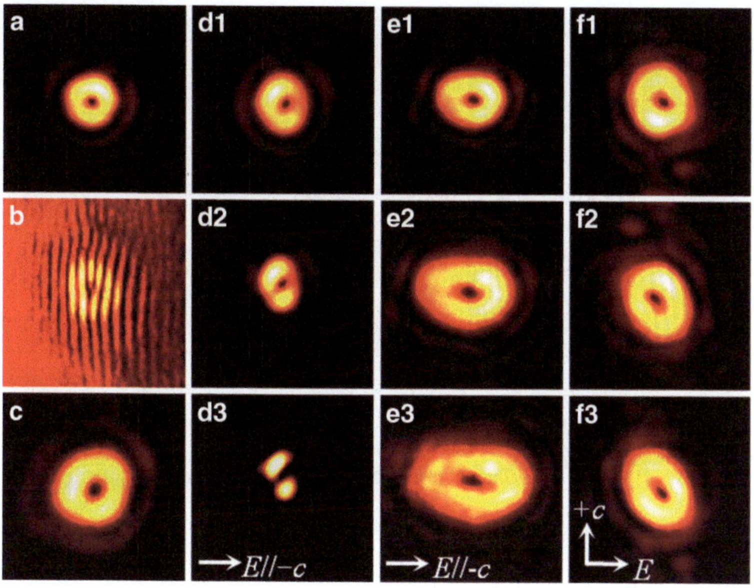

Fig. 5.11 Experimental results of a single-charged vortex under different bias conditions. (**a**) Intensity distribution and (**b**) interferogram of the input vortex beam. (**c**) Linearly diffracted output beam. (**d–f**) Nonlinear output at $E_0 \parallel +c$ ($\theta_e = \theta_c = 0$), $E_0 \parallel -c$ ($\theta_c = \pi$, $\theta_e = 0$), and $E_0 \perp c$ ($\theta_c = 0$, $\theta_e = \pi/2$), respectively, where results from *top* to *bottom* are obtained at gradually increased voltages (after ref. [84])

Next, we shall show the nonlinear dynamics of higher-order vortices in the presence of the hybrid nonlinearity. It has been demonstrated that a higher-order vortex will break up into an array of vortices with unit charge under the anisotropic boundary conditions or anisotropic nonlinearities [82]. In a self-focusing medium, the azimuthal modulation instability of higher-order vortices will be much more severe than that of single-charged vortices, leading to a quick collapse of the input vortex. Here, we use the self-defocusing case at $E_0 \parallel -c$ for comparison. Figure 5.12a, b display the simulation results of a double-charged vortex beam at $E_0 \perp c$ and $E_0 \parallel -c$, respectively. It is obvious that, in both cases, the double-charged vortex decays into two separate single-charged vortices, which can be seen from the fork fringes in the interferograms as described in the bottom row of Fig. 5.12. However, although the nonlinear expansion of input beam under the NCB condition is smaller than that at $E_0 \parallel -c$, the separation of the two resultant dark cores under hybrid nonlinearity is more distinct. This aggravation of the nonlinear decay might be due to the enhancement of the anisotropic nonlinearity driven by the NCB field. In addition, the orientations of the singly charged vortex pairs for the two cases are somewhat different, mainly determined by the anisotropy of the nonlinearity [65]. With the hybrid nonlinearity, the bright stripe separating the two dark cores (see Fig. 5.12a) will experience self-focusing during propagation [65], which will result in further separation of the two singly charged vortices. However, for the conventional case with

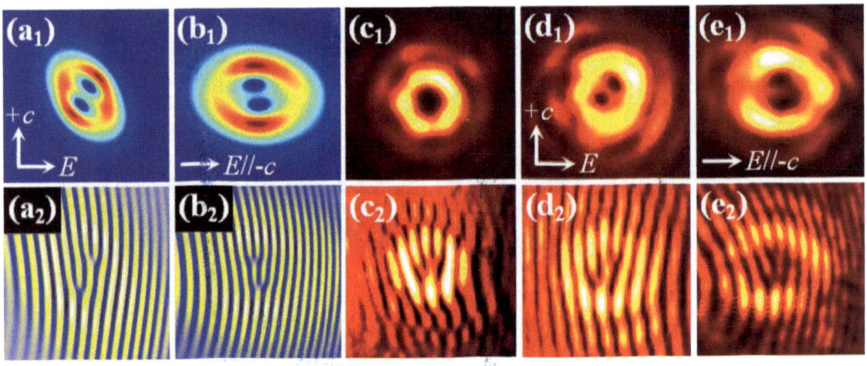

Fig. 5.12 Numerical (**a**, **b**) and experimental (**c**–**e**) results showing breakup of double-charged vortices. *Top* and *bottom* correspond to intensity patterns and interferograms, respectively. (**a**) and (**b**) Simulations at $z = 45$ for $E_0 \perp c\,(\theta_c = 0, \theta_e = \pi/2)$ and $E_0 \parallel -c\,(\theta_c = \pi, \theta_e = 0)$, respectively; (**c**) input vortex; (**d**) and (**e**) experimental results corresponding to (**a**) and (**b**) (after ref. [84])

self-defocusing, the bright stripe is almost parallel to the *c*-axis, which means it will experience less expansion during propagation to make sure the dark cores can be separated. The simulation results are also confirmed in experiment, as shown in Fig. 5.12. By launching a double-charged vortex beam (see Fig. 5.12c) into the crystal, the nonlinear decay of the input vortex is observed with the biased field perpendicular and antiparallel to the *c*-axis (see Fig. 5.12d, e), respectively. The experimental observations agree with the numerical simulations.

5.5 Reconfigurable Photonic Lattices

Wave propagation in periodic systems is of great interest for both its fundamental and technological applicability [86, 87]. Photonic lattices have served as an ideal platform for exploring various fundamental issues in discrete systems [40, 41]. The periodicity of the photonic lattice breaks the rotational symmetry of a normally isotropic medium, leading to a host of new phenomena [42–62] that have also been studied in a variety of other discrete systems ranging from photonic and phononic crystals to liquid surface waves, trapped Bose–Einstein condensates (BECs), and metamaterials [88–91]. In photorefractive crystals, photonic lattices can be optically induced by illuminating them with a periodic light pattern [49–62]. One of the advantages of such induced lattices is its reconfigurability, namely, the lattice structure and potential depth can be tuned at ease by varying the intensity of the lattice-inducing beam or the bias field. In this section, we shall discuss the lattice reconfiguration and associated discrete phenomena in the presence of hybrid nonlinearity [92–98].

5.5.1 Lattice Formation

By solving (5.4) and (5.5) with $I(x, y) = \cos^2(\pi x/\Lambda)\cos^2(\pi y/\Lambda)$, where Λ is the spatial period of the light field, the optically induced photonic lattices under different bias conditions can be obtained [93, 94]. The Brillouin zone (BZ) structure of the induced photonic lattices can be numerically visualized by using the method described in ref. [99]. In an experiment, such optically induced photonic lattices are created in a biased SBN:60 crystal by illumination with a periodic light pattern generated by using an optical amplitude mask, similar to that used in refs. [52, 55, 62]. The index distributions of the induced lattices are directly measured by the digital holography [70], and the BZ spectra are obtained by using the technique described in ref. [100]. Typical results of the index distributions and the BZ spectra are displayed in Fig. 5.13 for four lattice structures induced with a same square lattice beam but under different bias conditions. Clearly, under an identical bias scheme ($E_0 \parallel c$ or $E_0 \perp c$) different c-axis orientations lead to different lattice structures, whereas with the same c-axis orientation the induced lattice structure varies with the bias direction. Not only the shape and orientation of individual lattice site can be changed, but also the location of the index maxima can be shifted significantly with respect to that of the intensity maxima of the lattice inducing beam, as driven by the enhanced anisotropy and nonlocality of the hybrid nonlinearity [65]. The measured index profiles and BZ spectra are in good agreement with those obtained from simulation [93, 94].

5.5.2 Bloch Mode Transition

We now focus on the behavior of a probe beam propagating linearly through the above induced lattices. Taking Fig. 5.13c, d as examples, the period and orientation of these two induced lattices are quite different. It can be found that the first Brillouin zone (BZ) of the lattice in Fig. 5.13d happens to be overlapping with the second BZ of the lattice in Fig. 5.13c. Therefore, the high-symmetry M_1 point in one setting (Fig. 5.13c) corresponds to the X_1 point in another setting (Fig. 5.13d), as also illustrated in the band diagrams in Fig. 5.14a–c. This enables the excitation of Bloch modes associated with different high-symmetry points [101] in the first band with the same excitation scheme. Typical results are shown in Fig. 5.14d–g, where the Bloch modes are excited in these two lattices established under different NCB conditions. In our experiment, the probe beam is tilted at the same angle to excite the Bloch modes at M_1/X_1 point marked by a solid (green) dot in Fig. 5.14a–c. With both the probe beam and the lattice beam kept constant, a transition between Bloch modes at M_1 (Fig. 5.14d, e) and X_1 (Fig. 5.14f, g) is realized by rotating the biased crystal 45° about the z-axis. The measured intensity and phase (via interference) structures confirmed such a transition, in excellent agreement with calculated results.

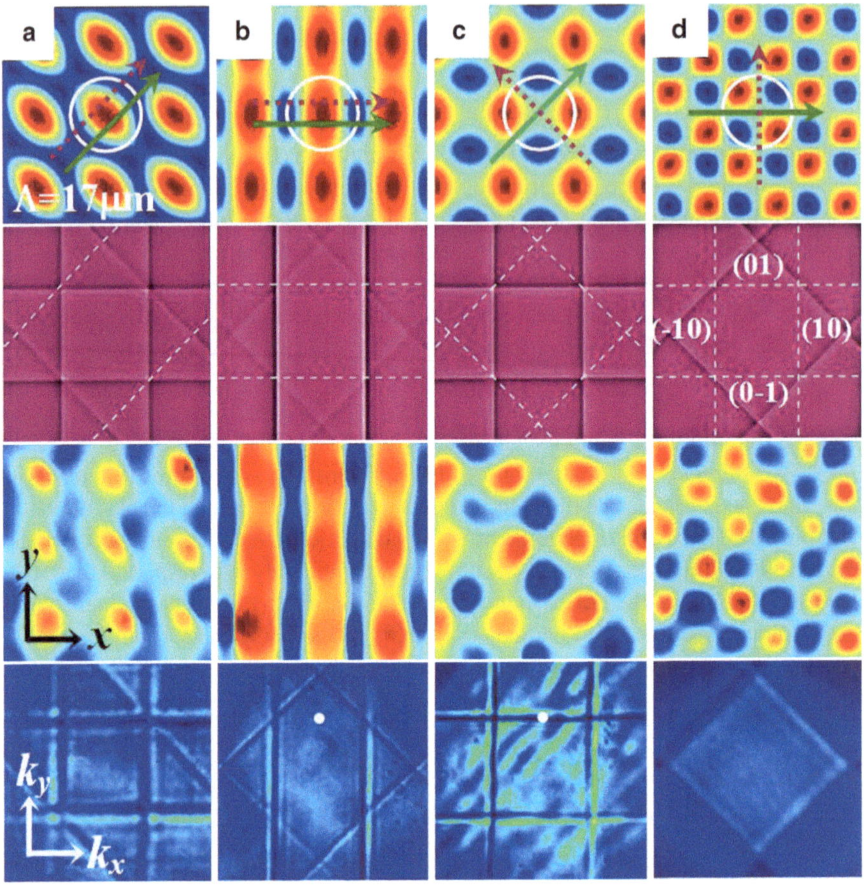

Fig. 5.13 Numerical (*upper two rows*) and experimental (*bottom two rows*) results of lattice and Brillouin zone (BZ) reconfiguration. First and third rows show the refractive index distributions, and second and fourth rows show the corresponding BZ spectra. *Solid* and *dashed arrows* in first row represent the directions of *c*-axis and bias field, and the *white circle* corresponds to an intensity spot of the lattice-inducing beam. *Dashed lines* in second row indicate the missing Bragg-reflection lines in BZ of a square lattice. From (**a**) to (**d**): $\theta_c = \theta_e = 0$; $\theta_c = \theta_e = \pi/4$; $\theta_c = \pi/4$, $\theta_e = 3\pi/4$, and $\theta_c = 0$, $\theta_e = \pi/2$ (after ref. [94])

5.5.3 *Diffraction Managements*

Since beam diffraction in 2D lattices depends on the position of its Bloch momentum vector within the BZ, we show next the transition between normal and anomalous diffraction using the above overlapping BZs. The region of normal diffraction for one lattice (Fig. 5.13d) but of anomalous diffraction along k_x for other lattices (Fig. 5.13a–c) can be identified (marked by a black circle in Fig. 5.14).

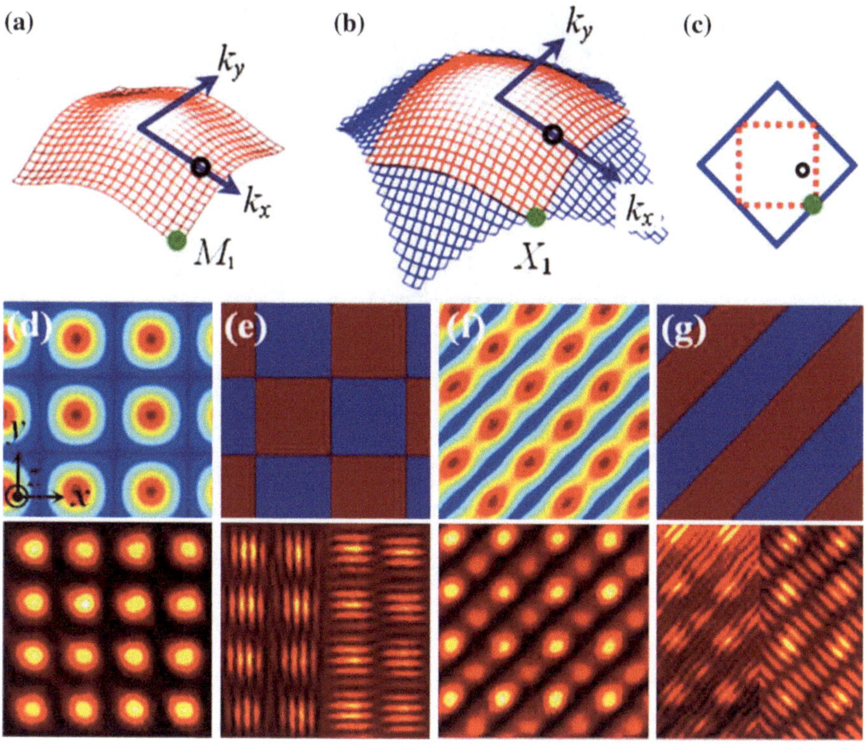

Fig. 5.14 Demonstration of Bloch-mode transition by lattice reconfiguration. (**a**) and (**b**) illustrate the first band diagram corresponding to lattices of Fig. 5.13c, d, respectively; (**c**) shows the boundary of the first BZ for (**a**) and (**b**); (**d–g**) are calculated (*top*) and experimentally observed (*bottom*) intensity (**d**, **f**) and phase (**e**, **g**) structures of Bloch modes excited at the *solid green spots* marked in (**a–c**) for lattices illustrated (**d**, **e**) in Fig. 5.13c and (**f**, **g**) in Fig. 5.13d (after ref. [95])

By probing with a tilted Gaussian beam (waist located before the front facet of the crystal) whose transverse **k**-vector corresponds to the black circle in Fig. 5.14, we observe different 2D diffraction patterns coming from different lattices as shown in Fig. 5.15. Clearly, the beam exhibits normal diffraction in both transverse directions in Fig. 5.15d but anomalous diffraction along k_x in Fig. 5.15a–c. Such anomalous diffraction has been demonstrated previously either by adjusting the angle in 1D lattices [43] or the nonlinearity in 2D lattices [101], but here the transition between linear normal and anomalous diffraction is realized with the same input tilt simply by reconfiguring the 2D photonic lattices.

5.5.4 *Refraction Managements*

Next we show that by lattice reconfiguration a probe beam can also excite Bloch modes from different bands, permitting the transition between normal (positive) and anomalous (negative) refraction [44]. An example is illustrated in Fig. 5.16,

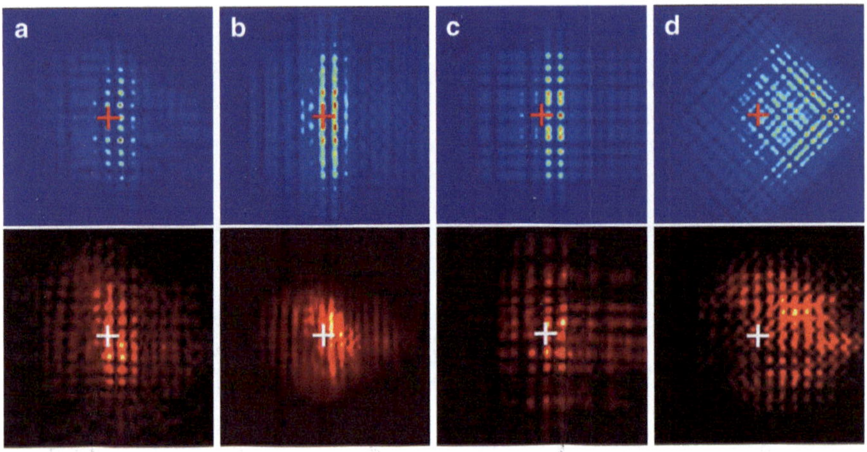

Fig. 5.15 Demonstration of 2D normal and anomalous diffraction by lattice reconfiguration. (**a–d**) Numerical (*top*) and experimental (*bottom*) results of output diffraction patterns of the same probe beam (excitation position marked by the circles in **k**-space of Fig. 5.14a–c) from the four lattices shown in Fig. 5.13a–d, respectively. The *crosses* indicate the center of input beam (after ref. [95])

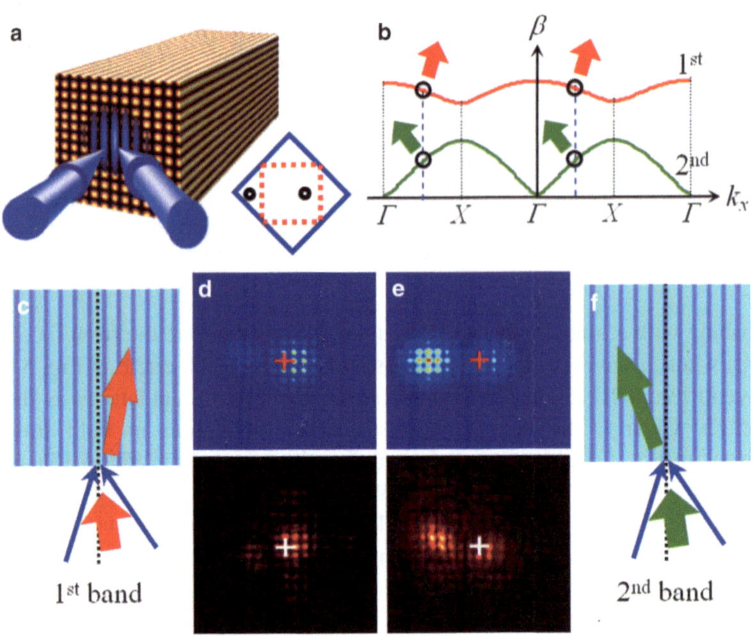

Fig. 5.16 Demonstration of 2D positive and negative refraction by lattice reconfiguration. (**a**) Excitation scheme with locations in **k**-space of the two input beams marked by *circles* in *inset*. (**b**) Direction of refracted light (*marked by arrows*) at the first and second band diffraction curves. (**c, f**) Illustration of light refraction from the first and second bands. (**d–e**) Numerical (*top*) and experimental (*bottom*) results of the output probe beam from lattices in Fig. 5.13a, c, respectively. The *crosses* indicate the center of the two beams at input (after ref. [95])

where the two-beam excitation technique [51] is employed to selectively excite Bloch modes. The angle between the two input beams is set to be twice of the Bragg angle, while the added input direction of the two interfering beams is tilted by a half Bragg angle. Under this condition, the k_x component of one of two beams is positioned inside the first BZ, but that of the other beam is outside (in lattices of Figs. 5.13a, c). In addition, the interference maxima of the two beams overlap with the intensity maxima of the lattice inducing beam. However, as seen in Fig. 5.13, reconfiguration of the lattices could result in a change from on-site to off-site excitation (e.g., from Fig. 5.13a–c). This in turn leads to a change of excitation of Bloch modes from the first to second band (Fig. 5.16b), and therefore a change of apparent refraction of the probe beam from anomalous (Fig. 5.16c, d) to normal (Figs. 5.16e, f). Two cases are shown in Fig. 5.16d, e, corresponding to outputs from lattices of Fig. 5.13a, c, respectively. For all two cases, the input direction of the two beams (or the direction of total energy flow) is initially tilted towards left, and it bends further to left in Fig. 5.16e, f due to dominant excitation of the second band Bloch modes by both beams in the periodic structure. However, it bends back to right (anomalous refraction) in Fig. 5.16c, d due to dominant excitation of the first band Bloch modes (Fig. 5.16b). From the direction of the energy flow depicted in Fig. 5.16c, it can be seen that the anomalous refraction just represents a negative refraction of energy flow, similar to that observed in other periodic systems [102].

5.5.5 Ionic-Type Photonic Lattices

Typically, all photonic lattices investigated theoretically or demonstrated experimentally are composed of either periodic positive or periodic negative potentials, not both [49–62]. Here, as shown in Fig. 5.13, under the conventional bias conditions ($E_0 \parallel c$), a square lattice solely composed of positive index changes is induced (see Fig. 5.13a). Simply by rotating the biased crystal by 45°, a non-Bravais lattice structure with alternating positive and negative index changes is induced (Fig. 5.13b), which is akin to an ionic crystal structure with alternating positive and negative charges. Therefore, we name such photonic lattice structures ionic-type photonic lattices. In the presence of hybrid nonlinearity ($E_0 \perp c$), another two different ionic-type lattices can be established (Fig. 5.13c, d). The three-dimensional display of the calculated refractive index distribution of Fig. 5.13c is shown in Fig. 5.17a, from which we can clearly see the egg-crate lattice structure.

By comparing the BZs in Fig. 5.13 with the textbook-calculated BZ for the square lattice, one can notice that some expected Bragg reflection lines indicated by the dashed lines in the second row of Fig. 5.13 are missing. This implies that Bragg reflection along some particular directions either is too weak to be observable or simply cannot occur (band-gap closed). Such reconfiguration of BZ spectra in ionic lattices introduces a new way for band-gap "engineering." For instance, one can design a particular photonic structure with its band-gap closed in a desired direction for control of light propagation. Figure 5.17b–e demonstrates a typical example.

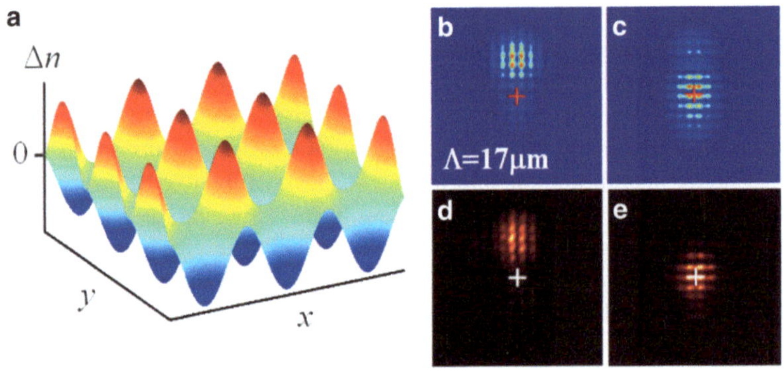

Fig. 5.17 (**a**) 3D display of the index distribution of an ionic-type lattice shown in Fig. 5.13c. (**b–e**) Numerical (*top*) and experimental (*bottom*) output patterns of a tilted probe beam with its excitation location in ***k***-space marked by a *white dot* in Figs. 5.13b, c. Here (**b**, **d**) and (**c**, **e**) correspond to excitation in lattices of Fig. 5.13b, c showing no Bragg-reflection and strong Bragg-reflection, respectively. The *crosses* indicate the input position of the probe beam (after ref. [94])

A Gaussian probe beam tilted at the same input angle is launched into the lattices shown in Fig. 5.13b, c, where the excitation locations of the probe beam in ***k***-space marked by a white dot. Both numerical (Fig. 5.17b, c) and experimental (Fig. 5.17d, e) results show that the output location of the probe beam is dramatically different in the two different lattices. In particular, no Bragg-reflection (Fig. 5.17b, d) is observed in the lattice of Fig. 5.13b, but strong Bragg-reflection (Fig. 5.17c, e) is observed for the same probe beam in the lattice of Fig. 5.13c.

To understand the underlying physics of the BZ spectra reconfigurations as well as the "band-gap engineering" in the optically induced ionic-type photonic lattices, we consider a general ionic-type photonic lattice composed of both positive and negative potentials, as shown in Fig. 5.18, which is a non-Bravais lattice. Assume that the dielectric constant distribution of the lattice is given by

$$\varepsilon(\mathbf{r}) = \varepsilon_0 + \sum_i \Delta\varepsilon(\mathbf{r} - \mathbf{r}_i), \tag{5.10}$$

where ε_0 is the dielectric constant of the uniform background medium, $\mathbf{r}_i$ denotes the location of the ith lattice site, and $\Delta\varepsilon$ is the magnitude of periodic variation of the dielectric constant with $\Delta\varepsilon < 0$ and $\Delta\varepsilon > 0$ representing the positive and negative lattice potentials, respectively. We point out that the novelty of the non-Bravais 2D ionic-type lattices arises from the periodic "dips" and "humps" in the refractive index pattern, which contains at least one dip and one hump in each lattice site and cannot be "washed out" or turned into simple lattices (composed of only "dips" or "humps" but not both) merely by resetting the background ε_0. Different from the 1D ionic-type lattices, which are considered to be simple lattices, by resetting the background at the maxima or minima of the dielectric constant, the background has to be chosen properly in the 2D case so that the lattice potentials

Fig. 5.18 Schematic diagram of an ionic-type non-Bravais photonic lattice, where the *dashed* and *solid circles* represent negative and positive lattice potentials, respectively, the *dashed square* marks a unit cell, the *white dots* indicate the centers of the unit cells, and *P* is an observing point. Other coordinates are defined in the text (after ref. [94])

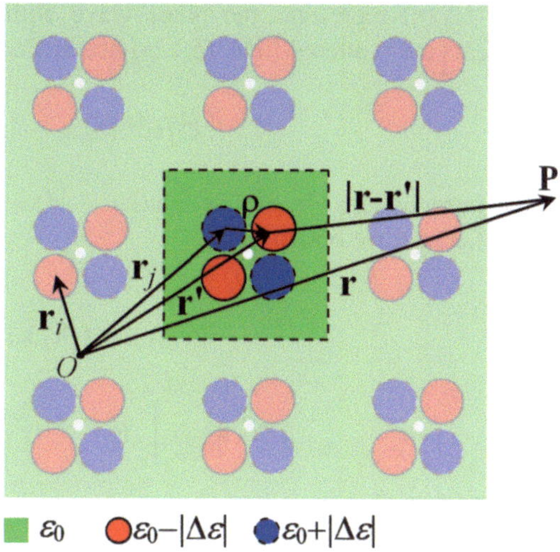

can be totally separated as described by (5.10). We mention that a chessboard lattice structure cannot be considered as ionic-type, and it contains single scattering center in each unit cell. It will turn into a simple lattice composed of only positive or negative potentials by setting the background at an appropriate level.

To analyze the BZ spectrum [99, 100] and the Bragg reflection in photonic structures, the geometrical structure factor (GSF) from solid state physics is often used as an efficient tool [86, 103, 104]. To deduce the GSF of the ionic-type photonic lattice structures as shown in Fig. 5.18 for a monochromatic light, here we assume the dielectric constant varies slowly with space and start with the Helmholtz equation by following the scattering theory for inhomogeneous media [103]

$$\nabla^2 U(\mathbf{r}) + k^2 \varepsilon(\mathbf{r})\, U(\mathbf{r}) = 0, \tag{5.11}$$

which can be rewritten as

$$\nabla^2 U(\mathbf{r}) + k^2 U(\mathbf{r}) = -4\pi F(\mathbf{r})\, U(\mathbf{r}), \tag{5.12a}$$

$$F(\mathbf{r}) = \frac{1}{4\pi} k^2 \left[\varepsilon_0 - 1 + \sum_i \Delta\varepsilon(\mathbf{r} - \mathbf{r}_i) \right], \tag{5.12b}$$

where $U(\mathbf{r})$ is the amplitude of the light field propagating through the lattice, and k is the wave number. Assuming a plane wave propagating in the otherwise uniform medium is scattered by lattice potentials, we can write $U(\mathbf{r}) = U^{(i)}(\mathbf{r}) + U^{(s)}(\mathbf{r})$, where $U^{(i)}(\mathbf{r})$ represents the plane wave with $(\nabla^2 + k^2)U^{(i)}(\mathbf{r}) = 0$ and $U^{(s)}(\mathbf{r})$ represents the scattered field, (5.12a) becomes

$$\left(\nabla^2 + k^2\right)U^{(s)}(\mathbf{r}) = -4\pi F(\mathbf{r})U(\mathbf{r}). \tag{5.13}$$

By introducing a Green function $G(\mathbf{r} - \mathbf{r}') = \exp(ik|\mathbf{r} - \mathbf{r}'|)/|\mathbf{r} - \mathbf{r}'|$ into (5.13), in the far field we obtain

$$U^{(s)}(\mathbf{r}) \approx f(\mathbf{k}', \mathbf{k})\frac{e^{ikr}}{r}, \tag{5.14a}$$

$$f(\mathbf{k}', \mathbf{k}) = \int_V F(\mathbf{r}')\,U(\mathbf{r}')e^{-i\mathbf{k}'\cdot\mathbf{r}'}d^3\mathbf{r}', \tag{5.14b}$$

where $\mathbf{r}'$ is the position of the scatterer, V is the volume of the lattice structure, and $\mathbf{k}$ and $\mathbf{k}'$ are the wave vectors of the incident and scattered fields, respectively. Considering the periodicity of the lattice structure of N cells, the amplitude of the scattered light field can be expressed by

$$f(\mathbf{k}', \mathbf{k}) = \frac{k^2(\varepsilon_0 - 1)}{4\pi}\int_V U(\mathbf{r}')e^{-i\mathbf{k}'\cdot\mathbf{r}'}d^3\mathbf{r}' + \frac{Nk^2}{4\pi}\int_{V_{\text{cell}}} \Delta\varepsilon(\mathbf{r}')U(\mathbf{r}')e^{-i\mathbf{k}'\cdot\mathbf{r}'}d^3\mathbf{r}', \tag{5.15}$$

where V_{cell} is the volume of the lattice unit cell. Taking the first order Born approximation [103] and $U^{(i)}(\mathbf{r}) = \exp(i\mathbf{k}\cdot\mathbf{r})$, (5.15) can be written as

$$f(\mathbf{k}', \mathbf{k}) = \frac{k^2(\varepsilon_0 - 1)}{4\pi}\delta(\mathbf{k} - \mathbf{k}') + \frac{Nk^2}{4\pi}\int_{V_{\text{cell}}} \Delta\varepsilon(\mathbf{r}')e^{i(\mathbf{k}-\mathbf{k}')\cdot\mathbf{r}'}d^3\mathbf{r}'. \tag{5.16}$$

Suppose now that there are n scatterers in an unit cell, and the center location of the jth scatterer is denoted by $\mathbf{r}_j$, then the dielectric constant distribution in the unit cell can be written as $\Delta\varepsilon = \Sigma\Delta\varepsilon_j$. For $\Delta\mathbf{k} = \mathbf{k} - \mathbf{k}' \neq 0$ and $\boldsymbol{\rho} = \mathbf{r}' - \mathbf{r}_j$ we obtain

$$f(\mathbf{k}', \mathbf{k}) = \frac{Nk^2}{4\pi}\sum_j^n f_j e^{i\Delta\mathbf{k}\cdot\mathbf{r}_j}, \tag{5.17a}$$

$$f_j = \int_{V_{\text{cell}}} \Delta\varepsilon_j(\boldsymbol{\rho})e^{i\Delta\mathbf{k}\cdot\boldsymbol{\rho}}dV. \tag{5.17b}$$

By writing $\mathbf{r}_j$ as $\mathbf{r}_j = x_j\mathbf{a}_1 + y_j\mathbf{a}_2 + z_j\mathbf{a}_3$ ($\mathbf{a}_1$, $\mathbf{a}_2$, $\mathbf{a}_3$ are the lattice vectors) and considering Bragg condition $\Delta\mathbf{k} = m\mathbf{b}_1 + n\mathbf{b}_2 + l\mathbf{b}_3$ (m, n, l are integers, and

$\mathbf{b}_1$, $\mathbf{b}_2$, $\mathbf{b}_3$ are the reciprocal lattice vectors), we can define the GSF for photonic lattices as

$$\& = \sum_{j}^{n} f_j e^{i2\pi\left(mx_j + ny_j + lz_j\right)}. \tag{5.18}$$

From (5.18), it can be seen that the GSF for photonic lattices shares a similar form to that of atomic lattices [86]. However, the amplitude of the scattered field f_j is always positive for atomic lattices (since the scattering of X-rays is attributed to the electron clouds of the atoms, therefore X-ray cannot experience the difference between positive and negative scattering potentials), whereas for the ionic-type photonic lattices, f_j can be positive or negative indicating there is a π phase difference between the scattered light fields from negative and positive optical potentials. In order to highlight this unique property, we introduce a modified GSF for ionic-type photonic lattices in the form of

$$\& = \sum f_{j+} e^{i2\pi\left(mx_{j+} + ny_{j+} + lz_{j+}\right)} + \sum f_{j-} e^{i2\pi\left(mx_{j-} + ny_{j-} + lz_{j-}\right)}, \tag{5.19}$$

where the "+" and "−" signs are for positive and negative lattice potentials, respectively.

Next, let us use the above modified GSF to analyze the missing lines in the BZ spectra as shown in Fig. 5.13, which are attributed to the coexistence of the positive and negative potentials in ionic-type lattices. By setting the origin at the center of the white circles shown in the top panels of Fig. 5.13b–d, we can use the GSF to analyze the Bragg reflection in the optically induced photonic lattices. Specifically, the lattice structure shown in Fig. 5.13b contains a positive and a negative potential with identical potential depth (i.e., $f_{j+} = -f_{j-}$) centered at $(x_-, y_-) = (0, 0)$ and $(x_+, y_+) = (½, 0)$, respectively. Then, from (5.19), it is easy to show that if m is an even number, the GSF & would be 0, which leads to zero amplitude of scattered light. Therefore, the Bragg reflection lines corresponding to **k**-space planes defined by $(m, n) = (0, \pm1)$, $(0, \pm2)$, $(\pm2, \pm1)$, $(\pm2, 0)$ are all missing in Fig. 5.13b. Likewise, for the lattice shown in Fig. 5.13c, Bragg reflection lines corresponding to $(m, n) = (\pm1, \pm1)$ are missing as in this case $\& = 0$ if $m + n$ ends up an even number. The lattice in Fig. 5.3d has a unit cell containing two negative potentials at $(x_{1-}, y_{1-}) = (-½, ½)$ and $(x_{2-}, y_{2-}) = (½, -½)$, and two positive potentials at $(x_{1+}, y_{1+}) = (½, ½)$ and $(x_{2+}, y_{2+}) = (-½, -½)$. The GSF in this case is found to be 0 (thus cancelled Bragg reflection) for the planes of $(m, n) = (0, 1)$, $(1, 0)$, $(-1, 0)$ and $(0, -1)$ (see Fig. 5.13d). Missing lines are also found for the nonionic-type lattice of Fig. 5.13a, but they result from a different mechanism of the anisotropic photorefractive nonlinearity. In this latter case, the induced index changes (waveguides) are highly elliptical and somewhat connected to adjacent lattice sites, thus the 2D lattice virtually turns into a quasi-1D lattice along a particular direction (−45° diagonal in Fig. 5.13a) for which Bragg reflection is weak or absent.

This anisotropy of the lattice potential should also play a role for the missing lines in Fig. 5.13b, as explained in ref. [100]. However, our numerical calculation indicates that the ionic nature of the lattice structure accounts more for these missing lines in Fig. 5.13b, as these lines can still be visible if we simply remove the positive lattice potentials. We thus emphasize that missing of the Bragg reflection lines in BZ spectra and associated band-gap closures are fundamentally new features in ionic-type photonic lattices. Note that the lattice structure shown in Fig. 5.13d represents a typical separable (or egg-crate) lattice structure in BEC, which has been investigated theoretically for dynamical Bloch oscillation of cold atoms [105, 106]. We expect that such Bragg reflection control could occur in other ionic-type photonic lattices with different symmetries including triangular, honeycomb, and quasi-crystal structures. Furthermore, the concept of ionic-type lattices may have direct impact on photonic crystals and other discrete systems beyond optics.

5.6 Nonlinear Discrete Light Behaviors

As shown in Sect. 5.5, hybrid nonlinearity enables the ability to achieve reconfigurable photonic lattices to modulate the linear beam dynamics. In the nonlinear regime, it is well known that discrete (gap) solitons can be formed by balancing normal (anomalous) diffraction with self-focusing (self-defocusing) nonlinearity [47, 48, 61, 62]. Considering the highly anisotropic discrete diffraction behaviors [107] and the enhanced anisotropy of the hybrid nonlinearity, one would expect novel nonlinear beam dynamics to occur in optically induced photonic lattices under NCB conditions [93, 96–98].

5.6.1 Transition Between Discrete and Gap Soliton-Trains

Similar to Sect. 5.4.1, as the simplest example, we first demonstrate the orientation-induced transition between a 1D discrete soliton under self-focusing nonlinearity and a 1D gap soliton under self-defocusing nonlinearity without the need of reversing the bias field [96]. Again, the same bias conditions as those used in Sect. 5.4.1 are employed, with $(\theta_e - \theta_c) = \pi/2$. By solving (5.7) with $I = |B(y,z)|^2 + I_{l0}\cos^2(\pi y/d)$ and $B(y,z) = b(y)\exp(i\beta z)$, the soliton solutions at different beam orientations are obtained as shown in Fig. 5.19a, b. As expected, discrete and gap solitons can be supported in the same crystal under identical bias conditions. The propagation constants β for the discrete solitons lie in the total internal reflection (or semi-infinite) gap, while those for the gap solitons reside in the first Bragg-reflection bandgap. All the discrete solitons are uniform in phase, while the gap solitons are always accompanied by staggered phase structures.

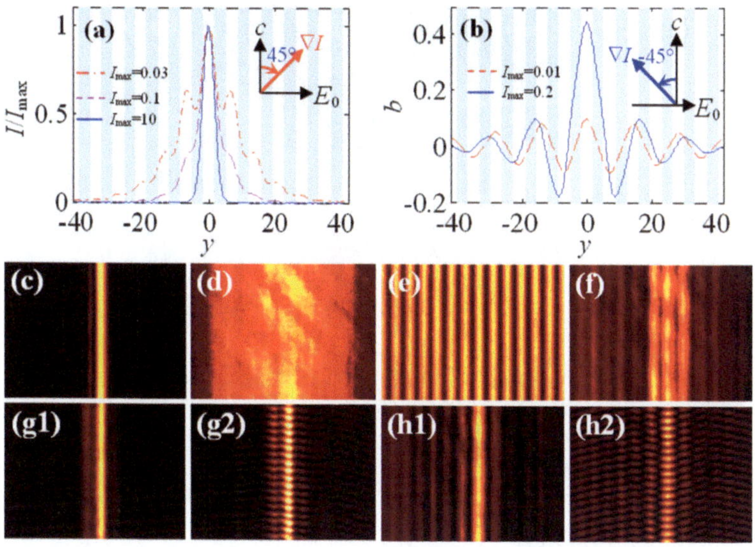

Fig. 5.19 Numerical (**a**, **b**) and experimental (**c–h**) demonstrations of orientation-induced transition between 1D discrete and gap solitons under self-focusing (**a**, **g**) and self-defocusing nonlinearity (**b**, **h**). (**a**, **b**) Soliton profiles at different peak intensities under bias conditions as indicated in the *upper right insets*; (**c**, **d**) input and linear output probe beams without lattice; (**e**) lattice beam pattern; (**f**) linear discrete diffraction of the probe beam; (**g**, **h**) are showing output soliton beams (1) and their interferograms with tilted plane waves (5.2) (after ref. [96])

Experimental demonstrations of these orientation-induced transitions between discrete and gap solitons are shown in Fig. 5.19c–h, where the photonic lattices are optically induced in an SBN:60 crystal under hybrid nonlinearity ($E_0 \perp c$). The lattice beam and the Gaussian probe beam both can be rotated freely in the transverse plane. For the convenience of comparison, the experimentally observed beam patterns are all rotated in transverse plane by 45°. It can be seen that without nonlinearity, the probe beam undergoes discrete diffraction (Fig. 5.19f). However, in the presence of nonlinearity, the balance between self-focusing or self-defocusing and discrete diffraction can leads to stable self-trapped states (Fig. 5.19g, h). Here, the interferograms clearly illustrate that the discrete and gap solitons possess uniform and staggered phase structures, respectively. These observations are in good agreement with the theoretical expectations.

Similarly, if the 1D lattice beams are replaced by the 2D square lattice beams these orientation-induced transitions can be extended directly to discrete [108] and gap soliton trains [55]. Figure 5.20 depicts typical numerical and experimental results. The normalized power curves of the soliton trains at different orientations are depicted in Fig. 5.20a, where the power is defined over one period along the train direction [55, 108]. Figure 5.20b corresponds to the soliton solutions in the semi-infinite gap (curve labeled with b in Fig. 5.20a) under self-focusing nonlinearity, which is bifurcated from the Γ_1 point. Figure 5.20c corresponds to the gap soliton trains residing in the first photonic gap (curve labeled with c in Fig. 5.20a) under self-defocusing

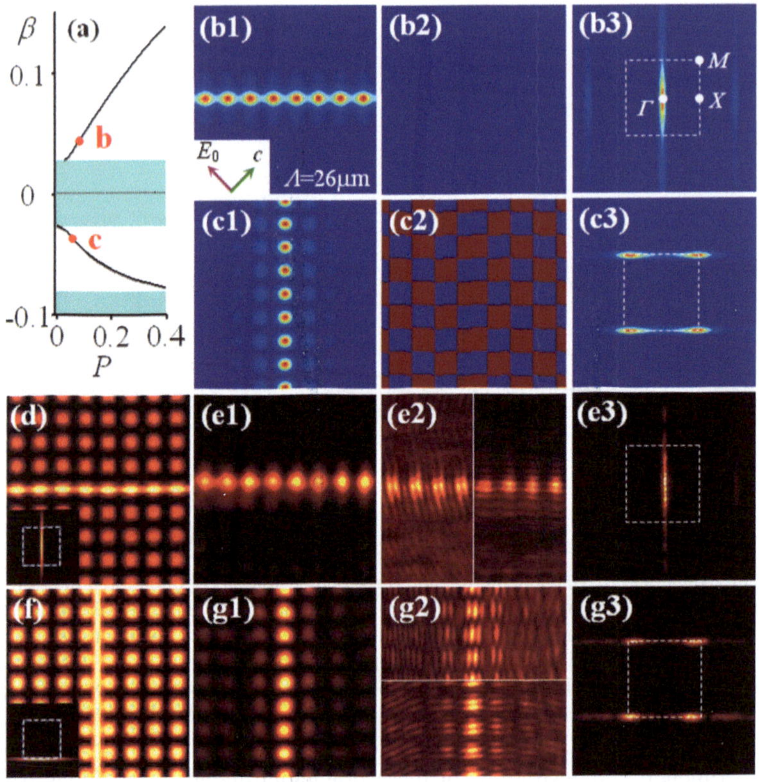

Fig. 5.20 Theoretical (**a–c**) and experimental (**d–g**) demonstrations of orientation-induced transition between discrete and gap soliton trains bifurcated from the Γ-point under self-focusing nonlinearity (**b**, **d**, **e**) and those bifurcated from the M-point under self-defocusing nonlinearity (**c**, **f**, **g**). (**a**) Existence curves for soliton trains bifurcated from the band-edge of the first Bloch band (Bloch bands are *shaded*); (**b**, **c**) Soliton solutions at corresponding marked points in (**a**), showing the soliton intensity patterns (1), phase structures (2), and Fourier spectra (3); (**d**, **f**) Superimposed input intensity pattern of the stripe and lattice beams (*insets* show the spectra of the probe beam excitation); (**e**, **g**) Output intensity pattern of the soliton trains (1), interferogram from two orthogonal directions (2), and soliton output spectra (3). *Added squares* in (**b–g**) depict the boundary of the first BZ (after ref. [97])

nonlinearity. These are bifurcated from the M_1 point. It is clear that although the intensity patterns of these soliton trains look somewhat similar, the phase structures and spectra indicate that they have different origins under different nonlinearities.

In an experiment, to obtain the semi-infinite gap soliton trains, the stripe beam is oriented horizontally and launched collinearly with the lattice beam (without input tilt) into the lattice as shown in Fig. 5.20d. In this case, the stripe beam experiences a self-focusing nonlinearity balancing the normal diffraction to form a discrete soliton train (Fig. 5.20e1), whose uniform phase structure (Fig. 5.20e2) and **k**-space spectrum (Fig. 5.20e3) indicate its constitute modes from the Γ point in the first band. To obtain the first photonic gap soliton train under the same bias condition, the stripe beam is oriented vertically and launched off-site with an input tilt at a Bragg angle as shown in Fig. 5.20f. In this case, the self-defocusing nonlinearity experienced by the probe beam balances anomalous diffraction to form a gap soliton train (Fig. 5.20g). The interferogram shown in Fig. 5.20g illustrates the "staggered" phase along the two orthogonal directions, representing the characteristic phase structure of Bloch modes from the M_1 points [101]. Due to nonlinear spectrum reshaping, the output spectrum of the gap soliton differs dramatically from the input spectrum, with most of the energy concentrated in regions around the four M points in the first BZ [55]. These observations perfectly match the theoretical results as shown in Fig. 5.20b and c .

5.6.2 Elliptical Discrete Solitons

In Sect. 5.4.2, it was shown that the enhanced anisotropy of hybrid nonlinearity results in elliptical solitons with nontrivial ellipticity. Here, we study elliptical discrete solitons in the presence of both a periodic optical potential and hybrid nonlinearity [93]. By solving (5.4) and (5.8) with $I = |b(x,y)|^2 + \cos^2(\pi x/\Lambda)\cos^2(\pi y/\Lambda)$, the elliptical soliton solutions at different bias conditions can be obtained. Typical results under $E_0 \perp c$ are shown in Fig. 5.21a–d, from which it is clear that the ellipticity of the solitons will be dramatically increased due to the enhanced anisotropy. This is in contrast to the discrete solitons formed under conventional bias conditions. In addition, the orientations of the elliptical solitons can be altered by changing the relative orientations of lattice beam, bias direction, and the c-axis. Figure 5.21e–h presents an experimental demonstration of such elliptical solitons. It is obvious that under linear condition, the input beam undergoes discrete diffraction. While in the presence of nonlinearity, however, the probe beam can indeed evolve into an elliptical self-trapped state. It should be noted that the soliton propagation constants of the elliptical discrete solitons reside in the semi-infinite gap and the soliton profiles possess uniform phase structures.

5.6.3 Saddle Solitons

In a 2D square lattice, the high-symmetry X-point in the first Bloch band, as shown in Fig. 5.22, is akin to a saddle point in a diffraction spectrum, where normal and anomalous diffractions co-exist along orthogonal directions [107]. Specifically, along the M–X (Γ–X) direction, the diffraction curve is convex (concave), indicating normal (anomalous) diffraction. Therefore, to form a 2D soliton at such a saddle point, orientation-dependent hybrid nonlinearity is required to balance the normal and anomalous diffractions simultaneously. This is somewhat similar to nonlinear X-waves [109] and light bullets [110] balancing beam diffraction and pulse dispersion at the same time. Synchronous compensation of normal

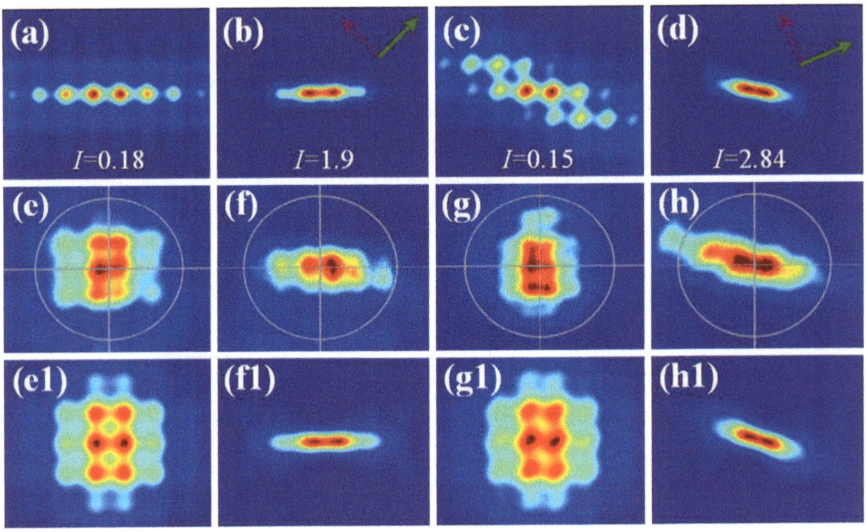

Fig. 5.21 Numerical and experimental demonstrations of elliptical discrete solitons under hybrid nonlinearity. Soliton solutions with different peak intensities at $\theta_c = \pi/4$, $\theta_e = 3\pi/4$ (**a**, **b**) and $\theta_c = \pi/8$, $\theta_e = 5\pi/8$ (**c**, **d**); (**e**, **g**) linear discrete diffraction of probe beam; (**f**, **h**) soliton output. The *bottom row* presents the numerical simulations for the experimental results shown in the *middle row*, and the bias conditions for (**e–f**) and (**g–h**) are corresponding to that for (**a–b**) and (**c–d**), respectively (after ref. [93])

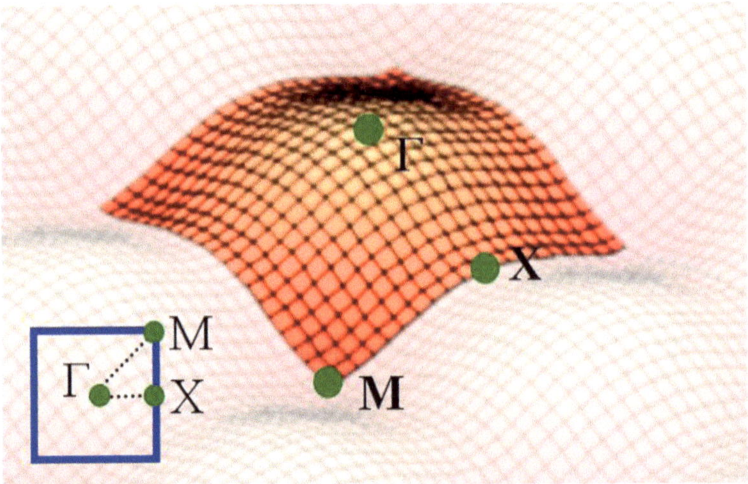

Fig. 5.22 Extended first Bloch band with high symmetry points marked of a typical square photonic lattice (after ref. [98])

and anomalous diffractions at the X point is impossible in conventional nonlinear media. However, quasi-1D saddle solitons have been successfully excited by balancing the anomalous diffraction along the Γ–X direction, while it is an extended plane wave in the orthogonal M–X direction. The propagation constants of these 1D soliton trains could reside within the first Bloch band, thus named "in-band" or "embedded" solitons [58]. In the following, we shall demonstrate the transition between self-focusing and self-defocusing quasi-1D (in-band) saddle solitons [97] and the 2D saddle solitons in the presence of hybrid nonlinearity [98].

5.6.3.1 Transition Between Self-Focusing and -Defocusing In-Band Solitons

By employing a numerical procedure and experimental setting similar to the one used in Sect. 5.6.1, the orientation-induced transition between quasi-1D saddle solitons under self-focusing and self-defocusing nonlinearities is demonstrated, as shown in Fig. 5.23. From the power curves of the soliton trains depicted in Fig. 5.23a, it is obvious that such quasi-1D saddle solitons are indeed bifurcated from the sub-band-edges (corresponding to the interior X point). When the propagation constant is embedded in the Bloch band (e.g., the solutions at the points b and c in Fig. 5.23a), the solutions represent in-band or embedded solitons [58]. Figure 5.23b, c depicts the intensity patterns, phase structures, and **k**-space spectra of typical soliton solutions corresponding to the marked points in Fig. 5.23a. It is clear that there is no qualitative difference in intensity and phase structure between the two cases of embedded solitons. However, the orientation of the soliton stripe is rotated 90° in transverse plane as supported by two different nonlinearities (self-focusing and self-defocusing) and the spectra near the two X points are somewhat different due to different initial excitations.

Experimental demonstrations of such transitions are shown in Fig. 5.23. To selectively excite the in-band solitons with a self-focusing nonlinearity, the stripe beam is oriented horizontally, but launched with an input tilt at a Bragg angle into the lattice as shown in Fig. 5.23d. Both the phase structure (Fig. 5.23e2) and the **k**-space spectrum (Fig. 5.23e3) clearly illustrate the excitation of Bloch modes at the X point of the first band. In this case, the normal diffraction of the stripe beam along y-direction is balanced by the self-focusing nonlinearity, so the soliton is quasi-one-dimensional in x-direction. Likewise, the in-band soliton train supported by the self-defocusing nonlinearity can be generated by off-site excitation of a vertically oriented stripe beam without input tilt as shown in Fig. 5.23f. Again the excitation of modes from the X points is evident in both phase and spectrum measurements, but now the soliton is quasi-one-dimensional in y-direction, and the anomalous diffraction of the stripe beam is balanced by the self-defocusing nonlinearity [101]. As the soliton power increases, the embedded soliton will turn into a semi-infinite gap soliton under self-focusing nonlinearity (moving up to out of the band along b branch in Fig. 5.23a) and into a photonic gap soliton under the self-defocusing nonlinearity (moving down to out of the band along c branch in Fig. 5.23a) [58]. These solitons have also been experimentally observed.

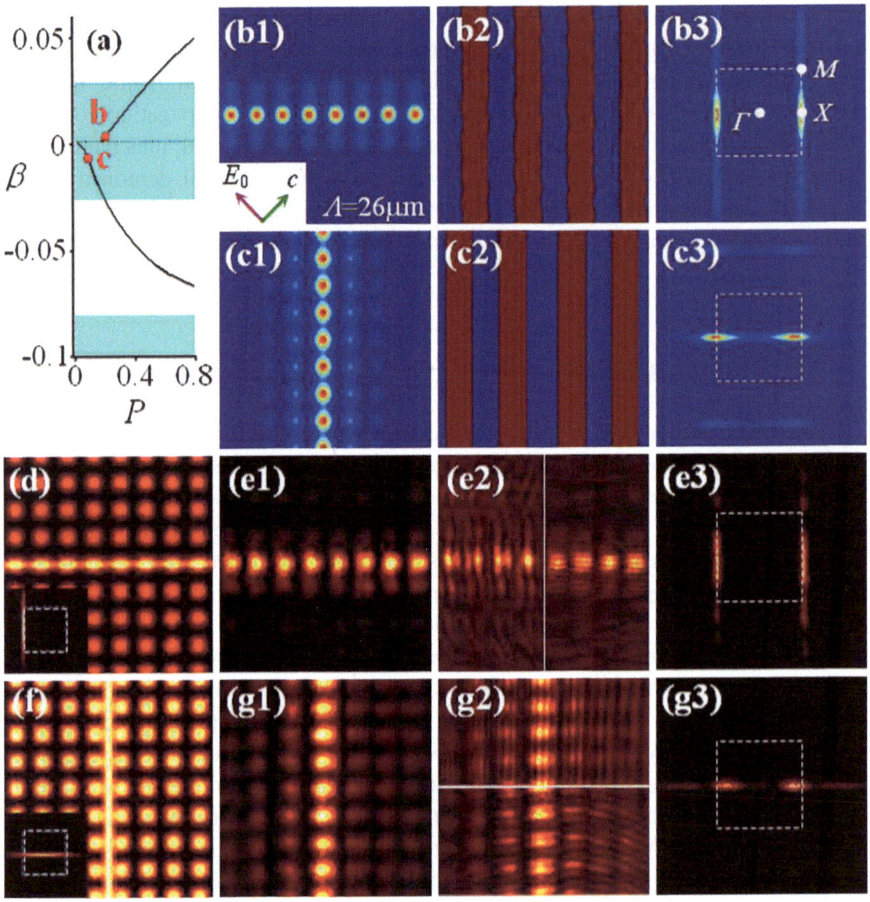

Fig. 5.23 Theoretical (**a–c**) and experimental (**d–g**) demonstrations of orientation-induced transition between in-band soliton trains bifurcated from the sub-band-edge X point under self-focusing (**b**, **d**, **e**) and -defocusing (**c**, **f**, **g**) nonlinearity. (**a**) Existence curves of the soliton-train solutions; (**b**, **c**) soliton-trains at corresponding marked points in (**a**), showing intensity patterns (1), phase structures (2), and Fourier spectra (3); (**d**, **f**) superimposed input intensity pattern of the stripe and lattice beams (*insets* show the spectra of the probe beam excitation); (**e**, **g**) output of the soliton trains (1), interferogram from two orthogonal directions (2), and soliton output spectra (3). *Added squares* in (**b–g**) depict the boundary of the first BZ (after ref. [97])

5.6.3.2 Two-Dimensional Saddle Solitons

By solving (5.4) and (5.8) with different initial conditions from those used in Sect. 5.6.2, 2D saddle solitons residing in the first Bragg reflection gap can be obtained as shown in Fig. 5.24. From Fig. 5.24d–f, it can be clearly seen that the phase structure of the central soliton region (Fig. 5.24e) fits that of the Bloch

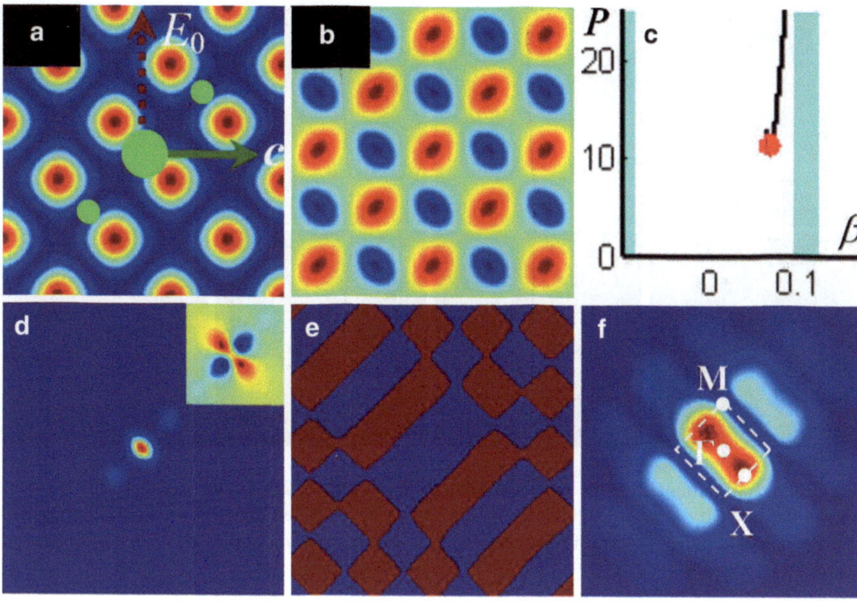

Fig. 5.24 (**a**) Orientations of crystalline *c*-axis, bias field, and square lattice-inducing beam; (**b**) refractive index distribution of the induced ionic-type lattice; (**c**) existence curve of 2D saddle solitons; (**d–f**) intensity pattern (associated index change shown in *inset*), phase structure, and Fourier spectrum of the saddle soliton at the marked point in (**c**). *Added squares* in (**f**) mark the boundary of the first BZ (after ref. [98])

mode at the X-point, and that the power spectrum (Fig. 5.24f) is concentrated mostly at the two X-points within the first BZ. This indicates that the saddle solitons must be bifurcated from the interior X-points of the first band, although far away from the soliton center staggered phase characteristic to band-edge M-point modes are evident. Apparently, these 2D saddle solitons differ from previously observed quasi-1D embedded or gap solitons (also from first-band X-points), which are localized only in one direction [58], and from the reduced symmetry solitons created solely by self-focusing nonlinearity (but from second-band X-points) [111]. We emphasize that such a 2D saddle soliton does not arise from single (self-focusing or -defocusing) nonlinearity or single (normal or anomalous) diffraction, but rather it results from a perfect balance between bi-diffraction and hybrid nonlinearity.

Experimental observations of such 2D saddle solitons are depicted in Fig. 5.25. To excite the 2D saddle soliton originated from the X-points, the relative orientation between the *c*-axis, the bias field, and the lattice beam is as depicted in Fig. 5.24a. The probe beam is reconfigured into three in-phase spots and launched into the lattice without any input tilt (illustrated by three green dots in Fig. 5.24a), with the central spot having a higher intensity. To make sure on-site excitation in the induced index lattice (which has an off-set with the intensity pattern as seen in Fig. 5.24a, b), the three beam spots are aimed at the intensity minima of the lattice beam. The experimentally observed intensity pattern of the saddle soliton is

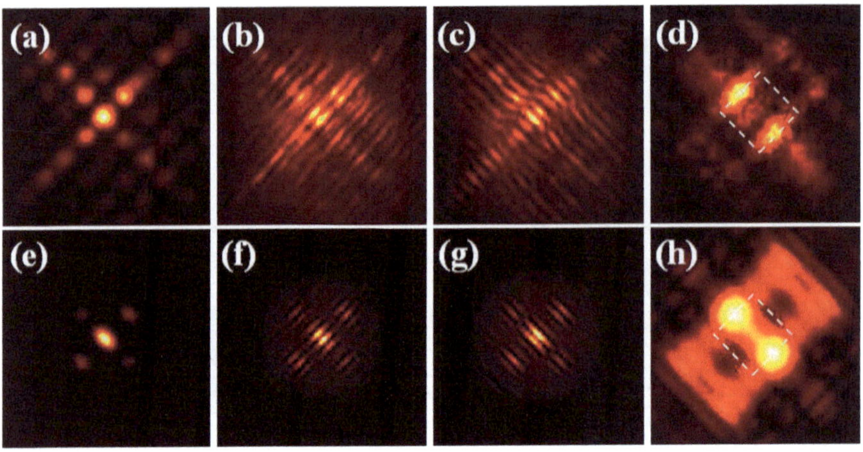

Fig. 5.25 Experimental (**a–d**) and numerical (**e–h**) results of 2D saddle solitons. The *top row* shows the intensity pattern (**a**), interferograms (**b**, **c**) with a tilted plane wave at two orthogonal directions, and Fourier spectrum (**d**) of the soliton. The *second row* shows corresponding numerical results (after ref. [98])

shown in Fig. 5.25a. From the interference patterns between the soliton beam and the reference plane wave tilted to two different directions (see Fig. 5.25b, c), we can tell that the soliton possesses uniform and "staggered" phase structure along two orthogonal directions. This kind of phase structure resembles that of the Bloch modes at the first-band X-points, as found in our numerical solution of Fig. 5.24e. And the **k**-space spectrum of the soliton also shows that most of its power concentrates at the two X-points of the first BZ (Fig. 5.25d). These results are also corroborated with our numerical simulations (shown in Fig. 5.25e–h). This confirms that the 2D saddle solitons are indeed due to a perfect balance between bi-diffraction and hybrid nonlinearity at the interior X-points of the first band. In addition, we mention that when the intensity of the lattice-inducing beam is reduced, the band-gap of the induced photonic lattice becomes narrower or not fully open, as found before for "backbone" photonic lattices [112]. In this case, quasi-localized saddle soliton solutions can exist with their propagation constant residing even in the Bloch band. However, these 2D in-band modes are only quasi-localized. Long distance propagation reveals that they are not stable [98].

5.7 Summary and Closing Remarks

In summary, we have reviewed recent progresses on the nonlinear beam dynamics in both continuous and discrete regimes under hybrid nonlinearity. A comprehensive theoretical model has been developed based on Kukhtarev's band transport model to describe the nonlinear beam dynamics as well as optically induced photonic lattices driven by the hybrid nonlinearity in NCB photorefractive crystals.

The direct experimental visualization of the complex nonlinear refractive index changes confirms the validity of the theoretical model in the first place. In homogenous bulk crystals, nonlinear beam dynamics including creation of elliptical optical solitons, suppression of the breakup of optical vortices, and orientation-induced transitions between bright and dark solitons have been demonstrated. In discrete media, hybrid nonlinearity enables the creation of a new type of photonic lattices with alternating positive and negative refractive index potentials, named ionic-type photonic lattices. These offer new opportunities for band-gap engineering and control of the flow of light by reconfiguring the lattice structures and BZs. Some typical examples include band-gap closure and Bragg reflection suppression, soliton transitions between different band-gaps, and interplays between normal and anomalous diffraction or positive and negative refraction under identical excitation conditions. Moreover, the enhanced anisotropy and nonlocality of hybrid nonlinearity result in some novel nonlinear discrete localized states including elliptical discrete solitons and saddle solitons by simultaneously balancing normal and anomalous diffractions along different directions.

In closing this chapter, we mention that the hybrid nonlinearity is directly originated from the anisotropic nature of photorefractive effect. Photorefractive nonlinearity has played a major role in experimentally exploring nonlinear beam dynamics in both homogenous and discrete regimes in the past two decades. Although elliptical solitons and their anomalous interactions due to the weak photorefractive anisotropy under conventional bias conditions have been studied, most of the previous studies were based on an isotropic and local approximation of the photorefractive model. Therefore, hybrid nonlinearity supported by the NCB conditions brings about opportunities for exploring novel nonlinear beam dynamics under strong anisotropy and nonlocality in both continuum and discrete regimes. More generally, the novel concept of hybrid nonlinearity opens a door to study the nonlinear beam dynamics in anisotropic nonlinear systems in optics and beyond. Specifically, this raises many intriguing questions, such as, would the asymmetric diffusive nonlinearity in photorefractive materials lead to self-bending of the anisotropic self-trapping states? If so, then how would solitons move in presence of hybrid nonlinearity? Is it possible to use the enhanced nonlocality of hybrid nonlinearity to fulfill wave mixing without diffusive nonlinearity? How does hybrid nonlinearity impact the modulation instability? Would other nonlinear phenomena, such as spontaneous pattern formation, self-pumped phase conjugation, exist under hybrid nonlinearity? In the presence of photonic lattices, would the enhanced anisotropy lead to symmetry breaking light propagation behaviors? What's the influence of the anisotropy on the nonlinear optical surface states? Our setting will help explore these concepts and understand the true universal nature of anisotropic nonlinear media. We expect more interesting and exciting results that are yet to come.

Acknowledgments This work was supported by US National Science Foundation, the Air Force Office of Scientific Research, and by the NPU Foundation for Fundamental Research, the Doctorate Foundation of NPU, the 973 Program, the 111 Project, NSFC, and PCSIRT in China. We are indebted to many of our collaborators over the years including F. Xiao, X. Gan, Q. Liu, Y. Gao, X. Tan, D. Yang, X. Wang, and J. Yang. We thank A. Miller for proofreading the chapter.

References

1. M.J. Ablowitz, P.A. Clarkson, *Solitons, Nonlinear Evolution Equations, and Inverse Scattering* (Cambridge University, Cambridge, 1991)
2. F. Abdullaev, V.V. Konotop, *Nonlinear Waves: Classical and Quantum Aspects* (Kluwer, New York, 2004)
3. E. Infeld, G. Rowlands, *Nonlinear Waves, Solitons, and Chaos* (Cambridge University, Cambridge, 1990)
4. J. Pedlosky, *Waves in the Ocean and Atmosphere: Introduction to Wave Dynamics* (Springer, Berlin, 2010)
5. A.S. Davydov, *Solitons in Molecular Systems* (Kluwer, Netherlands, 1991)
6. Y.R. Shen, *The Principles of Nonlinear Optics* (Wiley, New York, 2003)
7. R.W. Boyd, *Nonlinear Optics* (Academic, New York, 2008)
8. R.A. Rupp, F.W. Drees, Appl. Phys. B **39**, 223 (1986)
9. J.P. Huignard, A. Marrakchi, Opt. Commun. **38**, 249 (1981)
10. J. Feinberg, Opt. Lett. **7**, 486 (1982)
11. R.Y. Chiao, E. Garmire, C.H. Townes, Phys. Rev. Lett. **13**, 479 (1964)
12. G.I. Stegeman, M. Segev, Science **286**, 1518 (1999)
13. H. Hasegawa, Y. Kodama, *Solitons in Optical Communications* (Oxford University, New York, 1995)
14. Y.S. Kivshar, G.P. Agrawal, *Optical Solitons: From Fibers to Photonic Crystals* (Academic, San Diego, 2003)
15. M.D.I. Castillo, P.A. Marquez, J.J. Sanchez-Mondragon, S. Stepanov, V. Vysloukh, Appl. Phys. Lett. **64**, 408 (1994)
16. A. Barthelemy, S. Maneuf, C. Frcehly, Opt. Commun. **55**, 201 (1985)
17. J.E. Bjorkholm, A.A. Ashkin, Phys. Rev. Lett. **32**, 129 (1974)
18. M. Peccianti, C. Conti, G. Assanto, A. De Luca, C. Umeton, Nature **432**, 733 (2004)
19. C. Rotschild, O. Cohen, O. Manela, M. Segev, T. Carmon, Phys. Rev. Lett. **95**, 213904 (2005)
20. P. Günter, J.P. Huignard, *Photorefractive Materials and Their Applications I* (Springer, Berlin, 1988)
21. M. Segev, B. Crosignanim, A. Yariv, B. Fischer, Phys. Rev. Lett. **68**, 923 (1992)
22. G.C. Duree Jr., J.L. Shultz, G. Salamo, M. Segev, A. Yariv, B. Crosignani, P. Di Porto, E.J. Sharp, R.R. Neurgaonkar, Phys. Rev. Lett. **71**, 533 (1993)
23. W. Królikowski, B. Luther-Davies, C. Denz, IEEE J. Quantum Elect. **39**, 3 (2003)
24. M. Segev, G.C. Valley, B. Crosignani, D. Di Porto, A. Yariv, Phys. Rev. Lett. **73**, 3211 (1994)
25. D.N. Christodoulides, M.I. Carvalho, J. Opt. Soc. Am. B **12**, 1628 (1995)
26. A. Guo, M. Henry, G.J. Salamo, M. Segev, G.L. Wood, Opt. Lett. **26**, 1274 (2001)
27. M. Shih, P. Leach, M. Segev, M.M. Garrett, G. Salamo, G.C. Valley, Opt. Lett. **21**, 324 (1996)
28. M. Taya, M.C. Bashaw, M.M. Fejer, M. Segev, G.C. Valley, Phys. Rev. Lett. **52**, 3095 (1995)
29. Z. Chen, M. Mitchell, M. Shih, M. Segev, M.H. Garrett, G.C. Valley, Opt. Lett. **21**, 629 (1996)
30. Z. Chen, M. Segev, D.W. Wilson, R.E. Muller, P.D. Maker, Phys. Rev. Lett. **78**, 2948 (1997)
31. Z. Chen, M. Shih, M. Segev, D.W. Wilson, R.E. Muller, P.D. Maker, Opt. Lett. **22**, 1751 (1997)
32. Z. Chen, M. Segev, T.H. Coskun, D.N. Christodoulides, Opt. Lett. **21**, 1436 (1996)
33. M. Mitchell, Z. Chen, M. Shih, M. Segev, Phys. Rev. Lett. **77**, 490 (1996)
34. Z. Chen, M. Mitchell, M. Segev, T.H. Coskun, D.N. Christodolides, Science **280**, 889 (1998)
35. G.S. Garacía-Quirino, M.D.I. Castillo, V.A. Vysloukh, J.J. Sánchez-Mondragón, S.I. Stepanov, G. Lugo-Martínez, G.E. Torres-Cisneros, Opt. Lett. **22**, 154 (1997)
36. W. Krolikowski, S.A. Holmstrom, Opt. Lett. **22**, 369 (1997)

37. M. Shih, Z. Chen, M. Segev, T.H. Coskun, D.N. Christodoulides, Appl. Phys. Lett. **69**, 4151 (1996)
38. M. Shih, M. Segev, G. Salamo, Phys. Rev. Lett. **78**, 2551 (1997)
39. W. Krolikowski, M. Saffman, B. Luther-Davies, C. Denz, Phys. Rev. Lett. **80**, 3240 (1998)
40. D.N. Christodoulides, F. Lederer, Y. Silberberg, Nature **424**, 817 (2003)
41. F. Lederer, G.I. Stegeman, D.N. Christodoulides, G. Assanto, M. Segev, Y. Silberberg, Phys. Rep. **463**, 1 (2008)
42. D. Mandelik, H.S. Eisenberg, Y. Silberberg, R. Morandotti, J.S. Aitchison, Phys. Rev. Lett. **90**, 053902 (2003)
43. H.S. Eisenberg, Y.S. Silberberg, R. Morandotti, J.S. Aitchison, Phys. Rev. Lett. **85**, 1863 (2000)
44. T. Pertsch, T. Zentgraf, U. Peschel, A. Bräuer, F. Lederer, Phys. Rev. Lett. **88**, 093901 (2002)
45. H. Trompeter, W. Krolikowski, D.N. Neshev, A.S. Desyatnikov, A.A. Sukhorukov, Y.S. Kivshar, T. Pertsch, U. Peschel, F. Lederer, Phys. Rev. Lett. **96**, 053903 (2006)
46. K. Shandarova, C.E. Rüter, D. Kip, K.G. Makris, D.N. Christodoulides, O. Peleg, M. Segev, Phys. Rev. Lett. **102**, 123905 (2009)
47. H.S. Eisenberg, Y. Silberberg, R. Morandotti, A.R. Boyd, J.S. Aitchison, Phys. Rev. Lett. **81**, 3383 (1998)
48. D. Mandelik, R. Morandotti, J.S. Aitchison, Y. Silberberg, Phys. Rev. Lett. **92**, 093904 (2004)
49. N.K. Efremidis, S. Sears, D.N. Christodoulids, J.W. Fleischer, M. Segev, Phys. Rev. E. **66**, 046602 (2002)
50. Z. Chen, K. MacCarthy, Opt. Lett. **27**, 2019 (2002)
51. C.R. Rosberg, D.N. Neshev, A.A. Sukhorukov, Y.S. Kivshar, W. Krolikowski, Opt. Lett. **30**, 2293 (2005)
52. I. Makasyuk, Z. Chen, J. Yang, Phys. Rev. Lett. **96**, 223903 (2006)
53. T. Schwartz, G. Bartal, S. Fishman, M. Segev, Nature **446**, 52 (2007)
54. X. Wang, Z. Chen, P.G. Kevrekidis, Phys. Rev. Lett. **96**, 083904 (2006)
55. C. Lou, X. Wang, J. Xu, Z. Chen, J. Yang, Phys. Rev. Lett. **98**, 213903 (2007)
56. D.N. Neshev, T.J. Alexander, E.A. Ostrovskaya, Y.S. Kivshar, H. Martin, I. Makasyuk, Z. Chen, Phys. Rev. Lett. **92**, 123903 (2004)
57. J.W. Fleischer, G. Bartal, O. Cohen, O. Manela, M. Segev, J. Hudock, D.N. Christodoulides, Phys. Rev. Lett. **92**, 123904 (2004)
58. X. Wang, Z. Chen, J. Wang, J. Yang, Phys. Rev. Lett. **99**, 243901 (2007)
59. O. Cohen, G. Bartal, H. Buljan, T. Carmon, J.W. Fleischer, M. Segev, D.N. Christodoulides, Nature **433**, 500 (2005)
60. J.W. Fleischer, T. Carmon, M. Segev, N.K. Efremidis, D.N. Christodoulides, Phys. Rev. Lett. **90**, 023902 (2003)
61. J.W. Fleischer, M. Segev, N.K. Efremidis, D.N. Christodoulides, Nature **422**, 147 (2003)
62. H. Martin, E.D. Eugenieva, Z. Chen, D.N. Christodoulides, Phys. Rev. Lett. **92**, 123902 (2004)
63. C. Crognale, L. Rosa, J. Lightwave Technol. **23**, 2175 (2005)
64. P. Zhang, J. Zhao, C. Lou, X. Tan, Y. Gao, Q. Liu, D. Yang, J. Xu, Z. Chen, Opt. Express **15**, 536 (2007)
65. P. Zhang, S. Liu, C. Lou, Y. Gao, J. Zhao, J. Xu, Z. Chen, Appl. Phys. B **95**, 559 (2009)
66. N.V. Kukhtarev, V.B. Markov, S.G. Odulov, M.S. Soskin, V.L. Vinetskii, Ferroelectrics **22**, 949 (1979)
67. A.A. Zozulya, D.Z. Anderson, Phys. Rev. A **51**, 1520 (1995)
68. A. Yariv, P. Yeh, *Optical Waves in Crystals* (Wiley, New York, 1984)
69. P. Zhang, Y. Ma, J. Zhao, D. Yang, H. Xu, Appl. Opt. **45**, 2273 (2006)
70. J. Zhao, P. Zhang, J. Zhou, D.X. Yang, D.S. Yang, E. Li, Chin. Phys. Lett. **20**, 1748 (2003)
71. P. Zhang, J. Zhao, S. Liu, Y. Gao, C. Lou, J. Xu, Z. Chen, CLEO/QELS, OSA Technical Digest, CTuE5 (2008)

72. V.I. Petviashvili, Sov. J. Plasma Phys. **2**, 257 (1976)
73. A.A. Zozulya, D.Z. Anderson, A.V. Mamaev, M. Saffman, Phys. Rev. A **57**, 522 (1998)
74. A.A. Zozulya, D.Z. Anderson, A.V. Mamaev, M. Saffman, Europhys. Lett. **36**, 419 (1996)
75. L.M. Pismen, *Vortices in Nonlinear Fields: From Liquid Crystals to Superfluids, From Non-equilibrium Patterns to Cosmic Strings* (Oxford University, New York, 1999)
76. D.G. Grier, Nature **424**, 810 (2003)
77. A.S. Desyatnikov, Y.S. Kivshar, L. Torner, Prog. Opt. **47**, 291 (2005)
78. G.A. Swartzlander Jr., C.T. Law, Phys. Rev. Lett. **69**, 2503 (1992)
79. V. Tikhonenko, J. Christou, B. Luther-Daves, J. Opt. Soc. Am. B **12**, 2046 (1995)
80. D. Briedis, D.E. Petersen, D. Edmundson, W. Krolikowski, O. Bang, Opt. Express **13**, 435 (2005)
81. A.V. Mamaev, M. Saffman, A.A. Zozulya, Phys. Rev. Lett. **77**, 4544 (1996)
82. A.V. Mamaev, M. Saffman, A.A. Zozulya, Phys. Rev. Lett. **78**, 2108 (1997)
83. G. Duree, M. Morin, G. Salamo, M. Segev, B. Crosignani, P. Di Porto, E. Sharp, A. Yariv, Phys. Rev. Lett. **74**, 1978 (1995)
84. X. Gan, P. Zhang, S. Liu, Y. Zheng, J. Zhao, Z. Chen, Opt. Express **17**, 23130 (2009)
85. N.R. Heckenberg, R. McDuff, C.P. Smith, A.G. White, Opt. Lett. **17**, 221 (1992)
86. J.S. Blakemore, *Solid State Physics* (Cambridge University, Cambridge, 1985)
87. J.D. Joannopoulos, R.D. Meade, J.N. Winn, *Photonic Crystals: Molding the Flow of Light* (Princeton University, New Jersey, 1995)
88. M. Lu, C. Zhang, L. Feng, J. Zhao, Y. Chen, Y. Mao, J. Zi, Y. Zhu, S. Zhu, N. Ming, Nat. Mater. **6**, 744 (2007)
89. X. Hu, C.T. Chan, Phys. Rev. Lett. **95**, 154501 (2005)
90. E.A. Ostrovskaya, Y.S. Kivshar, Phys. Rev. Lett. **90**, 160407 (2003)
91. Y. Liu, G. Bartal, D.A. Genov, X. Zhang, Phys. Rev. Lett. **99**, 153901 (2007)
92. P. Zhang, C. Lou, S. Liu, F. Xiao, J. Zhao, J. Xu, Z. Chen, Opt. Photonics News **19**, 25 (2008)
93. P. Zhang, J. Zhao, F. Xiao, C. Lou, J. Xu, Z. Chen, Opt. Express **16**, 3865 (2008)
94. P. Zhang, S. Liu, C. Lou, F. Xiao, X. Wang, J. Zhao, J. Xu, Z. Chen, Phys. Rev. A 81, 041801 (R) (2010)
95. P. Zhang, C. Lou, S. Liu, J. Zhao, J. Xu, Z. Chen, Opt. Lett. **35**, 892 (2010)
96. P. Zhang, S. Liu, J. Zhao, C. Lou, J. Xu, Z. Chen, Opt. Lett. **33**, 878 (2008)
97. Y. Hu, C. Lou, S. Liu, P. Zhang, J. Zhao, J. Xu, Z. Chen, Opt. Lett. **34**, 1114 (2009)
98. Y. Hu, C. Lou, P. Zhang, J. Xu, J. Yang, Z. Chen, Opt. Lett. **34**, 3259 (2009)
99. S. Liu, P. Zhang, X. Gan, F. Xiao, J. Zhao, Appl. Phys. B **99**, 727 (2010)
100. G. Bartal, O. Cohen, H. Buljan, J.W. Fleischer, O. Manela, M. Segev, Phys. Rev. Lett. **94**, 163902 (2005)
101. D. Träger, R. Fischer, D.N. Neshev, A.A. Sukhorukov, C. Denz, W. Królikowski, Y.S. Kivshar, Opt. Express **14**, 1913 (2006)
102. C. Luo, S.G. Johnson, J.D. Joannopoulos, J.B. Pendry, Phys. Rev. B **65**, 201104 (2002)
103. M. Born, E. Wolf, *Principles of Optics* (Cambridge University, Cambridge, 1999)
104. M. Straub, M. Ventura, M. Gu, Phys. Rev. Lett. **91**, 043901 (2003)
105. A.R. Kolovsky, H.J. Korsch, Phys. Rev. A **67**, 063601 (2003)
106. D. Witthaut, F. Keck, H.J. Korsch, S. Mossmann, New J. Phys. **6**, 41 (2004)
107. J. Hudock, N.K. Efremidis, D.N. Christodoulides, Opt. Lett. **29**, 268 (2004)
108. Z. Chen, H. Martin, E.D. Eugenieva, J. Xu, A. Bezryadina, Phys. Rev. Lett. **92**, 143902 (2004)
109. C. Conti, S. Trillo, P. Di Trapani, G. Valiulis, A. Piskarskas, O. Jedrkiewicz, J. Trull, Phys. Rev. Lett. **90**, 170406 (2003)
110. Y. Silberberg, Opt. Lett. **15**, 1282 (1990)
111. R. Fischer, D. Träger, D.N. Neshev, A.A. Sukhorukov, W. Krolikowski, C. Denz, Y.S. Kivshar, Phys. Rev. Lett. **96**, 023905 (2006)
112. N.K. Efremidis, J. Hudock, D.N. Christodoulides, J.W. Fleischer, O. Cohen, M. Segev, Phys. Rev. Lett. **91**, 213906 (2003)

Chapter 6
Theory of Polariton Solitons in Semiconductor Microcavities

O.A. Egorov, D.V. Skryabin, and F. Lederer

6.1 Introduction

The research into the physics of localized structures (solitons) in nonlinear systems covers many branches of physical sciences, including optics, fluid dynamics, particle physics, and biology [1–3]. Depending on the physical setting, localization can occur in one, two, or three dimensions. One of the qualitative principles underpinning the localized structure formation is that the dispersion induced broadening is compensated by the nonlinearity of an appropriate sign and if loss is present as in non-Hamiltonian systems, then it has to be compensated by an external energy source [3]. Most generally, if dispersive spreading happens as for quantum mechanical particles with positive/negative effective mass, then a focusing/defocusing, or in other words, an attractive/repulsive nonlinearity, is required to compensate for the dispersion [2]. One of the active topics of recent research has been the so-called cavity solitons in weakly coupled, wide aperture semiconductor microcavities [4–11]. Apart from their interesting physics, these structures have been proposed for use in several information processing schemes [9–14]. However, applications of cavity solitons in weakly coupled microcavities are limited due to their slow response times and the relatively weak nonlinearity. Slow (nanosecond) carrier dynamics is dominant in weakly coupled vertical cavity surface-emitting lasers (VCSELs) and it does not catch up with the fast (picosecond) photon lifetime in the cavity [14, 19]. It is also challenging to trap photons in small volumes and thus dense packing of conventional light-only cavity solitons is

O.A. Egorov (✉) • F. Lederer
Institute of Condensed Matter Theory and Solid State Optics, Friedrich Schiller Universität Jena, Max-Wien-Platz 1, 07743 Jena, Germany
e-mail: oleg.egorov@uni-jena.de

D.V. Skryabin
Centre for Photonics and Photonic Materials, Department of Physics, University of Bath, Bath BA2 7AY, UK

Z. Chen and R. Morandotti (eds.), *Nonlinear Photonics and Novel Optical Phenomena*, Springer Series in Optical Sciences 170, DOI 10.1007/978-1-4614-3538-9_6,

problematic. In the strong coupling regime, the photons are replaced by polaritons [15], which are the exciton–photon quasiparticles [16–20]. The effective wavelength of these polaritons is smaller than that of photons and their lifetime is comparable to the one of photons and, hence, fast. Their nonlinear interaction, which is of excitonic origin, is 2–3 orders of magnitude stronger [19–22]. Polaritons are also known to exist in other condensed matter, atomic, and optical systems [23–28]. In the last few years, extensive studies of polaritons in strongly coupled microcavities have been largely motivated by the studies regarding polariton Bose–Einstein condensation [29, 30] and the transition to a superfluid regime [31, 32]. Recently, these exciton-polaritons have also been observed at room temperatures, see, e.g., [33, 34], which has further boosted their potential for practical applications. The strong repulsive interaction of polaritons has been shown to lead to low-threshold optical bistability [35–37] and parametric wave mixing [38–45]. In the latter case, the simultaneous matching of momenta and frequencies of the pump, signal, and idler beams can be achieved provided that the momentum of the pump exceeds a critical value associated with the so-called "magic angle" marking the point where the effective polariton mass changes its sign [44]. Besides Coulomb-induced polariton–polariton repulsion, there are two other nonlinear mechanisms which govern the dynamics of intracavity polaritons and lead to the saturation of the coupling strength between excitons and photons when the exciton density increases [18, 21, 22, 42]. The first effect is phase-space-filling due to the Pauli exclusion principle and the second one is the renormalization of the exciton orbital wave function induced by the presence of other excitons [21]. Both saturation mechanisms are important for GaAs- based devices with small exciton binding energy and they have been shown to result in polaritonic bistability [36, 37]. Solitonic and self-localization effects with polaritons in bulk media have attracted attention since the 1970s until now [24, 46, 47]. In the context of microcavity polaritons, the existence of vortices in exciton-polariton condensates [48] has been discussed. In a recent experimental paper [49], the authors claim the observation of localized structures or cavity solitons in a strongly coupled semiconductor microcavity. Further experimental investigations of the coherently driven semiconductor microcavities provide the evidence of polariton droplets moving on top of the background created by the signal and idler beams participating in parametric conversion [50–52]. In a series of our own papers, different types of resting and moving self-localized states of exciton-polaritons in coherently driven semiconductor microcavities have been reported [53–57]. This chapter reviews the findings of the latter. In Sect. 6.2, we describe the equations used to model the polariton dynamics. Then we analyze the linear dispersion of polaritons and study the bistability conditions of the homogeneous solutions (HSs). In Sect. 6.3, we derive the simplified nonlinear model in the polaritonic basis. Section 6.4 discusses stationary and moving dark cavity polariton solitons. Then, in Sect. 6.5, the formation of moving bright polariton solitons is considered for the pump beams tilted beyond the "magic angle." Finally, in the Sect. 6.6 we describe bright solitons existing close to the upper-polariton branch due to the saturation of the exciton–photon coupling.

6.2 Mathematical Model

Below we are dealing with a semiconductor microcavity Fig. 6.1a, where polaritons may form due to the coupling of quantum well (QW) excitons and microcavity photons. We start by introducing a standard model [18–20, 31, 53, 54, 57] for the case of l identical QWs and a sufficiently large exciton density to saturate the photon–exciton coupling:

$$\partial_t E - i\nabla^2 E + (\gamma - i\Delta)E = if(|\Psi|^2)\Psi + E_p e^{ik_p x}, \tag{6.1}$$

$$\partial_t \Psi - id\nabla^2 \Psi + (\gamma - i\Delta)\Psi + il^{-1}|\Psi|^2\Psi = if(|\Psi|^2)E. \tag{6.2}$$

Here, E and Ψ are the averages of the photon and exciton creation or annihilation operators. The normalization is such that $(\Omega_R/g)|E|^2$ and $(\Omega_R/g)|\Psi|^2$ are the photon and exciton numbers per unit area. Ω_R is the Rabi frequency and g is the exciton–exciton interaction coefficient. $\Delta = (\omega - \omega_{c,0})/\Omega_R$ is the detuning of the pump frequency ω from the identical cavity and exciton resonances ($\omega_c = \omega_0$). The cavity is driven by a plane pump wave with the amplitude E_p, k_p is the transverse momentum of the pump wave determined by the angle of incidence. The time t is measured in units of Ω_R^{-1}. The photon and exciton decay rates, γ, are assumed equal. Full details of the rescaling into physical units can be found in [53, 54]. A unit of t corresponds to 0. 25 ps and a unit of x to $\sim$ 1μm if typical parameters of the polariton experiments with InGaAs/GaAs QWs are used. The exciton dispersion coefficient $d = \hbar\omega_c n^2/m_{exc}c^2$ is normalized to the photon diffraction coefficient, where m_{exc} is the effective mass of excitons and n is the effective refractive index of the cavity. Realistically, it amounts to $d \sim 10^{-4}$–10^{-5}.

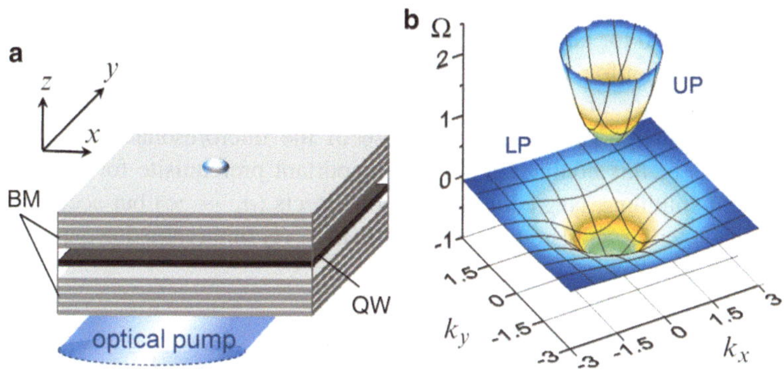

Fig. 6.1 (**a**) Sketch of the microcavity driven by a coherent optical pump. The semiconductor quantum well (QW) is sandwiched between two Bragg mirrors (BM). (**b**) Polariton dispersion of the planar semiconductor microcavity operating in the strong coupling regime. LP and UP depict lower and upper-polariton branches, respectively, for a single QW configuration (l = 1) and infinite exciton mass d = 0

The term determining the photon–exciton coupling $f(|\Psi|^2) \equiv [l/(1+|\Psi|^2/n_s l)]^{1/2}$ accounts for the saturation of the exciton–photon interaction with increasing exciton densities [18]. Here, $n_s = N_s g/\Omega_R$ is the normalized exciton saturation density. Starting with an effective Hamiltonian for the interacting electrons and holes one arrives at (6.1, 6.2) with saturation effects included [20, 22, 42]. This model is valid for moderate saturations and fits well the experimental measurements of the Rabi splitting [18]. The saturation density can be approximated as $N_s \approx 0.117/\pi a_{\mathrm{exc}}^2$ where a_{exc} is the exciton radius [21]. For InGaAs/GaAs quantum wells, it corresponds to the dimensionless quantity $n_s \simeq 0.5-1.0$.

First, we briefly describe the linear polariton dispersion, which is defined as the dependency of the frequency Ω (or energy) from the transverse momentum components k_x and k_y. Assuming that $E, \Psi \sim e^{-\gamma t} e^{-i\Omega T + ik_x x + ik_y y}$ and dropping the pump term and nonlinear effects, we find the eigenvalue problem:

$$i\Omega_{\pm}(k)\vec{p}_k = \begin{pmatrix} ik^2 & -i\sqrt{l} \\ -i\sqrt{l} & idk^2 \end{pmatrix} \vec{p}_k, \tag{6.3}$$

where $\vec{p}_k = \{e_k, \psi_k\}$ is the polariton basis vector and $\Omega_{\pm}(k)$ are the eigenfrequencies. e_k and ψ_k describe the content of the photonic and excitonic components of polaritons (Hopfield coefficients) [15]. The solution of the eigenvalue problem (6.3) yields the linear polariton dispersion relation

$$\Omega_{\pm}(k) = \frac{(1+d)k^2}{2} \pm \sqrt{l + \frac{1}{4}(1-d)^2 k^4}, \tag{6.4}$$

where $k^2 = k_x^2 + k_y^2$ and $\Omega_{\pm}$ are the relative frequencies of the upper (UP) and lower (LP) polariton branches, respectively, see Fig. 6.1b. In the strong coupling regime, the Rabi splitting exceeds the line-width ($\Omega_+(k=0) - \Omega_-(k=0) \gg \gamma$). Provided that $d \neq 0$ for large momenta, the LP branch extends into the gap between LP and UP existing for $d = 0$.

The nonlinearity of polaritons is accounted for by the terms proportional to the exciton density. The bistable response of the microresonator is one of the consequences of this nonlinearity and an important prerequisite for the existence of localized solutions. Neglecting saturation effects ($n_s \to \infty$) but accounting for the exciton–exciton repulsion, one finds that $E(E_p)$ is multivalued provided that $W(\Delta, k_p) > 0$, where

$$W(\Delta, k_p) \equiv \left[\left(\Delta - \frac{\left(\Delta - k_p^2\right) l}{\gamma^2 + \left(\Delta - k_p^2\right)^2} \right) - \sqrt{3}\gamma \left(1 + \frac{l}{\gamma^2 + \left(\Delta - k_p^2\right)^2} \right) \right]. \tag{6.5}$$

This cumbersome condition determines two intervals of Δ, where polariton bistability can be observed [53]. It simplifies for $\gamma = 0$ and normal incidence of the pump $k_p = 0$ giving two bistability intervals $\Delta>\sqrt{l}$ and $-\sqrt{l}<\Delta<0$. These two intervals overlap with either the LP or the UP band of the dispersion relation, see (6.4) and Fig. 6.2. The first interval belongs to the LP branch, and, thus, we call it the LP bistability [31, 35, 40]. The second one is located within the UP band.

The energy gap between the upper- and lower-polariton branches shrinks for larger polariton densities due to the saturation of the photon–exciton coupling ($n_s \neq \infty$). More precisely the LP branch shifts upward, i.e., experiences an additional blue shift, whereas the UP branch is red shifted [36]. Note that exciton–exciton interaction leads to the blue shift of both branches. The red shift of the UP branch due to saturation of $f(|\Psi|^2)$ can be sufficiently strong to compete with and surpass the blue shift due to exciton–exciton interaction. As a result, the bistability can appear in a new frequency interval Δ near the UP branch [57]. Note that unlike the conventional bistability of the upper polaritons (UP), this saturation bistability exists even for a vanishing exciton-exciton interaction term in (6.2). The interplay of two types of bistabilities of the UP branch can give rise to multistability of HSs.

6.3 Effective Equations in the Polaritonic Basis

To understand the physics behind the polariton soliton formation and to identify domains of their existence, we proceed by expanding the linear polariton dispersion around the pump momentum, and derive a single equation for the amplitude of both the lower (and upper) branch polaritons containing an effective nonlinearity and multiple dispersion terms. In this derivation, we simplify the nonlinear saturation of the coupling considering only the first density-dependent term in the Taylor expansion $f(|\Psi|^2) \approx l^{1/2} - |\Psi|^2/2n_s l^{1/2}$. We also assume that $\{E, \Psi\} \simeq \int a(t,k)\vec{p}_k e^{ikx} dk$, where $a(t, k)$ is the Fourier amplitude of the kth component. We assume that the spectrum of the polariton wavepacket is centered around k_p and expand $\Omega_\pm(k)$ up to the fourth order in $k - k_p$. The resulting equation for the amplitude of the polariton wave packet

$$A(t,x) = \int_{-\infty}^{\infty} a(t,k) e^{i(k-k_p)x} d(k - k_p)$$

is

$$i\partial_t A + iD_1\partial_x A + D_2\partial_x^2 A - iD_3\partial_x^3 A - D_4\partial_x^4 A + i\gamma A + \delta A - \xi\,|A|^2 A = i\eta E_p, \quad (6.6)$$

where $\delta = \Delta - \Omega_\pm(k_p)$ is the effective frequency detuning, γ the loss, $D_1 = \partial_k \Omega_\pm|_{k_p}$ the transverse group velocity, and $D_2 = (1/2)\partial_k^2\Omega_\pm|_{k_p}$, $D_3 = (1/6)\,\partial_k^3\,\Omega_\pm|_{k_p}$, $D_4 = (1/24)\partial_k^4\Omega_\pm|_{k_p}$ are the dispersion coefficients (Fig. 6.2b). The effective nonlinearity reads as $\xi = |\psi_{k_p}|^2(|\psi_{k_p}|^2 + Real(e_{k_p}^* \cdot \psi_{k_p})n_s^{-1})/l(|e_{k_p}|^2 + |\psi_{k_p}|^2)$ and

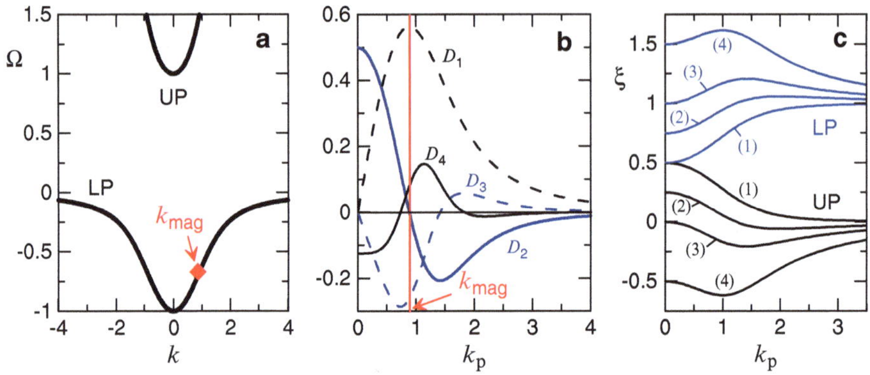

Fig. 6.2 (**a**) Polariton dispersion: lower (LP) and upper (UP) polariton branches in the strong coupling regime. (**b**) Dispersion coefficients of the lower polaritons (LP) vs pump momentum k_p for $d = 0$. (**c**) The effective nonlinear coefficient ξ of the upper (UP) and lower (LP) polariton branches vs pump momentum for $l = 1$ and the saturation densities $n_s = \infty$ (1), $n_s = 2.0$ (2), $n_s = 1.0$ (3), $n_s = 0.5$ (4)

$\eta = e_{k_p}/(|e_{k_p}|^2 + |\psi_{k_p}|^2)$. The essential difference between the weak and strong coupling regimes, relevant to our study, is not the mere existence of the LP branch, but the fact that it exhibits an inflection point, $\partial_k^2\Omega_- = 0$ at $k = k_{mag}$, where the second-order dispersion changes sign (Fig. 6.2a, b). Note that in general, the effective transverse mass of polaritons is a 2×2 tensor. The mass component corresponding to the direction of the pump momentum changes its sign from positive to negative for $k > k_{mag}$, while the mass in the orthogonal direction stays positive. Normalizing the excitonic component of the linear polariton eigenmode such that $|\psi_{k_p}|^2 \equiv 1$ results in an equivalence of the polariton and exciton densities ($|A|^2$).

The effective nonlinear coefficient for the lower (-) and upper (+) polariton branches is expressed as

$$\xi^{\pm}(k_p) = \frac{\left(k_p^2 \mp \sqrt{4l + k_p^4}\right)\left(\left(k_p^2 \mp \sqrt{4l + k_p^4}\right) + 2l/n_s\right)}{l\left(\left(k_p^2 \mp \sqrt{4l + k_p^4}\right)^2 + 4l\right)} \tag{6.7}$$

and its dependence on the pump momentum is shown in Fig. 6.2c. Due to the opposing actions of the two types of nonlinear shifts, the effective nonlinearity of the upper-polariton branch changes its sign for some values of momenta provided that the saturation density of the exciton–photon interaction is small enough: $n_s < n_{crit} = l/(\sqrt{l + k_p^4/4} - k_p^2/2)$.

For normal pump incidence, $k_p = 0$, the model (6.6) is identical to the one previously studied in the context of light-only cavity solitons in the case of Kerr nonlinearity, see, e.g., [4, 7, 58] and reference therein. From these studies, it is known that the points where the HSs destabilize are the potential bifurcation points

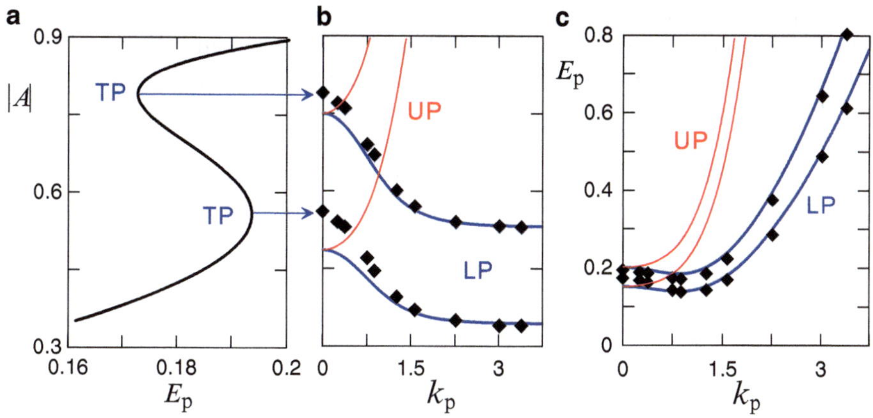

Fig. 6.3 (**a**) Bistable response of both lower (LP) and upper (UP) polaritons for the positive effective detuning δ = 0. 3 and a defocusing nonlinearity. (**b**) k_p-dependence of the $|A|$ values corresponding to the turning points (TP) of the bistability loop. (**c**) Pump amplitudes required for bistability. The full lines in (**b**, **c**) have been obtained using (6.6) and the diamonds depict results of the original model [(6.1) and (6.2)]

for solitons to emerge [7]. The HS response may be either monostable or bistable (see Fig. 6.3a). In the latter case, two steady-state HS solutions (stable and unstable) converge and disappear simultaneously at the limit or turning points (TP) of HS solutions. This is a saddle-node bifurcation. In accordance with this model, the operating frequency has to be slightly above the corresponding resonance $\delta > \gamma\sqrt{3}$ for the defocusing nonlinearity ($\xi > 0$) or slightly below for the $\delta < -\gamma\sqrt{3}$ focusing case ($\xi < 0$) [58]. As a result, one can find two intervals of polaritonic bistability associated with the lower and upper-polaritonic branches in agreement with the conditions for bistability (6.5) without saturation effects. Figure 6.3b, c shows the example of HS bistability domains for both the lower and the upper-polariton branches calculated in both the original model [(6.1) and (6.2)] and in the simplified model in the polaritonic basis (6.6). There is a reasonable agreement between the results of both models for the lower-polariton branch. It is quite unexpected that the pump amplitude required for bistability reaches the minimum for the nonzero moment k_p slightly beyond the "magic angle" (see Fig. 6.3c). It can be explained by the increase of the effective nonlinear coefficient with k_p for the lower polaritons (LP), see Fig. 6.2c. Note that a sufficiently strong nonlinear frequency shift can disturb the linear polariton basis and, therefore, the simplified model (6.6) ceases to be valid. Such distortion explains the substantial discrepancies between the bistability domains obtained in the original and in the simplified models for the upper polariton (UP) branch.

Before proceeding with comprehensive numerical simulations of cavity polariton soliton (CPS), we discuss three important regimes anticipated from the model (6.6).

Regime 1: For a moderate inclination of the pump beam $k_p \approx 0$, the second-order dispersion is positive ($D_2 > 0$) and prevails against all higher order dispersion terms (Fig. 6.2b). The effective nonlinearity is defocusing ($\xi^{\pm} > 0$) for the lower- as well as the upper-polariton branch for a small saturation nonlinearity $n_s \to \propto$. Dark CPSs are expected to be stable for these parameters (see Sect. 6.4 and [53]). The effective nonlinearity term ξ^- of the lower branch polaritons increases for strong saturation effects. Therefore, we assume that the saturation of the photon–exciton coupling does not change qualitatively the nonlinear solutions but reduces the intensity thresholds required for their excitations.

Regime 2: For a large inclination of the holding beam beyond the "magic angle," the leading order dispersion term of the lower-polariton branch changes the sign $D_2 < 0$. In this case, moving bright CPSs can be found for both 1D and 2D configurations (see Sect. 6.5 and [54, 55, 56]).

Regime 3: For normal incidence of the pump beam $k_p \approx 0$ and a strong saturation nonlinearity, the effective nonlinearity (6.7) changes the sign at the upper-polariton branch (Fig. 6.2c). The nonlinearity becomes focusing at this branch giving rise to a new domain of bistability associated with saturation effects and the formation of the stable bright cavity polariton solitons (see Sect. 6.6 and [57]).

6.4 Dark Polariton Solitons

In this chapter, we consider solitons with frequencies close to the bottom of the LP branch for zero and small pump momenta, where the effective polariton mass is positive along any direction in the cavity plane (regime 1). The saturation of the exciton–photon coupling does not have any qualitative impact on the results discussed in this chapter and therefore can be disregarded, $n_s \to \infty$. Since any type of cavity solitons is located on a homogeneous background, it is important that the linear stability of the background is investigated first. The spectrum of small perturbations around the background field in the bistable regime has been previously reported, e.g., in [31]. For sufficiently small pump momenta, the lower state of the polariton bistability loop can become unstable evidenced by the growth of perturbations with finite transverse momenta [modulational instability (MI)], thereby generating a growing modulation of the initially flat polariton state, while the upper state is generally stable, see Fig. 6.4a.

The soliton branches are known to emanate from the points of MI [7]. First, we consider the case when the MI point of the lower state of polaritons falls within the bistability interval of the pump field strength. Applying the Newton method to solve the stationary ($\partial_t = 0$) version of (6.1) and (6.2) with $\partial_x^2 + \partial_y^2 = \partial_r^2 + \frac{1}{r}\partial_r$ ($r^2 = x^2 + y^2$) we have found, perhaps surprisingly, a family of small amplitude bright solitons splitting from the MI point, see the dashed red line in Fig. 6.4a. These solitons become more intense for smaller values of E_p further away from the MI point, see Fig. 6.4d. The increase in intensity with the drop of the pump power is a ubiquitous signature of the underlying dynamical instability (see below).

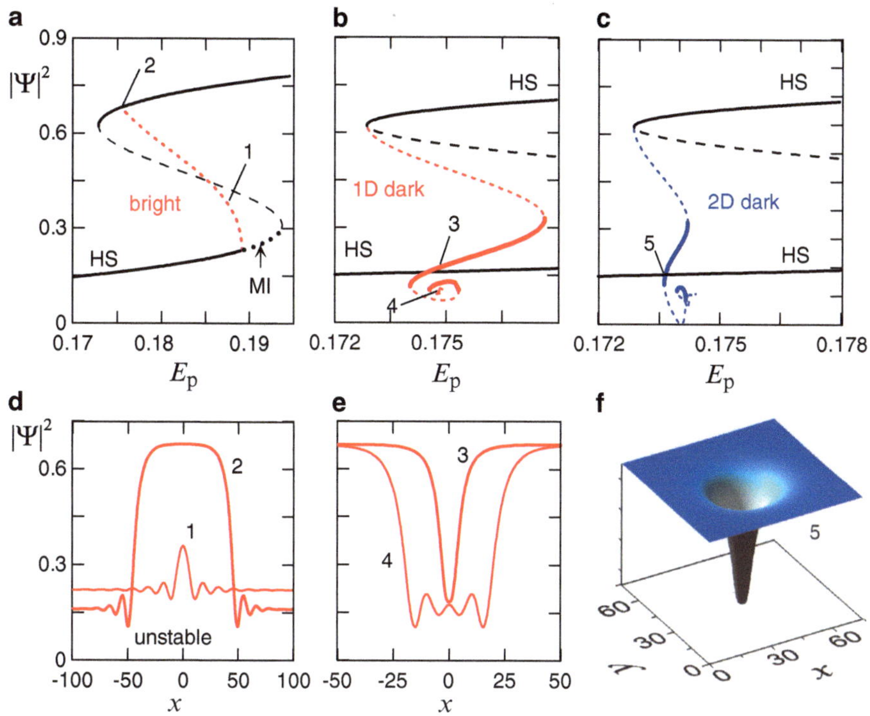

Fig. 6.4 Branches of unstable bright (**a**), 1D dark (**b**), and 2D dark (**c**) cavity polariton solitons (shown is the squared modulus of the excitonic component) as functions of pump amplitude E_p for $\Delta = -0.7$, $\gamma = 0.\ 1$. (**d–f**) Exciton density distribution $|\Psi(x, y = 0)|^2$ across the bright (**d**), 1D dark (**e**),and 2D dark (**f**) soliton profiles for the points marked by 1, 2, 3, 4, and 5 in figures (**a–c**). *Full and dashed lines* in (**a–c**) mark stable and unstable solutions, respectively

The bright CPS branch terminates at $E_p = 0.\ 1748$. At this particular point (Maxwell point), the lower and upper homogeneous states (HSs) can be connected by a zero velocity front. When the pump approaches the Maxwell point, the soliton broadens and its peak intensity tends toward the intensity of the upper homogeneous state. We also performed a stability analysis of these bright solitons. The linear perturbations around the radially symmetric solitons have been assumed in the general form $\epsilon_+(r)e^{iJ\theta+\kappa t} + \epsilon_-^*(r)e^{-iJ\theta+\kappa^* t}$, where $J = 0, 1, 2, \ldots$ [59]. The bright solitons have been found unstable with respect to the perturbation with the azimuthal index $J = 0$. This instability tends to transform the soliton into moving fronts. The front velocity becomes very small for E_p close to the Maxwell point (MP) and, hence, this class of bright solitons is only weakly unstable there. Pinning of the fronts to the inhomogeneities of the cavity can provide a further stabilizing effect.

Because of the defocusing nature of the polaritonic nonlinearity the dark solitons, see, e.g., [7], are expected not only to exist, but also to be genuinely stable. Dark cavity solitons in the weak coupling regime, have been previously reported in numerical modeling and experiments, see, e.g., [5, 8, 60]. In our case, the dark polariton solitons

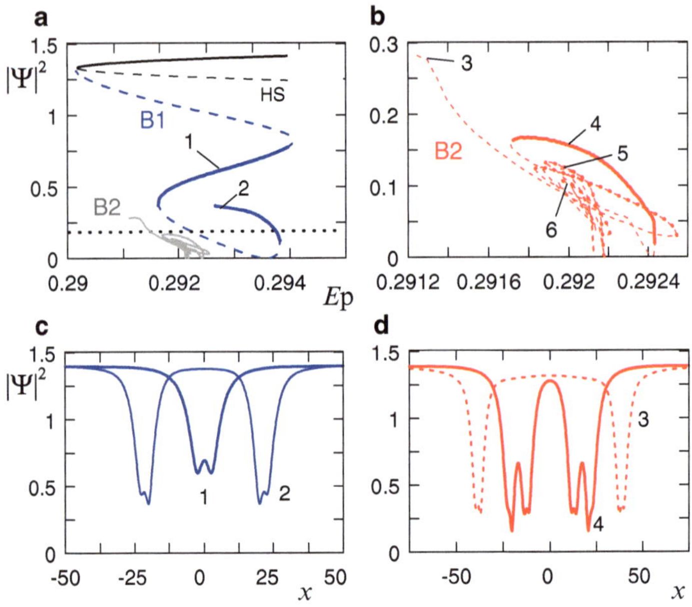

Fig. 6.5 (**a**) Amplitude of the homogeneous state (HS) (*black line*) and min $|\Psi(x, y)|^2$ for 2D dark CPSs shown as functions of E_p: $\Delta = -0.5$, $\gamma = 0$. 1.$B1$ and $B2$ mark two branches of dark CPSs. (**b**) Zoom of the area from (**a**) showing the B2 dark CPSs. (**c**, **d**) Exciton density distribution $|\Psi(x, y = 0)|^2$ across B1 (**c**) and B2 (**d**) CPSs for the points marked by 1, 2, 3 and 4 in panels (**a**) and (**b**). *Full and dashed lines* mark stable and unstable solutions, respectively

detach from the left part of the bistability loop, see Fig. 6.4b, c. Close to this point, the notch on the homogeneous background is very shallow and it deepens as E_p approaches the MP. Near the MP, the dark solitons become very broad and, similarly to the bright ones, can be qualitatively considered as a superposition of fronts. Note that the relaxation of the fronts toward the upper HS happens without oscillations however, the relaxation toward the lower state is oscillatory, see Fig. 6.4e. Thus, pinning of the two fronts one to another and, hence, stabilization of solitons is possible only for the dark structures (Fig. 6.4e). The stable branches of dark CPSs are shown by full lines in Fig. 6.4b, c. The unstable ones correspond to the instabilities with $J = 0$ (for the 2D case). Note, also, that the existence domain of 1D stable dark solitons wider than that for 2D ones (compare Fig. 6.4b, c).

For detunings Δ close to the exciton resonance ($\Delta = 0$), the MI point shifts toward the left edge of the bistability interval to pass it eventually (compare Figs. 6.4a and 6.5a). As a result the branch of bright solitons originating at this point disappears, see Fig. 6.5 ($\Delta = -0.5$). However, we have found two distinct branches of the dark solitons marked as B1 and B2 in Fig. 6.5a, b. The B1 branch bifurcates subcritically from the folding point of the upper HS and becomes stable after the turning point. Close to this point the B1 solitons have a single well shape, while later they transform into ring-like structures Fig. 6.5c. Note that close to the turning points the B1 solitons become unstable suffering from an oscillatory (Hopf) instability [53].

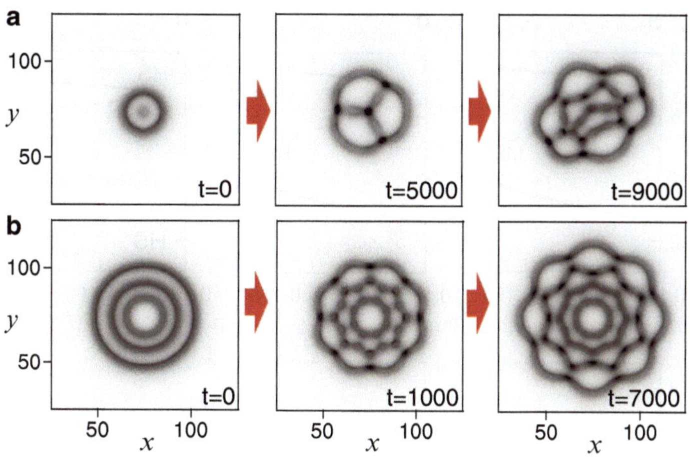

Fig. 6.6 Development of the symmetry-breaking instabilities of the dark CPSs marked as 5 (row **a**) and 6 (row **b**) in Fig. 6.5b

The B2 solitons exhibit a ring shape over the entire existence domain, see Fig. 6.5d. It is interesting that the B2 branch does not bifurcate from the HS. B2 solitons are linearly stable in the narrow interval marked by 4 in Fig. 6.5b. However, they are mostly unstable with respect to the perturbations breaking the radial symmetry, e.g., with $J \neq 0$. Examples of this instability with $J = 3$ and $J = 8$ are shown in Fig. 6.6.

Dark polariton solitons start to drift provided that the pump has a nonzero transverse momentum ($k_p \neq 0$). In first approximation, the velocity of motion is proportional to k_p [61].

To prove that drifting structures are indeed solitons, we consider (6.1) and (6.2) with $\partial_y = 0$ and seek moving solitons in the form: $E(t,x) = \tilde{E}(\tilde{x})e^{ik_px}$, $\Psi(t,x) = \tilde{\Psi}(\tilde{x})e^{ik_px}$, where $\tilde{x} = x - vt$ and v is the velocity (yet to be determined). It turns out that v is close, but not equal, to D_1, see Fig. 6.2b and (6.6). For the single QW configuration ($l = 1$) and negligible saturation effects ($n_s = \infty$), $\tilde{E}$ and $\tilde{\Psi}$ obey

$$(2k_p - v)\partial_{\tilde{x}}\tilde{E} - i\partial_{\tilde{x}}^2\tilde{E} + (\gamma - i\Delta + ik_p^2)\tilde{E} = i\tilde{\Psi} + E_p,$$
$$-v\partial_{\tilde{x}}\tilde{\Psi} + (\gamma - i\Delta)\tilde{\Psi} + i|\tilde{\Psi}|^2\tilde{\Psi} = i\tilde{E}. \qquad (6.8)$$

The soliton solutions of the above system were found using a modification of the Newton method allowing to treat v as an unknown variable (see Fig. 6.7a, d). The branch of moving dark solitons bifurcates subcritically from the left fold of the bistability loop. The solitons themselves have oscillating tails at their rear, also known from other moving cavity solitons [62].

In a 2D setting, the nonzero pump momentum breaks the equivalence between the x- and y- directions. While along the x-coordinate, one should expect the cross section of the 2D soliton to be similar to that of 1D solitons described above, along the

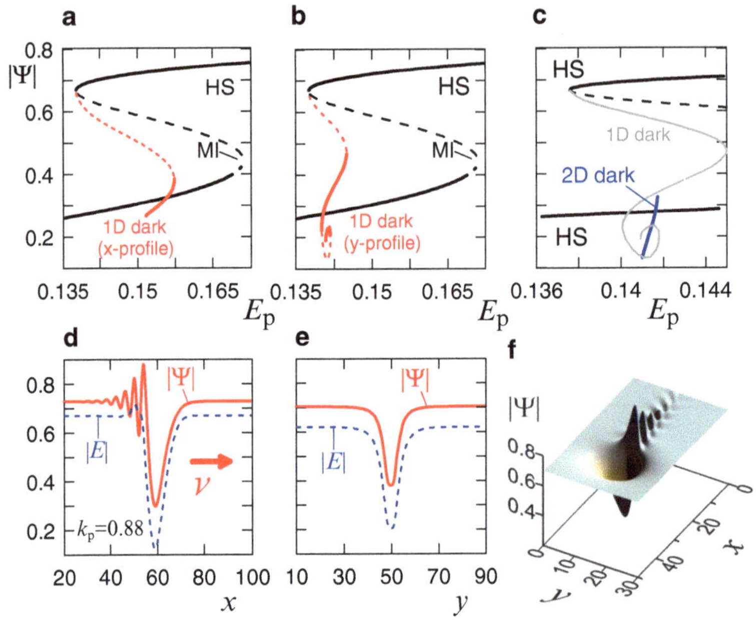

Fig. 6.7 Minima of the excitonic components of the 1D dark moving solitons in x-direction (**a**), stationary dark solitons localized in y-direction (**b**),and 2D moving dark CPS (**c**) vs E_p at the inflection point of the polariton dispersion $k_p \equiv k_{mag} \approx 0.88$ and $\Delta = -0.38$. Figures (**d**)–(**f**) show the soliton profiles for the branches given in (**a**)–(**c**), respectively

y-direction the soliton tails are expected to be symmetric, since the corresponding projection of the pump momentum is zero. Indeed if we now look for 1D solitons in the form $E(t,x,y) = \tilde{E}(y)e^{ik_px}$, $\Psi(t,x,y) = \tilde{\Psi}(y)e^{ik_px}$ we find a branch of stable solitons, see Fig. 6.7b, e. Thus, the shape of the stable 2D solitons, found through the modeling of (6.1) and (6.2) and shown in Fig. 6.7c, f, can be understood as a hybridization of 1D dark solitons independently localized in the x- and y-directions.

Overall, the semi-rigorous criterion for stable and moving dark solitons to exist in the present system is $k_p \leq k_{mag}$. For larger pump momenta, the effective mass becomes negative, so that the repulsive polariton–polariton interaction favors the stability of bright solitons, which we are going to describe in the succeeding chapters.

6.5 Bright Polariton Solitons

The effective polariton mass and the second-order dispersion coefficient D_2 change their signs from the positive to the negative one at $k_p > k_{mag}$. For k_p equal and close to k_{mag}, the third-order dispersion expressed by the coefficient D_3, see (6.6) and Fig. 6.2b, becomes the leading linear effect influencing polariton dynamics. As we increase k_p further, the value of D_3 drops and the polariton dispersion is determined by the competition of the D_2 and D_4 terms. Both $D_2 < 0$ and $D_4 > 0$ favor the

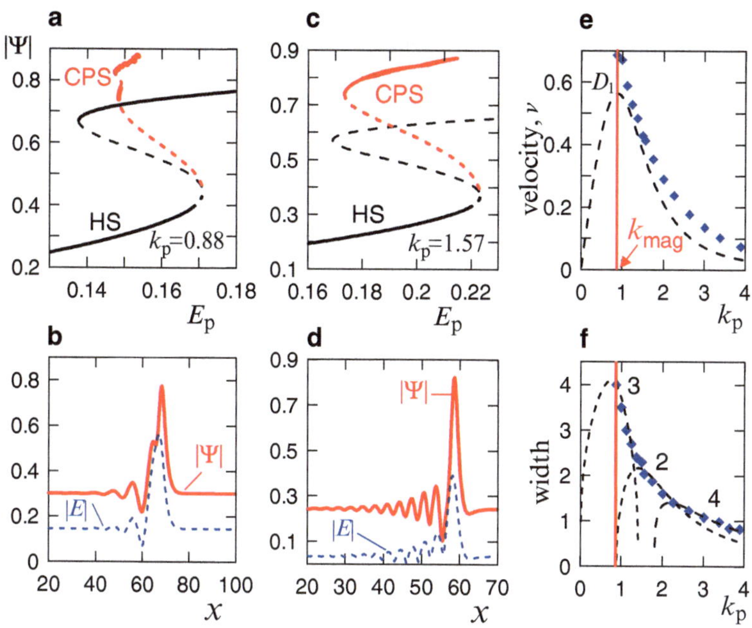

Fig. 6.8 Maxima of the excitonic components of the solitons vs E_p (**a**, **c**) and the soliton profiles (**b**, **d**) for different k_p. The *full/dashed lines* in the planes (**a**) and (**c**) correspond to stable/unstable solutions, respectively. E_p and Δ values are: (**a**, **b**) 0. 149 and − 0. 38 (**c**, **d**) 0. 19 and − 0. 05. (**e**) *Diamonds* indicate the numerically computed soliton velocity vs k_p. The *dashed line* shows $D_1(k_p)$. (**f**) *Diamonds* show the soliton width vs k_p. *Dashed lines* show the appropriately scaled $\sqrt[m]{D_m}$, where m = 2, 3, 4

existence of bright solitons for a repulsive nonlinearity. In this regime, $D_1 \neq 0$ and, hence, any solitons found here are expected to move.

Moving bright solitons have been calculated by using (6.8). Plots illustrating the dependence of the maximal soliton amplitude on the pump field and the associated bistability curves for the homogeneous ($\partial_{x,\,y} = 0$) solutions are shown in Fig. 6.8a, c. Figure 6.8b, d show typical transverse profiles of these bright solitons. They emerge for k_p slightly below k_{mag} and the intervals of their stability expand with increasing k_p. Note that (6.6) with the third-order dispersion term D_3 supports both bright and dark localized solutions irrespectively of the sign of nonlinearity [62]. Similarly, the bright and dark polariton solitons can coexist provided that k_p is close to k_{mag}.

The maximum of the soliton velocity v occurs at $k = k_{mag}$, and decreases rapidly with increasing k_p, see Fig. 6.8e. This is in remarkable contrast to what happens if dispersion is parabolic, where the soliton velocity continuously increases with k_p. For our choice of parameters v must be multiplied by $\simeq 4 \times 10^6$ m/s to give the physical velocity. It implies that a soliton with $v = 0.\ 25$ will traverse across a typical distance of ~ 100μm in ~ 100 ps. This is 40 times larger than the polariton lifetime. Hence, the solitons have enough time to be excited, formed, and experimentally observed.

To characterize the influence of the different dispersion orders on solitons, we display the soliton width vs the momentum k_p, see Fig. 6.8f. Changing k_p we have kept constant the detuning of the polariton frequency from the pump frequency, δ, fixed at δ = 0. 3, by adjusting the detuning of the pump field from the cavity resonance, Δ, which explicitly enters into the full model. For large k_p the excitonic part dominates over the photonic component, see Fig. 6.8. That is why the polariton solitons can be much narrower than the pure photonic cavity solitons in the weak coupling regime and may attain widths well below the ones allowed by the photonic dispersion. The soliton width is expected to scale with the dominant dispersion coefficients as $\sim \sqrt[m]{D_m}$ (m = 2, 3, 4). Figure 6.8f compares the numerically found soliton width (diamonds) with the scaling given by the different dispersion orders (dashed lines). Third-order dispersion describes well the soliton width for relatively small k_p where $D_2 \simeq 0$. A further inclination brings D_2 on the top, while the third-order dispersion vanishes. A further increase of k_p brings fourth-order dispersion into the play, which starts to compete with D_2.

In order to use polariton solitons as information bits, it is desirable to have them robustly localized in both transverse dimensions. The generalization of the 1D solutions toward the 2D case is not straightforward because the effective mass of polaritons has different signs along x and along y, with the latter being perpendicular to the pump momentum.

To find 2D bright solitons, we have taken the 1D soliton and extended it to infinity along the y- direction, see Fig. 6.9a, b. Then we multiplied this solitonic stripe with a broad, but finite top hat function in y-direction, and initialized (6.1) and (6.2) with this profile. The stripe edges start moving with velocity V_f along y forming moving fronts, Fig. 6.9d, where $V_f \neq 0$ does not result in the motion of the soliton center of mass (since two edges move in opposite directions), while $v \neq 0$ does. The front in Fig. 6.9d is analogous to the fronts connecting the upper and lower branches of the bistable HS [4]. For these fronts there exists the well-known MP, e.g., a special value of the pump, $E_p = E_{MP}$, such that the front rests [4]. For $E_p > E_{MP}$ the upper state is invading the lower one, and it is vice versa for $E_p < E_{MP}$, see Fig. 6.9c. The fronts observed in the present work, however, connect the 1D soliton to the lower branch of the HS; hence, their MP is different and shifts away, $E_{MP1} \neq E_{MP}$. Multihump 1D polariton solitons also exist [55] and can be connected by a front, see Fig. 6.9e. The MP in this case is again different, see Fig. 6.9c.

We have also performed a similar set of simulations with moving fronts designed to match practical experiments. In order to achieve this we have added the term $E_0(x, y, t)e^{ik_0 x - i\omega_0 t}$ representing a pulse, seeding a localized excitation to the equation for E. Using an elliptically shaped Gaussian beam elongated along the y-axis and having a 5ps duration, we have observed that the soliton profiles along x are easily generated, while their edges in the y-direction are either converging, so that the beam is shrinking ($E_p < E_{MP1}$), or diverging ($E_p > E_{MP1}$), so that the beam is expanding, see Fig. 6.10a, b. Remarkably, in a narrow window of the pump amplitudes left from the MP the shrinking in y-direction is suppressed, so that the emerging structures remain localized in both spatial coordinates over long propagation distances. Performing tedious numerical simulation of (6.1) and (6.2) over the time

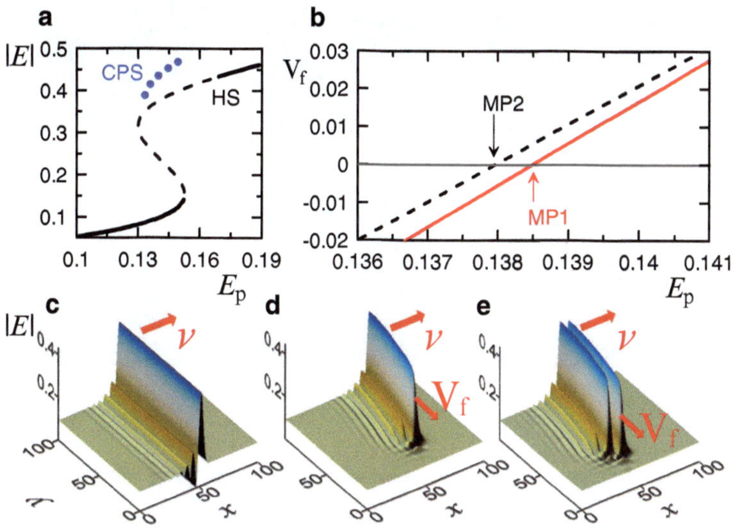

Fig. 6.9 (**a**) Bistability loop of the homogeneous solution (HS). The *dashed line* corresponds to the unstable HS. The *dotted line* marks the maximum intensity of the 1D CPSs localized along x, as shown in (**b**). (**b**) Soliton stripe moving in x-direction with velocity $v = 0.56$. (**c**) Velocities V_f of the single- (*solid line*) and double- (*dashed line*) hump fronts. $V_f = 0$ at the Maxwell points MP1 and MP2. Moving fronts connecting 1D single- (**d**) and double- (**e**) hump CPSs and HS background. Parameters: $E_p = 0.139$, $\Delta = -0.25$, $k_p = 1.2$, $\gamma = 0.1$

spans exceeding 2.5 ns, we have found that the 2D solitons with one and two humps represent stable attractors for a generic class of initial conditions within a finite interval of pump intensities, see Fig. 6.10c. The seed momenta applied in our simulation were $k_0 = 0$ (seed orthogonal to the cavity plane) and $k_0 = k_p$ (seed is collinear with the pump). Both choices have led to the excitation of 2D CPSs, albeit for different powers. The $k_0 = k_p$ case appeared to be more efficient. The branches of the single- and double-hump stable 2D bright solitons terminate at the respective MPs, see Fig. 6.11. Thus, perhaps surprisingly, stable 2D polariton solitons may exist even if the polariton mass along the orthogonal directions has opposite signs.

While the MP argument has been useful in finding 2D solitons, it relates to a specific value of E_p and cannot explain completely why solitons do exist within a finite interval of E_p. Other physical mechanisms are likely to be involved in the soliton formation. It was proposed in our recent paper [56] that parametric four-wave mixing of polaritons plays an important role for the localization of polaritons in the direction transverse to the pump momentum (y). The peculiar shape of the LP dispersion allows for phase matching of the pump, signal, and idler waves giving rise to strong parametric generation effects [19, 38, 39, 44]. In nonlinear optics it is well appreciated that the nonlinear phase shift resulting from parametric wave mixing is not simply proportional to the polariton density of pump, but involves phases and intensities of all the participating waves, thereby modifying the soliton existence conditions. This problem is currently considered by us and will be reported elsewhere.

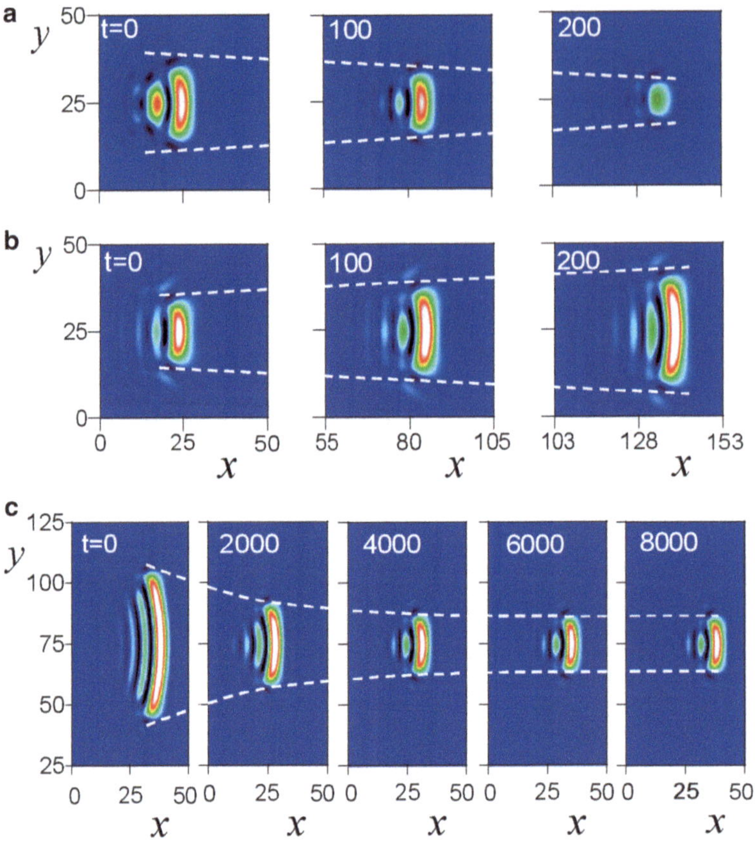

Fig. 6.10 (**a**) Shrinking of the initially localized excitation observed for $E_p = 0.136$. (**b**) Spreading of the initial excitation observed for $E_p = 0.141$. (**c**) Long-term dynamics showing the dynamical robustness and confirming the attractor properties of 2D CPSs, $E_p = 0.1378$. Other parameters as in Fig. 6.9

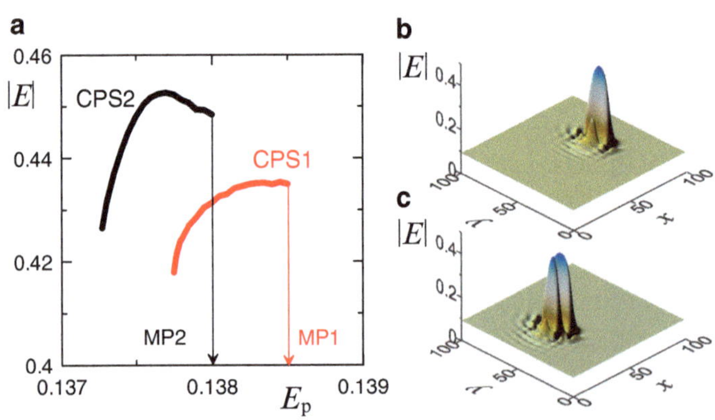

Fig. 6.11 (**a**) Maxima of E for different soliton solutions vs the pump E_p. CPS1 and CPS2 correspond to single- and double-hump solitons, respectively. Profiles of the photonic components of the single- (**a**) and two-hump (**b**) 2D stable bright polariton solitons for $E_p = 0.1378$ and $E_p = 0.1373$, respectively. Parameters: $\Delta = -0.25$, $k_p = 1.2$

6.6 Bright Solitons Near the Upper-Polariton Branch: Saturation and Finite Exciton Mass Effects

In the previous two sections, discussing properties of the solitons, we have disregarded the effect of saturation of the exciton–photon coupling coefficient f (Rabi frequency) with increasing exciton density by setting $n_s = \infty$, see (6.1) and (6.2). Near the LP branch the dynamics of the system is only quantitatively affected, because it simply enhances the repulsive interaction of the polaritons. However, for the UP branch, the saturation effect introduces an effective attractive polariton–-polariton interaction (focusing nonlinearity), which may counterbalance or even overcome the repulsive one. Indeed decreasing the parameter $f(|\Psi|^2)$, one shifts the LP branch upward and the UP branch downward, while for $n_s = \infty$, an increase of the exciton density effectively causes a blue shift of both branches. Thus, for upper branch polaritons accounting for finite values of n_s is expected to result in some qualitative impact on the properties of the homogeneous and solitonic polariton states. In particular, if the focusing nonlinearity becomes sufficiently strong, then bright solitons near the UP branch with small and zero velocities can be expected to exist in both 1D and 2D settings. In this section, we describe these solitons following our recent work [57].

Changing the detuning parameter Δ, so that the pump frequency appears in the vicinity of the UP branch and solving (6.1) and (6.2) with $\partial_t = 0$ in 1D and 2D cases, we have found bright solitons bifurcating from the left fold of the bistability loop, Fig. 6.12. The photonic and excitonic components of these solitons have the bell-like shape. The tails of the structures are nonoscillatory, if the excitonic dispersion is disregarded, $d = 0$. However, it turned out that the effect of the finite exciton mass d is important here, because the tails of the LP branch bend upward for large k and $d \neq 0$ and become resonant with the UP branch, see Fig. 6.13a, thereby opening a channel for energy transfer between the two branches. One of the manifestations of the $d \neq 0$ effects is that both homogeneous and soliton solutions undergo MI for $\Delta > 0$ against perturbations with large momenta. Physically, this instability leads to the generation of almost free excitons [57, 63].

To understand the nature of this excitonic instability, we consider in more detail how this instability appears for the HSs. The photonic field on the right-hand side of (6.2) serves as an external force driving the coherent excitons. Therefore, the excitonic component of the stationary HS solution $|\Psi_{st}|$ can be expressed as a function of its photonic part E_{st} (Fig. 6.13b). Both nonlinear terms in the excitonic equation introduce the negative nonlinear shift of the excitonic detuning, which can be compensated by $\Delta > 0$. Hence, the dependence of $|\Psi_{st}|$ on $|E_{st}|$ can become bistable. Assuming $n_s = \infty$, the condition of this intrinsic bistability becomes trivial [58]: $\Delta > \gamma\sqrt{3}$. To derive a simple expression for the growth rate of the perturbation with large momenta, we have disregarded the coupling between the excitonic and photonic components and performed a linear stability analysis

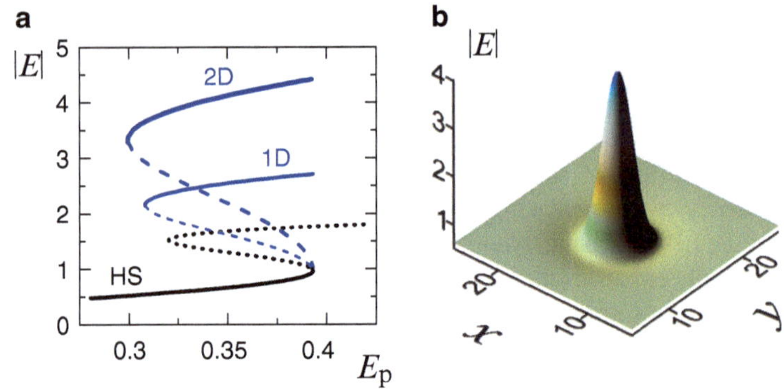

Fig. 6.12 (**a**) Branches of stable bright CPSs originating from the upper-polariton branch in both 1D and 2D configurations. (**b**) 2D profile of bright CPS. Parameters: $l = 16$, $\gamma = 0.\ 1$, $\Delta = 3.\ 7$, $n_s = 0.\ 75$

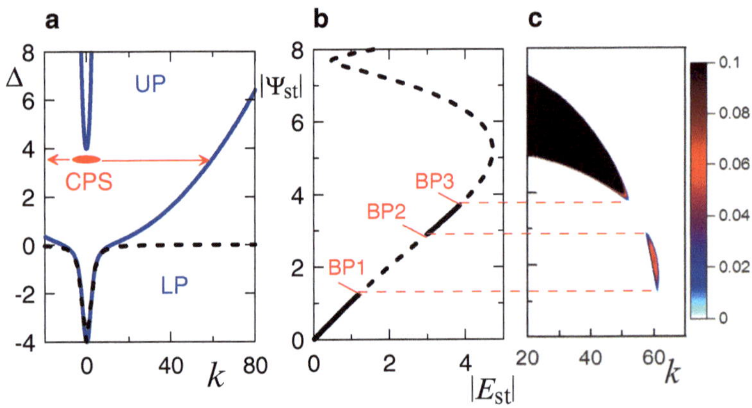

Fig. 6.13 (**a**) Dispersion relation of the lower- (LP) and upper- (UP) polariton branches for nonzero exciton dispersion (finite mass of excitons) $d = 0.\ 001$. The *dashed line* depicts the case of vanishing dispersion $d = 0$. (**b**) Dependence of the excitonic component on the photon component of stationary solution given by (6.2). *Solid lines* correspond to stable solutions. (**c**) Contour plot of the real part of the growth rate $Re\lambda$ of unstable perturbations in the plane $(|\Psi_{st}|, k)$ given by (6.10) for a finite exciton mass $d = 0.\ 001$. Parameters: $l = 16$, $\gamma = 0.\ 1$

using (6.2) only. Looking for the solution close to the stationary one (E_{st}, Ψ_{st}) in the form $\Psi_{st} + \delta\Psi e^{\lambda(k)t+ikx}$, we have:

$$\lambda^{\pm}(k) = -\gamma - A_1 \pm \sqrt{A_2(k) - A_3(k)}, A_1 = \frac{1}{2}|\Psi_{st}|^2 / \left(n_s l + |\Psi_{st}|^2\right),$$

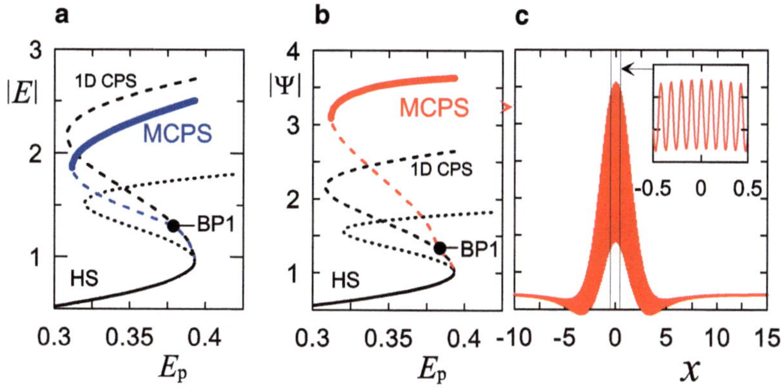

Fig. 6.14 (**a**, **b**) Branches of both uniform (CPS) and modulated (MCPS) one-dimensional cavity polariton solitons for nonzero exciton dispersion (finite mass) $d = 0.001$. HS depicts the homogeneous solution, and BP1 is a bifurcation point for MCPS. (**c**) Amplitude profiles of MCPSs. Parameters: $l = 16$, $\gamma = 0.1$, $\Delta = 3.7$, $n_s = 0.75$

$$A_2(k) = A_1^2\gamma^2 + A_1^2\left(\Delta - dk^2 - 3|\Psi_{st}|^2 l^{-1} - 2n_s\right)^2,$$
$$A_3(k) = \left((2 + A_1)|\Psi_{st}|^2 l^{-1} - (\Delta - dk^2)(1 + A_1)\right)^2. \qquad (6.9)$$

The stationary solution (E_{st}, Ψ_{st}) becomes modulationally unstable (Reλ $(k) > 0$) provided that its amplitude belongs to the intervals $|\Psi_{BP1}| < |\Psi_{st}| < |\Psi_{BP2}|$ and $|\Psi_{BP3}| < |\Psi_{st}|$ where BP1, BP2, BP3 denote bifurcation points in Fig. 6.13b. It follows from (6.10) that the period of the unstable perturbations scales as $k^{-1} \sim d^{1/2}$. Thus, for $d \sim 10^{-4}$ the period is much less than the soliton width. Therefore, our analysis is valid not only for the HS but also for the soliton solutions. The equation for the photonic component (6.1) practically does not contribute to this instability mechanism, since the momentum of unstable excitons is very large preventing them from efficiently coupling to photons.

The growth of this perturbation on the soliton background develops into a stable pattern forming a soliton with a sub-wavelength spatial modulation of the excitonic component, see Fig. 6.14. The formation of the periodic sub-wavelength pattern does not conflict with the parabolic dispersion approximation used for photons, since only the excitonic component of CPS is modulated (Fig. 6.14c). The branch of spatially modulated solitons bifurcates from smooth solitons at the point (BP1) where the MI sets in (Fig. 6.14a, b). Modulated solitons are essentially the bound states of polariton solitons and free excitons. Equations (6.1) and (6.2) neglect phonon effects and the relaxation of excitons with large momenta (k) toward the ground state [63]. We anticipate that these relaxation mechanisms can further stabilize the solitons due to the shortening of the exciton lifetime responsible for MI.

In many aspects the 2D case is not only similar to the above-discussed 1D case, but also exhibits some important distinct features. In particular, unlike in the 1D

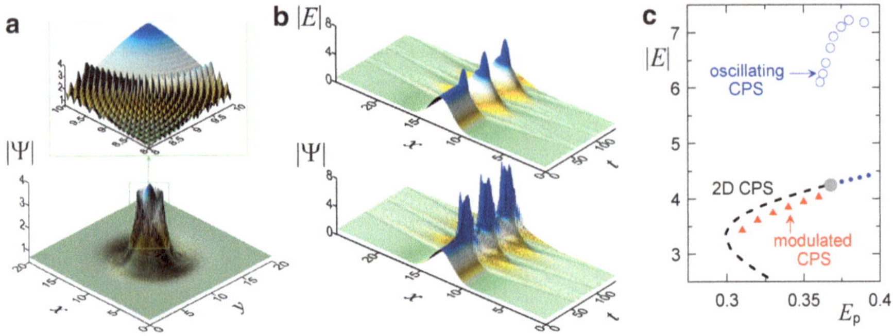

Fig. 6.15 (**a**) The modulus of the excitonic components of two-dimensional MCPSs for $d = 0.\ 001$ and $E_\mathrm{p} = 0.\ 35$.(**b**) Dynamics of two-dimensional oscillating CPSs for the parameters $E_p = 0.37, d = 0.001$. (**c**) The amplitude branches of unstable two-dimensional CPSs, modulated CPSs (*triangles*) and oscillating CPSs (*open circles*) for a finite exciton mass $d = 0.\ 001$. The *thick dot* in (**c**) depicts the point where the modulation starts to develop at the center of the 2D-modulated CPS (bifurcation BP3 in the Fig. 6.13b). Other parameters as in Fig. 6.14

case, the excitonic component of the 2D solitons is not modulated across the entire soliton profile. In particular, the area of the 2D solitons near their centers can stay free of spatial modulation, see Fig. 6.15a. This is because 2D solitons have larger amplitudes, so that the maxima of their excitonic component reach the window of stability, $|\Psi_{\mathrm{BP2}}| < |\Psi_{\mathrm{st}}| < |\Psi_{\mathrm{BP3}}$, see Fig. 6.13b, c.

Increasing the pump amplitude further, the modulated solitons undergo secondary instabilities. This instability induces the exponential growth of a periodic pattern at the soliton center (Fig. 6.15b). However, the soliton does not get destroyed, but experiences a high-amplitude periodic oscillation. This scenario is reminiscent soliton explosions in the Ginzburg–Landau model [3] and the excitability mediated by the localized structures in a nonlinear optical cavity operating in the weak coupling regime [64]. Direct numerical simulations show that bright soliton solutions of (6.1) and (6.2) can exist for large values of d, which can even be in the order of 1. If $d > d_{\mathrm{cr}}$ (for our set of parameters $d_{\mathrm{cr}} \approx 0.\ 003$), then the solitons do not experience MI. The diagram summarizing the existence of the stationary and oscillatory 2D solitons is shown in Fig. 6.15c.

6.7 Summary

We have reviewed the properties of bright and dark polariton solitons having zero and nonzero transverse momenta and existing in the spectral proximity of either the lower or the upper-polariton branch. In particular, we have shown the existence of 2D bright solitons for the case when the effective polariton masses along orthogonal directions have opposite signs. We demonstrated that accounting for the finite exciton mass and the saturation of the exciton–photon coupling is crucially important for the formation and the properties of solitons originating from the upper-polariton branch. Microcavity polariton solitons described in this contribution exhibit picosecond excitation times and can be observed for pump powers a few orders of magnitude lower than those required for pure photonic solitons in the weak coupling regime of the semiconductor microcavities. Thus, the hybrid, half-light, half-matter, polariton solitons have potentially significant advantages in all-optical signal processing applications over the light-only cavity solitons.

Acknowledgements Financial support by the Federal Ministry of Education and Research (PhoNa)as well as from the UK EPSRC project EP/D079225/1 and the Deutsche Forschungsgemeinschaft (Research unit 532) is acknowledged.

References

1. A. Scott, *Nonlinear Science* (Oxford University Press, Oxford, 1999)
2. Y. Kivshar, G. Agrawal, *Optical Solitons: From Fibers to Photonic Crystals* (Academic Press, San Diego, 2001)
3. N. Akmediev, A. Ankiewicz (eds.) *Dissipative Solitons* (Springer, Berlin, 2005)
4. N.N. Rosanov, *Spatial Hysteresis and Optical Patterns* (Springer, Berlin, 2002)
5. D. Michaelis, U. Peschel, F. Lederer, Multistable localized structures and superlattices in semiconductor optical resonators. Phys. Rev. A **56**, 3366(R) (1997)
6. V.B. Taranenko, I. Ganne, R. Kuszelewicz, C.O. Weiss, Spatial solitons in a semiconductor microresonator. Appl. Phys. B **72**, 377 (2001)
7. U. Peschel, D. Michaelis, C.O. Weiss, Spatial solitons in optical cavities. IEEE J. Quant. Electron. **39**, 51 (2003)
8. Ye. Larionova, C.O. Weiss, O. Egorov, Dark solitons in semiconductor resonators. Opt. Express **13**, 8308 (2005)
9. M. Brambilla, L.A. Lugiato, F. Prati, L. Spinelli, W.J. Firth, Spatial soliton pixels in semiconductor devices. Phys. Rev. Lett. **79**, 2042 (1997)
10. S. Barland, J.R. Tredicce, M. Brambilla, L.A. Lugiato, S. Balle, M. Giudici, T. Maggipinto, L. Spinelli, G. Tissoni, T. Knödl, M. Miller, R. Jäger, Cavity solitons as pixels in semiconductor microcavities. Nature **419**, 699 (2002)
11. Y. Tanguy, T. Ackemann, W.J. Firth, R. Jäger, Realization of a semiconductor-based cavity soliton laser. Phys. Rev. Lett. **100**, 013907 (2008)
12. G.S. McDonald, W.J. Firth, Spatial solitary-wave optical memory. J. Opt. Soc. Am. B **7**, 1328 (1990)
13. F. Pedaci, S. Barland, E. Caboche, P. Genevet, M. Giudici, J.R. Tredicce, T. Ackemann, A.J. Scroggie, W.J. Firth, G.L. Oppo, G. Tissoni, R. Jäger, All-optical delay line using semiconductor cavity solitons. Appl. Phys. Lett. **92**, 011101 (2008)
14. T. Ackemann, W.J. Firth, G.L. Oppo, Fundamentals and applications of spatial dissipative solitons in photonic devices. Adv. Atom. Mol. Opt. Phys. **57**, 324 (2009)
15. J.J. Hopfield, Theory of the contribution of excitons to the complex dielectric constant of crystals. Phys. Rev. **112**, 1555 (1958)
16. C. Weisbuch, M. Nishioka, A. Ishikawa, Y. Arakawa, Observation of the coupled exciton-photon mode splitting in a semiconductor quantum microcavity. Phys. Rev. Lett. **69**, 3314 (1992)
17. H. Haug, S.W. Koch, *Quantum Theory of the Optical and Electronic Properties of Semiconductots*, 2nd edn. (World Scientific Publishing Co. Pte. Ltd., Hong Kong, 1993)
18. R. Houdre, J.L. Gibernon, P. Pellandini, R.P. Stanley, U. Oesterle, C. Weisbuch, J.O′ Gorman, B. Roycroft, M. Ilegems, Saturation of the strong-coupling regime in a semiconductor microcavity: free-carrier bleaching of cavity polaritons. Phys. Rev. B **52**, 7810 (1995)

19. A.V. Kavokin, J.J. Baumberg, G. Malpuech, F.P. Laussy, *Microcavities* (Oxford University Press, New York, 2007)
20. B. Deveaud (ed.) *The Physics of Semiconductor Microcavities* (Willey-VCH, Weinheim, 2007)
21. S. Schmitt-Rink, D.S. Chemla, D.A.B. Miller, Theory of transient excitonic optical nonlinearities in semiconductor quantum-well structures. Phys. Rev. B **32**, 6601 (1985)
22. G. Rochat, C. Ciuti, V. Savona, C. Piermarocchi, A. Quattropani, P. Schwendimann, Excitonic Bloch equations for a two-dimensional system of interacting excitons. Phys. Rev. B **61**, 13856 (2000)
23. E. Burstein, C. Weisbuch (eds.) *Confined Electrons and Photons: New Physics and Applications* (Plenum, New York, 1995)
24. S.A. Moskalenko, D.W. Snoke, *Bose-Eistein Condensation of Excitons and Biexcitons* (Cambridge University Press, Cambridge, 2000)
25. J.P. Reithmaier, G. Sek, A. Löffler, C. Hofmann, S. Kuhn, S. Reitzenstein, L.V. Keldysh, V.D. Kulakovskii, T.L. Reinecke, A. Forchel, Strong coupling in a single quantum dot-semiconductor microcavity system. Nature **432**, 197 (2004)
26. D.G. Lidzey, D.D.C. Bradley, M.S. Skolnick, T. Virgili, S. Walker, M.S. Whittaker, Strong exciton-photon coupling in an organic semiconductor microcavity. Nature **395**, 53 (1998)
27. A. Wallraff, D.I. Schuster, A. Blais, L. Frunzio, R.-S. Huang, J. Majer, S. Kumar, S.M. Girvin, R.J. Schoelkopf, Strong coupling of a single photon to a superconducting qubit using circuit quantum electrodynamics. Nature **431**, 162 (2004)
28. G. Khitrova, H.M. Gibbs, M. Kira, S.W. Koch, A. Scherer, Vacuum Rabi splitting in semiconductors. Nat. Phys. **2**, 81 (2006)
29. J. Kasprzak, M. Richard, S. Kundermann, A. Baas, P. Jeambrun, J.M.J. Keeling, F. M. Marchetti, M.H. Szymanska, R. Andre, J.L. Staehli, V. Savona, P.B. Littlewood, B. Deveaud, L.S. Dang, Bose-Einstein condensation of exciton polaritons. Nature **443**, 409 (2006)
30. R. Balili, V. Hartwell, D. Snoke, L. Pfeiffer, K. West, Bose-Einstein condensation of microcavity polaritons in a trap. Science **316**, 1007 (2007)
31. I. Carusotto, C. Ciuti, Probing microcavity polariton superfluidity through resonant rayleigh scattering. Phys. Rev. Lett. **93**, 166401 (2004)
32. A. Amo, J. Lefrere, S. Pigeon, C. Adrados, C. Ciuti, I. Carusotto, R. Houdre, E. Giacobino, A. Bramati, Superfluidity of polaritons in semiconductor microcavities. Nat. Phys. **5**, 805 (2009)
33. R. Houdre, R.P. Stanley, U. Oesterle, M. Ilegems, C. Weisbuch, Room-temperature cavity polaritons in a semiconductor microcavity. Phys. Rev. B **49**, 16761 (1994)
34. S. Christopoulos, G. Baldassari, A.J.D. Grundy, P.G. Lagoudakis, A.V. Kavokin, J.J. Baumberg, G. Christmann, R. Butte, E. Feltin, J.-F. Carlin, N. Grandjean, Room-temperature polariton lasing in semiconductor microcavities. Phys. Rev. Lett. **98**, 126405 (2007)
35. A. Baas, J.Ph. Karr, H. Eleuch, E. Giacobino, Optical bistability in semiconductor microcavities. Phys. Rev. A **69**, 023809 (2004)
36. A. Tredicucci, Y. Chen, V. Pellegrini, M. Börger, F. Bassani, Optical bistability of semiconductor microcavities in the strong-coupling regime. Phys. Rev. A **54**, 3493 (1996)
37. D. Bajoni, E. Semenova, A. Lemaitre, S. Bouchoule, E. Wertz, P. Senellart, S. Barbay, R. Kuszelewicz, J. Bloch, Optical bistability in a GaAS-based polariton diode. Phys. Rev. Lett. **101**, 266402 (2008)
38. P.G. Savvidis, J.J. Baumberg, R.M. Stevenson, M.S. Skolnick, D.M. Whittaker, J.S. Roberts, Angle-resonant stimulated polariton amplifier. Phys. Rev. Lett. **84**, 1547 (2000)
39. D.M. Whittaker, Classical treatment of parametric processes in a strong-coupling planar microcavity. Phys. Rev. B **63**, 193305 (2001)
40. A. Baas, J.-Ph. Karr, M. Romanelli, A. Bramati, E. Giacobino, Optical bistability in semiconductor microcavities in the nondegenerate parametric oscillation regime: analogy with the optical parametric oscillator. Phys. Rev. B **70**, 161307(R) (2004)

41. C. Ciuti, P. Schwendimann, B. Deveaud, A. Quattropani, Theory of the angle-resonant polariton amplifier. Phys. Rev. B **62**, 4825(R) (2000)
42. C. Ciuti, P. Schwendimann, A. Quattropani, Theory of polariton parametric interactions in semiconductor microcavities. Semicond. Sci. Technol. **18**, 279 (2003)
43. M. Wouters, I. Carusotto, Parametric oscillation threshold of semiconductor microcavities in the strong coupling regime. Phys. Rev. B **75**, 075332 (2007)
44. D.M. Whittaker, Effects of polariton-energy renormalization in the microcavity optical parametric oscillator. Phys. Rev. B **71**, 115301 (2005)
45. D.N. Krizhanovskii, S.S. Gavrilov, A.P.D. Love, D. Sanvitto, N.A. Gippius, S.G. Tikhodeev, V.D. Kulakovskii, D.M. Whittaker, M.S. Skolnick, J.S. Roberts, Self-organization of multiple polariton-polariton scattering in semiconductor microcavities. Phys. Rev. B **77**, 115336 (2008)
46. I.B. Talanina, M.A. Collins, V.M. Agranovich, Polariton solitons in semiconductors. Solid State Communicat. **88**, 541 (1993)
47. D.V. Skryabin, A.V. Yulin, A. Maimistov, Localized polaritons and second-harmonic generation in a resonant medium with quadratic nonlinearity. Phys. Rev. Lett. **96**, 163904 (2006)
48. K.G. Lagoudakis, M. Wouters, M. Richard, A. Baas, I. Carusotto, R. Andre, L.S. Dang, B. Deveaud-Pledran, Quantized vortices in an exciton-polariton condensate. Nat. Phys. **4**, 706 (2008)
49. Ye. Larionova, W. Stolz, C.O. Weiss, Optical bistability and spatial resonator solitons based on exciton-polariton nonlinearity. Opt. Lett. **33**, 321 (2008)
50. A. Amo, D. Sanvitto, F.P. Laussy, D. Ballarini, E. del Valle, M.D. Martin, A. Lemaitre, J. Bloch, D.N. Krizhanovskii, M.S. Skolnick, C. Tejedor, L. Vina, Collective fluid dynamics of a polariton condensate in a semiconductor microcavity. Nature **457**, 291 (2009)
51. D. Sanvitto, A. Amo, F.P. Laussy, A. Lemaitre, J. Bloch, C. Tejedor, L. Vina, Polariton condensates put in motion. Nanotechnology **21**, 134025 (2010)
52. A. Amo, D. Sanvitto, L. Vina, Collective dynamics of excitons and polaritons in semiconductor nanostructures. Semicond. Sci. Technol. **25**, 043001 (2010)
53. A.V. Yulin, O.A. Egorov, F. Lederer, D.V. Skryabin, Dark polariton solitons in semiconductor microcavities. Phys. Rev. A **78**, 061801(R) (2008)
54. O.A. Egorov, D.V. Skryabin, A.V. Yulin, F. Lederer, Bright cavity polariton solitons. Phys. Rev. Lett. **102**, 153904 (2009)
55. D.V. Skryabin, O.A. Egorov, A.V. Gorbach, F. Lederer, One-dimensional polariton solitons and soliton waveguiding in microcavities. Superlatt. Microstr. **47**, 5 (2010)
56. O.A. Egorov, A.V. Gorbach, F. Lederer, D.V. Skryabin, Two-dimensional localization of exciton polaritons in microcavities. Phys. Rev. Lett. **105**, 073903 (2010)
57. O.A. Egorov, D.V. Skryabin, F. Lederer, Polariton solitons due to saturation of the exciton-photon coupling. Phys. Rev. B **82**, 165326 (2010)
58. L.A. Lugiato, R. Lefever, Spatial dissipative structures in passive optical systems. Phys. Rev. Lett. **58**, 2209 (1987)
59. D.V. Skryabin, W.J. Firth, Dynamics of self-trapped beams with phase dislocation in saturable Kerr and quadratic nonlinear media. Phys. Rev. E **58**, 3916 (1998)
60. O. Egorov, F. Lederer, Rotating cavity solitons in semiconductor microresonators. Phys. Rev. E **75**, 017601 (2007)
61. S. Fedorov, D. Michaelis, U. Peschel, C. Etrich, D.V. Skryabin, N. Rosanov, F. Lederer, Effects of spatial inhomogeneities on the dynamics of cavity solitons in quadratically nonlinear media. Phys. Rev. E **64**, 036610 (2001)
62. O. Egorov, F. Lederer, K. Staliunas, Subdiffractive discrete cavity solitons. Opt. Lett. **32**, 2106 (2007)
63. F. Tassone, C. Piermarocchi, V. Savona, A. Quattropani, P. Schwendimann, Bottleneck effect in the relaxation and photoluminescence of microcavity polaritoms. Phys. Rev. B **56**, 7554 (1997)
64. D. Gomila, M.A. Matias, P. Colet, Excitability mediated by localized structures in a dissipative nonlinear optical cavity. Phys. Rev. Lett. **94**, 063905 (2005)

Chapter 7
Observation of "True" Optical Vortices in a Laser System

S. Barland, E. Caboche, P. Genevet, X. Hachair, M. Giudici, F. Pedaci, and J.R. Tredicce

7.1 Introduction

Nonlinear dynamics in optics developed mainly during the 1980s in the last century. Probably the interest in the subject begun in 1975 when Haken [1] demonstrated that Maxwell–Bloch equations in the single mode approximation were isomorphous to the Lorenz model equations [2]. Thus, a laser would show deterministic chaos for appropriate values of the control parameters. Unfortunately, almost no laser is able to operate at those parameter values because pumping and loss rate for the field were too high in order to reach Lorenz instabilities. It was only in 1982 that chaotic operation of a laser was experimentally demonstrated [3]. Since then, a series of papers have appeared in the literature dedicated to the unpredictable dynamical behavior in optical systems [4], and particularly in lasers and lasers systems [5]. Furthermore, a new classification of lasers was made based on their dynamical behavior instead of the characteristics of the material, or the type of pumping mechanism [6, 7]. At that time, almost all of the work was dedicated to the temporal behavior of lasers ignoring the spatial coordinates. Only a few papers considered the temporal instabilities in multimode lasers, which is equivalent to considering the spatial variation of the amplitude of the field along the direction of propagation of the light [8–10]. All this activity confirms the important role that lasers, or in a more general view, optical systems, play on the experimental studies in nonlinear dynamics. They are an excellent test-bench for more general theories because instabilities occur at frequency ranges enough fast to allow a fine and constant control of parameter values as well as an easy way to explore all the interesting control parameter space in reasonable times. They also had the advantage that there exist reliable models based on basic principles that are able to reproduce most of the experimental results.

S. Barland • E. Caboche • P. Genevet • X. Hachair • M. Giudici • F. Pedaci • J.R. Tredicce (✉)
CNRS, Institut Nonlineaire de Nice (UMR6618), Universite de Nice-Sophia Antipolis,
1361 Route des Lucioles, Valbonne 06560, France
e-mail: jorge.tredicce@inln.cnrs.fr

Z. Chen and R. Morandotti (eds.), *Nonlinear Photonics and Novel Optical Phenomena*,
Springer Series in Optical Sciences 170, DOI 10.1007/978-1-4614-3538-9_7,

The interest on spatiotemporal instabilities in lasers grew by the end of the 1980s. It was again some theoretical papers at the origin of developing the activity on transverse instabilities in lasers. Though the appearance of spatial structures in the intensity distribution had been very well known since the 1960s, the interpretation of the observed patterns was mainly done on the basis of a linear theory based on the empty cavity modes [11–15]. Modulation instabilities (also called Turing instabilities) [16, 17], cooperative frequency locking among transverse modes [18], and optical vortices [19] appear as possible observable dynamical behavior in broad area lasers. Pattern formation and cooperative frequency locking were soon observed in different experimental setups [20–23], but it was difficult instead to reach the conditions for which "true" optical vortices would spontaneously appear in the transverse intensity distribution. However, several papers appear claiming the observation of optical vortices [24, 25].

Here, we call "true" optical vortices those points in space where the local solution takes the value corresponding to an unstable fixed point, but it is free to move in the transverse plane. In lasers above threshold, the unstable fixed point corresponds to the zero solution of the electromagnetic field. It is well known that all transverse modes (except the fundamental or Gaussian one) of an optical cavity have at least one point where the electromagnetic field vanishes and the circulation of the phase around such point is 2π. These are also properties of "true" optical vortices. However, these points would not be able to have an intrinsic dynamics and move around because they belong to a fixed structure, like the nodes of a standing wave. In order to reach the second condition it is necessary that the correlation length of the structure in the transverse plane be much smaller than the spatial size of the system and of course of the boundary conditions. If the laser is able to construct such solution, then we can say that probably we can observe spatiotemporal chaos or even fully developed optical turbulence. Unfortunately, most lasers prefer to show simple or more or less complex structures but always imposing long-range correlations [26] because different spatial structures covering the whole space are strongly competitive among them [27]. One cavity mode tends to win, due in general to its higher gain, and the winning-takes-all dynamics destroy the possibility of creating a complex intensity distribution with a small correlation length. In conclusion, observing experimentally "true" optical vortices in laser systems is a kind of nightmare requiring a high Fresnel number, a strong nonlinearity, and almost equal probability for a high number of transverse modes. Furthermore, the observation of any type of localized structure, a structure that exist in a spatial region decorrelated from the boundaries and other region of space, requires a bistable behavior between a homogeneous state and a pattern [28].

In this chapter, we describe a laser system complying with the necessary conditions to observe optical vortices. In Sect. 7.2, we describe the experimental setup. The experimental results and their analysis are "explored" in Sect. 7.3.

7.2 Experimental Setup

The experimental setup shown in Fig. 7.1 is based on two nominally identical vertical cavity surface emitter lasers (VCSELs) put into a face-to-face configuration [30]. The distance between the two VCSELs can be varied between 6 and 60 cm. The diameter of each VCSEL is 200 μm. They are electrically pumped by a stabilized power supply and stabilized in temperature by a Peltier. Two identical lenses are placed between the two lasers in self-conjugate planes such that the near-field distribution of each VCSEL is imaged into the other. This scheme insures that we can maintain a high Fresnel number despite the relatively long length of the composed cavity [29, 30]. A 20% beam splitter placed almost exactly in the middle gives rise to two output beams. One of them is sent to a fast detector in order to observe temporal oscillations if present and two CCD cameras give the near-field profiles of the two beams. To insure that the position of the lenses is such that the self-imaging condition is obtained, we superpose the emission of each VCSEL separately with its reflection from the mirror of the other.

The two VCSELs provided by ULM photonics operate at a wavelength of 980 nm; thus, they are bottom emitter devices [31]. This feature plays an important role in allowing a relatively uniform pumping current density over almost the whole transverse section of the laser. In fact, one of the electrodes is covering the full

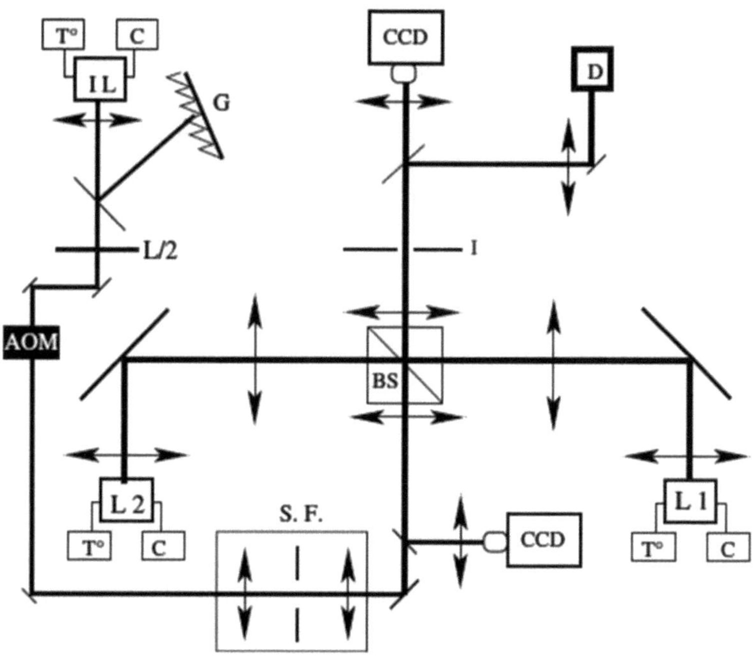

Fig. 7.1 Schematic of the experiment. *L1* laser pumped above threshold, *L2* laser pumped below transparency, *IL* incoherent writing laser, *bs* beam splitter, *D* high-bandwidth detector, *CCD* charge coupled device camera, *T°* temperature controller, *C* power supply, *L/2* half-wave plate, *G* grating, *AOM* acoustooptic modulator, *S.F.* spatial filter, *I* iris (reprinted from [29])

surface and the other is a metallic ring through which the electromagnetic beam is coming out. Usually, the ring electrode would generate a greater current density on the border than at the center of the active material. This inhomogeneity is not important in small area devices but would generate strong gradients in broad area's ones. The presence of gradients would affect any spatial structure that is not fixed by the boundary conditions and, if such structures exist, they will disappear so fast that it would be impossible to observe them with the state-of-the-art CCD cameras. The nominal threshold current for the VCSELs we used is 400 mA. One of the two lasers is pumped above threshold, while the current density on the other one is low enough to not reach the transparency condition. Thus, the second laser will behave as an absorber rather than an amplifier or an emitter. In other words, our optical system can be described as a laser with saturable absorber in a compound cavity. However, to get interesting dynamical behavior in a laser with saturable absorber, it is necessary that the resonances of both active and passive media be close enough. In VCSELs, and in semiconductors in general, the resonance frequency is a function of the temperature and therefore also a function of the pumping current due to Joule effect. In our experimental setup, the mismatch between the cavity resonances is then controlled by the difference in the operating temperature of each substrate. The strength of the coupling between the two VCSELs depends also on the loss rate of the electromagnetic field which, in turn, depends on the reflectivity of the intracavity beam splitter. A 20% reflectivity beam splitter allowed a good coupling and at the same time enough output intensity to be detected easily by the detection system. A diaphragm in the path of one of the output beams is used in order to detect the temporal behavior of the intensity at a point of the pattern with the fast detector. The bandwidth was limited at 6 GHz by the LeCroy Wavemaster 8600A oscilloscope. Furthermore, an external beam 15 μm in diameter provided by an edge emitter semiconductor laser can be used to ignite localized structures in the system. This so-called "writing beam" is controlled both in frequency and amplitude. Its frequency is selected by an intracavity grating while its intensity can be switched on and off by an acoustooptic modulator inserted on its optical path.

7.3 Experimental Results

As we describe in Sect. 7.2, the main control parameter of the device is the detuning between the resonances frequencies of the two VCSELs. Such difference in frequency depends on the temperature of the substrate and the pumping current. We choose to keep both the temperatures constant and change the pumping currents in order to study the behavior of the system as the detuning changes. In Fig. 7.2 we show the local intensity output as a function of the current (I1) of the emitter while keeping constant the current (I2) of the absorber. By sweeping I1 back and forth, we can observe different behavior in different regions of parameter space:

1. Region A of Fig. 7.2 corresponds to an increase in intensity of the spontaneous emission.

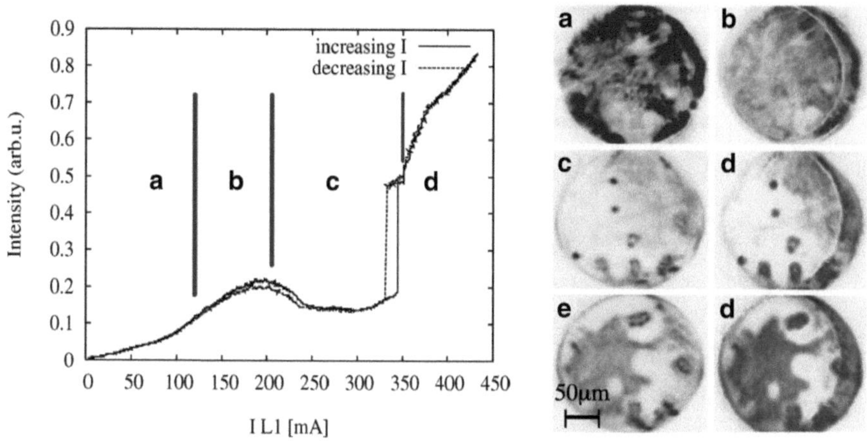

Fig. 7.2 *Left*: Local intensity output emitted by the system as a function of I1 keeping all other parameters constant. (A) System below threshold, (B) lasing action takes place, (C) absorption by L2, yielding bi-stable behavior, (D) pattern formation. *Right*: Near field of both devices. *Dark areas* correspond to high intensities. (**a** and **b**) Near-field image of L1 (L2), before the interaction (IL1 = 180 mA), (**c** and **d**) near field of L1 (L2), in the absorption zone (IL1 = 358 mA), (**e** and **f**) near field of L1 (L2), when the pattern is developed (IL1 = 365 mA). L2 is slightly shifted on the left (reprinted from [30])

2. Region B, coherent emission is obtained but the two VCSELs do not interact because the detuning is big enough such that the absorber is almost transparent to the radiation emitted.
3. Region C, the detuning becomes small enough and the absorber begins to interact with the emitter. The output intensity decreases, and then the absorber saturates and there is a discontinuous increase of the intensity. Sweeping back the control parameter, we observe a region of bistability. It is in this region where localized structures may appear.
4. Region D, a modulational instability develops inducing the appearance of patterns.

As a matter of fact, the appearance of bright localized structures was reported in this system and also in an usual laser with saturable absorber [29, 30, 32, 33]. Such single peak localized structures are usually called "cavity solitons" (CS). Left panel in Fig. 7.2 shows typical intensity distributions including the presence of CS for those taken in region C (left panel c and d).

Usually, CS have been observed in VCSEL amplifiers and VCSEL laser with injected signal [34–37]. If the diameter of the VCSEL is larger than 100 μm, two or more CS do not interact among them if they are separated by a distance of the order of 20–25 μm. Furthermore, they can be switched on and off independently by a coherent or an incoherent beam [36, 37] at relatively high rate because the switching time is of the order of 50 ps [38]. Thus, CS may constitute the basis for

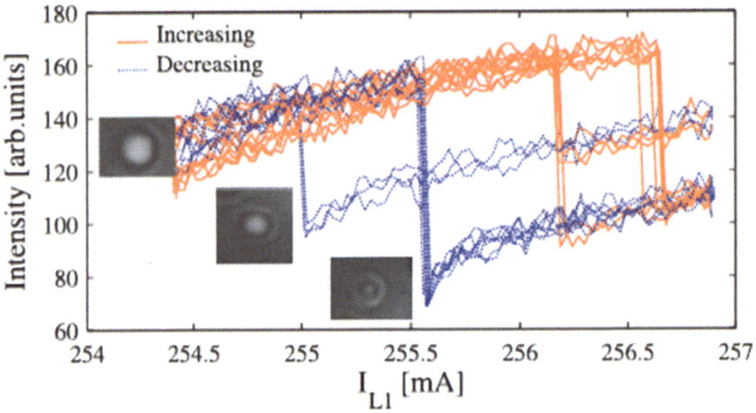

Fig. 7.3 Interferometric intensity signal of an LS as a function of I1. The interferometric intensity signal is obtained by integrating the phase profile of the monitored LS. The *dashed blue traces* correspond to decreasing values of I1, while *continuous red traces* correspond to increasing values of I1. Variation of the interferometric intensity signal corresponds to a mode hopping of the LS. The figure shows the multistability of the CS emission frequency. The phase profiles corresponding to each branch of the hysteresis loop are shown in the *insets*. The spatial region monitored is centered on the LS considered and has a diameter of 10 μm. *Bright zones* represent high intensities (reprinted from [47])

a fast all optical memory [34–38]. On the other hand, CS are very sensitive to gradients, both intensity and phase ones, then their position, motion and speed can be controlled by artificially adding gradients to the system [39–41]. This property of CS, which is a general property of localized structures in nonlinear dynamical systems, opens the way to some interesting applications like an all optical shift register or a force microscope [42–45].

A broad area laser with saturable absorber or face-to-face VCSELs do not require an external injected signal and therefore it is a simpler device easy to miniaturize. Such optical systems belong to the class of the so-called CS Lasers (CSL). Another example of CSL is a broad area laser with frequency selective feedback [46]. The main physical difference imposed by the presence of an external field is that the phase symmetry is broken. The phase is coupled to the intensity, and if there are not temporal instabilities, the phase of the output field is determined by the injected field. It is worthwhile to notice that all localized structures, for which it was proven that the correlation length in the transverse plane was much smaller than the size of the system, are a single intensity peak structure when the phase symmetry is broken.

Instead for a CSL, like the two face-two-face VCSEL, the system is phase symmetric. In principle, localized structures may appear for the same parameter values with different phases, and even frequencies being incoherent among them as it was shown in [47]. Furthermore, single peak structures may coexist with other multipeak localized structures. In Fig. 7.3 it is shown not only that different

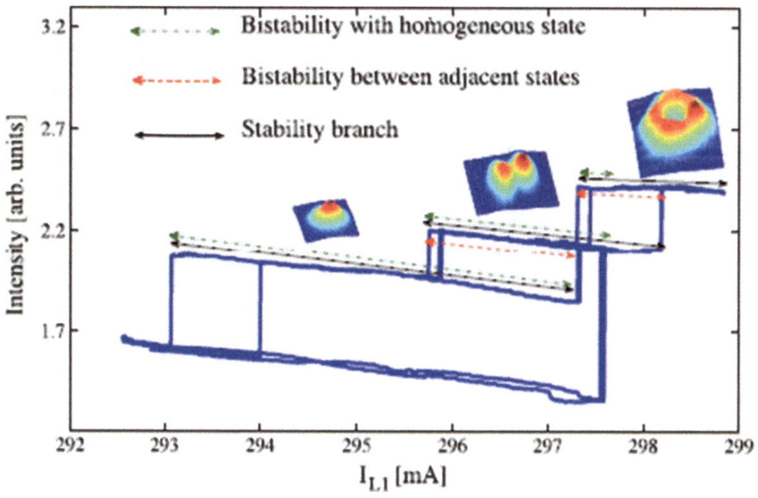

Fig. 7.4 Bifurcation diagram showing the spontaneous switching of the system between different solutions. The ring can coexist with an "off" state and/or with a two-humps state (reprinted from [49])

localized structures appear but also that there is a marked multistability among them [48]. Each type of structure is distinguished by both the intensity distribution and its operating frequency as transverse modes in a laser.

In a well-defined region of parameter values, three types of structures are observed which coexist with the homogeneous zero solution. In Fig. 7.4, bifurcation diagram shows the multistability among the homogeneous, a single (CS), a two peak, and a ring structures [49]. By increasing the current I1 from 290 mA, we begin with the homogeneous solution. This solution loses stability at around 297.5 mA and a two peak structure appears which loses stability at 298.2 mA jumping to the upper branch characterized by a ring-like intensity distribution. Sweeping back I1, we reach the two peak branch at 297.3 mA which remains stable till 296 mA. Then, a single peak structure appears. Increasing the current again, the single peak structure remains stable until 297.2 mA.

Thus, the ring-like structure is bistable with the homogeneous solution and the two humps structure but we cannot reach its branch from the single peak structure. This type of bifurcation diagram for localized structures has been already discussed in the literature [28, 50, 51]. In Fig. 7.5, we can see how such rings can be generated in different spatial locations [49]. Adding a Mach–Zender interferometer at the output path of the device, it is possible to observe the interference of each localized structure with itself or with adjacent ones. A detailed explanation of how the interference patterns are obtained is found in [37].

The result of the interference of a ring with itself when both arms of the Mach–Zender interferometer are perfectly aligned is shown in the right panel of Fig. 7.6. The reference beam for the interference is taken from the region of the pattern where the ring is located. Interference appears only where there is mutual

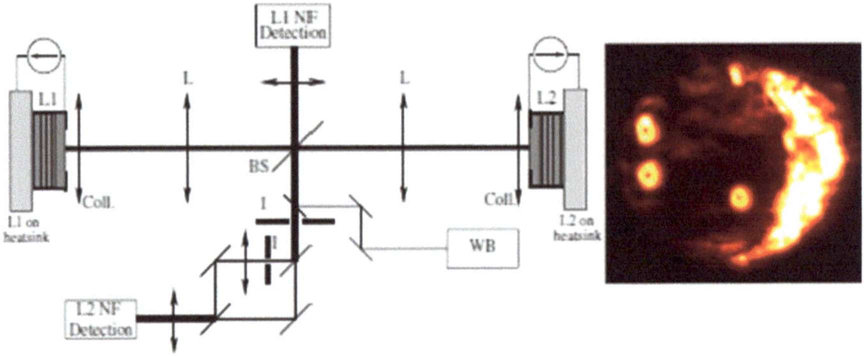

Fig. 7.5 *Left*: Experimental setup. Two broad area semiconductor lasers (L1 and L2) are coupled by imaging them onto each other via collimating optics (Coll) and lenses (L). Part of the emitted beam is extracted from the compound cavity via a beam splitter (BS) for near-field (NF) detection allowing interferometric measurements and spatial filtering (I). A tiny beam from a tunable laser can be used to apply a local perturbation to the system (WB). *Right*: Spontaneously formed intensity rings in the near field. The two devices (200 m diameter) are laterally shifted with respect to each other. In the spatial region where they overlap, the absorption can locally saturate and lead to the formation of several bright bistable rings (reprinted from [49])

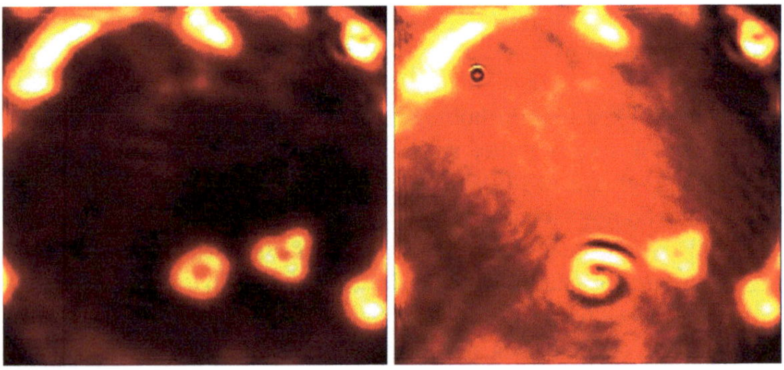

Fig. 7.6 *Left panel*: Near-field intensity of the system, showing two bright spatial structures sitting on the dark homogeneous background corresponding to non saturated absorption. *Right*: Interferometric measurement. When both arms of the interferometer are aligned (a small part of the ring structure being superimposed with the whole ring structure and surrounding region), the phase profile of the ring structure is indicated by spiral-like interference pattern (reprinted from [49])

coherence between the reference beam and the spatial region with which it overlaps. A remarkable feature is that the interference pattern appears only where the ring structure itself is, demonstrating that this structure is de-correlated from the rest of the pattern. In particular, the neighboring structure does not interfere at all with the reference beam. This absence of mutual coherence between the fields emitted in distinct spatial areas shows the lack or extreme weakness of any coherent coupling between these two areas.

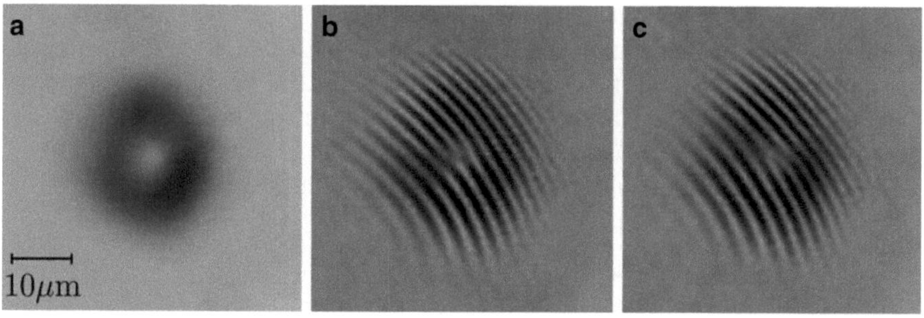

Fig. 7.7 (**a**) Near-field intensity of a localized vortex. When a part of it is magnified and interferes with the whole vortex, fringes appear if both beams are tilted with respect to each other. Their orientation and periodicity is set by the tilt angle. (**b** and **c**) The dislocation of the fringe pattern indicates the presence of a phase defect and the direction of the dislocation gives the sign of the charge (reprinted from [49])

The other striking feature is the presence of a phase discontinuity originating in the center of the ring, which contains, therefore, a phase defect revealed by the typical spiral of the phase.

While the chirality of the phase profile of the ring structure is readily apparent in Fig. 7.6, it can also be conveniently detected by performing an identical measurement, but tilting the two beams with respect to each other. In this case, interference fringes are detected, as shown in Fig. 7.7. The presence of a dislocation in the pattern (Fig. 7.7b, c) reveals the existence of a phase defect at the core of the ring since the circulation of the phase around the center is 2π. Even though the chirality of the ring structure appears to be very robust, we occasionally observed spontaneous switching between the two chirality states shown in Fig. 7.7b, c. This switching can be attributed to exchange of charge with the surrounding zero solution (as was observed at the boundaries of the system in [39]).

Thus, the observed ring structures in this face-to-face VCSELs device have the following properties: (1) there is a zero of the electromagnetic field at the core of the structure; (2) there is a circulation of the phase around the core equal to 2π; and (3) there is no correlation among this structure and other regions of the pattern.

Furthermore, these ring structures can be switched on and off independently at different locations and they move under the presence of any intensity of phase gradient.

In other words, we are in the presence of what we defined as a "TRUE OPTICAL VORTEX" at the beginning of this manuscript. It is worthwhile to note that this phenomenon is strongly related to the absence of any phase reference in this system. In the majority of experiments regarding localized structures in optics which involve some form of coherent energy input, the phase symmetry is broken and ring structures are forbidden. In fact, a system with injected field will fix the phase of the whole localized structures not allowing for not vanishing circulation of the phase. On the other hand, it is important to also note that the intensity vanishes for the homogeneous solution from which we can construct the ring-like structure.

This property of the homogeneous solution is also an important feature because (1) it will not allow the generation and propagation of phase waves [52] which would not allow the generation of localized structures and (2) it provides the source of pairs of defects from which the system construct the ring structure with just one defect at the core.

We thank L. Gil and G. Tissoni for a very helpful discussion.

References

1. H. Haken, Phys. Lett. **53A**, 77 (1975)
2. E.N. Lorenz, J. Atmos. Sci. **20**, 130 (1963)
3. F.T. Arecchi, R. Meucci, G.P. Puccioni, J.R. Tredicce, Phys. Rev. Lett. **49**, 1217 (1982)
4. F.T. Arecchi, N.B. Abraham, L. Lugiato (eds.), *Instabilities and Chaos in Quantum Optics*, vol. II (Plenum, New York, NY, 1988)
5. N.B. Abraham, P. Mandel and L.M. Narducci, *Dynamical Instabilities and Pulsations in Lasers,* Progress in Optics, **25**, pp. 1–190 (1988)
6. F. Arecchi, G. Lippi, G. Puccioni, J. Tredicce, Opt. Commun. **51**, 308 (1984)
7. J.R. Tredicce, F.T. Arecchi, G.L. Lippi, G.P. Puccioni, J. Opt. Soc. Am. B **2**, 173–183 (1985)
8. N.J. Halas, S.N. Liu, N.B. Abraham, Phys. Rev. A **28**, 2915 (1983)
9. L.W. Casperson, Phys. Rev. A **21**, 911 (1980)
10. L.W. Casperson, Phys. Rev. A **23**, 248 (1981)
11. H. Kogelnik, T. Li, Appl. Opt. **5**, 1550 (1966). See, for example
12. H. Kogelnik, in *Lasers: A Series of Advances*, ed. by A.K. Levine, vol. 1 (Marcel, New York, NY, 1966), p. 295
13. A. Yariv, *Optical Electronics*, 3rd edn. (Holt, Rinehart and Winston, New York, NY, 1985)
14. A.E. Siegman, *Lasers* (University Science Books, Mill Valley, CA, 1986)
15. O. Svelto, *Principles of Lasers* (Plenum, New York, NY, 1982). Chapter 4
16. A.M. Turing, Phil. Trans. R. Soc. London, Ser. B **237**, 36 (1952)
17. L.A. Lugiato, R. Lefever, Phys. Rev. Lett. **58**, 2209 (1987)
18. L.A. Lugiato, G.-L. Oppo, M.A. Pernigo, J.R. Tredicce, L.M. Narducci, D.K. Bandy, Opt. Commun. **68**, 63 (1988)
19. P. Coullet, L. Gil, F. Rocca, Opt. Commun. **73**, 403 (1989)
20. Z. Chen, N.B. Abraham, S. Balle, E.J. D'Angelo, J.R. Tredicce, Chaos Solitons Fractals **4**, 1489 (1994)
21. C. Green, G.B. Mindlin, E.J. D'Angelo, H.G. Solari, J.R. Tredicce, Phys. Rev. Lett. **65**, 3124 (1990)
22. R. López Ruiz, G.B. Mindlin, C. Pérez García, J. Tredicce, Phys. Rev. A **47**, 500 (1993)
23. J.R. Tredicce, E.J. Quel, A.M. Ghazzawi, C. Green, M.A. Pernigo, L.M. Narducci, L.A. Lugiato, Phys. Rev. Lett. **62**, 1274 (1989)
24. G. Slekys, C.O. Weiss, D.Y. Tang et al., J. Opt. Soc. Am. **B11**, 2089–2094 (1994)
25. A.V. Mamaev, M. Saffman, A.A. Zozulya, Phys. Rev. Lett. **77**, 4544 (1996)
26. G. Huyet, C. Martinoni, J.R. Tredicce, S. Rica, Phys. Rev. Lett. **75**, 4027 (1995)
27. G. Huyet, C. Mathis, H. Grassi, J.R. Tredicce, N.B. Abraham, Opt. Commun. **111**, 488 (1994)
28. P. Coullet, C. Riera, C. Tresser, Phys. Rev. Lett. **84**, 3069 (2000)
29. P. Genevet, S. Barland, M. Giudici, J.R. Tredicce, Phys. Rev. A **79**, 033819 (2009)
30. P. Genevet, S. Barland, M. Giudici, J.R. Tredicce, Phys. Rev. Lett. **101**, 123905 (2008)
31. M. Grabherr, R. Jäger, M. Miller, C. Thalmaier, J. Heerlein, R. Michalzik, K. Ebeling, Photon. Technol. Lett. **10**, 1061 (1998)

32. T. Elsass, K. Gauthron, G. Beaudoin, I. Sagnes, R. Kuszelewicz, S. Barbay, Eur. Phys. J. D **59**, 91 (2010)
33. T. Elsass, K. Gauthron, G. Beaudoin, I. Sagnes, R. Kuszelewicz, S. Barbay, Appl. Phys. B **98**, 327 (2010)
34. S. Barland, J. Tredicce, M. Brambilla, L. Lugiato, S. Balle, M. Giudici, T. Maggipinto, L. Spinelli, G. Tissoni, T. Knödel, M. Miller, R. Jäger, Nature **419**, 699 (2002)
35. X. Hachair, F. Pedaci, E. Caboche, S. Barland, M. Giudici, J. Tredicce, F. Prati, G. Tissoni, R. Kheradmand, L. Lugiato, I. Protsenko, M. Brambilla, IEEE J. Sel. Top. Quant. Electron. **12**, 339 (2006)
36. S. Barbay, Y. Ménesguen, X. Hachair, L. Leroy, I. Sagnes, R. Kuszelewicz, Opt. Lett. **31**, 1504 (2006)
37. X. Hachair, S. Barbay, T. Elsass, I. Sagnes, R. Kuszelewicz, Opt. Exp. **16**, 9519 (2008)
38. X. Hachair, L. Furfaro, J. Javaloyes, M. Giudici, S. Balle, J. Tredicce, Phys. Rev. A **72**, 013815 (2005)
39. F. Pedaci, P. Genevet, S. Barland, M. Giudici, J.R. Tredicce, Appl. Phys. Lett. **89**, 221111 (2006)
40. E. Caboche, F. Pedaci, P. Genevet, S. Barland, M. Giudici, J.R. Tredicce, Phys. Rev. Lett. **102**, 163901 (2009)
41. W.J. Firth, A.J. Scroggie, Phys. Rev. Lett. **76**, 1623 (1996)
42. F. Pedaci, S. Barland, E. Caboche, P. Genevet, M. Giudici, J.R. Tredicce, T. Ackemann, A.J. Scroggie, W.J. Firth, G.-L. Oppo, G. Tissoni, R. Jäger, Appl. Phys. Lett. **92**, 011101 (2008)
43. F. Pedaci, G. Tissoni, S. Barland, M. Giudici, J. Tredicce, Appl. Phys. Lett. **93**, 111104 (2008)
44. Y. Tanguy, N. Radwell, T. Ackemann, R. Jäger, Phys. Rev. A **78**, 023810 (2008)
45. Y. Tanguy, T. Ackemann, R. Jager, Phys. Rev. A **74**, 053824 (2006)
46. Y. Tanguy, T. Ackemann, W.J. Firth, R. Jäger, Phys. Rev. Lett. **100**, 013907 (2008)
47. P. Genevet, M. Turconi, S. Barland, M. Giudici, J.R. Tredicce, Eur. Phys. J. D **59**, 109114 (2010)
48. P. Genevet, L. Columbo, S. Barland, M. Giudici, L. Gil, J.R. Tredicce, Phys. Rev. A **81**, 053839 (2010)
49. P. Genevet, S. Barland, M. Giudici, J.R. Tredicce, Phys. Rev. Lett. **104**, 223902 (2010)
50. W.J. Firth, L. Columbo, A.J. Scroggie, Phys. Rev. Lett. **99**, 104503 (2007)
51. S. Barbay, X. Hachair, T. Elsass, I. Sagnes, R. Kuszelewicz, Phys. Rev. Lett. **101**, 253902 (2008)
52. L. Columbo, L. Gil, J. Tredicce, Opt. Lett. **33**, 995 (2008)

Chapter 8
Scale-Free optics

E. DelRe and C. Conti

8.1 Introduction

The most fascinating images and patterns emerge when light diffracts from minute structures [1]. Even the image of an otherwise featureless hole produces enthralling ripples that spread out to invest space and form what is known as an Airy pattern. It is a basic fact that diffraction becomes dominant when the size of the feature becomes micrometric, and the transmitted wave has an angular spread $\Delta\theta$ that depends on the size of the aperture d measured in units of the optical wavelength λ, i.e., $\Delta\theta \simeq \lambda/d$. From a practical perspective, diffraction represents a major obstacle to imaging of finer details, and a great research effort is continuously exerted to overcome it. In fact, diffraction spreads the optical wave and blurs the spatial information encoded in the optical beam. Consider an image composed of separate pixels of characteristic size d and spacing l. Light emitted from the single pixels will blur after a propagation of $L_z \sim ld/\lambda$, i.e., when light from one pixel superimposes with light from an adjacent one. The result is a progressive loss of the initial encoded information. Diffraction entails limitations to all imaging applications that range from astronomical observations to high-resolution optical microscopy. Efforts aimed at taming the effects of diffraction on waves can be loosely fitted under the general heading of diffraction compensation. The basic idea is to have transmitted light not propagate in a homogeneous medium, where diffraction must intrinsically produce spreading, but in an inhomogeneous medium, in which the index of refraction n is spatially resolved. For example, in an optical waveguide, diffraction is compensated by an effective lensing effect distributed along the propagation direction of light [2]. The problem is that exact diffraction

E. DelRe (✉) • C. Conti
Dipartimento di Fisica, Università di Roma Sapienza and IPCF-CNR, Roma, Italy
e-mail: eugenio.delre@uniroma1.it; claudio.conti@uniroma1.it

Z. Chen and R. Morandotti (eds.), *Nonlinear Photonics and Novel Optical Phenomena*, Springer Series in Optical Sciences 170, DOI 10.1007/978-1-4614-3538-9_8,

compensation will take place only for specific waveforms, the guided modes, with specific shapes and sizes that depend on the features of the waveguide, precisely on the amplitude of the index of refraction modulation Δn and on the spatial scale d measured, once again, in units of λ [3]. The modes only form a discrete set of guided solutions that, hence, fail to allow the propagation of an image of arbitrary size. Even encoding the optical waveform in an extended family of modes in a multimode fiber simply transfers the distortive mechanism from the standard phase slippage of angled plane-wave components at the heart of diffraction to inter-modal phase slippage typical of mode dispersion.

Diffraction compensation can also be enacted through nonlinearity, the basic example being the optical spatial soliton [4–17]. In this case, the index of refraction inhomogeneity Δn that gives rise to distributed lensing, termed self-lensing, is produced by the propagating light beam itself. As for the case of guided modes, here the soliton waveforms are determined by the peak intensity I_p and size of the beam w_0 measured in units of the optical wavelength λ, giving rise to soliton existence conditions that amount to strict laws that form rigid constraints on the waves that can actually propagate without spreading.

The matter can be summarized in the basic observation that since diffraction is dependent on the size of the wave measured in units of λ, also its compensation will make use of effects that depend on the spatial scales measured in units of λ. Put differently, diffraction introduces an intrinsic spatial scale into the picture, the optical wavelength, that breaks scale-invariance symmetry and thwarts efforts at transmitting high-resolution images. For visible beams with $\lambda \sim 0.5$ μm, this implies that the direct observation of features below several micrometers is in itself a challenge, to the point that even the once simple task of observing electronic circuitry has become impractical with optical microscopes and requires more exacting techniques.

Limits to imaging caused by the wave nature of light become even more dramatic for subwavelength scales, where the very nature of electro-magnetic radiation does not allow the propagation of waves [3]. Hence the simple fact that the optical wave has a limited bandwidth, with a spectrum peaked at λ, introduces an even more fundamental limit associated to the evanescent wave spectrum, so that no plane wave component of transverse (with respect to the propagation direction) wave vector $|\mathbf{k}_{\perp}| > 2\pi n/\lambda$ will actually propagate from the emitting/transmitting object. This implies that an imaging system will simply not be able to detect features below $w_0 \sim \lambda/2n$. For visible light and imaging done in water solution, this means that features smaller than approximately 200 nm cannot be detected even with all diffraction compensating techniques in force.

In recent achievements, a profoundly innovative approach to the entire issue has been proposed and experimentally demonstrated, termed scale-free optics [18]. In scale-free optics, the issue of diffraction and evanescent wave formation is circumvented by using nonlinearity to deprive the optical propagation mechanism of its spatial scale λ. We here describe and review scale-free optics with particular focus on experiments and theory relative to effects observed in nanodisordered ferroelectrics [19]. The subject is still in its infancy and rests on nonlinear optical effects in complex solids [20], which is itself a new branch of optical research. Results hold the promise of a new general setting for optics where diffraction can be canceled and where subwavelength beams can propagate, in crystals whose optical response can be greatly varied through thermal history [21].

8.2 The Development of the Idea

The basic idea underlying scale-free optics is that when light propagates in a medium in which nonlinearity introduces an intensity-independent response that amounts to an anti-diffraction, conditions can be found for which diffraction is completely canceled. Seen in physical terms, this happens when the nonlinearity introduces an independent spatial scale L that cancels the optical wavelength scale in the propagation equation. The result is that there is no spatial scale that breaks scale-invariance. Accordingly, no diffraction intervenes and, in principle, no limitations associated to evanescent waves hold, so that also subwavelength features will propagate into the optical system. The idea was originally formulated in [22], where a set of optical waves in the form of nondiffracting, spreading, and converging solutions were predicted that supported scale invariance. In fact, the governing nonlinear model did not contain the wavelength scale λ in the diffractive propagation. Unfortunately, this prediction remained such because at the time the required material susceptibility was inaccessible: the nonlinear response was experimentally insufficient for the predicted scale-free regime. In fact, the effect emerged from the interplay between diffraction and a diffusive effect. This diffusive effect, at accessible temperatures, amounted to a canceling spatial scale $L \ll \lambda$ and, congruently, no scale-free effects were ever observed. It was the recent development of a new generation of out-of-equilibrium photorefractive ferroelectric crystals that profoundly changed the scenario [23]. It was found that relevant compositional disorder in composite optical quality samples [24] introduces a glass-forming liquid dielectric response [19]. This allowed the observation of an anomalously enhanced electro-optic susceptibility with $L \sim \lambda$. It was thus possible to observe scale-free optics in newly engineered Cu-doped lithium-enriched potassium-tantalate-niobate crystals (KTN:Li) [18]. The theoretical framework and numerical studies were then extended to encompass a full nonparaxial model, and this has allowed the prediction of subwavelength beam propagation, a striking phenomenon that still awaits experimental confirmation [25]. Making use of out-of-equilibrium nanodisordered ferroelectrics supercooled to their ferroelectric Curie point T_C, scale-free optics has opened the way both to programmable optical effects and to optical response in highly non-ergodic contexts, where the response of single materials depends strongly on history effects, such as previous thermal cycles. In a recent set of experiments, this has allowed the observation of the optical Kovacs effect [26]. Finally, the very notion of scale-free optics suggests an environment that supports optical phenomena that are independent the wavelength, i.e., on the color of the beams. This has been recently confirmed by experiments in KTN:Li, where nonlinear effects occur simultaneously for propagating beams at two very different wavelengths ("achromatic optics") [27].

As opposed to diffraction compensation, which is a ubiquitous concept in wave mechanics and in optics, diffraction cancelation first predicted in [22] demonstrated in [18] appears as a truly innovative and original solution. An interesting and related experimental result has also recently been reported in a slow-light apparatus in 2009, where it was shown that linear conditions can be found in which, on consequence of strong coupling to diffusing atoms, light can be made to suffer anomalous diffraction, with some effects in common with paraxial scale-free optical effects [28].

8.3 Scale-Free Optics: A General Picture

In order to grasp the core idea behind scale-free optics, we consider the propagation of an optical wave in the paraxial scalar approximation. The slowly varying part of the optical field A (i.e., $|A|^2 = I$ is the optical intensity) obeys the paraxial wave equation

$$2ik\partial_z A + \nabla_\perp^2 A + \frac{2k^2}{n}\Delta n A = 0, \tag{8.1}$$

where $k = (\omega/c)n$ is the wavenumber, ω is the optical angular frequency, n is the material unperturbed index of refraction, Δn the nonlinear index of refraction perturbation, z is the propagation direction of the beam and $\perp \equiv (x, y)$ are the two transverse coordinates. The second term in (8.1) $\nabla_\perp^2 A$ is the paraxial diffraction operator. In a standard diffraction compensation scheme, this term is balanced for each z by the third term, where Δn is a scalar point dependent function that is either fabricated in the material, as in an optical waveguide, or produced by the wave itself in response to the local light intensity distribution I. In both cases, diffraction compensation entails a boundary value problem that leads to specific linear or nonlinear modal solutions (a situation that does not warrant for diffraction cancelation). We note that the diffraction operator term scales with $A/{w_0}^2$, where w_0 is the characteristic transverse spatial scale of A. The idea behind scale-free optics is the use of a Δn that instead of being a function in space, such as a $\Delta n(x, y, z)$, is, like the diffraction term, itself an operator. If this operator ΔnA scales itself with $A/{w_0}^2$ it amounts to a "diffraction" term that, combined with the original diffraction term, can lead to diffraction cancelation. For example, if $2(k^2/n)\Delta nA = -(L/\lambda)^2\nabla_\perp^2 A$, then the result is what amounts to a profoundly altered linear-like propagation in an effective homogeneous medium of the type

$$2ik\partial_z A + (1 - (L/\lambda)^2)\nabla_\perp^2 A = 0. \tag{8.2}$$

Evidently, for $L = \lambda$ this implies that diffraction simply ceases to exist. However, to appreciate the physics involved, in cases in which, for example, $L < \lambda$ (but not $L \ll \lambda$), then the optical propagation can be described by the standard propagation in a homogeneous material

$$2ik_{eff}\partial_z A + \nabla_\perp^2 A = 0, \tag{8.3}$$

where $k_{eff} = n_{eff}(2\pi/\lambda)$ and $n_{eff} = n/(1 - (L/\lambda)^2)$. So what we expect is to observe something analogous to what occurs in metamaterials, i.e., that the optical wave behaves according to diffractive laws governed by an effective index of refraction. In our case, of course, this change in index of refraction strongly depends on the spatial scale L introduced by the nonlinear interaction giving rise to the operational nature of Δn, and the effect itself is limited to the diffractive properties of the localized beams.

The requirements for scale-free optics and diffraction cancelation can be summarized by the statement that the diffraction term and the nonlinear term in (8.1) be proportional. This means (1) that the nonlinear interaction be nonlocal and introduce a spatial scale L that mimics the optical wavelength scale λ as regards to diffraction and (2) that the nonlinearity be intensity independent, exactly like diffraction. In principle, these two conditions can be met in a great variety of materials through various nonlocal responses, such as photothermal, reorientational, and so forth. In what follows, we will describe how these two conditions can be met in nanodisordered photorefractive ferroelectric crystals [18].

8.4 Diffusive Photorefraction in Supercooled Ferroelectric KTN:Li

As described above, scale-free optics can, in principle, occur in a great variety of materials and through a number of physical processes. Here, we focus on scale-free optics in nanodisordered ferroelectrics, i.e., on the system in which scale-free effects have been recently demonstrated [18]. Scale-free optics in KTN:Li requires photorefraction, a nonlinear optical mechanism that is common in ferroelectrics with deep in-band donor impurities. However, in distinction to the greater part of photorefractive optics, the response harnessed here must emerge on consequence of the thermal motion of the photogenerated charges, as they diffuse in the sample itself, and not of their drift caused by an external applied electric bias field. In order for the diffusive response to give rise to diffraction cancelation, the electro-optic response must be quadratic, as occurs in centrosymmetric crystals or in crystals that are disordered below the wavelength scale (the case of KTN:Li, where the system is in fact a glass-forming liquid of polar nanoregions—PNRs). This is in distinction to common conditions for electro-optic experiments, where the samples are in a poled ferroelectric state and have a linear electro-optic response. A last, but fundamental ingredient is that the electro-optic response must be anomalously enhanced, since the space-charge electric fields produced by charge diffusion are invariably miniscule, and the quadratic response is in itself generally weak. To achieve this the material must be engineered to support an accessible non-ergodic phase that allows a dielectric anomaly to be optically used, a condition that can be obtained if the critical phase-transition temperature is reached at T_C without having the crystal undergo an all-out phase transition, which brings with it opacity and scattering [9]. In what follows, we describe these different ingredients and how they combine to give rise to a system that supports scale-free optics.

8.4.1 Diffusion-Driven Space-Charge Field

The basic process that constitutes the photorefractive nonlinearity is the formation of a light-induced electric field **E**, which changes the local index of refraction by affecting the underlying crystal through the electro-optic effect, which in our case boils down to the reorientation of large polar nanoregions (PNRs). The electric field is caused by the diffusion of mobile electrons excited into the conduction band by the propagating light. This diffusion, which drives electrons from the more illuminated areas of the crystal, is halted once a sufficiently strong space-charge field has formed. It is this charge field that ultimately affects the PNR and alters the sample index of refraction.

The process is described using the band-transport model, which hinges on the rate equation

$$\frac{\partial N_d^+}{\partial t} = (\beta + sI)(N_d - N_d^+) - \gamma N N_d^+, \tag{8.4}$$

where N_d is the concentration of donor impurities, N_d^+ that of ionized donor impurities, N the concentration of electrons in the conduction band, β the thermal ionization constant, s the optical cross-section of the donors, I the optical intensity, and γ is the electron recombination constant. Considering steady-state conditions, (8.4) leads to

$$\gamma N N_d^+ = (\beta + sI)(N_d - N_d^+), \tag{8.5}$$

so that

$$N = \frac{\beta + sI}{\gamma} \frac{(N_d - N_d^+)}{N_d^+}. \tag{8.6}$$

These charges reach thermal equilibrium and obey the transport law for the current density **J**

$$\mathbf{J} = q\mu N\mathbf{E} + k_B T \mu \nabla N, \tag{8.7}$$

where q is the electron charge, μ is the electron mobility, k_B is the Boltzmann constant, and T is the temperature of the sample. Equation (8.7) expresses the fact that the local current density has two separate contributions, the $q\mu N\mathbf{E}$ term, which is the drift term, and the second term $k_B T\mu \nabla N$, which is thermal diffusion term. In turn, the charge density ρ is

$$\rho = q(N_d^+ - N_a - N), \tag{8.8}$$

where N_a is the concentration of acceptor impurities (normally $N_a \ll N_d$) that play a fundamental role in rendering the spatial charge separation efficient and semipermanent. Using (8.6) and (8.8) leads to (neglecting the thermal ionization)

$$N = \frac{s}{\gamma} I \frac{\left(\frac{N_d - N_a}{N_a} - \frac{\rho}{N_a q}\right)}{1 + \frac{\rho}{N_a q}}. \tag{8.9}$$

Using the Gauss law $\rho = \nabla \cdot (\varepsilon \mathbf{E})$ this leads finally to the expression

$$N = \frac{s}{\gamma} I \frac{\left(\delta - \frac{\nabla \cdot (\epsilon \mathbf{E})}{q N_a}\right)}{1 + \frac{\nabla \cdot (\epsilon \mathbf{E})}{N_a q}}, \tag{8.10}$$

where $\delta = (N_d - N_a)/N_a$. In standard conditions leading to, for example, photorefractive solitons, the spatial scales involved imply that $\nabla \cdot (\varepsilon \mathbf{E}) \ll 1$, so that $N \simeq (s/\gamma)I\delta$. In our present case, however, the situation is noticeably different, since the sample is in its critical state in proximity of its ferroelectric phase-transition temperature, so that $T \simeq T_C$. In this condition, the sample manifests a dielectric anomaly, accessible through the PNRs, for which ε grows greatly, and we are actually in the opposite case of $\nabla \cdot (\varepsilon \mathbf{E}) \gg \delta$ (and, hence, also $\nabla \cdot (\varepsilon \mathbf{E}) \gg 1$), so that

$$N = -\frac{s}{\gamma} I. \tag{8.11}$$

The last step is to consider that the crystal is itself not inserted in any form of circuit so that $\mathbf{J} = 0$ at steady state. From (8.7), this means that

$$q\mu N\mathbf{E} = -k_B T\mu \nabla N, \tag{8.12}$$

i.e., that

$$\mathbf{E} = -\frac{k_B T}{q} \frac{\nabla N}{N}. \tag{8.13}$$

Using (8.11) in (8.13) we finally conclude that

$$\mathbf{E} = -\frac{k_B T}{q} \frac{\nabla I}{I}. \tag{8.14}$$

We note that this expression is based on the assumption that the thermal ionization of donors (described by β) is negligible. The general expression will have a $I + I_d$ at the denominator of (8.14), where $I_d = \beta/s$. Whereas it is generally true that I_d is negligibly small compared to any optical intensity normally used in experiments, yet this assumption will forcibly break down at the beam tails, where I will decay to zero. In general, I_d is some orders of magnitude smaller than peak values of I, so it affects not truly the beam tails, but the very lateral areas of beams that are indistinguishable from noise. This is in direct distinction to what happens for space-charge fields that support photorefractive solitons, where the saturation parameter I_d plays such a fundamental role that it is generally artificially enhanced by using a co-propagating plane wave superimposed on the soliton beam itself. This said, we should, hence, consider the validity of (8.14) to extend to all optical intensities above the thermal noise level, which is the equivalent dark illumination of the band-transport mechanism in the specific crystal. In the case of KTN:Li, no experimental evidence of I_d has been observed in scale-free optics experiments. However, such studies could in principle be carried out, simply by artificially enhancing the I_d.

The expression of the space-charge field in (8.14) forms one of the first tassels in the mosaic leading to scale-free optics. In fact, the field is independent of the actual peak intensity of the beam, which factors out in the ratio. This important fact stems, in this context, from the fact that the mobile electrons described by N are both the mechanism driving diffusion, through ∇N, and the mechanism giving rise to conduction, which is proportional to N and appears at the denominator in the strength of the electric field formed.

8.4.2 Electro-Optic Response in the Nanodisordered Dipolar Glass Formed by the PNR

The second tassel in the attainment of scale-free optics in KTN:Li is the quadratic electro-optic response. It is this response that turns the diffusion-driven electric field of (8.14) into an operator that approximately achieves the condition for diffraction cancelation discussed in (8.2).

A ferroelectric is generally characterized by a strong low frequency susceptibility, which in turn means that a quasi-static electric field $\mathbf{E}$ readily generates a relevant quasi-static crystal polarization $\mathbf{P}$. This polarization, which stems from the underlying dipolar nature of the crystal lattice structure typical of ferroelectrics, alters in an anisotropic way the index of refraction of the crystal, an effect that is known as the electro-optic effect. The most general expression for this effect is

$$\Delta n_{ij} = -\frac{1}{2} n^3 g_{ijkl} P_k P_l, \tag{8.15}$$

where g_{ijkl} is the quadratic electro-optic tensor of the crystal lattice, the indices represent the three spatial axes (i.e., $i = 1, 2, 3 \equiv x, y, z$), and summation over repeated indices is assumed. Δn_{ij} is, in general, a matrix that, fixing the relationship between the high-frequency polarization of the crystal and the optical field, enters into the optical propagation equations (the Helmholtz equation) and determines the evolution of the optical field in all its components (i.e., its polarization). In a paraelectric, the absence of spontaneous polarization leads to the so-called quadratic electro-optic effect, which is none other than the expression in (8.15).

In a poled ferroelectric phase, the presence of a large spontaneous polarization makes (8.15) be dominated by linear terms in the external electric field **E**, giving rise to what is known as the linear electro-optic effect. In our present case of a dipolar glass composed of PNRs, we have that each component of the polarization is caused by the combination of the polarization of the single dipoles and of the paraelectric region of the crystal in between the regions, both affected by the optically induced **E** of (8.14). In general, the response can be described introducing the effective relative dielectric constant ε_r

$$P_k = \epsilon_0(\epsilon_r - 1)E_k = \epsilon_0\chi E_k, \tag{8.16}$$

along with the effective susceptibility χ. In our case of a supercooled dipolar glass, the susceptibility will be thus the combination

$$\chi = \chi_{PNR} + \chi_P, \tag{8.17}$$

where χ_{PNR} is the susceptibility associated with the PNRs and χ_{P} that associated with the paraelectric surroundings. As we will discuss in what follows, we will operate in conditions in which the dipolar glass is excited and provides an anomalously enhanced susceptibility, so that in our system $\chi_{\mathrm{PNR}} \gg \chi_{\mathrm{P}}$. The PNRs are equivalent to randomly oriented random birefringent crystals, so that the result is that averaging on a given volume V (larger than the size of the PNR themselves but smaller than the optical wavelength) (8.15) leads to

$$\Delta n_{ij} = -\frac{1}{2}n^3(g_{ijxx}P_x^2 + g_{ijyy}P_y^2). \tag{8.18}$$

which entails, on the basis of simple symmetry considerations, that

$$\Delta n_{xx} = -\frac{n^3}{2}\left(g_{11}P_x^2 + g_{12}P_y^2\right), \tag{8.19}$$

and

$$\Delta n_{yy} = -\frac{n^3}{2}\left(g_{11}P_y^2 + g_{12}P_x^2\right), \tag{8.20}$$

where $g_{11} = g_{xxxx}$ and $g_{12} = g_{xxyy}$ of the specific crystal used.

For the x-component (and analogously for the y-component), we have:

$$P_x^2 = \rho_{PNR}^2\langle p_x^2\rangle, \tag{8.21}$$

where ρ_{PNR} is the PNR density and p_x the x-component of their local polarization. As mentioned, the brackets denote an average on a volume V much larger than the PNR but smaller than the wavelength.

The lattice in the PNRs is in general different from the cubic lattice of the host crystal, and we have

$$p_x = p_{0x} + \alpha_{xj}E_j, \tag{8.22}$$

where α_{ij} is the PNR polarizability tensor.

As the PNR are randomly oriented in the volume V, one readily finds that

$$\langle p_x^2 \rangle = \langle p_y^2 \rangle = p_0^2 + \alpha_0^2 E^2, \tag{8.23}$$

where p_0 is the spontaneous polarization and $\alpha_0 > 0$ is the mean microscopic polarizability of the PNR, such that the microscopic polarization due to the PNR will be written as

$$P_x^2 = P_y^2 = P_0^2 + \epsilon_0^2 \chi_{PNR}^2 E^2 \tag{8.24}$$

with $P_0 = \rho_{\mathrm{PNR}} p_0$ and the mean low frequency susceptibility of the non-ergodic phase is given by

$$\chi_{PNR} = \rho_{PNR} \alpha_0. \tag{8.25}$$

χ_{PNR} accounts for the nonlinear optical effects that are due to the PNR.

Finally, the index perturbation for the two polarizations reads as $\Delta n_{xx} = \Delta n_{yy} \equiv \Delta n_{\mathrm{PNR}}$

$$\Delta n_{PNR} = -\frac{n^3}{2}(g_{11} + g_{12})\epsilon_0^2 \chi_{PNR}^2 E^2, \tag{8.26}$$

where we have dropped a constant term with P_0^2.

8.4.3 Enhanced Susceptibility and Electro-Optic Response in Out-of-Equilibrium Ferroelectrics

The last tassel in the design of a system capable of supporting scale-free optics is the achievement of a huge electro-optic response using the anomalously large χ_{PNR} when the ferroelectric is supercooled into a highly non-ergodic dielectric phase. In fact, the electric field caused by charge diffusion for a beam with a micrometric size is very weak. Using (8.14), we find that $|\mathbf{E}| \sim$ 10–50 V/cm. A sufficient electro-optic response would require values of $\chi \sim 10^5$, and these can only be observed for ferroelectrics undergoing a phase transition. Unfortunately, a ferroelectric undergoing a phase transition cannot support optical propagation, because the ferroelectric domains reach beyond the nanometric scales and scatter the optical beam ultimately giving rise to a strong absorption of the sample [9]. This limit can be qualitatively grasped considering the strong constraint imposed by equilibrium physics on the susceptibility and the correlation length in the material r_c, which both diverge as the system is brought to its phase transition at T_C. In fact, equilibrium imposes that

$$\chi \propto \frac{1}{(T - T_C)^{\delta}}, \tag{8.27}$$

and

$$r_c \propto \frac{1}{(T - T_C)^{\delta/2}}, \tag{8.28}$$

where δ is a critical parameter. As the dielectric anomaly is approached to enhance the response for $T \simeq T_C$, r_c also diverges and, when it reaches the macroscopic scales of the optical wavelength, light is scattered and the material becomes incapable of supporting optical propagation. The point is that ferroelectrics with compositional disorder are known to support also a non-ergodic phase, which does not obey the constraints of equilibrium [19]. In fact, whereas in standard conditions, crystal behavior is the macroscopic manifestation of all the possible microstates compatible with the various external constraints on the system, and its dynamics consists in the rapid exploration of the microstates and has no memory of previous history, when a ferroelectric crystal with compositional disorder, i.e., with lattice sites occupied at random by different chemical elements that do not greatly alter the lattice structure, is rapidly cooled to its Curie temperature, the crystal develops a disordered mosaic of electrically polarized domains, a dipolar glass, whose surface energy and volume energy balance out halting further crystal structural changes. The crystal is now in a condition in which it is inhomogeneous (at the nanoscale) and cannot occupy all the microstates compatible with its thermal equilibrium. This out-of-equilibrium or glass-forming liquid state can never reach equilibrium and can break the bond between χ and r_c. It is in this phase that it is possible to harness the dielectric anomaly associated to the critical temperature T_C without suffering the limits of optical scattering.

Working with supercooled ferroelectrics at the Curie point guarantees that the dielectric anomaly can be accessed without scattering, but the actual details of the mechanism leading to χ_{PNR} as a function of the cooling rate and the details of the thermal history are phenomenological, as at present no theory for glasses, let alone dipolar glasses, exists. Below, in describing the experimental apparatus, we will provide the details of how to tap the non-ergodic phase of KTN:Li.

8.5 Scale-Free Nonlinearity in Nanodisordered Ferroelectrics

8.5.1 Scaling and Approximate Diffraction Cancelation: The Basic Scale-Free Equation

We are now in a position to establish in what way the diffusive electric field of (8.14) produces, through the electro-optic response of (8.26), an effect capable of diffraction cancelation. Clearly, we do not achieve in this manner a truly anti-diffraction operator $\propto \nabla^2 A$, since the response will amount to a term $\propto (|\nabla I|/I)^2 A$. However, both terms are independent of peak intensity, and both have the same fundamental scaling in terms of the spatial extent of the beam, i.e., both are $\propto 1/{w_0}^2$. These two facts guarantee a diffraction cancelation for arbitrary waveforms, and an exact mathematically rigorous cancelation for the solutions that we discuss in what follows. For all cases, in conditions in which the nonlinear length scale $L \simeq \lambda$, scale-free optics is observed.

To put together the different pieces in our physical mosaic, we begin from the paraxial nonlinear wave equation of (8.1) and insert into the nonlinear response term Δn the electro-optic response of the PNR described by (8.26) using the diffusive photoinduced electric field of (8.14). Thus, a linearly polarized beam with amplitude A obeys

$$i\partial_z A + \frac{1}{2k}\nabla_\perp^2 A + \frac{k}{n}\Delta n A = 0, \tag{8.29}$$

where Δn is expressed by

$$\Delta n = \Delta n_{PNR} = -g\frac{n^3}{2}\epsilon_0^2\chi_{PNR}^2\left(\frac{K_B T}{q}\right)^2\frac{(\partial_x I)^2 + (\partial_y I)^2}{I^2}, \tag{8.30}$$

and $g \equiv g_{11} + g_{12}$ (that depends on the specific PNR-supporting ferroelectric used). The nonlinear propagation equation is

$$i\partial_z A + \frac{1}{2k}\nabla_\perp^2 A - K\frac{\left(\partial_x|A|^2\right)^2 + \left(\partial_y|A|^2\right)^2}{|A|^4}A = 0, \tag{8.31}$$

where

$$K = kg\left(\frac{n\epsilon_0\chi_{PNR}K_B T}{\sqrt{2}q}\right)^2. \tag{8.32}$$

The sign of g and, hence, the focusing/defocusing nature of the effect depends on the specific lattice structure: for KTN:Li $g_{11} + g_{12} > 0$ ($g_{11} = 0.\ 16\ \mathrm{m^4C^{-2}}$, $g_{12} = -0.02\ \mathrm{m^4C^{-2}}$) [29], and the effect is self-focusing. Since $g > 0$, (8.33) is also

$$2ik\frac{\partial A}{\partial z} + \nabla_\perp^2 A - \frac{L^2}{\lambda^2}\frac{\left(\partial_x|A|^2\right)^2 + \left(\partial_y|A|^2\right)^2}{4|A|^4}A = 0. \tag{8.33}$$

The spatial scale L that is used to factor out the wavelength scale is

$$L = 4\pi n^2\epsilon_0\sqrt{g}\chi_{PNR}(K_B T/q). \tag{8.34}$$

8.5.2 Scale-Free Analytical Solutions

Interestingly, this model admits analytical self-trapped solutions [22] when

$$8kK = \frac{L^2}{\lambda^2} \geq 1. \tag{8.35}$$

For the PNR susceptibility, this implies that

$$\chi_{PNR} \geq \chi_{thr} = \frac{q}{2kn\epsilon_0 K_B T\sqrt{g_{11}+g_{12}}} \simeq 10^5. \tag{8.36}$$

In other words, there exists a critical value for the nonlinear optical response due to the PNR for which $L = \lambda$. It is this threshold value that is simply inaccessible in ferroelectrics at equilibrium, as discussed previously [30]. In our experiments, we achieve this anomalously large response supercooling KTN:Li to its T_C.

We note that the diffraction-free (diverging n_{eff} or zero effective wavelength $\lambda_{eff} = \lambda/n_{eff}$) solutions of (8.33) are scale free, because there is no relationship between the peak intensity and the waist. In other words, the waves with a given waist exist for any power level. The situation is found in the Gaussian solution for $8kK = 1$ ($\chi_{\text{PNR}} = \chi_{thr}$, $L = \lambda$)

$$A = A_0 \exp\left(-\frac{x^2+y^2}{w_0^2}\right)\exp\left(-2i\frac{z}{kw_0^2}\right). \tag{8.37}$$

More precisely, in (8.37) the waist w_0 of the soliton and its amplitude A_0 are free independent parameters.

8.5.3 Scale-Free Solutions in Conditions of L > λ

When $\chi_{\text{PNR}} > \chi_{thr}$ ($L > \lambda$), the very nature of the propagation equation changes, because the relative sign of the effective diffraction term switches. In these conditions, we once again find self-trapped solutions given by

As in the previous case, also here A_0 and w_0 free parameters (i.e., the existence curve is flat). Remarkably, as χ_{PNR} grows the beam appears to lose its radial symmetry, developing a square-like profile (see Fig. 8.1).

$$A = A_0 \frac{e\left[-\frac{4iz}{kw_0^2(8kK-1)}\right]}{\cosh\left(\frac{2x}{w_0}\right)^{\frac{1}{8kK-1}} \cdot \cosh\left(\frac{2y}{w_0}\right)^{\frac{1}{8kK-1}}}. \tag{8.38}$$

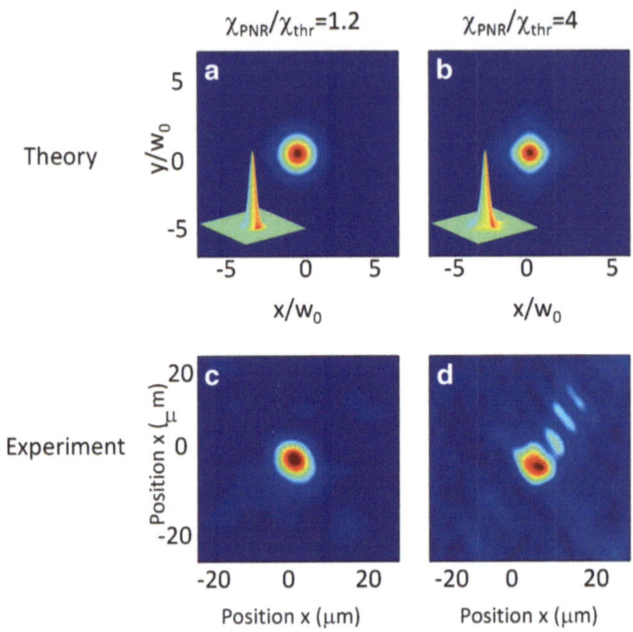

Fig. 8.1 Squaring-off of the beam profile as the cooling rate is increased for $L > \lambda$ compared to experimental results

8.6 The Observation of Scale-Free Optical Propagation

Scale-free optical propagation has been demonstrated in nanodisordered lithium-enriched potassium-tantalate-niobate (KTN:Li) doped with copper impurities. The experimental setup is schematically illustrated in Fig. 8.2, and is relatively simple. The idea is to detect the propagation dynamics of a diffracting laser beam within the crystal. Beams are obtained from standard low power continuous lasers, such as a He–Ne laser, and the initially diffracting condition is achieved by appropriately shrinking the laser beam down to micrometric widths at the input of the sample. The propagation dynamics can be detected in two different and complementary ways, i.e., through a top-view CCD or through an on-axis CCD (see Fig. 8.2). The top-view detection is made possible by the enhanced scattering that intervenes as the crystal is brought into its non-ergodic phase, whereas the on-axis intensity distributions are detected imaging the input and output planes of the crystal through an appropriate optics.

Perhaps the most complicated element in the apparatus is the control of the thermal history $T(t)$ of the crystal. This is because the giant electro-optic effect required to reach the scale-free regime is only achieved when the ferroelectric is brought into a non-ergodic phase, as described below. To do this, the crystal must be supercooled to its phase-transition temperature.

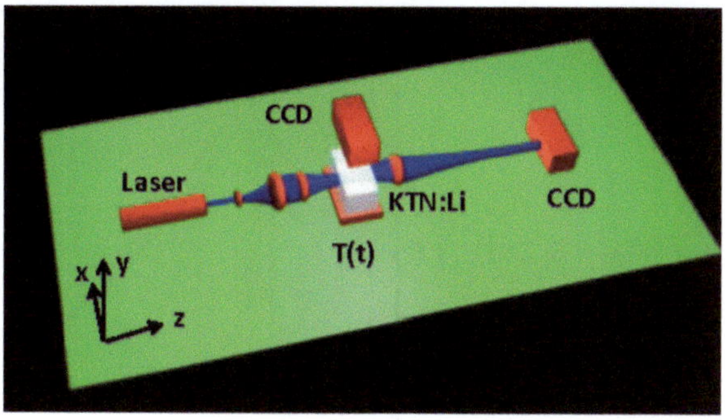

Fig. 8.2 Schematic of experimental setup used to observe scale-free optics

8.6.1 The KTN:Li

The heart of the apparatus is the newly engineered optical quality nonlinear photorefractive crystal, Cu-doped KTN:Li. It is a ferroelectric designed to simultaneously host a glass-forming liquid-like phase, maintain an elevated degree of optical transparency, absent in typical glassy media [21], and have a nanostructured electro-optical and nonlinear optical response. Scale-free behavior is observed only for a sufficiently high χ_{PNR}, a requirement that can be achieved supercooling the sample into its non-ergodic phase in proximity of its Curie point T_C. In a commonly accepted model, on approaching the ferroelectric phase transition, the Cu-doped KTN:Li behaves like a disordered distribution of polarizable dipoles that are formed by the emergence of the Nb ions from the center of the crystal lattice unit cells. These form the PNRs, and their size is controlled by the cooling rate α to the final operating room temperature. It is in fact observed that changing α affects the size of the PNRs (see Fig. 8.4 below) and consequently the KTN:Li nonlinearity, so that at the same operating temperature (T), the permanent structure of the PNR provides different and tunable nonlinear responses, a fact which is phenomenologically described by a crystal susceptibility χ_{PNR} that depends on α.

The non-ergodic phase required for the observations can be pinpointed dielectrically by measuring the capacitance of the sample as a function of the temperature T. Non-ergodicity is signaled by strong thermal hysteresis in the relative dielectric constant [19]. An example of thermal hysteresis in KTN:Li is shown in Fig. 8.3.

We note that when a ferroelectric with compositional disorder is brought into a non-ergodic phase achieved through supercooling to the Curie temperature, it effectively turns into a material that is fundamentally different from a standard ferroelectrics undergoing a phase transition [9]. If the crystal is slowly cooled to T_C, its interaction with light is mediated by large ferroelectric domains with a strong spontaneous polarization that is rigidly fixed by the principal axes of the crystal lattice, evident in the transmission results shown in Fig. 8.4 (note the 45° orientation of the domain walls). The scale-free regime, in turn, occurs when the crystal is rapidly cooled to T_C (supercooled state). Dielectric dynamics are those of a glass-forming liquid, and

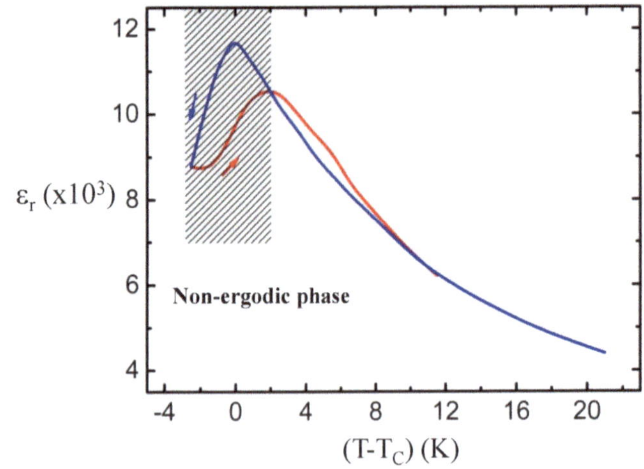

Fig. 8.3 Dielectric characterization of the non-ergodic phase: Low-frequency average dielectric permittivity ε_r for fast (*blue line*) and slow (*red line*) temperature rates. The *shaded region* where the two curves considerably differ represents the region in which the non-ergodic behavior of the PNR is more pronounced [19]

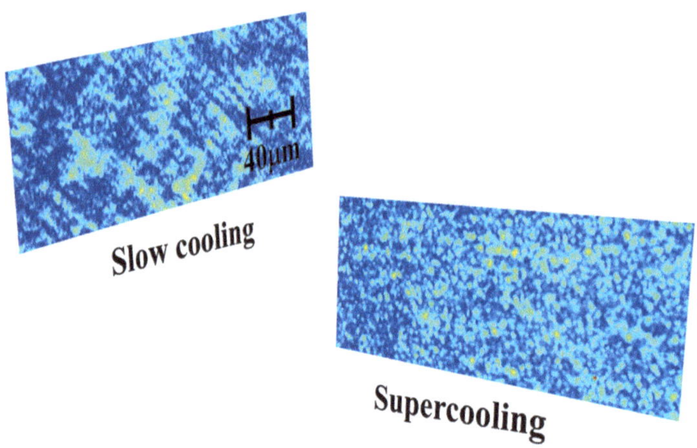

Fig. 8.4 Comparison of transmission through the KTN:Li sample at T_C for slow and fast cooling rates. For slow cooling, the transmission indicates the formation of large ferroelectric clusters and domains with geometrically fixed boundaries. For fast cooling (supercooling), transmission radically changes, indicating a more homogeneous and disordered dielectric state. The specific sample used in these experiments is a $K_{1-x}Ta_{1-y}Nb_yO_3$:Li_x composite with $x = 0.003$, $y = 0.36$, a Curie point at $T_C = 14.5\,^\circ$C, and a non-ergodic phase for $T < T_B = 17\,^\circ C$

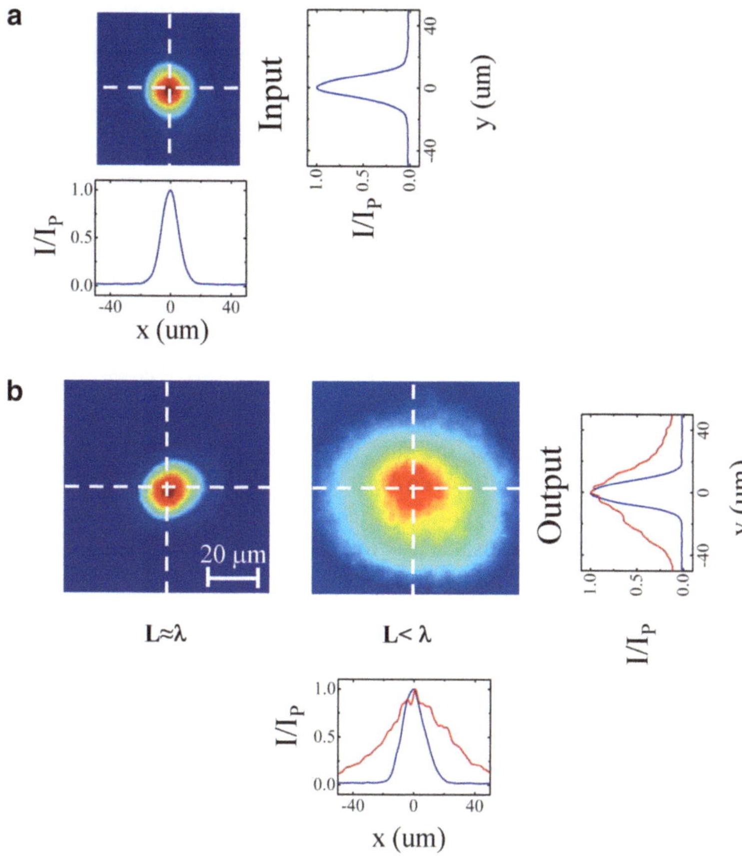

Fig. 8.5 Scale-free propagation. Input intensity distribution (**a**) fails to spread according to the standard diffraction laws for a fast cooling rate at the output (**b**) when $L \simeq \lambda$, and suffers a partial spreading in the slow cooling rate case, i.e., when $L < \lambda$

the interaction with light is mediated by PNRs that lock into a disordered network. Here the size of the PNR depends on the cooling rate (i.e., the previous thermal history of the sample) and not on the operating temperature, and optical transmission is fundamentally different, as shown in Fig. 8.4.

8.6.2 *Scale-Free Propagation and Supercooling*

The analysis of the spreading of the intensity distribution of the beam in propagating through the crystals shows the basic signatures of a scale-free optical propagation. In Fig. 8.5, for example, we report how, for a scale-free regime, $L \simeq \lambda$, spreading due to diffraction is absent, compared to diffraction reported for $L/\lambda \simeq 0.4$ ($n_{\text{eff}} \simeq 1.2n$). In the case reported, the scale-free regime is achieved by cooling the KTN:Li sample

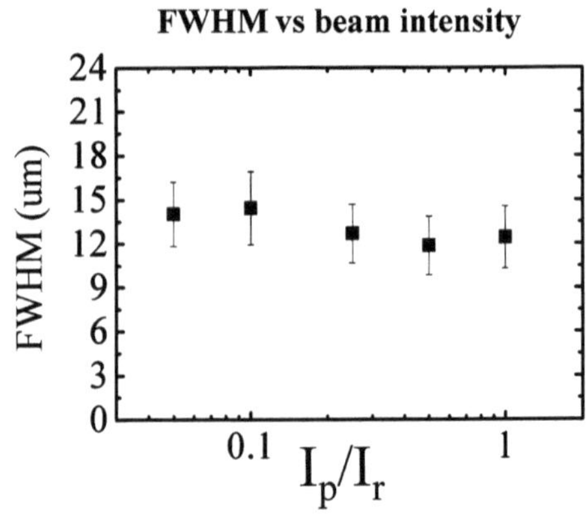

Fig. 8.6 A typical result of output intensity distribution for a scale-free beam (i.e., in conditions in which supercooling is enacted so that $L \simeq \lambda$) that depends only on the input beam FWHM, and not on the peak intensity or on any other saturation parameter

from 35° C to $T_C = 14.5°$ C at the fast rate of $\alpha_f = 0.07°$ C/s and the output intensity distribution has a full-width-half-maximum (FWHM) $\Delta x(L_z) \simeq \Delta y(L_z) \simeq 15$ μm equal to the input one, after $L_z = 6$ mm of propagation in the z-direction: the beam does not spread because diffraction is fundamentally absent ($\lambda_{\rm eff} \simeq 0$). In turn, when the KTN:Li sample is slowly cooled with $\alpha_s = 0.\ 015°$ C/s, the intensity distribution of the input beam widens to $\Delta x(L_z) = 43$ μm and $\Delta y(L_z) = 40$ μm at the output only slightly less than what is expected from the standard diffraction of a Gaussian beam.

8.6.3 Violation of Scale-Dependent Soliton Laws

Absence of beam spreading is not in itself evidence of scale-free optical propagation. Diffraction can be readily compensated through nonlinear self-focusing, as occurs for spatial solitons. What is a direct signature of scale-free optics is a violation of scale-dependent scaling laws that must intervene when diffraction is compensated. If these laws are in fact violated, diffraction has been canceled. Experiments in KTN:Li have indeed shown that non-spreading beams of a given width are observed over a wide range of peak beam intensities, as illustrated in Fig. 8.6. It can be noted that there are physical systems in which the value of peak intensity is not important, but the ratio between this value and a second quantity is relevant. This occurs in all saturated nonlinearities, such as the Kerr-saturated model valid for photorefractive screening solitons. On the contrary, in the scale-free case, no saturating parameter is changed.

8.6.4 Instability

One of the most striking phenomena associated to scale-free optics is the observation of beam instability. In a standard soliton-supporting nonlinearity, beams with specific symmetries, such as plane waves or stripe beams lead to a characteristic

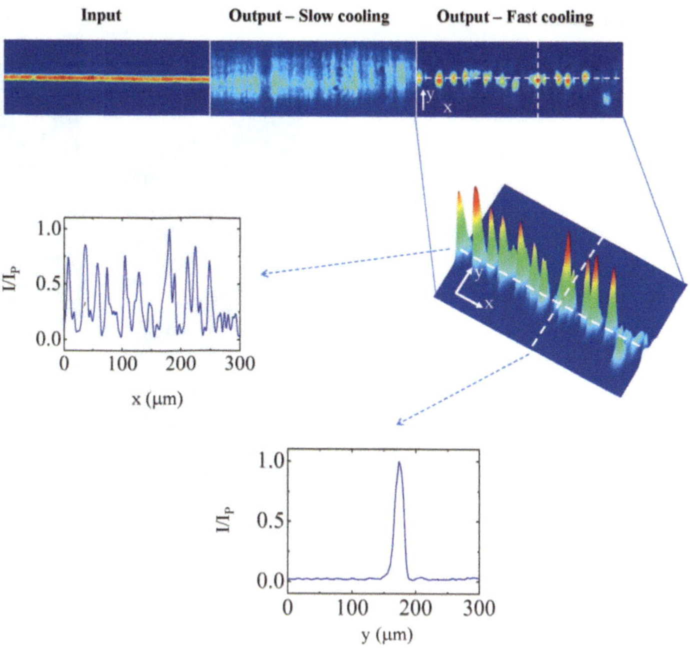

Fig. 8.7 Scale-free modulation instability, once again directly dependent on the relationship between L and λ

breakup into patterns formed by solitons. Experiments in KTN:Li in the scale-free regime have led to the unexpected breakup of a stripe into a "gas" of different localized beams of different sizes and intensities, with no specific relationship between the two, in direct response to inhomogeneities in the input beam pattern. This nonselective instability is in fact a true sign of a scale-free optical propagation, since it underlines the basic absence of soliton scaling laws [31]. The experimental demonstration of scale-free instability is shown in Fig. 8.7.

8.6.5 *Intensity-Independent Beam Interaction*

An important aspect of scale-free propagation is that although the phenomenon is the product of a peak intensity-independent mechanism, it is profoundly nonlinear in nature. For one, the effect depends on the shape of the optical wave. This nonlinearity seems absent once we consider the entire matter in the form of a diverging effective index of refraction or a vanishing effective wavelength, but these are not applicable to plane waves, but to localized beams. Experiments that highlight the nonlinear nature of diffraction cancelation are beam–beam interaction phenomena, which involve beam attraction (see Fig. 8.8), crossing (Fig. 8.9), and even beam spiraling (Fig. 8.10), three interaction phenomena that are identical to those normally associated to solitons.

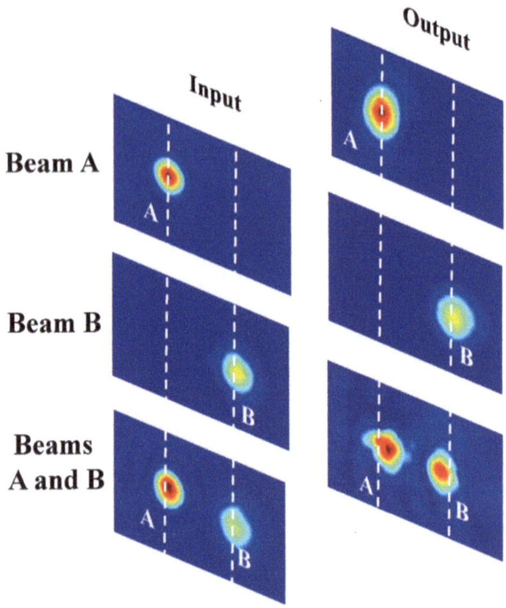

Fig. 8.8 Scale-free beam attraction

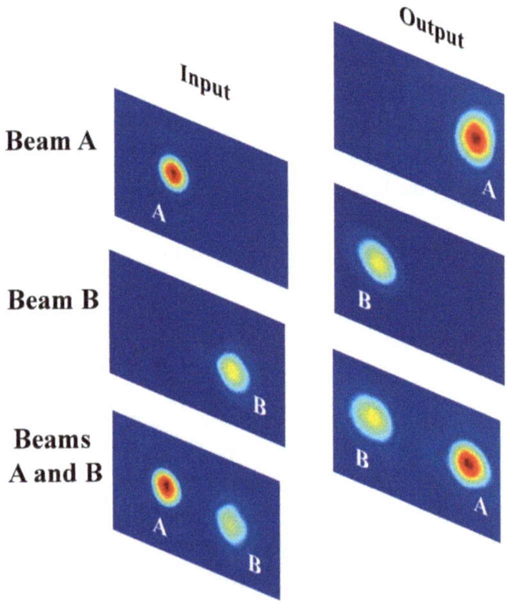

Fig. 8.9 Scale-free beam crossing

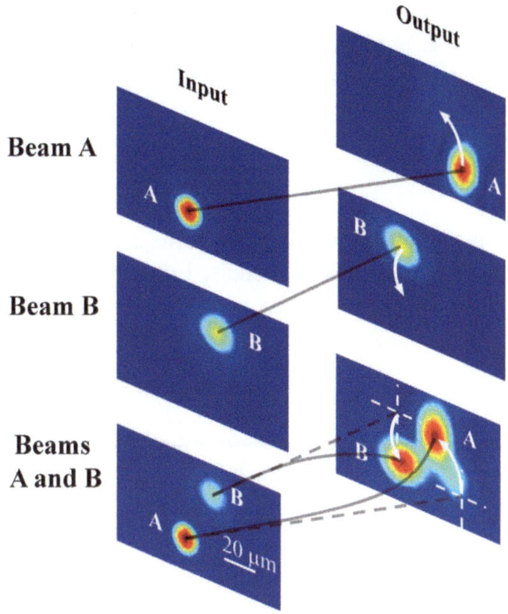

Fig. 8.10 Scale-free beam spiraling

8.7 Subwavelength Beam Propagation

Scale-free optics affects the diffraction of optical waves, effectively reducing the spreading and distortion of beams due to diffraction without incurring in the limits of linear and nonlinear diffraction compensation. The factoring out of the wavelength from the diffractive dynamics suggests one basic question: does the factoring out of λ extend into the nonparaxial regime and, hence, affect the limits to propagating waves? Put differently, can scale-free optics be used to allow the propagation of subwavelength beams with impact, for example, on super-resolution in optical microscopy? The issue is not a farfetched proposition because what we term nonparaxiality and the propagation spectrum as opposed to the evanescent wave spectrum all depend on the use of the wavelength as a fundamental scale!

Remarkably, scale-free solutions exist also for a model which is valid beyond the paraxial approximation and contains (in the linear limit) also the evanescent waves. The model here considered is the Helmholtz equation, with the simplification, with respect to Maxwell's equations, that vectorial coupling is neglected, an assumption that can, in most cases, be shown to lead to very small corrections.

The Helmholtz equation is obtained for a monochromatic beam starting from the vectorial wave equation

$$\nabla \times \nabla \times \mathbf{E} = \left(\frac{\omega n}{c}\right)^2 \mathbf{E}, \tag{8.39}$$

with $n^2 = (n_0 + \Delta n)^2 \cong n_0^2 + 2n_0\Delta n$ and Δn given in (8.30). For a linearly polarized beam $\mathbf{E} = \hat{\mathbf{x}}E_0 A \exp(i\beta z)$, with $|A|^2 = I$ the beam intensity ($E_0^2 = 2Z_0/n_0$), and neglecting vectorial effects ($\nabla\nabla \cdot \mathbf{E} \cong 0$), one has the Helmholtz equation

$$\nabla_\perp^2 A + \left[\left(\frac{\omega n}{c}\right)^2 - \beta^2\right] A = 0. \tag{8.40}$$

β is the overall wave vector in the z direction, and not its nonlinear perturbation as in standard paraxial models. Equation (8.40) admits an exact Gaussian solution, which is scale free, i.e., with an arbitrary amplitude A_0 and waist w_0:

$$A = A_0 \exp\left(-\frac{x^2}{w_0^2} - \frac{y^2}{w_0^2}\right) \tag{8.41}$$

with the wavector β given by

$$\beta = \sqrt{\left(\frac{\omega n_0}{c}\right)^2 - \frac{4}{w_0^2}}, \tag{8.42}$$

when the condition $L = \lambda$ is satisfied.

A solution exists (i.e., β is real) as long as

$$w_0 > \frac{2c}{\omega n_0} = \frac{\lambda}{\pi n_0} \tag{8.43}$$

as expected.

The point is that more general solutions exist for $L > \lambda$, as in the paraxial case. These are given by

$$A = A_0 \frac{e^{i\beta z}}{\cosh\left(\frac{2x}{w_0}\right)^{\gamma^2} \cosh\left(\frac{2y}{w_0}\right)^{\gamma^2}}, \tag{8.44}$$

where

$$\gamma = \frac{1}{\sqrt{\frac{L^2}{\lambda^2} - 1}} \tag{8.45}$$

and the wavector is

$$\beta = \sqrt{\left(\frac{\omega n_0}{c}\right)^2 - \frac{4\gamma^2}{w_0^2}}. \tag{8.46}$$

The corresponding lower limit for the waist is, hence,

$$w_0 > \gamma \frac{\lambda}{\pi n_0}. \tag{8.47}$$

Thus, for $L > \lambda$, the lower limit for the waist is scaled by a factor γ: a beam smaller than the wavelength can propagate in the medium without distortion (for $L > \sqrt{2}\lambda$) irrespective of its intensity.

8.8 Conclusions

Scale-free optical propagation appears as a general and new paradigm for the propagation of undistorted optical beams for imaging and microscopy, with the promise to allow the propagation of beams with no limits associated to evanescent waves, with obvious impact on super-resolved microscopy.

Acknowledgments This contribution is based on work carried out with Prof. Aharon J. Agranat of the Applied Physics Department at the Hebrew University of Jerusalem, and with Elisa Spinozzi, now at the University of Rome La Sapienza. The research leading to these results has received funding from the Italian Ministry of Research (MIUR) through the "Futuro in Ricerca" FIRB-grant PHOCOS—RBFR08E7VA, and from the European Research Council under the European Community's SeventhFramework Program (FP7/2007-2013)/ERC Grant Agreement No. 201766. Partial funding was received through the SMARTCONFOCAL project of the Regione Lazio.

References

1. M. Born, E. Wolf, A. Bhatia, *Principles of Optics* (Cambridge University Press, Cambridge, 1999)
2. D. Marcuse, *Theory of Dielectric Optical Waveguides* (Academic Press, New York, 1974)
3. A. Yariv, *Quantum Electronics*, 3rd edn. (Wiley, New York, 1988)
4. P.G. Drazin, R.S. Johnson, *Solitons: An Introduction* (Cambridge University Press, Cambridge, 1989)
5. Y.S. Kivshar, G.P. Agrawal, *Optical Solitons* (Academic Press, San Diego, 2003)
6. M. Segev, B. Crosignani, A. Yariv, B. Fischer, Spatial solitons in photorefractive media. Phys. Rev. Lett. **68**, 923–926 (1992)
7. M. Segev, G.I. Stegeman, Self-trapping of optical beams: spatial solitons. Phys. Today **51**, 43–48 (1998)
8. E. DelRe, M. Segev, *Self-focusing and Solitons in Photorefractive Media*, in Topics in Applied Physics, vol. 114 (Springer, Berlin, 2009), pp 547–572
9. E. DelRe, M. Tamburrini, M. Segev, R. Della Pergola, A.J. Agranat, Spontaneous self-trapping of optical beams in metastable paraelectric crystals. Phys. Rev. Lett. **83**, 1954–1957 (1999)
10. E.A. Ultanir, D. Michaelis, F. Lederer, G.I. Stegeman, Stable spatial solitons in semiconductor optical amplifiers. Opt. Lett. **28**, 251–253 (2003)

11. W.E. Torruellas et al., Observation of two-dimensional spatial solitary waves in a quadratic medium. Phys. Rev. Lett. **74**, 5036–5039 (1995)
12. P.J. Reece, E.M. Wright, K. Dholakia, Experimental observation of modulation instability and optical spatial soliton arrays in soft condensed matter. Phys. Rev. Lett. **98**, 203902 (2007)
13. C. Conti, M. Peccianti, G. Assanto, Observation of optical spatial solitons in a highly nonlocal medium. Phys. Rev. Lett. **92**, 113902 (2004)
14. J.W. Fleischer, M. Segev, N.K. Efremidis, D.N. Christodoulides, Observation of two-dimensional discrete solitons in optically induced nonlinear photonic lattices. Nature **422**, 147–150 (2003)
15. C. Conti, G. Ruocco, S. Trillo, Optical spatial solitons in soft matter. Phys. Rev. Lett. **95**, 183902 (2005)
16. S. Trillo, W. Torruealls (eds.), *Spatial Solitons* (Springer, Berlin, 2001)
17. A.D. Boardman, A.P. Sukhorukov (eds.), *Soliton Driven Photonics* (Kluwer Academic, Dordrecht, 2001)
18. E. DelRe, E. Spinozzi, A.J. Agranat, C. Conti, Scale-free optics and diffractionless waves in nanodisordered ferroelectrics. Nat. Photon. **5**, 39–42 (2011)
19. A.A. Bokov, Z.-G. Ye, Recent progress in relaxor ferroelectrics with perovskite structure. J. Mater. Sci. **41**, 31–52 (2006)
20. E. Dagotto, Complexity in strongly correlated electronic systems. Science **309**, 257–262 (2005)
21. N. Ghofraniha, C. Conti, G. Ruocco, F. Zamponi, Time-dependent nonlinear optical susceptibility of an out-of-equilibrium soft material. Phys. Rev. Lett. **102**, 038303 (2009)
22. B. Crosignani, A. Degasperis, E. DelRe, P. Di Porto, A.J. Agranat, Phys. Rev. Lett. **82**, 1664 (1999)
23. P.B. Ishai, C.E.M. de Oliveira, Y. Ryabov, Y. Feldman, A.J. Agranat, Glassforming liquid kinetics manifested in a KTN:Cu crystal. Phys. Rev. B **70**, 132104 (2004)
24. A. Agranat, R. Hofmeister, A. Yariv, Characterization of a new photorefractive material: KLTN. Opt. Lett. **17**, 713–715 (1992)
25. C. Conti, A.J. Agranat, E. DelRe, Subwavelength optical spatial solitons and three-dimensional localization in disordered ferroelectrics: Toward metamaterials of nonlinear origin. Phys. Rev. A **84**, 043809 (2011)
26. E. DelRe, J. Parravicini, G. Parravicini, A.J. Agranat, C. Conti, Wavelength-insensitive negative optical permittivity without nanofrabrication in transparent nonlinear dipolar glasses under extreme frustration, submitted to Physical Review Letters (2012)
27. J. Parravicini, F. Di Mei, C. Conti, A.J. Agranat, and E. DelRe, Diffraction cancellation over multiple wavelengths in photorefractive dipolar glasses. Opt. Exp. **19**, 24109 (2011)
28. O. Firstenberg, P. London, M. Shuker, A. Ron, N. Davidson, Elimination, reversal and directional bias of optical diffraction. Nat. Phys. **5**, 665–668 (2009)
29. A. Bitman, N. Sapiens, L. Secundo, A.J. Agranat, G. Bartal, M. Segev, Opt. Lett. **31**, 2849 (2006)
30. Y. Xu, *Ferroelectric Materials and Their Applications* (North-Holland, Amsterdam, 1991)
31. V. Folli, E. DelRe, C. Conti, Beam Instabilities in the Scale Free Regime. Phys. Rev. Lett. **108**, 033901 (2012)

Chapter 9
Spatially Dispersive Shock Waves in Nonlinear Optics

Christopher Barsi, Wenjie Wan, Shu Jia, and Jason W. Fleischer

9.1 Introduction

Shock waves occur when an object immersed in a medium travels faster than waves in that medium [1]. They are inherently nonlinear, as their speed depends on their density/intensity, and arise when higher-velocity components of the wave overtake lower-velocity ones. The result is a transition region, or front, between the undisturbed fluid downstream and the perturbed fluid upstream. Although there are many variants, shock waves come in two basic flavors: dissipative and nondissipative. Dissipative shock waves relieve their excess energy through heating, e.g., via viscosity, which results in a monotonically decreasing front connecting the high-pressure and low-pressure sides. Nondissipative shock waves have an oscillatory and expanding front, as the excess energy is relieved via the generation of dispersive waves. The most familiar examples of shock waves are dissipative, e.g., the sonic boom of a supersonic airplane, created because air molecules downstream of the plane cannot be "warned" of its arrival by information-carrying sound waves. Dispersive shock waves (DSWs) are in a sense more "exotic" because their fronts rely on wave-dynamical, rather than collisional, effects; in turn, the basic excitations require a degree of coherence not normally found in classical fluids. Examples include superfluids, plasmas, and laser systems [2]. In these systems, DSWs are a fundamental means of energy transport.

C. Barsi • W. Wan • S. Jia • J.W. Fleischer (✉)
Department of Electrical Engineering, Princeton University, Olden Street, Princeton, NJ 08544, USA
e-mail: jasonf@princeton.edu

Z. Chen and R. Morandotti (eds.), *Nonlinear Photonics and Novel Optical Phenomena*, Springer Series in Optical Sciences 170, DOI 10.1007/978-1-4614-3538-9_9,

The first consideration of shock waves in optics was modeled on the nonlinear behavior of classical fluids. Early work focused on temporal pulses in fibers, showing that short pulses can experience traditional wave steepening and breaking [3–6]. Later work revealed that spatiotemporal kink states can result from Raman effects [7–10], polarization dynamics [11, 12], and frequency-dependent gain or loss [13, 14]. In the spatial domain, kink shock waves were predicted in quadratic media [15] and in photorefractive crystals [16, 17], and observed in the latter [18]. In these cases, the model behavior was dissipative, with energy transfer between beams replacing viscosity as a means of sustaining a well-defined (vs. oscillatory) front.

The study of DSWs in optics gained popularity in the 1990s, as non-return-to-zero formats in telecommunications created step-like initial conditions that were ripe for their creation [19–21]. More recently, such waves were demonstrated in bulk photorefractive crystals [22] and in thermal fluids [23, 24]. Particularly exciting has been the extension to two transverse dimensions, which has enabled an examination of wavefront geometry in shock propagation and interaction [22, 25].

In the remainder of this chapter, we discuss recent work on spatially DSWs in optical beams. In Sect. 9.2, we introduce some basic theory and shock wave formalism. In Sect. 9.3, we present the experimental observation of DSWs in self-defocusing photorefractive crystals. In Sect. 9.4, we illustrate the use of DSWs in material characterization by using their propagation dynamics to measure optical anisotropy and nonlocal response. In Sect. 9.5, we consider DSWs in photonic lattices and show their importance in nonlinear tunneling dynamics. Conclusions are given in Sect. 9.6.

9.2 Basic Theory and Formalism

The particular system considered here is a spatial one in which a continuous optical wave propagates in a nonlinear medium, mainly along the z axis. To an excellent approximation, the slowly varying amplitude ψ of such a field can be described by the nonlinear Schrödinger equation (NLSE):

$$i\frac{\partial\psi}{\partial z}+\frac{1}{2k_0}\nabla_{\perp}^{2}\psi+\frac{k_0}{n_0}\Delta n\left(|\psi|^2\right)\psi=0, \tag{9.1}$$

where $k_0 = 2\pi n_0/\lambda$ is the wavenumber, λ/n_0 is the wavelength in a homogeneous medium of refractive index n_0, and Δn is a nonlinear index change which depends on the intensity $|\psi|^2$. For the spatial case, the transverse Laplacian describes beam diffraction, while in the temporal case it describes pulse spreading due to dispersion. In most of the discussion that follows, we consider a Kerr-like medium, for which the index change $\Delta n = n_2|\psi|^2$ ($n_2 < 0$ for defocusing).

As is well-known [26–29], (9.1) also describes the (macroscopic) ground-state wavefunction for a fully condensed quantum state: $i\hbar\partial_t\psi + (\hbar^2/2m)\nabla_{\perp}^{2}\psi +$

$g|\psi|^2\psi = 0$, where m is the mass of the underlying particle, and the nonlinear coefficient g represents the mean-field contribution of (s-wave) interactions. In this approximation, the dynamics is more properly described as wave mechanical rather than quantum, with $\hbar$ simply serving as a parameter which normalizes the wavefunction. Note in particular that wavepacket evolution in time corresponds to beam propagation in space. Therefore, all of the shock phenomena here can appear in condensed matter systems as well. Indeed, the fundamental DSW, in both one [30, 31] and two [25] dimensions, has been demonstrated in Bose–Einstein condensates.

A fluid interpretation of light propagation follows by applying the polar (Madelung) transformation [32] $\psi(x,z) = \sqrt{\rho(x,z)}\exp[iS(x,z)]$ to (9.1), where ρ is the intensity of the beam, and S is its coherent phase. Scaling $(x,z) \to k_0(x,z)$ then gives the Euler-like fluid equations [33, 34]:

$$\frac{\partial \rho}{\partial z} + \nabla_\perp \cdot (\rho \mathbf{u}) = 0. \tag{9.2}$$

$$\rho\left(\frac{\partial \mathbf{u}}{\partial z} + \mathbf{u}\cdot\nabla_\perp \mathbf{u}\right) = -\nabla_\perp P + \frac{1}{2}\rho\nabla_\perp\left(\frac{1}{\sqrt{\rho}}\nabla_\perp^2\sqrt{\rho}\right). \tag{9.3}$$

This set of equations is a nonlinear eikonal representation, expressing the conservation of intensity ρ and momentum $\rho\mathbf{u}$, where $\mathbf{u} = \nabla_\perp S$ is the direction of energy propagation, or equivalent fluid velocity. The self-defocusing nonlinearity gives rise to an effective pressure $P = |n_2|\rho^2/2n_0$ whose gradient drives the optical flow. Indeed, we note that the nonlinear contribution to the phase $S \sim |n_2| k_0\rho(\Delta z)/n_0$, so that (9.3) is self-consistent with the definition of velocity. The last term in (9.3), known as the "quantum pressure" in condensed matter systems, has the highest-order derivatives and regularizes the system. Without the quantum pressure, (9.2) and (9.3) represent ideal Eulerian flow, and there would be no limit to the amount of energy that could accumulate in small spatial scales. For example, shock waves would develop infinitely sharp fronts and instabilities would have perturbation growth at arbitrarily short wavelengths.

To gain insight into the dynamics of shock formation, let us consider the case of a small wave packet/hump propagating on a uniform background ρ_∞. For simplicity, we consider a (1 + 1)-D system [35], so that the intensity may be written as

$$\rho(x,z) = \rho_\infty + \delta\rho(x,z). \tag{9.4}$$

In a fluid sense, the background intensity is a reference density which sets an effective "sound" speed $c = \sqrt{|n_2|\rho_\infty/n_0}$. This appears clearly when we linearize (9.2) and (9.3) for small perturbations. Keeping only the lowest-order terms, i.e., neglecting the convective derivative and the quantum pressure, we obtain the

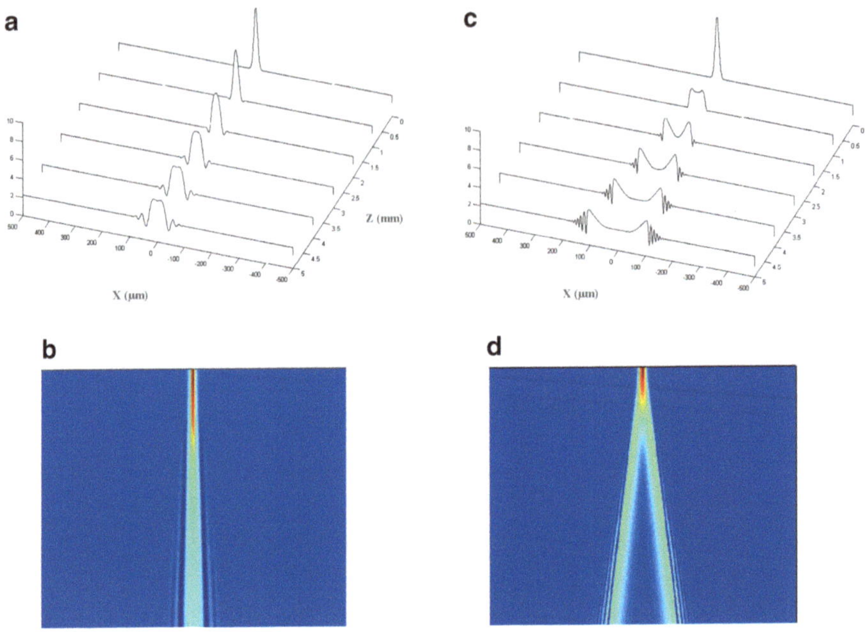

Fig. 9.1 Dispersive shock wave formation in nonlinear spatial optics. (**a**) Waterfall and (**b**) overhead view of linear propagation of hump-on-background. Simple diffraction yields some interference, but the localized beam retains its shape. (**c**, **d**) Same views as in linear case but for nonlinear evolution. The Gaussian hump splits into two and shocks the background

classical wave equation for the phase (optical stream function) S:

$$\frac{\partial^2 S(x,z)}{\partial z^2} - c^2 \frac{\partial^2 S(x,z)}{\partial x^2} = 0 \tag{9.5}$$

As before, $u = \partial S/\partial x$ and, through continuity, $\delta\rho = -\,(n_0/|n_2|)\partial S/\partial x$. This standard equation has the usual traveling wave solutions $\delta\rho(x,z) = [\delta\rho(x - cz, 0) + \delta\rho(x + cz, 0)]/2$, indicating that the initial perturbed intensity hump splits into two separated ones traveling with opposite directions.

This nonlinear splitting and repulsion contrasts sharply with linear dynamics. To see this explicitly, we numerically simulate a Gaussian perturbation $\delta\rho(x,0) = 2\eta\rho_\infty e^{-x^2/2a^2}$ using a split-step beam propagation code. Results are shown in Fig. 9.1. In the linear case (Fig. 9.1a,b), the hump retains its Gaussian profile as it propagates, creating small ripples as its tails interfere with the background field. In the nonlinear case, the perturbation splits into two smaller humps that walk off from each other (Fig. 9.1c,d). For weak nonlinearity, these humps travel at the linearized sound speed $u = u(\rho_\infty) = c$. For stronger nonlinearity, the two pieces will propagate with a velocity $u = u(\rho)$, which depends on the local intensity. Higher-intensity parts of the profile will travel at faster speeds, leading to wave steepening and eventual shock formation. We emphasize that any perturbation with intensity higher than the background, $\rho > \rho_\infty$, will travel faster than the local sound speed c, i.e., it will be supersonic.

In shock models with viscosity, e.g., Burgers-type descriptions [36, 37], dissipation counteracts wave steepening and leads to a monotonically decreasing front between the high- and low-pressure sides of the shock. For the case here, there is no absorption; instead, the increasing gradient triggers an increase in spatial dispersion. More accurately, self-phase modulation within the high-intensity region generates new (spatial) frequencies, which then disperse into the surrounding medium. This range of phases creates a series of interference fringes with varying periods, resulting in an oscillating wave 0 (Fig. 9.1c,d). The inner, nonlinear part of the front resembles a train of dark (or gray) solitons, whereas the outer part is a low-intensity region with oscillations that are effectively sound-like. In one dimension, for step-like initial conditions, the shock profile is given by a Jacobi elliptic function [25, 38–41]. As the derivation is somewhat involved, it is not repeated here.

9.3 Experimental Realization of DSWs in Spatial Optics

Experiments were performed using the set-up shown in Fig. 9.2. Light from a 532 nm laser was sent into an 8 mm × 8 mm × 8 mm SBN:75 ($Sr_{0.75}Ba_{0.25}Nb_2O_6$) photorefractive crystal and then imaged at the output. The initial profile was created using a Mach–Zehnder interferometer: one arm acted as a low-intensity plane-wave background, and a lens (cylindrical or spherical) placed in the other arm focused a central intensity hump onto the input face of the crystal. The nonlinear response of the SBN crystal was created by applying a voltage bias of −500 V across the crystal and taking advantage of the photorefractive screening effect [42]. A discussion of the photorefractive response, and the influence of various models of it on shock dynamics, is

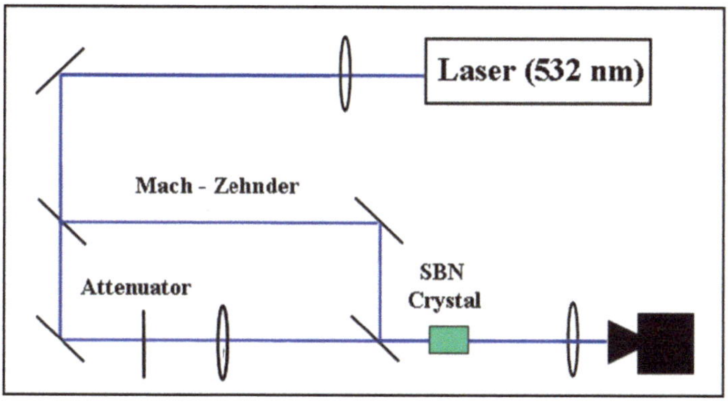

Fig. 9.2 Shock wave experimental setup

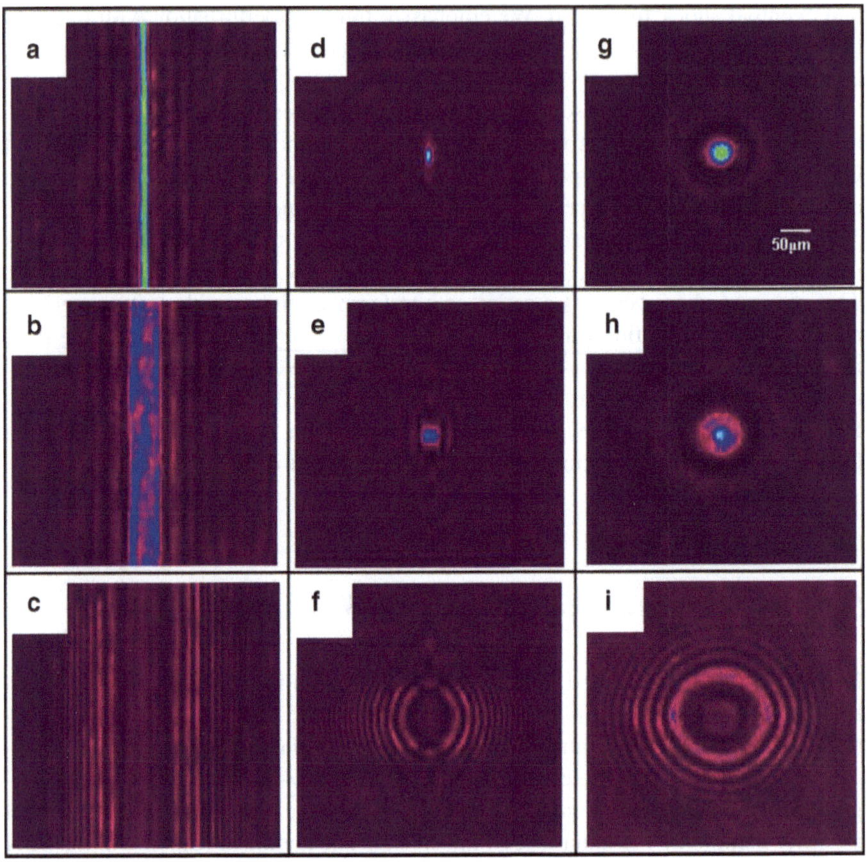

Fig. 9.3 Experimental pictures of superfluid-like optical spatial shock waves. (**a–c**) Input, (**d–f**) linear diffraction at output face, and (**g–i**) nonlinear shock waves at output face. *Left column*: 1D stripe. *Middle column*: 2D ellipse. *Right column*: 2D circle

deferred to Sect. 9.4. To keep the response as Kerr-like as possible, we held the applied voltage constant throughout the experiments and changed only the intensity of the central hump (using a variable attenuator) to probe nonlinearity. To maximize the response, the input light was polarized extraordinarily.

Typical shock waves are shown in Fig. 9.3. A stripe profile was created using a single cylindrical lens, an elliptical profile using two cylindrical lenses, and a circular shock using a spherical lens. The background beam was 10 mW and the hump-to-background intensity ratio was 20:1. In the linear case (Fig. 9.3d–f), the initial humps simply diffract, creating small ripples in the tails as the phase front curves. In the nonlinear case (Fig. 9.3g–i), self-defocusing forces the hump apart, depletes the central region, and shocks the background to create the oscillating front characteristic of DSWs.

The strength of shock formation was examined by adjusting the intensity of the central hump. As expected, the shocks become more violent with increasing

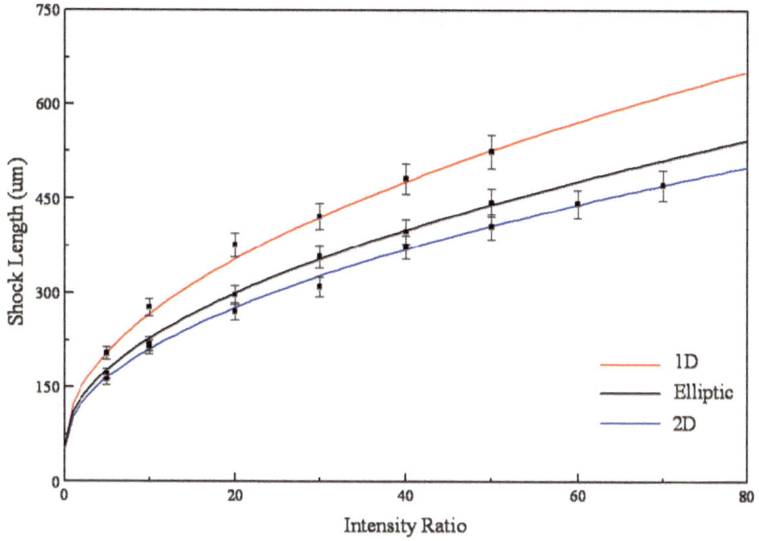

Fig. 9.4 Shock length, measured from the centerline to the end of oscillations, with respect to hump-to-background intensity ratio. *Top to bottom*: three solid curves are plotted of the functions $D_s = a_s(1 + b_s\sqrt{\rho/\rho_\infty})$ to fit the 1D stripe, 2D ellipse and 2D circle cases, respectively

hump-to-background contrast, displaying faster wave propagation and more oscillations within the front (stronger effective repulsion and higher nonlinear phase). In Fig. 9.4, we plot the measured front length (measured from the centerline to the end of oscillations) as a function of ρ/ρ_∞. The solid curves are best fits of the functions $D_\mathrm{s} = a_\mathrm{s}(1 + b_\mathrm{s}\sqrt{\rho/\rho_\infty})$, calculated independently for each shape s. Here, $a_\mathrm{s} = 54$ μm is a baseline distance set by the fixed background sound speed (background intensity and crystal properties), and the b-coefficients are 1.2, 1.0, 0.92 ± 0.04 for the stripe, ellipse and circle, respectively.

This square-root velocity scaling follows directly dimensional analysis of (9.3) (obtained by balancing the convective derivative with the nonlinear pressure). However, simple scaling arguments cannot determine the coefficient. More significantly, photorefractive media are saturable and known to deviate from Kerr-like responses for high intensity. This is probably why the measured stripe coefficient is higher than the $b = 1.0$ scaling predicted by recent 1D theory [25, 35, 38, 43].

To date, the only analytic treatment of dispersive, dissipationless shock waves in higher dimensions has been one-dimensional shocks propagating at an oblique angle [41]. The experimental results here show that geometry and the available expansion directions play a significant role in their dynamics.

9.4 Material Characterization

The basic DSWs shown above can be interpreted as nonlinear point-spread functions (though their use to describe image transmission is complicated by the non-shift-invariant nature of the spatially varying response). The intensity-dependent self-repulsion of the main central beam enhances diffraction, while the plane-wave background serves as a reference beam to highlight phase accumulation from propagation. Naturally, the output pattern depends on the particular nonlinear properties of the medium. Of course, the pattern also depends on the input, as demonstrated in Fig. 9.4. With knowledge of both the input and output, however, measurements of the output shock can be used to *characterize* a material's (possibly unknown) optical response.

Measuring nonlinear diffraction in this way is similar to z-scan techniques, commonly used to characterize both thin [44] and thick [45, 46] nonlinear samples. However, the latter methods require multiple measurements along the optical axis, cannot measure the full transverse index profile, and usually are limited by assuming Gaussian input beams and Gaussian-like output beams. For more complex nonlinear outputs, matching the shock parameters with model coefficients allows calibration of the underlying propagator. Further, the shock-formation characterization technique does not rely on symmetry considerations of the initial beam, so that the material properties can be studied by optimizing the arbitrary initial conditions.

We demonstrate the method here on two different types of nonlinear response: isotropic media with a nonlocal nonlinearity and uniaxial crystalline media with an anisotropic response.

9.4.1 DSWs with Nonlocal Nonlinearity

Many optical materials respond nonlocally, in that the index of refraction at a particular location is determined by the intensity not only at that point, but at nearby portions of the material as well. The spatial extent of the index contribution at that point is assumed to be determined only by the properties of the medium itself. Nonlocal phenomena appear in many fields, such as plasma physics [47] and BEC [48], and can arise in optics through physical processes such as atomic diffusion [49] and thermal self-action [50, 51]. The spatially extended response competes with small-scale perturbations, inhibiting dynamics that appear when wave coupling is local. For example, modulation instability can be suppressed [52] and incoherent solitons in instantaneous media are possible [53, 54]. Here, we consider the influence of nonlocality on the high-frequency, oscillatory nature of dispersive shock fronts [23, 24, 55, 56].

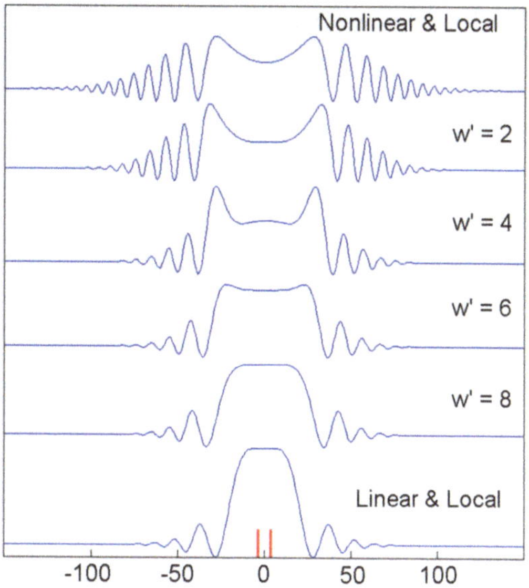

Fig. 9.5 Numerical simulation of dispersive shock waves as a function of nonlocal response width (w', normalized to the width of the input Gaussian hump). The *red bars* on the x-axis indicate the FWHM of the initial hump, which has a 25:1 hump-to-background intensity ratio

To begin, we return to the NLSE (9.1) but generalize the nonlinearity to a nonlocal response. Generically, this response can be modeled by a convolution operation: $\Delta n[\rho(\mathbf{r})] = n_2 \int R(\mathbf{r}')\rho(\mathbf{r} - \mathbf{r}')\mathrm{d}\mathbf{r}'$, where $R(\mathbf{r}')$ is the medium's response function normalized to unity, i.e., $\int R(\mathbf{r}')\mathrm{d}\mathbf{r}' = 1$ [52], and the integration takes place over the transverse dimensions of the system. For the thermal medium considered below, we assume a Gaussian response function $R(x) = (\pi w^2)^{-1/2}\exp(-x^2/w^2)$, where w represents the range of the nonlocality.

Numerical simulations of dispersive/diffractive shock waves in a nonlocal nonlinear medium are shown in Fig. 9.5. The input intensity profile is the same as in Fig. 9.1: a Gaussian hump on a constant, low-level background, and output profiles are shown as the nonlocal parameter w is increased. As in Fig. 9.3, the expanding wave consists of two repulsive humps whose fronts are characterized by oscillations. As the range of nonlocality increases, the central region broadens and the oscillations in the tails become damped. These effects are intuitively reasonable, because the convolution in Δn should both broaden and smooth out the resultant field. Several limiting forms of the response function are useful to consider. In the limit of a delta-function response, the nonlinearity reduces to the local Kerr case $\Delta n = n_2\rho$, and the problem reduces to the system described in Sect. 9.2, i.e., an initial hump that is "supersonic" compared to the background's effective sound speed $c = \sqrt{(|n_2|\rho_\infty/n_0)}$. In the limit of a response width narrow compared to the input, the nonlinearity can be expanded in a Taylor series, giving $\Delta n = n_2\rho + n_2\left(\langle x^2\rangle/2\right)\left(\mathrm{d}^2\rho/\mathrm{d}x^2\right)$, where $\langle x^2\rangle = w^2/2$ is the average variance of the nonlocality. The second term has the same dispersive order as diffraction and has a spatially dependent effect, weakening the effective repulsive pressure in the central part of the Gaussian beam while enhancing it in the tails. In the opposite limit of response

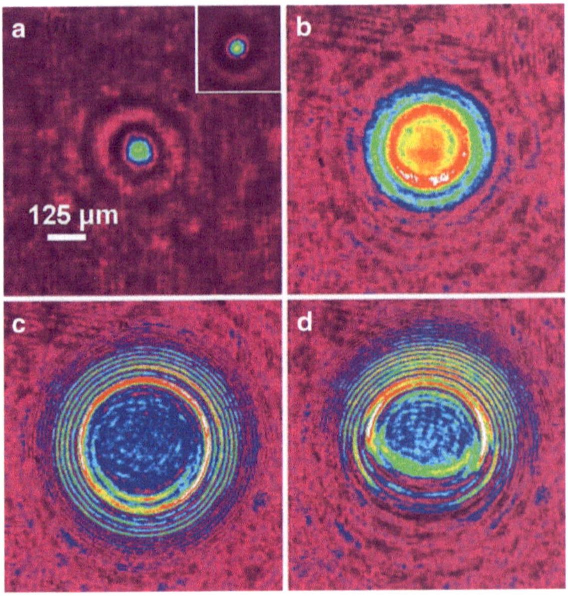

Fig. 9.6 Experimental evolution of diffractive shock wave after 1 cm of propagation in an ethanol + iodine liquid cell. (**a**) Linear case. *Inset*: input profile. (**b**) Initial shock formation in nonlinear, nonlocal case. (**c**) ~200 ms later, quasi-steady state (before convection of fluid) shock profile. (**d**) Steady state, with asymmetry due to convection

width much greater than input beam width, the overall response (nonlinearity plus nonlocality) approaches the linear, local case; here, the response of the medium is equally strong across the entire beam profile, so that the hump does not experience a nonlinear phase shift relative to the background.

We demonstrate these effects experimentally by considering the self-defocusing of CW spatial beams in a thermal liquid cell [24, 25]. More specifically, we project 532 nm laser light onto a 1 cm × 1 cm plastic cell containing ethanol doped with iodine. Physically, the iodine absorbs the green light, which then acts as a thermal source. The liquid develops a temperature gradient ΔT, which is proportional to the index change Δn via the thermo-optic coefficient, $\beta \equiv \partial n/\partial T$. Overall, the system is quite complex as it involves three coupled systems: thermal diffusion (Poisson equation) [53–55, 57, 58], fluid dynamics (Navier-Stokes equation), and nonlinear optical propagation (NLSE). However, for weak heating, convection can be ignored, so that the fluid can be taken as stationary, and the heating effects can be modeled accurately by the phenomenological convolution operation above [52, 59, 60].

As in the photorefractive case, a hump-on-background beam is incident onto the liquid. A shutter is placed in front of the liquid cell so as to observe the time dynamics of the system. By itself, the plane-wave background causes uniform heating and thus no relative index change, while the Gaussian hump produces a spatially dependent nonlinearity. The liquid medium responds to this profile in two different stages: relatively fast thermal diffusion followed by convection of the fluid itself (after ~0.5 s of CW heating). The temporal evolution of the output, shown in Fig. 9.6, demonstrates the development of the different stages of nonlocality. When the shutter is first open, the liquid has had no time to absorb the light, so the output

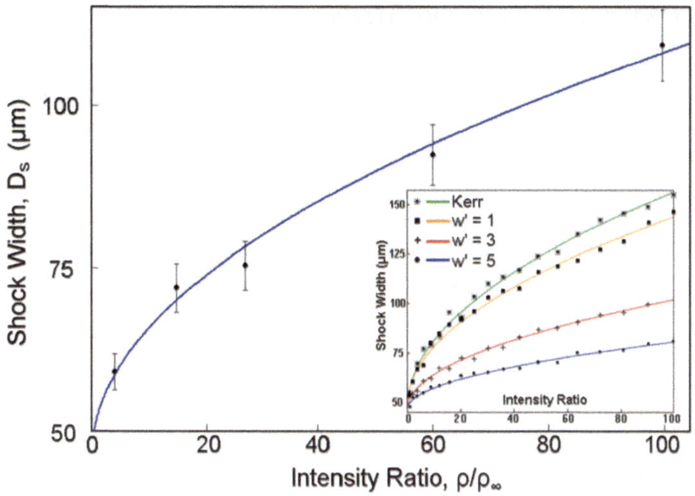

Fig. 9.7 Nonlocal shock length, measured horizontally from centerline to the end of oscillations, as a function of hump-to-background intensity ratio. *Black dots*: experimental measurements. *Solid line*: best fit of $D_s = a_s\left(1 + b_s\sqrt{\rho/\rho_\infty}\right)$, with $a_s = 46$ μm and $b_s = 0.13$. *Inset*: numerical simulation, showing consistency of scaling relation as a function of nonlocal response width w'

is simply a linearly diffracted beam (Fig. 9.6a), with some interference fringes due to interference between the plane wave and the curved Gaussian phase front. As time progresses, the nonlocal shock begins to form (Fig. 9.6b). The thermal gradient is weak, but not negligible, so that the beam diameter increases dramatically to about five times the diameter in the linear case. The next panel, Fig. 9.6c, shows the full beam expansion due to the increased temperature gradient, and hence nonlocality, in quasi-steady-state. The observed oscillations are signature features of DSWs [2, 23, 24], and are similar to previous observations in the context of thermal blooming [50, 51, 61, 62]. For longer times, the shock begins to move vertically and develop an asymmetry (Fig. 9.6d), due to convection of the liquid medium itself. At this point, the beam does not propagate simply in a static medium, as the light and fluid dynamics become coupled [63–65].

As before, increasing the hump-to-background ratio makes the shocks become more violent, with faster oscillations (higher self-phase modulation) in the tails. We again characterize the nonlinearity by measuring the shock width and fitting the resulting plot to the relation $D_s = a_s\left(1 + b_s\sqrt{\rho/\rho_\infty}\right)$. This is the same scaling as obtained in Figs. 9.2–9.4, but the fitting parameters $a_s = 46$ um and $b_s = 0.13$ indicate a rate of spreading much slower than in the local, Kerr case (Fig. 9.7, inset). For the liquid medium here, the b_s-coefficient suggests a nonlocal response of roughly twice the width of the input hump. Thus, shock propagation allows for characterization of the two material properties of the liquid: the nonlinear strength and the degree of nonlocality.

9.4.2 DSWs in Anisotropic Media

Many nonlinear optics experiments take advantage of anisotropic materials, as differences in response, e.g., between different polarizations, give another degree of freedom. The SBN crystal used in Sect. 9.3 is no exception: it is a uniaxial crystal with a much stronger nonlinear response along its crystalline c-axis than perpendicular to it. To maximize the response, all of the input beams were extraordinarily polarized. For solitons in SBN, for which self-focusing balances diffraction, circular inputs normally remain circular [66–69]. For the shock case, in which self-defocusing enhances diffraction, anisotropy appears. This birefringence can be seen most clearly by examining the 2D beam in Fig. 9.3, which shows that an initially circular input (Fig. 9.3g) leads to an output shock wave with a slightly elliptical shape (Fig. 9.3i).

There are two potential causes to the noncircular profile: anisotropic corrections to linear diffraction [70] or to the nonlinear response [71]. For SBN, the expected anisotropy in the diffraction operator is $\delta = (n_o/n_e)^2 - 1 \sim 2\text{–}3\%$, but this was not observed when the voltage was turned off (Fig. 9.3h). However, uniform bulk changes to the crystal (e.g., internal stress) when the nonlinearity is turned on cannot be ruled out [72]. Therefore, we consider both possibilities here.

The nonlinear index change of the photorefractive response is rather complex. The standard model is the one considered by Kuhktarev et al. in which photoexcitation is followed by charge transport and an electro-optic change in the index [71]. However, the model cannot be solved in its general form. Instead, two main limits have been studied: noncollinear wave mixing of broad beams [73–77] and the steady-state propagation of solitons [42, 78]. Neither limit is applicable to the dynamic expansion considered here. On the other hand, the different forms of the material response can be considered in a shock context, where the continuously evolving profile will quickly illustrate differences in the approximations used.

As shock waves are triggered by a spatially localized variation in the intensity, we base our discussion on the models for photorefractive screening solitons [42, 78]. For these steady-state beams, the combination of electron diffusion and drift leads to two intensity-dependent terms: a local, saturable term and a nonlocal, diffusive term that results in self-bending. Since this bending occurs in the direction of the applied voltage, i.e., along the c-axis, one-dimensional models are normally sufficient to describe material and soliton response (2D models can account explicitly for anisotropic boundary conditions [79, 80] and have predicted elliptical soliton formation [81], but circular solitons occur for nearly all experimental parameters [66–69]). A list of the most popular reduced models is given in Table 9.1.

We now wish to compare these theoretical models with experiment. To examine a pure point-like spread function, without coupling to a background, we consider a single Gaussian beam focused onto the input face of an SBN:75 crystal (Fig. 9.8). Here, a DSW is formed as the center of the beam breaks into its tails. In this example, the intensity was measured as before, but an additional interference measurement [82] was used to calculate the phase as well. This is necessary here, as the different models of the photorefractive response (Table 9.1) have different gradient terms,

Table 9.1 Models of the photorefractive effect and their effects in the reconstruction

1	$\Delta n \propto I$	Simplest model generates ellipticity near the focus of the Gaussian beam
2	$\Delta n \propto I, \tilde{D} = \mathrm{e}^{-\frac{z}{2k_0 n_0}\left(\frac{k_x^2}{1+\delta}+k_y^2\right)}$	Algorithmic modification (δ-term in Fresnel propagation kernel) yields accurate reconstruction
3	$\Delta n \propto \frac{I}{I+I_{\text{sat}}}$	Saturation provides circular reconstruction in intensity, but phase reconstruction is distorted
4	$\Delta n \propto \frac{\partial}{\partial x}[\log(I+I_{\text{sat}})]$	Intensity profile is distorted, and the phase contains spurious modulations
5	$\Delta n \propto \frac{\partial}{\partial x}[I]$	
6	$\Delta n \propto I + \kappa \frac{\partial}{\partial x}[\log(I)]$	Results are similar to 1), 1), 3) respectively, but with unwanted noise in both intensity and phase and incorrect phase curvature
7	$\Delta n \propto I + \kappa \frac{\partial}{\partial x}[I]$	
8	$\Delta n \propto \frac{I}{I+I_{\text{sat}}} + \kappa \frac{\partial}{\partial x}[\log(I+I_{\text{sat}})]$	

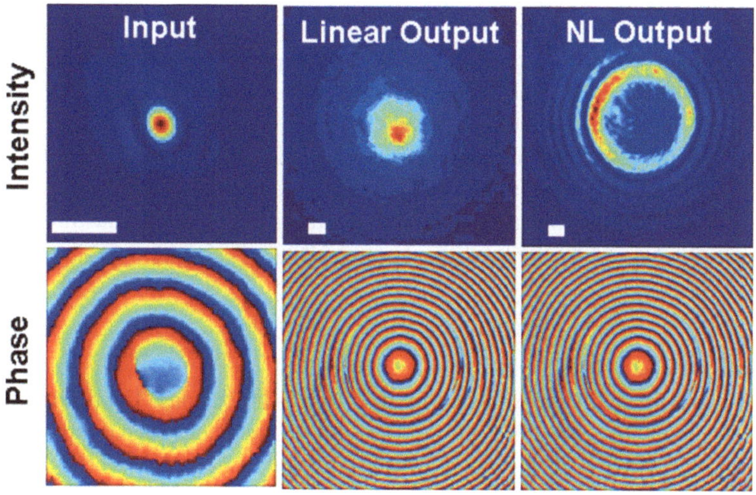

Fig. 9.8 Experimental measurements of linear and nonlinear Gaussian beam propagation. *Left*: experimental input. *Center*: linear (diffracted) output. *Right*: a dispersive optical shock wave is formed. Comparison of pure Kerr nonlinearity and Kerr + anisotropic diffraction term. The *c*-axis is horizontal in the figures. *Scale bar*: 50 μm

which manifest themselves in the phase. Because the full complex field is known, it can be substituted as an initial condition into numerical beam propagation algorithms that model (9.1).

There are two ways of making the comparisons: forward propagation [83] or backward propagation [84, 85]. In the former case, the experimental input is fed into a computer code, whose numerical output is compared with the measured output. The latter case starts with the output and runs in reverse. For ideal behavior, both methods are equivalent. In practice, however, back-propagation is more sensitive and more accurate, as the output profile has finer features, i.e., higher spatial frequencies, and the input profile provides a cleaner and more stringent target.

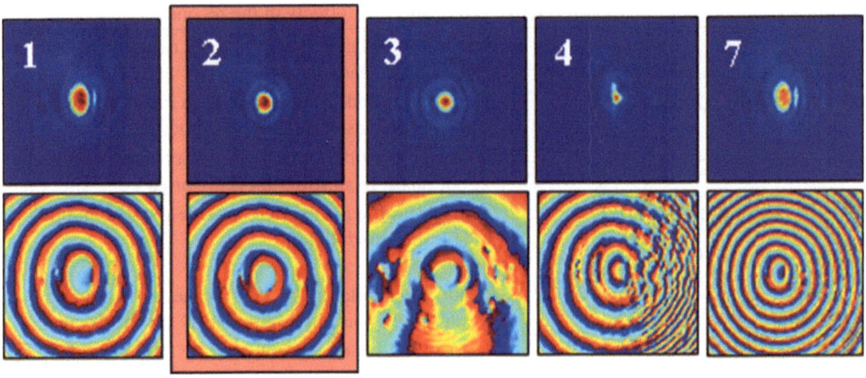

Fig. 9.9 Reconstruction of input amplitude (*top row*) and phase (*bottom row*) from measured output for different propagation models. *Numbers* in each panel indicate which model from Table 9.1 was used for the reconstruction. The *highlighted box* shows the best results, obtained using the new model (9.2) of anisotropic propagation with Kerr-dependent diffraction coefficients

A reverse propagation algorithm is effected simply by replacing the propagation distance z with $-z$ in the standard split-step Fourier method [84, 85]. As an initial model, we use the symmetric Kerr nonlinearity, $\Delta n = -|n_2|I$, which has been suitable for the experimental measurements made thus far (e.g., see Fig. 9.4). The coefficient γ is calculated by choosing the value that minimizes the sum-of squares error between the measured input intensity and the reconstructed input intensity, normalized by the total pixel count M^2:

$$\mathrm{SSE} = M^{-2} \sum_{m_x, m_y} \left(I_{m_x m_y}^{\text{in,reconstructed}} - I_{m_x m_y}^{\text{in,measured}} \right),$$

where m_x and m_y are the pixel counts in the x- and y-dimensions. The result is shown in Fig. 9.9. Note that even after error minimization, the reconstructed input is highly elliptical. As expected, the isotropic Kerr nonlinearity is not sufficient in describing photorefractive propagation.

Perhaps more surprising is the fact that traditional anisotropic models of the photorefractive response also do not capture the observed behavior, even though they are highly successful models for soliton propagation. Interestingly, some models provide a reasonable match for the intensity but fail in the phase, an indication of the latter's importance here.

To account for the ellipticity, we return to the anisotropic model of linear diffraction but parameterize the coefficients with a nonlinear change. More specifically, we include a voltage-dependent correction of 7% in the x-direction for the linear propagator, that is, in the Fourier domain

$$e^{-i\frac{\lambda \Delta z}{4\pi}\left(\frac{k_x^2}{n_e}+\frac{k_y^2}{n_e}\right)} \rightarrow \mathrm{e}^{-i\frac{\lambda \Delta z}{4\pi}\left(\frac{k_x^2}{n_e(1+\delta)}+\frac{k_y^2}{n_e}\right)},$$

where $\delta = 0.07$, and k_x and k_y are the wavevectors for the x- and y-axis. As shown in Fig. 9.9, the reconstructed input provides a good match with known one.

Despite this reasonable agreement with experiment, a first-principles derivation of the modified propagator has yet to be performed. Likewise, the modification induced by more complex beam profiles remains to be determined. The results here simply illustrate that the Gaussian-to-shock evolution can serve as a sensitive probe for testing and comparing theories with model material responses.

9.5 DSWs in Nonlinear Junctions and Arrays

Up to now, we have studied wave dynamics in homogeneous media, where we have shown that DSWs are fundamental to energy transport. It should not be surprising, then, than DSWs play a critical role in the presence of potentials as well. Here, we give two basic examples: a periodic potential and a single nonlinear junction.

9.5.1 *Wave Tunneling in Nonlinear Junctions*

It is clear that the introduction of a potential will affect wave propagation, as reflection and tunneling will create perturbations that launch waves. In the nonlinear case (for nonsolitonic conditions), energy transport will be dominated by shocks.

Even for a single interface, nonlinearity introduces a variety of new dynamics. Self-phase modulation modifies boundary conditions, leading to enhanced refraction and reflection [86, 87] as well as multistability [88], while cross-phase effects can give rise to splitting solutions for vector beams [89]. For double interfaces, i.e., tunneling junctions [90–92], the problem is compounded by the dynamics between the faces [93, 94]. In the latter case, most work has concentrated on a sequence of linear–nonlinear connections [86, 90–92]. More recently, potential wells fully immersed in the nonlinear medium, with nonlinear–nonlinear boundaries, have been considered [95–100].

Here, we consider the fundamental case of plane-wave scattering from a barrier potential [94, 101]. A schematic of the geometry is shown in Fig. 9.10. An ordinarily polarized sheet of light, created using a cylindrical lens, optically induces a refractive index defect (antiwaveguide) in an $8 \times 8 \times 8$ mm^3 SBN:60 photorefractive crystal [102, 103]. An extraordinarily polarized plane wave is then sent across this defect. In the experiments, the angle of incidence (and thus initial kinetic energy) is fixed to 1° and the barrier width is 200 μm.

To begin, we introduce a potential $V(x)$, constant along z, to an initial field that has constant intensity everywhere. (This setup is different than the traditional transmission–reflection geometry, discussed below, in which a plane wave is incident from the right side of the barrier and there is no initial field on the left side.) This type of profile occurs in many physical situations, such as flow over and through an obstacle, and facilitates an understanding of dynamics on the transmission side (as the background field did in the homogeneous case).

Fig. 9.10 Experimental scheme of nonlinear wave tunneling

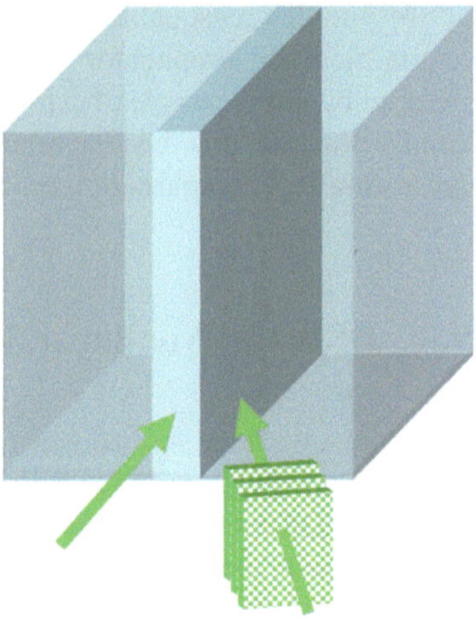

Tunneling dynamics with initial field everywhere are shown numerically in Fig. 9.11. As before, simulations were performed using a split-step beam propagation code, this time using a potential $V(x)$ in (9.1) [a force $\nabla_{\perp} V$ in (9.3)]. A plane wave is incident from right to left, giving rise to significant dynamics on both sides of the barrier potential. In the linear case (Fig. 9.11a,c), phase changes at the interfaces lead to interference fringes, but there is no evolution of the wavefronts themselves. By contrast, in the nonlinear case (Fig. 9.11b,d), the development of large phase gradients leads to large flow velocities. Energy is transported farther away from the interfaces, creating a distributed profile with much lower peak intensities, and there is significant evolution dynamics. The reflected and transmitted waves both self-steepen as they propagate, eventually breaking into right and left-going DSWs. These are the phase jumps/dark soliton trains predicted for nonlinear tunneling in a superfluid [94, 104].

Interestingly, DSWs appear in the shadow region behind the barrier as well. Here, the darkness of the shadow implies a higher refractive index (the medium is defocusing), drawing a back-flow towards the barrier and further encouraging the tunneling of light through it. From the rest frame of the angled wave, the barrier is being dragged through it. There is now a (dispersive) bow shock created at the front of the obstacle, while the dynamics behind the barrier is the optical version of a (1D) fluid wake. Outgoing shock waves develop at the trailing edge, while optical back-flow starts to follow the barrier due to the induced low-pressure area in the shadow.

Experimental observations of tunneling dynamics are shown in Fig. 9.12. In the linear case, there are interference fringes upon reflection and transmission, with the highest intensity peaks appearing at the barrier edges. In the nonlinear case, these peaks propagate faster, appearing a significant distance from the junction walls. Tunneling and back-flow are enhanced, leading to an intensity decay from

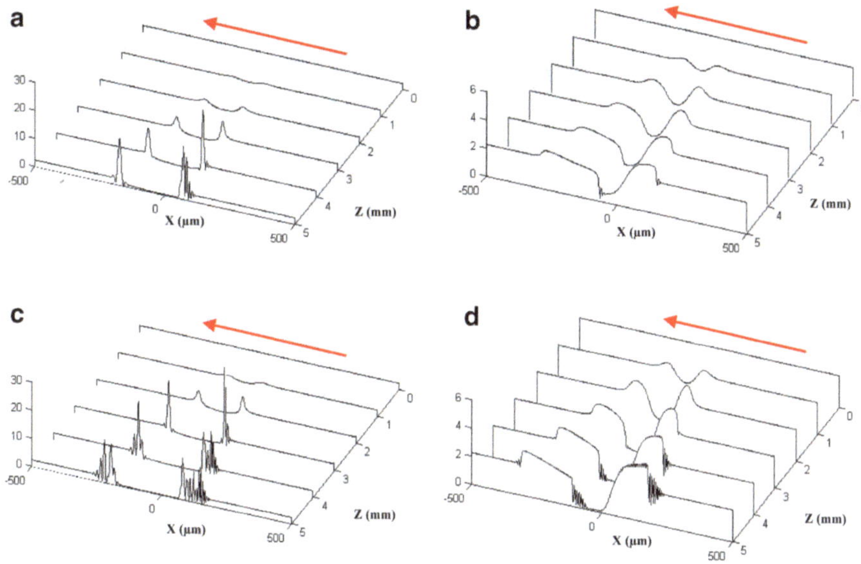

Fig. 9.11 Numerical simulation of full-field scattering by a barrier. Plane wave incidence onto a barrier centered in $x = 0$ with FWHM ~100 μm from *right* (*red arrow*). x is the spatial transverse axis, z is propagation axis. (**a**, **c**) Linear case. (**b**, **d**) Nonlinear case. Note the difference in scale. Shown are output pictures for (**a**, **b**) weak ($\Delta n/n_0 = 2 \times 10^{-4}$), (**c**, **d**) strong barrier heights ($\Delta n/n_0 = 6 \times 10^{-4}$)

the interfaces that is more gradual than in the linear case. As predicted [94], dark solitons in the fluid wake appear only if the barrier height exceeds a threshold value (~2×10^{-4} for 1° incidence) and these nonlinear effects become more pronounced as the barrier height is increased.

In tunneling experiments, a common metric of performance is the amount of energy transmission through the barrier. To measure this, we block half of the plane wave at the input, so there is only initial field energy on one side of the defect. As before, we fix the angle of incidence to 1°. Figure 9.13 shows the end result of tunneling, obtained by integrating the total energy transmitted through the barrier. In the linear case, there is the usual transmission for low potentials and total internal reflection for higher barriers, resulting in an exponentially decaying evanescent wave. In the nonlinear case, there are two distinct behaviors, depending on whether the initial incidence angle (beam kinetic energy) is above or below the barrier height [101]. For small barrier heights, linear transport is more efficient, as the self-defocusing nonlinearity (repulsive interaction) creates an enhanced pressure upon reflection. For stronger potentials, however, the self-repellant force of the tunneled light encourages further transport across the barrier, resulting in enhanced transmission. Paradoxically, this nonlinear effect increases for narrower and higher barrier potentials [94], as the nonlinear pressure within the sandwich layer becomes more pronounced.

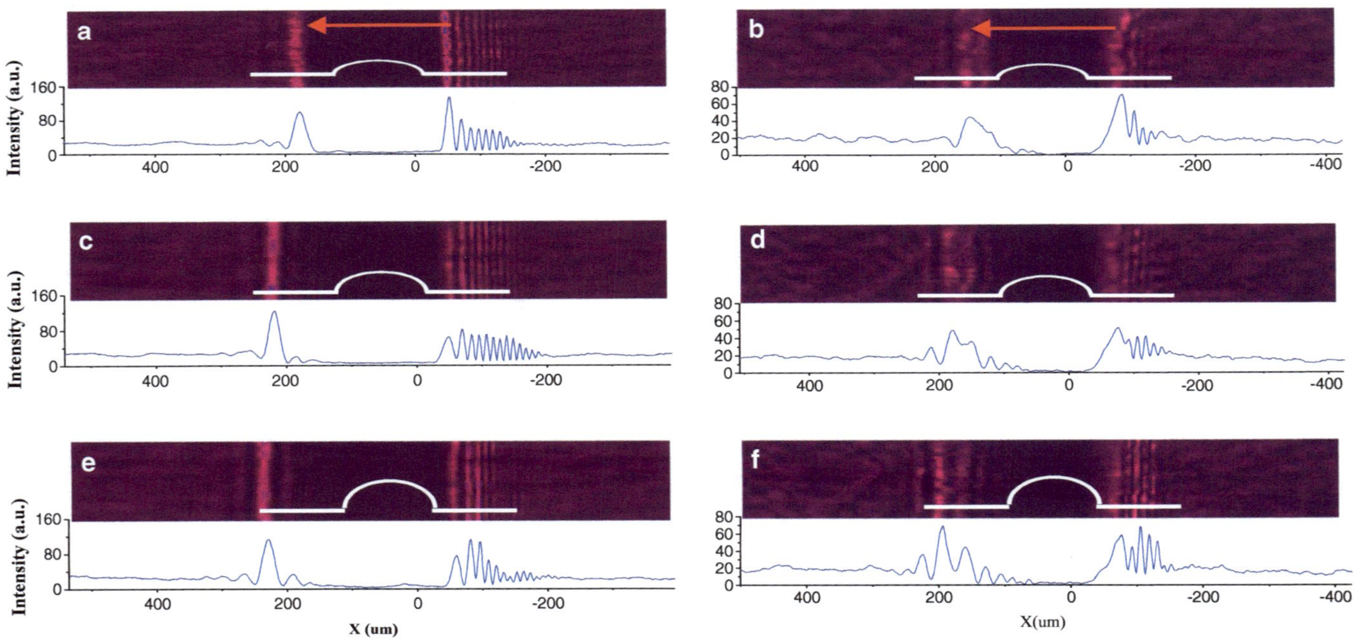

Fig. 9.12 Experimental observation of full-field tunneling through barrier. *Left column*: linear case; *right column*: nonlinear case. Shown are output pictures for (**a**, **b**) weak ($\Delta n/n_0 = 2 \times 10^{-4}$), (**c**, **d**) medium ($\Delta n/n_0 = 4 \times 10^{-4}$), and (**e**, **f**) strong barrier heights ($\Delta n/n_0 = 6 \times 10^{-4}$). In all cases, the input angle was 1°

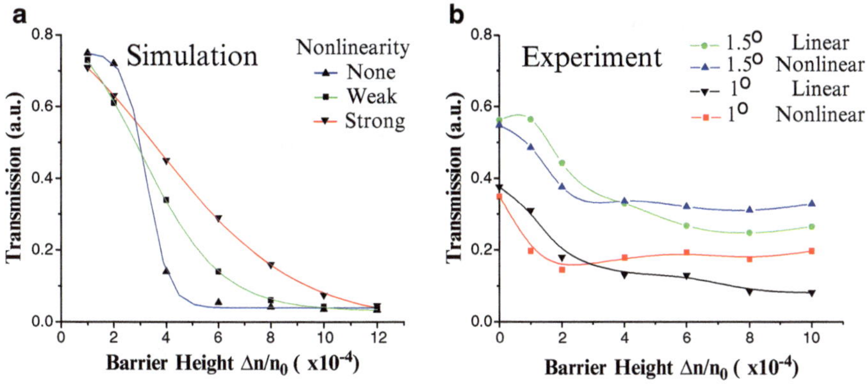

Fig. 9.13 Wave transmission/tunneling through a small barrier potential. (**a**) Numerical simulation of transmission through the junction. (**b**) Experimental results

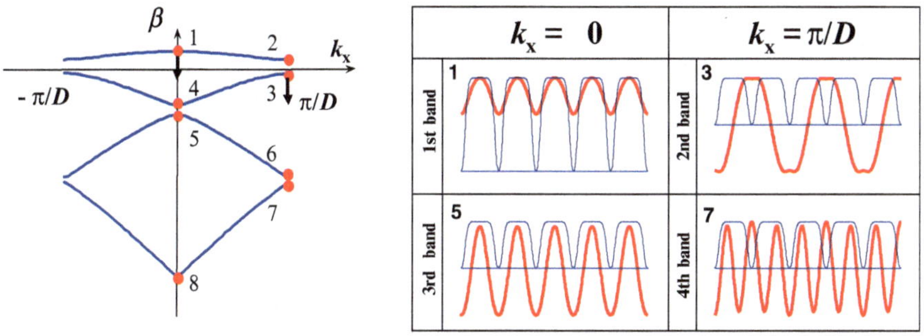

Fig. 9.14 Linear transmission spectrum, showing initial propagation constants and field profiles of representative modes. Modes numbered 1 and 3 are the initial background k-vectors, and the *arrows* indicate the effect of defocusing nonlinearity

9.5.2 DSWs in Nonlinear Arrays

When many junctions are arranged in an array, the geometry becomes a photonic lattice. For periodic potentials, the result is the familiar appearance of transmission bands, forbidden gaps, and Brillouin zones for wave dynamics.

Figure 9.14 shows the linear transmission spectrum and underlying Floquet-Bloch modes of the waveguide array. As the system is spatial, rather than temporal, the eigenvalue is the propagation constant $\beta = k_z(k_x)$ rather than the wave energy (frequency) ω. As in other photonic systems, the geometry of the transmission bands characterizes the dispersion (diffraction) of modes. Momentum regions with concave curvature (the odd-numbered modes in Fig. 9.14) experience normal diffraction while regions with convex curvature experience anomalous diffraction [105, 106].

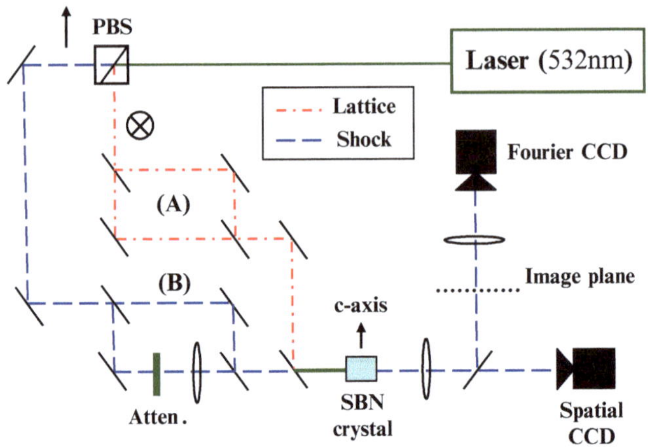

Fig. 9.15 Experimental setup of shock waves in nonlinear arrays. Light from a 532 nm laser gets split by a polarizing beam splitter. The ordinarily polarized beam passes through a Mach–Zehnder interferometer (A) to induce a waveguide array in an SBN:75 crystal. The extraordinarily polarized beam passes through a modified Mach–Zehnder interferometer (B) to create a Gaussian-on-background input shock profile. Light exiting the crystal is imaged into two CCD cameras, one for the intensity in position (**x**) space, one for the power spectrum in momentum (**k**) space (obtained by performing an optical Fourier transform). To excite a second-band Bloch background for a higher-band shock, the plane wave arm in (B) is blocked and the interferometer in (A) is partially polarized in the extraordinary direction

When nonlinearity is present, the propagation constant moves off the linear transmission bands. If β shifts into a gap, then the wave is decoupled from the linear modes and can be localized in its own self-induced defect [106]. Examples of such lattice solitons occur with self-focusing nonlinearity for modes 1 [107] and 3 [108] and self-defocusing nonlinearity for mode 2 [103] in Fig. 9.14. If β shifts in the opposite direction, e.g., self-focusing for modes 2 and 4 and self-defocusing for modes 1 and 5, then coupling with linear modes is facilitated. This is the regime of lattice shock waves.

To observe lattice shocks experimentally, we use the setup shown in Fig. 9.15 [109]. It consists of three basic parts: (1) an ordinarily polarized pair of plane waves to optically induce a lattice structure [102, 103], (2) an extraordinarily polarized hump-on-background profile, similar to the input in Fig. 9.2, and (3) an imaging system to observe the light exiting the crystal. At the exit face of the crystal, the output is imaged into two CCD cameras, one for the direct (near-field) intensity in position (**x**) space and one for the Fourier (far-field) intensity in momentum (**k**) space.

Experimental shock waves are shown in Fig. 9.16. For reference, a DSW in the homogeneous crystal (no induced array) is shown in Fig. 9.16a. The input, not shown, has a hump-to-background intensity ratio of 10:1. This is the same 1D shock wave discussed in Sect. 9.3.

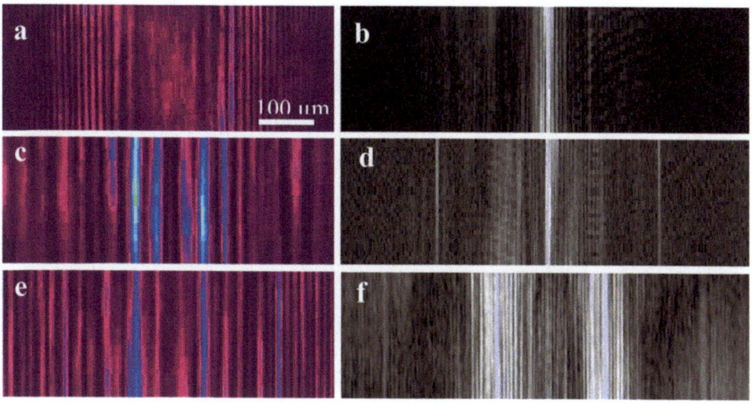

Fig. 9.16 Output pictures of dispersive shock waves in homogeneous and lattice systems. The Gaussian-on-background input intensity ratio is 10:1. (**a**, **b**) homogeneous system; (**c**, **d**) against fundamental, first-band background (mode 1); (**e**, **f**) against second-band, cosine background (mode 3). In (**c–f**), the period of the lattice is 30 μm. *Left* and *right columns* show intensity in position (**x**) space and power spectrum in momentum (**k**) space, respectively

Figure 9.16c shows the output of the same hump-on-background profile launched into a waveguide array with period $D = 30$ μm. In this case, the intensity ratio of hump to background remains 10:1 while the hump-to-lattice ratio is 1:1. Due to the lattice potential, the front does not propagate as fast, and the smaller-scale oscillations are lost. The slower speed arises from the periodic energy barrier as the wave tunnels from site to site, while the size of the waveguide (potential well) sets the spatial scale for oscillations.

Further insight into the basic shock behavior can be obtained from the power spectrum, shown in Fig. 9.16b,d. For the homogeneous shock, there is a large peak at $k = 0$, due to the plane-wave background, surrounded by two broader bands: an inner one representing large-scale envelope modulations and an outer one representing the small-scale oscillations in the shock tails. For the lattice shock, the central peak remains but two additional side peaks appear, one each at the double-Bragg angles $\pm 2k_B = \pm 2\pi/D$. The reasons for these can be seen in Fig. 9.17, where we propagate each of the input components separately through the nonlinear array. For the Gaussian input beam (Fig. 9.17a,b), the output intensity profile is a Gaussian $\exp[-x^2]$ modulated by the periodic lattice $\cos(k_B x)\exp[-x^2]$. The corresponding profile in k-space (obtained by a simple Fourier transform) is $\exp[-(k-k_B)^2/4] + \exp[-(k+k_B)^2/4]$. The nonlinear output spectrum of the plane-wave input is shown in Fig. 9.17c,d. As can be seen, this uniform input excites a broad spectrum of modes across the first band, with dominant peaks at $k = 0$ and $k = \pm 2k_B$. The latter peaks are a direct result of the lattice periodicity, aided by the nonlinearity pushing down the propagation constant (from mode 1 in Figs. 9.3–9.9 to mode 5). Note that it is mode 5 from the third band, rather than

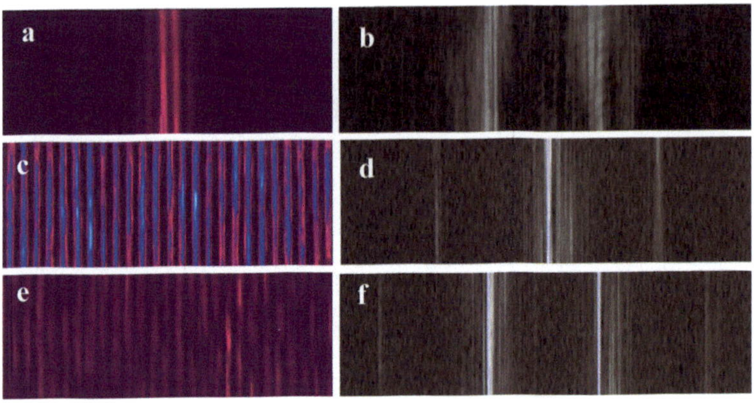

Fig. 9.17 Output pictures of different initial intensity profiles in optical lattices. (**a**, **b**) Bright Gaussian hump with width 15 μm, intensity ratio 1:1 to the lattice; (**c**, **d**) plane background wave, with intensity ratio 1:10 to the lattice; (**e**, **f**) cosine background wave, with intensity ratio 1:10 to the lattice. The period of the lattice is 30 μm. *Left* and *right columns* show intensity in position (**x**) space and power spectrum in momentum (**k**) space, respectively

mode 4 from the second band, because of the concave curvature of the band; also, excitation of this mode sharpens the edges within the waveguides (its dipole structure is phase-shifted), rather than contributing an intensity dip in the waveguide centers. The fact that the broad output peaks at $\pm k_B$ from the Gaussian input (Fig. 9.17b) do not appear in the discretized shock spectrum (Fig. 9.16d) means that the energy initially in the Gaussian hump has been effectively coupled to the background light.

Interpretation of the previous results is complicated by the fact that the background plane wave is not a pure eigenmode of the underlying array. To excite a cleaner lattice shock wave, we launch the initial Gaussian input against the cosine Floquet-Bloch mode at the edge of the first Brillouin zone (mode 3 in Fig. 9.14). This mode is excited by partially rotating the polarization of the array beams (Fig. 9.15a) in the extraordinary direction and recalibrating the intensity ratios. Compared with the previous shock wave (Fig. 9.16c), the Bloch shock wave (Fig. 9.16e) has much more intensity and higher spatial resolution in the shock tails. As shown by the power spectrum in Fig. 9.16f, this real-space behavior arises from wave coupling to higher-order modes. The two high peaks at $\pm k_B$ are from the initial cosine mode, with the spread around these peaks showing clear coupling to other modes within the band. In contrast with the broad first-band excitation in Fig. 9.16d, there is a sharp cutoff halfway through the Brillouin zone as the band curvature changes sign. As before, though, the nonlinearity couples the initial background mode to the next-higher mode modulo $2\pi/D$. These are the peaks appearing at an additional $\pm 2k_B$ in Figs. 9.16f and 9.17f.

9.6 Discussion and Conclusion

Similar to the soliton, the DSW is a basic unit for energy transport. However, solitons require a special balance between nonlinearity and dispersion/diffraction, while DSWs arise from the action of these forces in concert. This distinction makes soliton dynamics more particle-like, while the behavior of dispersive shocks is inherently wave-like. Indeed, DSWs are naturally dynamic creatures, with wave steepening and front spreading an integral part of their character.

When a wave encounters a potential in a self-defocusing medium, the subsequent radiation is dominated by dispersive shocks. This includes flow around material obstacles [41, 110–112] as well as self-induced potentials from collision and reflection [101, 113]. In the former case, the evolution of 1D shock waves propagating in two dimensions has led to the observation of fluid-like instabilities [114] and the onset of turbulence [41]. In the case of collisions, interacting fronts give rise to a nonlinear Huygens' principle, in which superposition of waves creates high-intensity sources for new shocks [22]. The result can lead to multi-phase flow [113] and, in arrays, self-focusing from mutual pressure [115–117]. When the potential allows tunneling, e.g., in nonlinear junctions, shock interaction at the boundary can lead to nontrivial transmission and hysteresis [101].

When external pressure leads to collapse, e.g., of a dark stripe, the resulting gradient catastrophe generates a fan of dark soliton filaments [118]. Similar soliton emission occurs in the temporal domain in supercontinuum generation [119] and optical rogue waves [120], in which dispersive waves and solitons created with opposite dispersion properties interact in optical fiber [121]. In the spatial domain, supercontinua have been observed in arrays [117, 122], where the four-wave mixing of Bloch modes leads to a wide spectrum of daughter shocks. Opposite shock-soliton behavior has also been observed for beams in self-focusing media, in which wave diffraction from a sharp edge was increased due to shock-generated solitons radiating outward from the edge [112].

The increased sensitivity to phase and intensity changes suggests that nonlinear optical systems hold much potential for sensing and imaging applications. Indeed, the basic DSW can be considered as a nonlinear point spread function. This aspect was used above in Sect. 9.4 as a probe for material characterization and can serve as a type of optical limiter. For use in imaging, however, care must be taken, as the intensity-dependent propagation is not shift-invariant and superposition does not hold. On the other hand, the dispersive (rather than dissipative) dynamics of the shock waves means that both energy and entropy are conserved. With suitable knowledge of the nonlinear propagator, interactions can be deconvolved, enabling new forms of dynamical imaging [85].

The ability to easily control the input, directly image the output, and create complex potentials gives optics a considerable advantage over other systems that support DSWs. Particular modes and waveforms can be excited nearly at will, with measurement possible in the full position-momentum (x,k) phase space. In this chapter, we have highlighted only the first generation of experiments, showing basic shock waves in homogeneous, interface, and lattice systems. Many more examples, including interactions, boundary layer flow, instabilities, and turbulence, remain to be demonstrated. Indeed, any (nonviscous) dynamics possible in fluids should arise in optics as well. Likewise, observations with light may back-react on the fluid systems which inspired them, particularly with regard to flow control. These will be subjects for future work.

References

1. G.B. Whitham, *Linear and Nonlinear Waves* (Wiley, New York, 1974)
2. V.I. Karpman, *Nonlinear Waves in Dispersive Media* (Pergamon, New York, 1974)
3. F. DeMartini, C.H. Townes, T.K. Gustafson, P.L. Kelley, Phys. Rev. **164**(2), 312 (1967)
4. D. Anderson, M. Lisak, Phys. Rev. A **27**(3), 1393 (1983)
5. R.L. Fork, C.V. Shank, C. Hirlimann, R. Yen, W.J. Tomlinson, Opt. Lett. **8**(1), 1 (1983)
6. G.Z. Yang, Y.R. Shen, Opt. Lett. **9**(11), 510 (1984)
7. Y.S. Kivshar, Phys. Rev. A **42**(3), 1757 (1990)
8. D.N. Christodoulides, Opt. Comm. **86**(5), 431 (1991)
9. G.P. Agrawal, C. Headley, Phys. Rev. A **46**(3), 1573 (1992)
10. Y.S. Kivshar, B.A. Malomed, Opt. Lett. **18**(7), 485 (1993)
11. M. Haelterman, A.P. Sheppard, Phys. Rev. E **49**(4), 3389 (1994)
12. S. Pitois, G. Millot, S. Wabnitz, Phys. Rev. Lett. **81**(7), 1409 (1998)
13. D.N. Christodoulides, M.I. Carvalho, Opt. Lett. **19**(4), 251 (1994)
14. D. Cai, A.R. Bishop, N. Grønbech-Jensen, B.A. Malomed, Phys. Rev. Lett. **78**(2), 223 (1997)
15. S. Darmanyan, A. Kamchatnov, F. Lederer, Phys. Rev. E **58**(4), R4120 (1998)
16. V.A. Vysloukh, V. Kutuzov, V.M. Petnikova, V.V. Shuvalov, J. Exp. Theor. Phys. **84**(2), 388 (1997)
17. M.I. Carvalho, A.G. Grandpierre, D.N. Christodoulides, M. Segev, Phys. Rev. E **62**(6), 8657 (2000)
18. B. Freedman, T. Carmon, M.I. Carvalho, M. Segev, D.N. Christodoulides, Phys. Rev. Lett. **91** (13), 133902 (2003)
19. Y. Kodama, S. Wabnitz, Opt. Lett. **20**(22), 2291 (1995)
20. M.G. Forest, K.T.-R. McLaughlin, J. Nonlinear Sci. **8**(1), 43 (1997)
21. M.G. Forest, J.N. Kutz, K.R.-T. McLaughlin, J. Opt. Soc. Am. B Opt. Phys. **16**(11), 1856 (1999)
22. W. Wan, S. Jia, J.W. Fleischer, Nat. Phys. **3**(1), 46 (2007)
23. N. Ghofraniha, C. Conti, G. Ruocco, S. Trillo, Phys. Rev. Lett. **99**(4), 043903 (2007)
24. C. Barsi, W. Wan, C. Sun, J.W. Fleischer, Opt. Lett. **32**(20), 2930 (2007)
25. M.A. Hoefer, M.J Ablowitz, I. Coddington, E.A. Cornell, P. Engles, V. Schweikhard, Phys Rev. A **74**(2), 023623 (2006)
26. V.L. Ginzburg, L.P. Pitaevskii, Zh. Eksp. Teor. Fiz. **34**, 1240 (1957)
27. L.P. Pitaevskii, Sov. Phys. JETP **13**(2) 451 (1961)
28. E.P. Gross, Nuovo Cimento **20**(3), 454 (1961)
29. G.A. El, V.V. Geogjaev, A.V. Gurevich, A.L. Krylov, Phys. D Nonlinear Phenomena **87** (1–4), 186 (1995)
30. Z. Dutton, M. Budde, C. Slowe, L.V. Hau, Science **293**(5530), 663 (2001)
31. N.S. Ginsberg, J. Brand, L.V. Hau, Phys. Rev. Lett. **94**(4), 040403 (2005)
32. E. Madelung, Z. Phys. **40**(3–4), 322 (1927)
33. S.A. Akhmanov, R.V. Khoklov, A.P. Sukhorukov, in *Laser Handbook*, ed. by F.T. Arecchi, E.O. Schultz-DuBois, M.L. Stitch (North-Holland, Amsterdam, 1972), p. 115

34. P.H. Roberts, N.G. Berloff, *Quantized Vortex Dynamics and Superfluid Turbulence*, ed. by C.F. Barenghi, R.J. Donnely, W.F. Vinen (Springer, Berlin, 2001), p. 235
35. B. Damski, Phys. Rev. A **69**(4), 043610 (2004)
36. J.M. Burgers, *The Nonlinear Diffusion Equation* (Reidel, Boston, 1974)
37. J. Fleischer, P.H. Diamond, Phys. Rev. E **61**(4), 3912 (2000)
38. A.V. Gurevich, A.L. Krylov, Zh. Eksp. Teor. Fiz. **92**(5), 1684 (1987)
39. G.A. El, A.L. Krylov, Phys. Lett. A **203**(17), 77 (1995)
40. A.M. Kamchatnov, A. Gammal, R.A. Kraenkel, Phys. Rev. A **69**(6), 063605 (2004)
41. M.A. Hoefer, B. Ilan, Phys. Rev. A **80**(6), 061601(R) (2009)
42. M. Segev, G.C. Valley, B. Crosignani, P. DiPorto, A. Yariv, Phys. Rev. Lett. **73**(24), 3211 (1994)
43. I. Kulikov, M. Zak, Phys. Rev. A **67**(6), 063605 (2003)
44. M. Sheik-Bahae, A.A. Said, T.-H. Wei, D.J. Hagan, E.W. Van Stryland, IEEE J. Quantum Electron. **26**(4), 760 (1990)
45. M. Sheik-Bahae, A.A. Said, D.J. Hagan, M.J. Soileau, E.W. Van Stryland, Opt. Eng. **30**(8), 1228 (1991)
46. J.A. Hermann, R.G. McDuff, J. Opt. Soc. Am. B (Opt. Phys.) **10**(11), 2056 (1993)
47. A. Litvak, V. Mironov, G. Fraiman, A. Yunakovskii, Sov. J. Plasma Phys. **1**, 31 (1975)
48. F. Dalfovo, S. Giorgini, L.P. Pitaevskii, S. Stringari, Rev. Mod. Phys. **71**(3), 463 (1999)
49. D. Suter, T. Blasberg, Phys. Rev. A **48**(6), 4583 (1993)
50. J.P. Gordon, R.C.C. Leite, R.S. Moore, S.P.S. Porto, J.R. Whinnery, J. Appl. Phys. **36**(1), 3 (1965)
51. S. Akhmanov, D. Krindach, A. Migulin, A, Sukhorukov, IEEE J. Quantum Electron. **4**(10), 568 (1968)
52. W. Krolikowski, O. Bang, J.J. Rasmussen, J. Wyller, Phys. Rev. E **64**(1), 016612 (2001)
53. C. Rotschild, O. Cohen, O. Manela, M. Segev, Phys. Rev. Lett. **96**(21), 213904 (2005)
54. C. Rotschild, B. Alfassi, O. Cohen, M. Segev, Nat. Phys. **2**(11), 769 (2006)
55. G. Assanto, T.R. Marchant, N.F. Smyth, Phys. Rev. A **78**(6), 063808 (2008)
56. A. Armaroli, S. Trillo, A. Fratalocchi, Phys. Rev. A **80**(5), 053803 (2009)
57. C. Sun, C. Barsi, J.W. Fleischer, Opt. Exp. **16**(25), 20676 (2008)
58. D. Buccoliero, A.S. Desyatnikov, W. Krolikowski, Y.S. Kivshar, J. Opt. A (Pure Appl. Opt.) **11**(9), 094014 (2009)
59. W. Krolikowski, O. Bang, N.I. Nikolov, D. Neshev, J. Wyller, J.J. Rasmussen, D. Edmundson, J. Opt. B (Quant. Semiclass. Opt.) **6**(5), S288 (2004)
60. A. Minovich, D.N. Neshev, A. Dreischuh, W. Krolikowski, Y.S. Kivshar, Opt. Lett. **32**(12), 1599 (2007)
61. E.A. McLean, L. Sica, A.J. Glass, Appl. Phys. Lett. **13**(11), 369 (1968)
62. P.M. Livingston, Appl. Opt. **10**(2), 426 (1971)
63. S.H. Lehnigk, B. Steverding, J. Fluid Mech. **66**(4), 817 (1974)
64. C.M. Vest, Phys. Fluid **17**(11), 1945 (1974)
65. E. Greenfield, C. Rotschild, A. Szameit, J. Nemirovsky, R. El-Ganainy, D.N. Christodoulides, M. Saraf, E. Lifshitz, M. Segev, New J. Phys. **13**, 053021 (2011)
66. M.F. Shih, P. Leach, M. Segev. M.H. Garrett, G. Salamo, G.C. Valley, Opt. Lett. **21**(5), 324 (1996)
67. H.X. Meng, G. Salamo, M. Segev, Opt. Lett. **23**(12), 897 (1998)
68. E. DelRe, A. Ciattoni, A.J. Agranat, Opt. Lett. **26**(12), 908 (2001)
69. E. DelRe, G. De Masi, A. Ciattoni, E. Palange, Appl. Phys. Lett. **85**(23), 5499 (2004)
70. A. Ciattoni, C. Palma, J. Opt. Soc. Am. A **20**(11), 2163 (2003)
71. N.V. Kukhtarev, V.B. Markov, S.G. Odoulov, M.S. Soskin, V.L. Vinetskii, Ferroelectrics **22** (1), 949 (1979)
72. J. Safioui, M. Chauvet, F. Devaux, V. Code, F. Pettazzi, M. Alonzo, E. Fazio, J. Opt. Soc. Am. B (Opt. Phys.) **26**(3), 487 (2009)
73. J. Feinberg, J. Opt. Soc. Am. **71**(1), 46 (1982)

74. N.V. Kukhtarev, E. Krätzig, H.C. Külich, R.A. Rupp, Appl. Phys. B Laser Opt. **35**(1), 17 (1984)
75. M.D. Ewbank, P. Yeh, J. Feinberg, Opt. Comm. **59**(5–6), 423 (1986)
76. D.A. Temple, C. Werder, J. Opt. Soc. Am. B (Opt. Phys.) **5**(8), 1800 (1988)
77. K. Nakagawa, T. Minemoto, J. Opt. Soc. Am. B (Opt. Phys.) **11**(9), 1639 (1994)
78. D.N. Christodoulides, M.I. Carvalho, J. Opt. Soc. Am. B (Opt. Phys.) **12**(9), 1628 (1995)
79. A.A. Zozulya, D.Z. Anderson, Phys. Rev. A **51**(2), 1520 (1995)
80. M.R. Belic, D. Vujic, A. Stepken, F. Kaiser, G.F. Calvoe, F. Agulló-López, M Carrascosa, Phys. Rev. E **65**(6) 066610 (2002)
81. M. Saffman, A.A. Zozulya, Opt. Lett. **23**(20), 1579 (1998)
82. I. Yamaguchi, T. Zhang, Opt. Lett. **22**(16), 1268 (1997)
83. A.R. Pandey, J.W. Haus, P.E. Powers, P.P. Yaney, J. Opt. Soc. Am. B (Opt. Phys.) **26**(2), 218 (2009)
84. M. Tsang, D. Psaltis, F.G. Omenetto, Opt. Lett. **28**(20), 1873 (2003)
85. C. Barsi. W. Wan, J.W. Fleischer, Nat. Phot. **3**(4), 211 (2009)
86. P. Smith, W. Tomlinson, IEEE J. Quantum Electron. **20**(1), 30 (1984)
87. M. Peccianti, A. Dyadyusha, M. Kaczmarek, G. Assanto, Nat. Phys. **2**(11) 737 (2006)
88. A.E. Kaplan, JETP Lett. **24**(3), 114 (1976)
89. F.W. Ye, Y.V. Kartashov, L. Torner, Opt. Lett. **32**(4), 394 (2007)
90. B. Bosacchi, L.M. Narducci, Opt. Lett. **8**(6), 324 (1983)
91. Th. Peschel, P. Dannberg, U. Langbein, F. Lederer, J. Opt. Soc. Am. B (Opt. Phys.) **5**(1), 29 (1988)
92. G.I. Stegeman, E.M. Wright, N. Finlayson, R. Zanoni, C.T. Seaton, IEEE J. Light. Tech. **6**(6), 953 (2002)
93. A.C. Newell, J. Math. Phys. **19**(5), 1126 (1978)
94. V. Hakim, Phys. Rev. E **55**(3), 2835 (1997)
95. M. Albiez, R. Gati, J. Fölling, S. Hunsmann, M. Cristiani, M.K. Oberthaler, Phys. Rev. Lett. **95**(1), 010402 (2005)
96. S. Fölling, S. Trotzky, P. Cheinet, M. Feld, R. Saers, A. Widera, T. Müller, I. Bloch, Nature **448**(7157), 1029 (2007)
97. Y. Linzon, R. Morandotti, M. Volatier, V. Aimez, R. Arez, S. Bar-Ad, Phys. Rev. Lett. **99** (13), 133901 (2007)
98. G. Dekel, V. Fleurov, A. Soffer, C. Stucchio, Phys. Rev. A **75**(4), 043617 (2007)
99. A. Barak, O. Peleg, C. Stucchio, A. Soffer, M. Segev, Phys. Rev. Lett. **100**(15), 1513901 (2008)
100. M. Peccianti, A. Dyadyusha, M. Kaczmarek, G. Assanto, Phys. Rev. Lett. **101**(15), 153902 (2008)
101. W. Wan, S. Muenzel, J.W. Fleischer, Phys. Rev. Lett. **104**(7), 073903 (2010)
102. N.K. Efremidis, S. Sears, D.N. Chistodoulides, J.W. Fleischer, M. Segev, Phys. Rev. E **66**(4), 046602 (2002)
103. J.W. Fleischer, T. Carmon, M. Segev, N.K. Efremidis, D.N. Christodoulides, Phys. Rev. Lett. **90**(2), 023902 (2003)
104. P. Engels, C. Atherton, Phys. Rev. Lett. **99**(16), 160405 (2007)
105. H.S. Eisenberg, Y. Silberberg, R. Morandotti, J.S. Aitchison, Phys. Rev. Lett. **85**(9), 1863 (2000)
106. J.W. Fleischer, G. Bartal, O. Cohen, T. Schwartz, O. Manela, B. Freedman, M. Segev, H. Buljan, N. Efremidis, Opt. Exp. **13**, 1780 (2005)
107. H.S. Eisenberg, Y. Silberberg, R. Morandotti, A.R. Boyd, J.S. Aitchison, Phys. Rev. Lett. **81** (16), 3383 (1998)
108. D. Mandelik, H.S. Eisenberg, Y. Silberberg, R. Morandotti, J.S. Aitchison, Phys. Rev. Lett. **90**(5), 053902 (2003)
109. S. Jia, W. Wan, J.W. Fleischer, Phys. Rev. Lett. **99**(22), 223901 (2007)
110. E.G. Khamis, A. Gammal, G.A. El, Y.G. Gladush, A.M. Kamchatnov, Phys. Rev. A **78**(1), 013829 (2008)

111. G.A. El, A.M. Kamchatnov, V.V. Khodorovskii, E.S. Annibale, A. Gammal, Phys. Rev. E **80** (4), 046317 (2009)
112. W. Wan, D.V. Dylov, C. Barsi, J.W. Fleischer, Opt. Lett. **35**(16), 2819 (2010)
113. M.A. Hoefer, M.J. Ablowitz, Phys. D Nonlinear Phenomena **236**(1), 44 (2007)
114. S. Jia, J.W. Fleischer, Rayleigh-Taylor instability in Nonlinear Optics. In: Lasers and Electro-Optics/International Quatum Electronics Conference, OSA Technical Digest (CD) (Optical Society of America, 2009), paper CtuHH4, 2009
115. G.P. Agrawal, Phys. Rev. Lett. **64**(21), 2487 (1990)
116. A.J. Stentz, M. Kauranen, J.J. Maki, G.P. Agrawal, R.W. Boyd, Opt. Lett. **17**(1), 19 (1992)
117. S. Jia, W. Wan, J.W. Fleischer, Opt. Lett. **32**(12), 1668 (2007)
118. C. Conti, A. Fratalocchi, M. Peccianti, G. Ruocco, S. Trillo, Phys. Rev. Lett. **102**(8), 083902 (2009)
119. J. Herrmann, U. Griebner, N. Zhavoronkov, A. Husakou, D. Nickel, J.C. Knight, W.J. Wadsworth, P.St.J. Russel, G. Korn, Phys. Rev. Lett. **88**(17), 173901 (2002)
120. D.R. Solli, C. Robers, P. Koonath, B. Jalali, Nature **450**(7172), 1054 (2007)
121. A. Efimov, A.V. Yulin, D.V. Skryabin, J.C. Knight, N. Joly, F.G. Omenetto, A.J. Taylor, P. Russell, Phys. Rev. Lett. **95**(21), 213902 (2005)
122. R. Dong, C.E. Rüter, D. Kip, O. Manela, M. Segev, C. Yang, J. Xu, Phys. Rev. Lett. **101**(18), 183903 (2008)

Chapter 10
Nonlinear Interaction of Intense Ultrashort Filaments

Heping Zeng and Jia Liu

10.1 Introduction

Rapid advance in laser technology has greatly increased the attainable high-peak-intensity range of ultrashort lasers. By using well-developed adaptive correction of the wave-front distortion with deformable mirrors, intense femtosecond pulses could be tightly focused into a nearly diffraction-limited point with a peak intensity up to 2×10^{22} W/cm^2 [1]. Such a high intensity could only be reached in vacuum without multiphoton ionization to detrimentally influence the pulse focusing and tight focusing produced a limited spatial region of high-peak intensity, while many applications require tightly guided high-peak intensities along sufficiently long interaction distances. To meet this requirement, a number of approaches have been investigated and guiding of intense pulses has been demonstrated in hollow capillary [2], relativistic channeling [3], plasma waveguide [4], and self-channeled filaments [5–9]. Among all these demonstrated approaches, plasma guiding of intense pulses was demonstrated as an efficient solution to avoid detrimental defocusing caused by further multiphoton ionization as charged particles within plasma typically have quite high ionization potentials and thus the already-existed plasma experiences negligible laser-induced changes to alter the pulse propagation dynamics. Although pulse propagation might be influenced by hydrodynamic plasma expansion, tightly guiding in plasma waveguide could reach up to 5×10^{17} W/cm^2 [10]. Obviously, such a tight focusing or guiding of intense laser pulses is quite difficult in air or other neutral media due to unavoidable

H. Zeng (✉)
State Key Laboratory of Precision Spectroscopy, East China Normal University,
Room A1215, Science Building, 3663 North Zhongshan Road, Shanghai 200062, China
e-mail: hpzeng@phy.ecnu.edu.cn

J. Liu
State Key Laboratory of Precision Spectroscopy, East China Normal University,
Room A1110, Science Building, 3663 North Zhongshan Road, Shanghai 200062, China

Z. Chen and R. Morandotti (eds.), *Nonlinear Photonics and Novel Optical Phenomena*,
Springer Series in Optical Sciences 170, DOI 10.1007/978-1-4614-3538-9_10,

defocusing from multiphoton-ionization-induced plasma. There always exists a counterbalance between Kerr self-focusing and plasma-defocusing in neutral media, bringing about robust self-guided channels that facilitate abundant self-action nonlinear processes, such as spatial self-cleaning [5], intensity-clamping [11], and self-phase locking of nonlinearly coupled multicolor pulses [6]. Filamentary propagation of intense femtosecond laser pulses has been so far demonstrated quite useful for super-continuum white-light generation [7], pulse self-compression down to a few cycles in duration [8, 9], harmonic generation [6], and so on. All the nonlinear processes not only enforce observable changes on the spatiotemporal dynamics of the intense ultrashort pulses during their nonlinear propagation but also prohibit a tight focusing or guiding in small cores. The focusable peak intensity could not increase to quite a high level since any possible tendency to increase the peak intensity to quite a high level would be defeated by the accompanied multiphoton-ionization-induced plasma defocusing. In air, the peak intensity in filamentary self-guided waveguides was found to be clamped down to about 5×10^{13} W/cm^2 [11]. As the input power increases much higher than the critical self-focusing threshold, multiple filaments occur [12] and as a result, filament-mediated nonlinear interactions take place in spatially separated channels. Recent experimental explorations on the interaction of multiple femtosecond filaments have already revealed quite a lot intriguing features of spatiotemporal light bullets with abundant self-action and cross-coupling nonlinearities [13–17]. In molecular gases, the spatiotemporal phase modulation induced by molecular alignment [18, 19] offers an additional degree of freedom to control filamentation [20–25] and filament interaction [26, 27]. Nonlinear filament interaction and its dynamic control may not only stimulate potential applications with nonlinearly coupled multiple filaments instead of single filaments but also solve current challenge on coalescence of multiple intense ultrashort pulses to beat the tightly guided peak intensity beyond the intensity-clamping limit.

On the other hand, current available nonlinear optical waveguides cannot guide high-peak-intensity ultrashort pulses. Standard optical fibers guide light via total internal reflection, and high-fidelity delivery of ultrashort laser pulses was limited so far less than the nanojoule level. A new paradigm of light guidance was realized in photonic-crystal fibers [28, 29] by creating photonic band gaps with periodic wavelength-scale lattice of microstructures in glass to trap light in hollow fiber core of refractive index lower than that of the cladding, wherein optical nonlinearities are reduced by a factor of 1,000 lower than the conventional silica-core fibers, and thus supports megawatt optical solitons of femtosecond pulses [30]. Peak intensities are still limited by the material damage at the input and high-peak-intensity photonic-crystal waveguides or nonlinear waveguide couplers are so far unrealized. Photonic-crystal waveguides could be also created in an all-optical way by recording interference pattern in photosensitive media [31], typically applicable to guide low-intensity beams. Photo-sensitivity in solid materials could be used to record dynamic structures of four-wave mixing for high-density disk storage [32] and three-dimensional (3D) photonic crystals [33]. It is also well-known that nonlinear interference of non-collinear beams could establish transient holographic gratings by virtue of nonlinear changes of the refractive index within a limited interaction length, which induced nonlinear diffraction rather than tightly guiding. In bulk quadratic nonlinear crystals, two-dimensional (2D) transient Kerr grating [34] could be formed on the basis of nonlinear wave-packet breakup involved in cascaded quadratic nonlinear couplings during second harmonic generation. By using the preformed 2D transient grating, synchronized super-continuum seed pulses were diffracted and the diffracted seeds were amplified in the corresponding diffraction directions, resulting in 2D up-converted multicolor arrays along the diffraction directions [34]. Laser-induced gratings could be extended to gaseous media (such as air) through multiphoton ionization and other higher-order optical nonlinear effects with self-projected high-intensity femtosecond laser pulses. At high-peak intensities, the transient Kerr gratings were mediated by filamentation in spatially localized region. Well-localized plasma channels could not only establish nonlinear waveguides to facilitate nonlinear interaction of the laser pulses but also enforce filamentation into different plasma microstructures that could last with a duration much longer than the excitation pulses. Periodic plasma microstructures could be used as plasma photonic components with ultrahigh intensity damage thresholds to sustain intense laser fields. In addition, nonlinear interaction of preformed plasma microstructures with intense laser pulses exhibits unique dynamics with multiphoton-sensitive responses of extremely high nonlinearities. All-optical control of the plasma channels is expected to stimulate intriguing studies of solitonary-like waves in nonlinear waveguides at ultrahigh peak intensities, including nonlinear filament interaction, fusion [13, 14] and spiraling [15–17] of interacting multiple filaments, and so on. In particular, the full potential of photonic crystal waveguides could be maximized to guide high-peak-intensity pulses over relatively long lengths by creating a wavelength-scale periodic plasma density modulation to change the local refractive index periodically in the surrounding gaseous phase media. This could be realized with spatial interference grating established by using non-collinearly overlapped intense femtosecond pulses to assist coalescence of multiple filaments into a lattice of strongly coupled parallel self-channels along the entire waveguide length, dubbed hereby as plasma lattices, which could be extended from 1D plasma gratings excited by two non-collinear interfering pulses to 2D plasma photonic crystals by using three non-collinear and non-coplanar interfering pulses.

In this review, we show that coalescence of interfering non-collinear intense femtosecond pulses assisted periodic wavelength-scale self-channeling into photonic-crystal plasma waveguides with encircling air molecules, wherein high-peak-intensity light bullets were tightly guided with strong self-channeled spatiotemporal couplings and survived as particle-like attraction and repulsion. Two non-collinearly intersected filaments were fused several millimeters long in their interaction region and then departed along their corresponding incident directions. Spatially localized Kerr self-focusing around constructive interference intensity peaks followed its counterbalance with local plasma defocusing and accordingly, the non-collinearly propagated intense ultrashort pulses in air were efficiently coalesced into regularly spaced filament bundles, one-dimensional (1D) or 2D plasma lattice with periodic plasma density modulation. Such a periodic modulation not only supported efficient coupling and coalescence of multiple intense femtosecond pulses at arbitrary incident angles but also influenced the nonlinear frequency conversions. Third harmonic generation was observed to enhance at least two orders of magnitude in the presence of preformed plasma lattices. As a result of the periodically distributed refractive index modulation, long-lifetime 1D and 2D plasma gratings were generated, which were clearly confirmed by the Bragg diffraction and in-line holographic imaging. We anticipate that our results are of significant importance in creating photonic-crystal waveguides for high-order harmonic generation and high-intensity relativistic optics.

10.2 Nonlinear Spatiotemporal Coupling of Interacting Filaments

If two intense femtosecond pulses are non-collinearly overlapped in air, their spatial interference redistributes the intensity according to the interference fringes and local intensity peaks are located around the constructive interference peaks. Self-focusing occurs at first around the interference intensity peaks, which further makes the local intensity increase, followed by an increased multiphoton ionization probability. Self-focusing around the interference peaks is then counterbalanced by multiphoton-ionization-induced plasma defocusing and higher-order terms of the nonlinear refractive index, giving rise to filamentation in periodically localized regions with periodic plasma density modulation along the interference fringes. As a consequence, wavelength-scale periodic lattice of plasma microstructures are created, where strongly coupled parallel self-channels are formed along a relatively long distance, manifesting as long-distance projection of the constructive and destructive interference fringes along the bisector of the non-collinearly overlapped filaments. In distinct contrast with transient holographic gratings, the plasma microstructures induce tightly guiding rather than nonlinear diffraction, and accordingly, input pulses will be eventually guided into a bundle of plasma self-channels of local reduction in the refractive index encircled by air, which have wavelength-scale structures similar to those of photonic-crystal fibers. Extremely high-peak intensities could be guided in the wavelength-scale plasma self-channels with strong spatiotemporal couplings, wherein the counter-balance among self-focusing, plasma defocusing, and beam diffraction could be roughly estimated by

$$n_2 I = \rho(I)/2\rho_c + (1.22\lambda_0)^2/(8\pi n_0 d^2), \tag{10.1}$$

with the nonlinear refractive index n_2, critical plasma density ρ_c above which the plasma becomes opaque, and self-channel diameter d. As the electron densities are dramatically increased within localized channels of diameters much smaller than the beam waist, the clamping intensity I is expected to increase dramatically. Intrinsically different from internal light reflection of the conventional fiber [35] or diffraction in photonic band gap fibers [28–30], intense pulse guiding in the plasma lattices is caused by the strong spatiotemporal couplings in the plasma self-channels. Such unique intrinsic features are anticipated to stimulate various

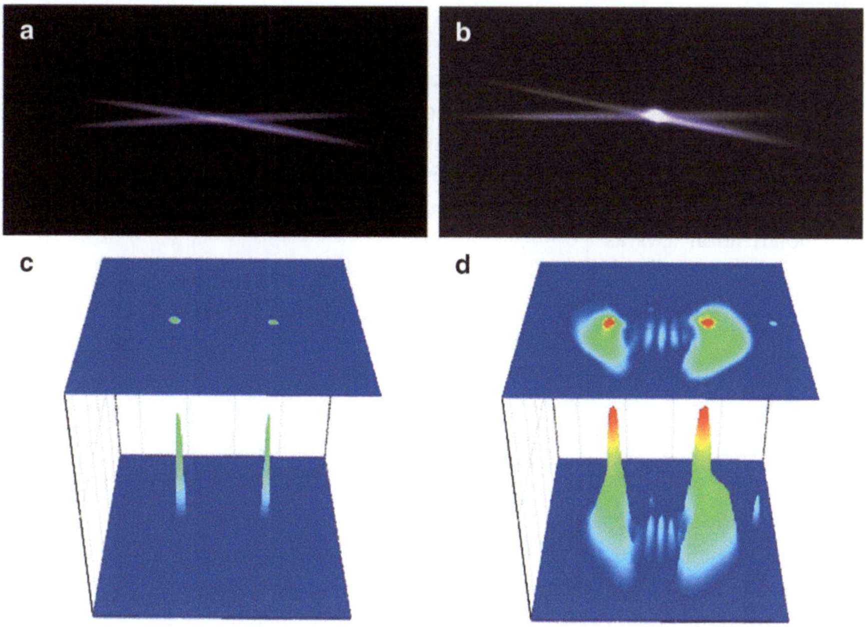

Fig. 10.1 Camera-captured pictures of the intersected non-collinear filaments from the top view as the intersecting filaments were (**a**) delayed far away and (**b**) synchronized with observable filament interaction, and the observed spatial distributions of the interacting filaments in the far-field region, indicating no filament interaction (**c**) and strong spatiotemporal couplings (**d**), respectively

technological and scientific breakthroughs in various fields, such as high precision frequency measurements [36], high-intensity nonlinear optics [6], ultrafast pulse compression [8, 9], control of spontaneous emissions [37], sharp bending of intense pulses [38], and zero-threshold white-light lasing [39].

The experiments were performed with a Ti:sapphire chirped pulse amplifier (1 kHz repetition rate, 50 fs, 800 nm, 2.0 mJ). The output laser pulse was equally split into a pump and probe beams that were then focused with two $f = 1{,}000$ mm high reflection coated concave mirrors, inducing two non-collinear filaments in air that slightly crossed nearby their foci with a variable crossing angle from 2° to 16°. Without any temporal overlapping, the induced filaments exhibited no observable couplings and as shown in Fig. 10.1a, independent self-guiding was accompanied by spatial self-cleaning [5] in the filament core over a typical length about ~40 mm, leading to a significant improvement of the spatial distribution of the output pulses in the far-field region (Fig. 10.1c). As these two incident ultrashort pulses with the same polarization were synchronized in the intersection region, the local field intensity in the overlapped region increased significantly due to their intensity interference, and the corresponding plasma density increased as well. Self-focusing around the interference peaks attracted surrounding laser energies to the plasma subchannels, resulting in fusion of the non-collinear filaments into plasma lattice.

Strong nonlinear interaction was observed when two filaments were synchronized. As expected, periodic distributions of the local intensity in the overlapping region could be produced due to the constructive and destructive interference of the incident pulses. Such intensity modulation of the interaction region could thus in turn modulate the local refractive index. The refractive index of air is closely associated to the electronic Kerr effect and ionization-induced plasma. The electronic Kerr effect produces a refractive index change along the excitation polarization axis as

$$n_{\mathrm{Kerr}\|} = n_2 I, \tag{10.2}$$

while the perpendicular component is proportional to

$$n_{\mathrm{Kerr}\perp} = \frac{1}{3} n_{\mathrm{Kerr}\|}, \tag{10.3}$$

where n_2 is the Kerr nonlinear coefficient related to the third-order susceptibility, I is the peak intensity of the laser pulse. The positive coefficient n_2 in air leads to a refractive index increase in the presence of intense pulse. However, the Gaussian beam distribution causes a convex lens-like wave-front bending, leading to the self-focusing of a laser beam. Meanwhile, ionization takes place with the increased pulse intensity, together with a burst of generation of electrons or plasma. At sufficiently high-peak intensities that multiphoton ionization of air molecules becomes observable, plasma generated during pulse propagation in air makes a significant reduction of the refractive index in the interaction regions according to the phenomenon expression

$$n \cong n_0 - \rho(r,t)/2\rho_c, \tag{10.4}$$

where the decrease of the refractive index n_0 caused by the local density of free electrons $\rho(r, t)$ is normalized with respect to that of the critical plasma density $\rho_c = \varepsilon_0 m_e \omega_0{}^2/e^2$ (m_e and e are the electron mass and charge, ω_0 is central frequency of the laser pulse). The refractive index variation induced by plasma is eventually balanced by that from Kerr nonlinearity. The Kerr and plasma refractive index variation and the balance itself are sensitive to the local peak intensity.

As shown in Fig. 10.1b, a new bright white fluorescence bulb emerged as a result of filament fusion [40]. Such fusion could sustain a finite propagation distance due to the energy loss from multiphoton ionization, diffraction, fluorescence, and so on. As shown in Fig. 10.2a, the fusion length decreased from 4.2 to 0.2 mm as the non-collinear crossing angle varied from 2.0° to 16.0°. The far-field distributions were modulated with butterfly-like expansion of each filament and thread-like fringes between the two filaments (Fig. 10.1d).

In order to investigate the details about the observed butterfly-like patterns, we chose five points and used a fiber-coupled spectrometer to measure the corresponding spectrum. As shown in Fig. 10.2b, the spectrum is gradually broadened from point A

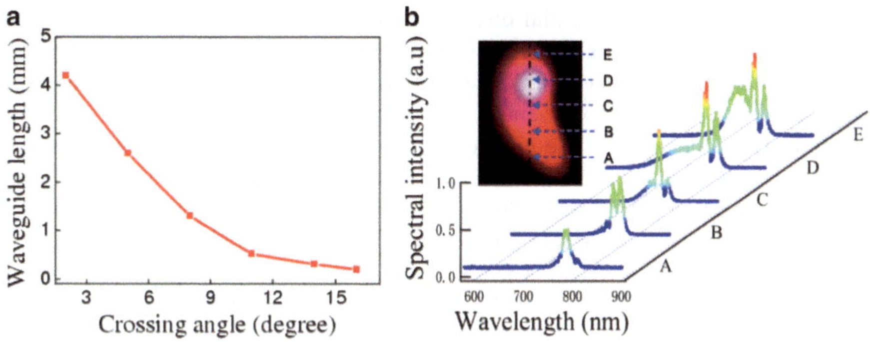

Fig. 10.2 (**a**) The length of the filament fusion as a function of the crossing angle between two non-collinearly intersected filaments. (**b**) The spectrum distribution at five different positions in the butterfly-like pattern

to point D and blue-shifted frequency components appear as spectral humps. Spectral breakups were observed in the upper side of the butterfly-like pattern. The observed spectral broadening and splitting could be ascribed to the strong spatiotemporal coupling in the interaction region, as well as the complicated self-phase and cross-phase modulations. As a whole, the butterfly-like patterns angularly dispersed in a spectral range from 650 to 830 nm, while the inner thread-like fringes ranged from 775 to 825 nm. The observed spectral difference implied that the far-field butterfly-like patterns and inner thread-like fringes should be originated from quite different nonlinear processes. As the coalesced filaments in the plasma lattice experienced fission into different propagation directions with quite different energy distributions, the spatiotemporal modulation and plasma density modulation experienced by the propagating pulses were completely different. Butterfly-like bending in the far-field region revealed the spatial inhomogeneous plasma density distribution in the crossing plane and planar plasma waveguides of small thickness vertical to crossing plane, while the mysterious inner thread-like fringes demonstrated self-guided propagation within the plasma waveguide even after the filament fission out of the plasma lattice.

10.3 Visualization of the Plasma Density Modulation

The wavelength-scale plasma density modulation and self-projection of the interfering pulses in the interaction region could be directly imaged by inserting therein a thin plate at a grazing angle, facilitating a direct imaging of the self-projected intensity interference fringes to explore the spatial periodicity of the guided laser intensity and visualize the filamentary propagation of the laser fields with a high spatial resolution determined by the 4-f optical imaging system to reveal the periodic plasma density modulation. The plasma microstructures were accompanied by spatially inhomogeneous fluorescence change of the ionized molecules and thus their spatial distribution could be captured by collecting the plasma fluorescence from ionized molecules into an imaging system, enabling a nearly diffraction-limited spatial resolution for the plasma microstructures. As plasma self-channels experienced hydrodynamic expansion and ultrafast exponential plasma decay after the interacting filaments, the ultrafast dynamics of the preformed plasma should be recorded in a holographic way with a time-delayed femtosecond probe pulse passing through the interaction region, which measured the dynamic change of the local refractive index with a spatial resolution determined by the 4-f optical imaging system. The holographic imaging provided spatial information on the ultrafast evolution of the plasma density modulation and an accurate lifetime measurement of the preformed plasma lattice. Throughout this review article, we used all the three methods to visualize the non-collinearly interfering filaments and the filament coalescence in the interaction region as well as the nonlinear interaction of preformed plasma lattices with intense femtosecond pulses, such as nonlinear Bragg diffraction from preformed plasma gratings, 1D and 2D plasma density modulations, nonlinear energy transfer among interacting filaments, filament-interaction-mediated spatiotemporal phase modulation, filamentation suppression in the weak interference fringes, multicolor filament interaction, third harmonic generation modulated by plasma lattice, and some other intriguing phenomena.

10.3.1 Direct Imaging of Plasma Channel at a Grazing Angle

The filament interaction region could be directly imaged with a setup schematically shown in Fig. 10.3a, where a thin plate was inserted at a grazing angle nearby the filament-crossing point. The thin plate reflected part of the laser fields [41], which were directly recorded by a CCD camera after a 4-f configuration optical imaging system. The recorded intensity distribution revealed the laser intensity distribution in the filament interaction region.

The inset in Fig. 10.3a presents the captured intensity distribution for the cases without and with filament interaction, respectively. No intensity modulation was observed when the filaments were intentionally delayed far away so that no interference took place, while periodic intensity modulation and spatial localization of the laser fields occurred inside the interaction region for synchronized intersecting filaments, which were observed to exhibit distinct dependence on the intensity interference fringes (Fig. 10.3b). The observed spatial intensity distribution changed its modulation depth as the incident laser intensity varied. Sharp intensity peaks became more observable as two incident lasers increased in intensity. Nevertheless, the observed central modulation depth became almost saturated as both laser pulses underwent filamentation. As shown in the insets of Fig. 10.3a, the observed intensity modulation was enveloped with a spatial intensity profile quite different from the individual incident filaments, indicating that the laser fields in the interaction region were completely redistributed with spatiotemporal

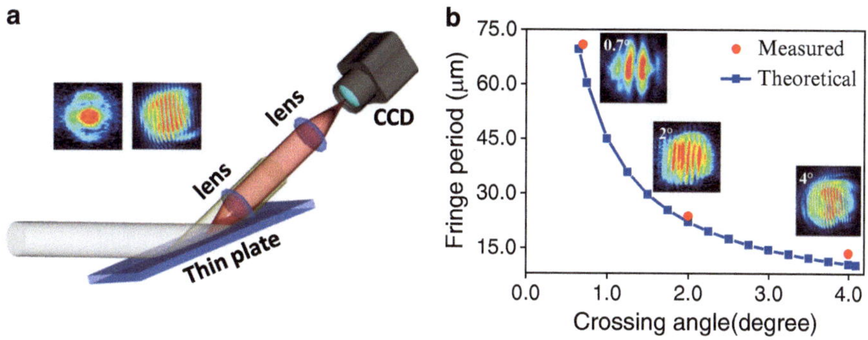

Fig. 10.3 (**a**) The schematic measurement of the interaction region by direct imaging of plasma channels with a thin plate inserted at a grazing angle. (**b**)The transverse fringe periods and the corresponding intensity distributions (*insets*) of the non-collinearly intersected filaments at different crossing angles from ref. [41]

cross-couplings and spatial modulation of the plasma density. In comparison with individual incident filaments, the thickness of the filamentary interference fringes was enlarged (vertical to the filament-crossing plane) while the width of the whole profile was almost the same (parallel with the filament-crossing plane), implying that there occurred spatially well-localized filamentation around each interference peaks. In the filament-crossing plane, laser energies were attracted in the fringe self-channels with more serious plasma defocusing around the interference peaks, while along the direction vertical to the filament-crossing plane, the filamentary fringe peaks were expanded with an enlarged thickness by plasma defocusing. As a whole, the counterbalance between Kerr self-focusing and plasma focusing equivalently functioned as periodically distributed cylindrical lens arrays around the interference fringes due to the inhomogeneous intensity distribution in the interaction region.

Assuming that the inserted thin plate was tilted at a grazing angle θ_g with respect to the bisector of the incident filaments of core diameter D, the captured image was $L \sim 2D/\sin(\theta_g)$ in length and accordingly, the captured intensity distribution actually came from the intensity integrated over a length L of the overlapped filaments. The measured spatial distribution illustrated the filamentary propagation along the filament bisector. The filamentary self-guiding with an observation length L clearly indicated self-projection of intensity interference along the bisector, with the spatial intensity modulation period determined by

$$\Lambda = \lambda_c / 2\sin(\theta/2), \tag{10.5}$$

where θ is the crossing angle of the non-collinear beams, λ_c is the central wavelength. Figure 10.3b shows the measured intensity modulation periods at different crossing angles θ (red dotted), which fits well with the interference fringe (blue-squared curve).

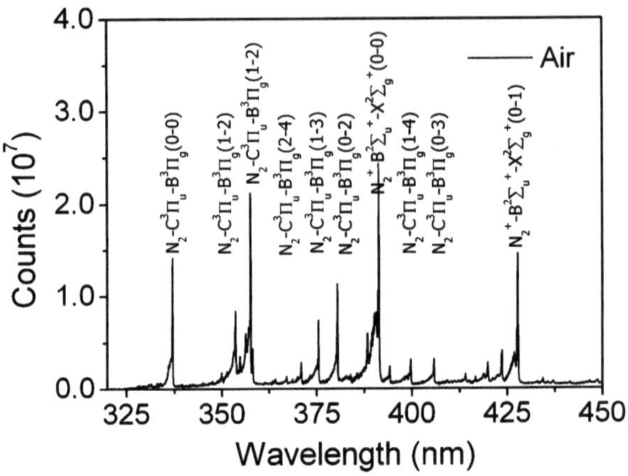

Fig. 10.4 Typical fluorescence spectrum of a linearly polarized laser-induced filament in air in the range of 330–450 nm. The corresponding transition band are marked near the spectral lines

10.3.2 Fluorescence Detection of the Femtosecond Filaments

Multiphoton ionization of air molecules occurs during intense femtosecond filamentation and some ionized molecules are excited to highly lying electronic states, which subsequently undergoes electronic transitions to emit characteristic fluorescence of plasma [42]. For filaments in air, the observed visible and near UV fluorescence were mainly assigned to the second positive band system of N_2 and the first negative system of N_2^+ [43–45]. The first negative system ($B^2\Sigma_u^+$–$X^2\Sigma_g^+$), from well marked sequences and degraded to shorter wavelengths, is the main system of N_2^+, originated from the laser-induced multiphoton ionization or tunneling ionization of neutral nitrogen molecule, while the second positive band system ($C^3\Pi_u$–$B^3\Pi_g$) comes from the transitions [42]:

$$N_2 + N_2 \rightarrow N_4^+,$$

$$N_4^+ + e \rightarrow N_2(C^3\Pi_u) + N.$$

Figure 10.4 depicts the typical filamentation fluorescence spectra, measured by a fiber-coupled gated imaging spectrometer (ANDOR, Mechelle 5000), with the transition bands marked near the corresponding spectral lines.

As shown in Fig. 10.1b, strong interaction/coupling with a bright white fluorescence bulb was observed, indicating a significant fluorescence enhancement in the interaction region. A fluorescence microscope with an optical imaging system and a UV-intensified CCD camera (DS-QilMc, Nikon), as schematically shown in Fig. 10.5a, was used to monitor the plasma structure near the interaction region, whose image was captured by a CCD camera and digitized by computer software. A band-pass filter from 250 to 380 nm was used to filter out the fluorescence from the neutrals and excited ions.

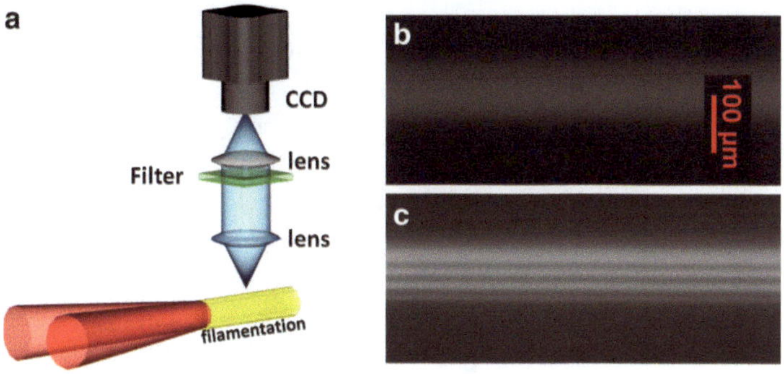

Fig. 10.5 (**a**) The schematic of the fluorescence detection of the femtosecond filament structures. Measured N_2 fluorescence profile of the interaction region (**b**) without and (**c**) with filament interactions

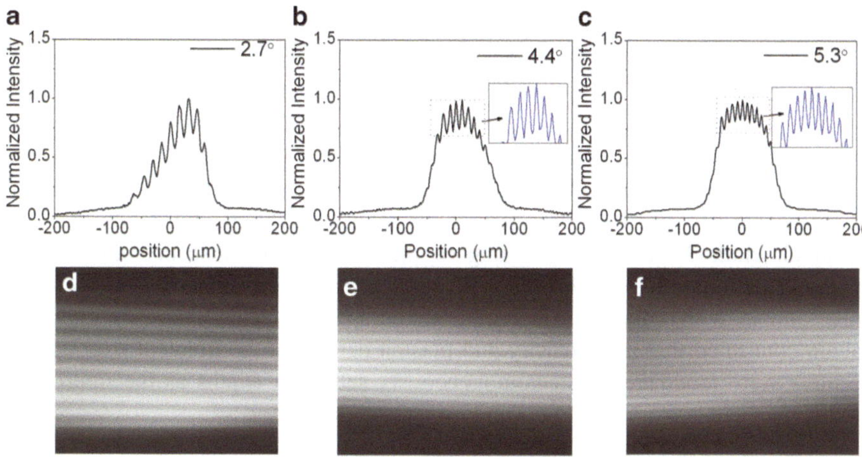

Fig. 10.6 The fluorescence images and the related transverse profiles of the observed 1D plasma channels with different forming cross angles of (**a**) 2.7°, (**b**) 4.4°, and (**c**) 5.3°

Figure 10.5b and c shows the measured fluorescence profiles of the interaction region without and with interactions. We observed an enormous enhancement of the fluorescence with a distinct periodic distribution in the overlapping region, justifying the existence of intensity and refractive index modulation inside the interaction region. As mentioned above, the spatial size of the plasma channel could be simply manipulated by changing the crossing angle of the non-collinear filaments. Figure 10.6 shows the fluorescence images and transverse profiles of the observed 1D plasma channels between two filaments with different crossing angles 2.7°, 4.4°, and 5.3°. The corresponding spatial periodic modulation period could be calculated from (10.5). The corresponding period of the 1D plasma lattices was measured to be 8.8, 10.4, and 17.0 μm, respectively. The measured periods fit well

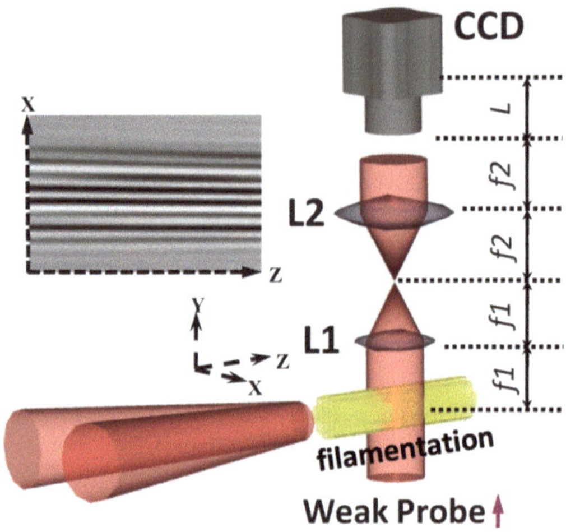

Fig. 10.7 The experimental setup for holographic recording of the plasma dynamics. A weak probe pulse propagates through the filament after a certain time delay. The magnification of the system is determined by f_2/f_1. *L1*, *L2* lenses

with the calculated interference fringe periods. As the crossing angle increased, the number of the plasma self-channels increased approximately as $N \sim [D/\Lambda]$ (D represents the typical core diameter of the interfering filaments), while the self-channel width decreased. According to the observed fluorescence images, the periodic modulation near the filament bisector exhibited a nearly homogeneous distribution, with a flattened envelope of the spatial profile. Those observations again confirmed that the intensity and plasma density at the filament overlapping centers were modulated with homogeneously saturated depths as the interfering filaments were coalesced into plasma self-channels.

10.3.3 Holographic Recording of the Plasma Dynamic Evolution

Holographic recording serves as an important tool to record fast processes by making holograms, and it allows one to catch even ultrafast events and retain all the involved information. For example, a hologram contains the information about the shape, size, and brightness of an object, or the pulse's amplitude and phase information in the case of laser pulse propagation. Up to now, in the field of high-intensity laser physics, the holographic imaging has been applied in the studies of ultrafast events like filamentation dynamics [46, 47], greasing the speed of light, and watching the molecular dynamics of chemical reactions [48].

Here, we implemented holographic imaging in capturing plasma wake generated by the filamentary propagation of intense femtosecond pulses in neutral medium. The setup of the in-line digital holography is shown in Fig. 10.7. Briefly, an 800-nm probe pulse, propagating perpendicularly to the filament, was used to record the structure of filament interaction region in air. Weak probe was used here to order to minimize detrimental influence induced by the probe pulse itself on the preformed plasma microstructures. The image was magnified by a 4-f configuration imaging system with a factor of $M = f_2/f_1$. The CCD camera was placed at the image plane (at the position $L = 0$ as shown in Fig. 10.7) to record the magnified image of the interaction region. If the CCD camera was placed at a distance L from the image plane, what captured was actually the interference between the transmitted and diffracted probe pulse after the interacting filaments [48], which could be used to retrieve the phase and refractive index change induced by the filamentation channel, and estimate the involved electron density distribution. As compared with the above two methods, the holographic recording technique provided a proficient approach to decode the time evolution of the plasma self-channels.

10.4 Plasma Waveguide Induced by Filament Interaction

As mentioned above, we observed strong spatiotemporal couplings in the interaction region when two intersected filaments were synchronized. The N_2 fluorescence imaging technique provided a straightforward observation of the periodic structure in the interaction region. As shown in Fig. 10.5c, a significant enhancement of the fluorescence and also notable periodic modulation in the overlapping region were observed as compared with that of a single filament. Apart from the fluorescence enhancement, modulated far-field distribution of butterfly-like spatial beam profiles and thread-like fringes were observed. Note that the thread-like fringes could not be originated from diffraction of the incident pulse, since diffraction should be a bright spot rather than thread-like fringes. The thread-like fringes were measured to exhibit a spectrum in the range from 775 to 825 nm. The thread-like fringes had quite similar structures with the observed periodic plasma microstructures, suggesting their intrinsic origin as self-guiding within the plasma self-channels, which could be understood as follows. Interfering pulses built up a local intensity modulation in the interaction region, Kerr self-focusing was at first launched around the constructive interference peaks [49], where filamentary propagation was reached via spatially localized counterbalance between self-focusing and plasma defocusing, leading to tight guiding of the incident laser pulses and a wavelength-scale periodic plasma density modulation. After a few millimeters of coalesced propagation, filamentary self-guiding could no longer maintain its counterbalance due to diffraction and other losses, the guided lasers within the plasma self-channels were partly projected in the far-field region as thread-like fringes as experimentally observed. Most of the pulse energies were diffracted out of the plasma self-channels at Bragg angles, i.e., along the directions almost the same as the incident pulses. Slight beam bending was observed due to inhomogeneous spatial distribution of the generated plasma density, which functioned equivalently as a plasma lens to deform the incident pulse wavefronts. Strong spatiotemporal couplings inside the plasma self-channels brought about significant cross-phase modulations and spatial-resolved changes of the diffracted pulses.

As a whole, the filament interaction processes functioned equivalently as X-type nonlinear waveguide couplers for high-peak intensity laser pulses. Two interfering pulses launched individual filamentation and then were coalesced into plasma self-channels along the intensity interference peaks at the input port of the plasma waveguide coupler. After a few millimeters of coalesced co-propagation with strong spatiotemporal couplings, the tightly guided laser pulses underwent filament fission and most of the laser energies were diffracted along the incident pulse directions and a small portion of guided pulses were projected along the plasma lattice, which could be regarded as the output port of the plasma waveguide coupler. Interestingly, the output pulses along the incident directions were demonstrated to have almost the same intensity ratio as the incident pulses. This presented a solid evidence that the coalesced filamentary propagation in the plasma self-channels came from self-projection of the interference intensity fringes, i.e., the filament interference maintained its fringe contrast during the whole coalesced propagation. As we had two beams at the input port, two main output beams and multiple fringes at the output port, the generated plasma microstructures could be used as two-to-multiple nonlinear waveguide couplers. The strong spatiotemporal and plasma-laser couplings within the plasma couplers may find interesting application in tailoring the spatiotemporal dynamics of the involved intense ultrashort pulses.

Due to the aperture size limitation, our CCD camera could only detect a small spatial range of the whole interaction region. In order to further confirm our clarification, we captured the fluorescence image of the whole interaction region by moving the optical imaging system and CCD camera along the bisector of the interacting filaments, as shown in Fig. 10.8a. Interestingly, the interference existed not only in the center of the interaction region but also in the starting and ending regions where spatial overlapping was not so obvious. The energy reservoir played a vital role in the formation and persistence of coalesced filament interaction and plasma microstructures [50], and most of the filament energy was distributed in its energy reservoir. Strong energy reservoir was generated with a surrounding energy attraction in the center of the filament interaction region, and weak but regularly separated intensity modulation also occurred in starting and ending regions of very small spatial overlapping due to the interference between the energy reservoirs. From this point of view, the observed non-collinear filament interference as shown in Fig. 10.8 differed a lot from the standard optical interference of weak lasers.

Non-collinear pulses at arbitrary incident angles could be efficiently coupled into the periodic plasma structure (1D lattice) by automatic balance between the incident beam wave-vectors and plasma lattice periodicity, as shown in Fig. 10.8c. This took place owing to

$$\vec{K}_1 + \vec{K}_2 = 2\pi/\Lambda, \tag{10.6}$$

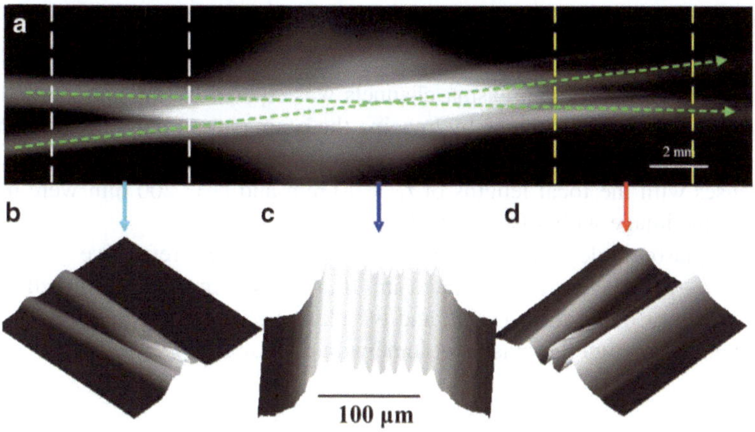

Fig. 10.8 (**a**) Measured fluorescence profiles of the interaction region of the intersected filaments. (**c**) The observed fringe-like spatial distribution of plasma fluorescence inside the filament interaction region. (**b**, **d**) The three-dimensional plot of the leading and trailing parts of the interaction to illustrate guiding of the incident pulse

where $\vec{K}_{1,2}$ represents the wave-vectors of the incident pulses and $2\pi/\Lambda$ is the wave-vector of the wavelength-scale plasma microstructures. As the plasma density modulation was formed by the incident pulses themselves, the wave-vector matching was adaptively reached for efficiently coupling the incident pulses under Bragg conditions.

Figure 10.8b and d shows the three-dimensional plots of the leading and trailing parts of the interaction region, which confirmed fusion and guiding of the input pulse along the corresponding bisector, consistent with the observed far-field profiles. However, after a few millimeters, the preformed plasma lattice was split into individual filaments (Fig. 10.8d) due to the collapse of the counterbalance between self-focusing and plasma defocusing. Interestingly, a slight bending of the incident pulses was observed at the input and output ports. The curved pulse propagations at the leading and trailing regions of the filament interaction manifested that laser–plasma interaction actually changed the plasma density gradually, resulting in a gradient plasma density and refractive index changes in the leading and trailing regions during the interference of energy reservoirs. This also differed dramatically from the standard optical interference, showing clearly that interfering filament interaction was accompanied by strong spatiotemporal and laser–plasma interaction, where highly nonlinear photosensitive responses of multiphoton-ionization-induced plasma microstructures played a vital role in filamentary self-projection of the intensity interference. It presents another solid evidence of plasma microstructures consistent with direct imaging and dynamic probing of interacting filaments and plasma density modulation.

10.5 Dynamics of the Plasma Channels

The time evolution of 1D plasma channels was studied by the above-mentioned holographic imaging technique by imaging the beam pattern of a third perpendicularly propagated weak probe beam with controllable time delays. In the experiment, two lenses with the focal lengths of $f_1 = 40$ mm and $f_2 = 200$ mm were used to magnify the image with a factor of $M = f_2/f_1 = 5$.

The plasma density modulation changes the refractive index as $\delta n(r,t) \cong -\rho(r,t)/2\rho_c$, as illustrated in (10.4). Assuming the preformed plasma microstructures had a thickness d, and the plasma changed the probe pulse vertically passing through the plasma volume with a phase delay

$$\delta\phi(r,t) \cong -\rho(r,t)d/2\rho_c. \tag{10.7}$$

The wavelength-scale microstructures are described by the spatial dependence in $\rho(r,t)$ and $\delta\phi(r,t)$, a time-delayed probe laser pulse passing through the plasma microstructures experienced a position-dependent phase shift, which interfered with the transmitted part and got an interference holography closely associated with the phase shift. As the probe pulse was changed in the time delay, the ultrafast dynamics of the preformed plasma density $\rho(r,t)$ could be thoroughly studied with a high spatial resolution.

The images of plasma channels at different time delays are shown in Fig. 10.9a–f. We clearly observed periodic plasma microstructures with a probe time delay of 0.25 ps (Fig. 10.9a). With such a small delay $\tau \sim 0.25$ ps, the interacting filaments were overlapped with a short length $c\tau \sim 0.075$ mm. The captured holographic imaging thus covered a short length. We could distinguish the interfering filament

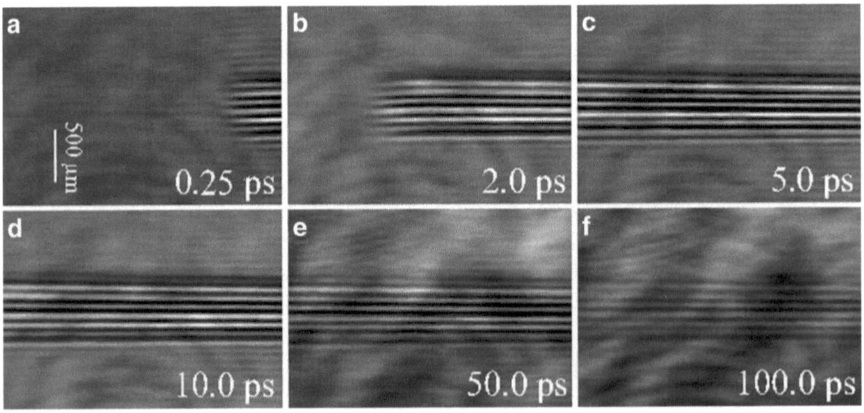

Fig. 10.9 The time evolution of the 1D plasma channels at different time delays (**a**) 0.25 ps, (**b**) 2 ps, (**c**) 5 ps, (**d**) 10 ps, (**e**) 50 ps, and (**f**) 100 ps

fronts and their interference self-projection as well. The interfering plasma wakes built at the filament fronts with parallel plasma self-channels projected along the interference peaks.

The plasma subfilament channels moved forward with almost the same plasma density modulation depth at larger time delays as the interacting filaments propagated in the interaction region (Fig. 10.9b). Gradient blurring of holographic image was observed, which was ascribed to the hydrodynamic plasma expansion that tended to make the plasma density spatially distributed homogeneously. In this way, the plasma density modulation became weaker as plasma expansion went further. The gradually blurred images in Fig. 10.9c–f represent the plasma channels at delays of 5, 10, 50, and 100 ps, from which we can conclude that the plasma density remained to be spatially modulated with a lifetime up to ~100 ps.

10.6 Two-Dimensional Plasma Waveguides

In this part, we demonstrate that more complicated 2D plasma lattices could be generated by simply adding another interaction filament, i.e., three intersected filaments could produce 2D photonic plasma lattice in air.

In the experiment, a 50 fs pulse of 1 kHz repetition rate and 2.4 mJ pulse energy from a Ti:sapphire laser system at 800 nm were equally split into three pulses (A, B, and C), and were independently focused with a lens ($f = 60$ cm) and two high-reflection concave mirrors ($f = 100$ cm), with changeable crossing angles from 3° to 6°. Each pulse produced a single filament as a result of the counterbalance between self-focusing and plasma defocusing. Two step-stage time-delay lines were used to finely control the relative time delays. As sketched in Fig. 10.10, pulses A and B were manipulated in the horizontal plane while pulses B and C in the vertical plane. As shown in Fig. 10.10b–e, for each of the two pulses, 1D intensity modulation of the interaction region was observed as previous observation [41]. Meanwhile, an interference induced 2D periodically modulated intensity was observed when all the three pulses were synchronized (A, B, and C), suggesting a localization of the laser fields inside the interaction region. The experimental observations clearly showed that with three or more pulses crossed nearby their common foci at different crossing planes, similar to the formation of the 1D plasma lattice, the interference patterns consequently induced 2D periodic plasma density modulation and the input pulses were eventually guided by the 2D plasma lattices.

Once again, the holographic imaging technique was used to observe the plasma lattice structures by recording the spatial distribution of a perpendicularly propagated weak probe beam. Moreover, by introducing the probe beam from below to top and from left to right through the interaction region, the difference of 1D and 2D plasma density modulation was clearly visualized, as shown in Fig. 10.11a–d. Figure 10.11a–d depicts the images of the 1D and 2D plasma gratings when the probe pulse was delayed ~5.0 ps after the plasma lattice

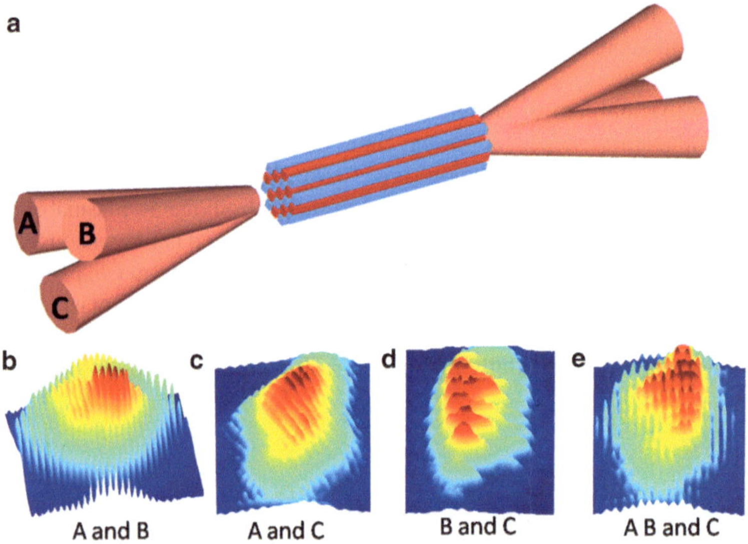

Fig. 10.10 (**a**) The sketch geometry for 2D plasma lattices generated by three non-collinearly intersected pulses A, B, and C. (**b–e**) The 3D distributions of the measured 1D and 2D intensity modulation patterns for different crossly overlapped pulses (marked beneath the patterns)

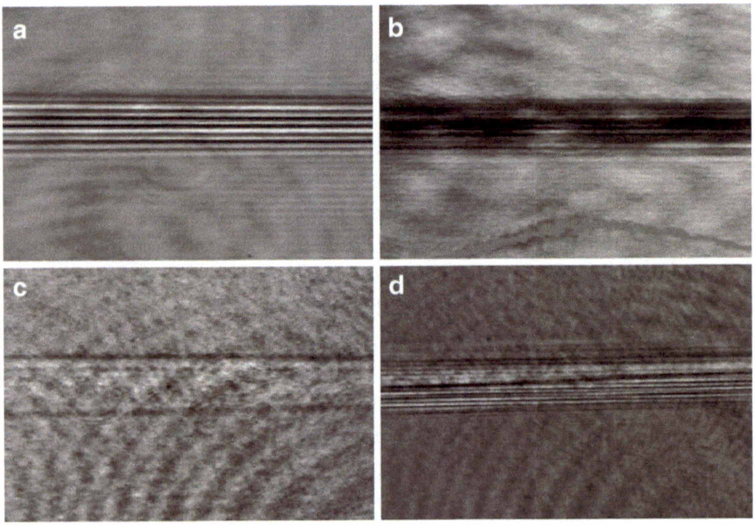

Fig. 10.11 The measured holographic images of the (**a**, **c**) 1D and (**b**, **d**) 2D plasma gratings from different views. (**a**, **b**) Top view (*bottom* to *top*), (**c**, **d**) Side view (*left* to *right*). The probe pulse was delayed ~5.0 ps after the plasma lattice formation

formation, in which (a) and (b) represent the top view (bottom to top) while (c) and (d) are the side view (left to right). For the case of two interacting filaments (A and B), owing to the refractive index modulation, the weak probe beam pattern clearly verified the 1D photonic lattice structure in the top view. However, in the side view, no periodic structures were observed, indicating no evident refractive index difference. When another synchronized pulse C was introduced, the photonic lattices were extended from 1D to 2D. The image of 2D plasma lattices in the top view became somehow vague (Fig. 10.11b), while some periodic structures emerged in the side view (Fig. 10.11d), clearly indicating the formation of a 2D plasma photonic lattice.

10.7 The Formation of Plasma Grating

As already discussed above, non-collinear intensity interference resulted in periodic intensity distribution in the interaction region, and the interference fringes were projected through filamentation and the corresponding multiphoton ionization created a spatially modulated plasma density [40, 41]. Consequently, local refractive index was modulated periodically according to the plasma density modulation. Such a dynamic refractive index modulation could function as a diffraction grating, similar to the classic optical diffraction grating with a periodic structure to split and diffract the incident light beams to several beams in different directions. Based on the diffraction characteristics, the gratings could be sorted as thin and thick ones. A thin grating exhibits little angular dependence and wavelength selectivity, with the thickness D and the period Λ of the grating follows $D/\Lambda < 10$, while a thick grating with $D/\Lambda > 10$ shows relatively strong angular and wavelength selectivity [51]. As the plasma grating period $\Lambda = \lambda_c/2\sin(\theta/2)$ could be easily tuned by changing the intersection angle θ of the non-collinear beams, the plasma grating could behave as either a thin or thick grating, where λ_c is the central wavelength. In our experiment, the typical plasma grating thickness was $D \sim 100$ μm, a thin grating was formed with the crossing angle $\theta < 4.6°$ in the case of $\lambda_c = 800$ nm. In the case of a thin grating, an incident light beam at wavelength λ was diffracted to an angle φ_m according to the condition

$$\Lambda[\sin(\theta+\varphi_{\mathrm{m}}) - sin(\theta)] = \mathrm{m}\lambda, \qquad (10.8)$$

where $m = 0, \pm1, \pm2\ldots$ denote the diffraction order, θ is the angle of the incidence onto the grating.

The characteristics of 1D and 2D plasma gratings could be revealed by the diffraction of a time-delayed TH pulse. As nonlinear frequency conversion such as third harmonics could be efficiently generated within an intense femtosecond filament in air [6], in which the phase of the fundamental-wave (FW) pulse and the generated TH pulse are nonlinearly locked within a long distance. The probe

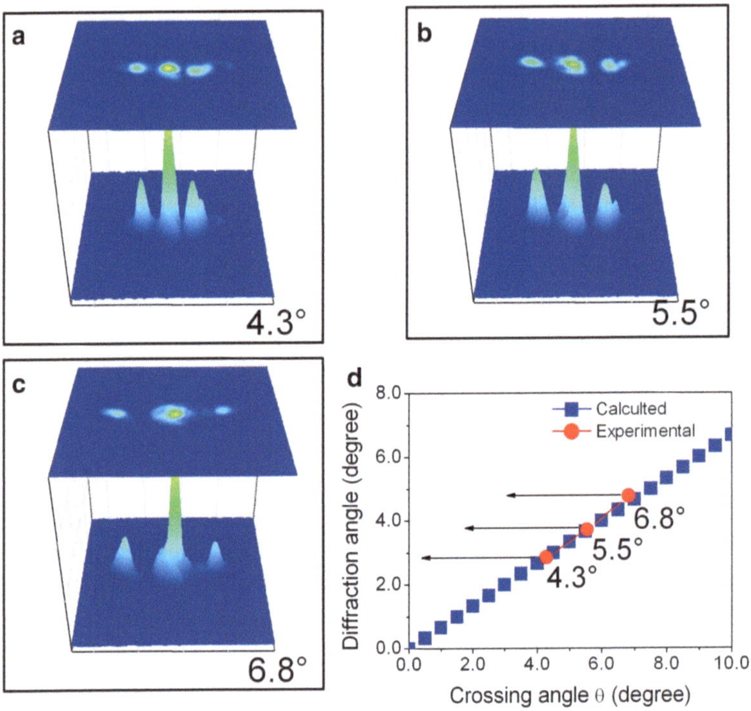

Fig. 10.12 The 3D distribution of the diffracted third harmonic pulse by 1D plasma grating with different crossing angles of (**a**) $\theta = 4.3°$, (**b**) $\theta = 5.5°$, and (**c**) $\theta = 6.8°$, respectively. (**d**) The comparison of the experimental result (*red-circular line*) and calculated (*blue-square line*) dependences of the +1 order diffraction angle on the crossing angle θ

pulse was generated with an intense pulse undergoing filamentary propagation through the preformed plasma gratings. The experiments were realized with three pulses (A, B, and C) independently focused to form three intersected filaments. Synchronized pulse A and B were manipulated in the horizontal plane to form a plasma grating. Pulse C was placed in the same plane of pulses A and B, with an incident angle 0° onto the grating. As 1D plasma density grating was formed by pulses A and B, the time-delayed TH pulse from pulse C was diffracted according to

$$\varphi_{\mathrm{m}} = 2\mathrm{m}\sin(\theta/2)\lambda_{\mathrm{TH}}/\lambda_{\mathrm{FW}}, \tag{10.9}$$

where $m = 0, \pm 1, \pm 2\ldots$ denotes the diffraction order, θ is the cross angle of pulses A and B that determines the grating period, φ_m is the corresponding diffraction angles, λ_{TH} and λ_{FW} are the wavelengths of the TH and FW pulses, respectively. By blocking the pulses A and B after their interaction region, as shown in Fig. 10.12a–c, photographs of the diffracted TH pulse from pulse C after a low-pass filter (filter out the FW pulse) were taken by a digital camera on a paper screen placed perpendicularly to the propagation direction of pulse C at a distance of ~150 cm away from the

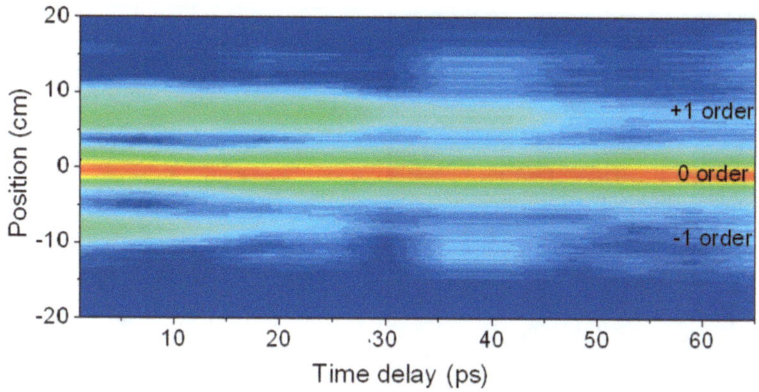

Fig. 10.13 The measured dynamic evolution of the intensity of the diffracted TH pulse generated from pulse C by the plasma grating created by pulses A and B [($\theta = 4.3°$)]

intersecting point. The crossing angles between the pulses A and B were set to be ~4.3°, 5.5°, and 6.8°, respectively. As determined by the diffraction condition, the ±1 orders moved apart from the TH pulse as the crossing angle θ increased, consistent with the decrease of the plasma grating period. Figure 10.12d shows a simulated dependence of the +1 order diffraction angle (blue-squared line) as a function of the crossing angle θ, which agreed well with experimental measurements (red-circular line). According to the dynamic diffraction of the time-delayed pulses, the preformed plasma gratings were evidenced to last a few tens picoseconds after the excitation pulses, in agreement with the plasma lifetime measured by using the above-mentioned holographic imaging technique. We emphasize that the plasma grating period in this work could be readily tuned by adjusting the cross angle between the incident filaments which meanwhile could be applicable to high-peak-intensity pulses, in contrast with the periodic plasma structures fabricated with a spatial-light modulator [52, 53] that sustained only low pulse energies.

The dynamic evolution of the plasma grating was characterized by recording the diffracted TH pulse as a function of the time delay of pulse C with respect to the plasma grating forming pulses A and B (4.3° crossed). Figure 10.13 displays the normalized integrated spatial distribution of the diffracted TH pulse when the time delay of the pulse C with respect to the plasma grating formed by pulses A and B was tuned from 1.0 to 65.0 ps. Before the formation of the plasma grating, no diffraction of the TH pulse was observed. The intensity of the ±1 order of the diffracted TH pulse decreased gradually as the time delay increased up to several tens picoseconds, confirming the long-lifetime plasma grating in contrast with transient Kerr gratings that exist only within the pulse duration [34].

2D plasma grating could be produced with three interacting filaments in different planes. 2D diffraction of the simultaneously generated TH pulses was resulted. The diffraction was caused by the corresponding plasma grating preformed by either the interfering pulses A and B, B and C, A and C, or three of them.

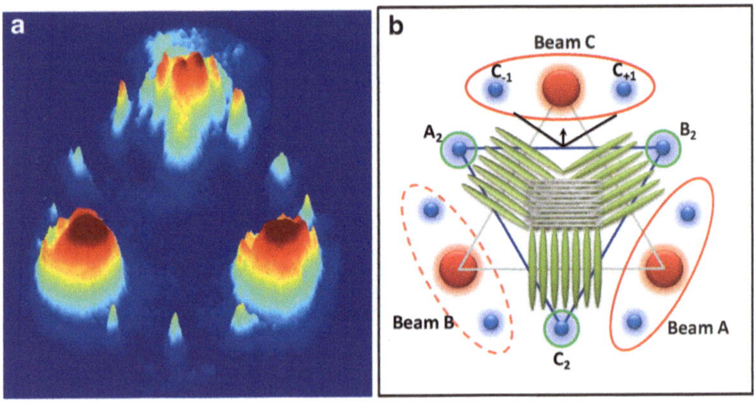

Fig. 10.14 (**a**) The measured diffraction pattern of the generated third harmonic pulses as a two-dimensional plasma grating created by the interaction of three femtosecond pulses. (**b**) The sketch map of the observed diffraction pattern from the two-dimensional plasma grating

As diffraction was observed under the situation that all the three pulses A, B, and C are synchronized, it behaved in a quite similar way as the transient Kerr grating observed in the quadratic nonlinear crystals [34]. Nevertheless, the observed Kerr nonlinearity of air molecules was dramatically enhanced in the presence of plasma density modulation. The TH pulses were diffracted and meanwhile amplified due to the strong interaction among the three intersecting filaments. Here, no low-pass filter was used to remove the FW pulses in the photograph. In addition to ±1 diffraction orders beside the FW pulse, as shown in Fig. 10.14a, three new TH pulses were observed as labeled by A_2, B_2, and C_2, which were due to the additional diffraction of the preformed 2D plasma grating.

Figure 10.14b schematically shows the observed TH diffraction pattern from the 2D plasma grating, which can be divided into two parts: the first part marked in the dashed ellipse was originated from the diffraction of the 1D plasma grating as discussed above, the second part labeled as A_2 (B_2, and C_2) was originated from the additional diffraction of the 2D plasma grating. For example, the TH pulse labeled C_{+1} and C_{-1} is the ±1 diffraction orders of the TH pulse (generated from the pulse C) diffracted by the 1D plasma grating formed by the pulses A and B. According to the diffraction condition of a thin plasma grating (10.9), the ±1 orders diffraction angle is expected to be $\pm2.67°$, which is in good agreement with the measured values of $\pm2.58°$. Since all the three pulses were temporally overlapped, two additional plasma gratings could be formed by the pulses A and C and pulses B and C, which could be treated as an equivalent plasma grating marked with silver lines. This equivalent plasma grating diffracted the generated TH pulse in pulse C and the +1 diffraction order led to the observed TH pulse labeled as C_2. Based on the measured diffracted TH pulse at C_2, the period of the equivalent plasma grating was estimated to be ~4.6 μm. The new TH pulses of A_2 and B_2 could be similarly understood to be the +1 order diffraction of the generated TH pulses in pulses A and

B by the correspondingly equivalent plasma gratings. These overall led to the observed 2D diffraction pattern of the generated TH pulses as shown in Fig. 10.14a, which were simultaneously amplified by the strong filament interactions. From this point of view, 2D plasma grating presented here is not only the superposition of three independent 1D plasma gratings but also should be treated as a complicated volume grating.

10.8 Plasma Grating for Efficient Energy Transfer Between Intense Femtosecond Filaments

Wavelength-scale plasma density modulation could form a dynamic plasma grating for efficient energy transfer from one filament to the other. This could be realized on the basis of the second-order diffraction from the preformed plasma grating with the experimental setup as schematically shown in Fig. 10.15. A 50-fs pulse of 1 kHz repetition rate, 800 nm and 2.2 mJ pulse energy from a Ti:sapphire laser system was equally split into two parts as the pump and probe pulses. A BBO crystal (type I, 29.2°-cut, and 500-μm thick) placed in the arm of pump was used for second harmonic (SH) generation. A dichotic mirror was used to separate the 800 and 400 nm pulses, and the SH pulse passed through a motorized delay line and then collinearly combined with the FW pulse. In the probe arm, a combination of a half-wave plate (HWP) and a neutral density attenuator (ND) was used to control the polarization and the energy of the probe pulse. The pump and probe pulses were separately focused by two lens with $f = 100$ cm, producing two filaments in air with the crossing angle variable from 2° to 4°. When the pump and probe pulses were synchronized, a 1D plasma grating was formed.

Different from previous studies, there was also frequency-doubled 400 nm pulses in the pump arm. Interestingly, an efficient energy transfer of the time-delayed SH pulse from the pump to the probe was observed to be accompanied with the formation of the 1D plasma grating. Figure 10.16 presents the photographs and

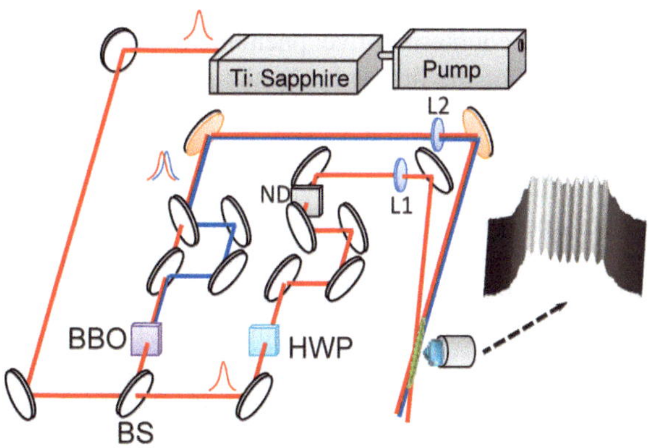

Fig. 10.15 The schematic of the experimental setup. *BS* beam splitter, *HWP* half-wave plate, *L1*, *L2* lenses, *ND* neutral density attenuator

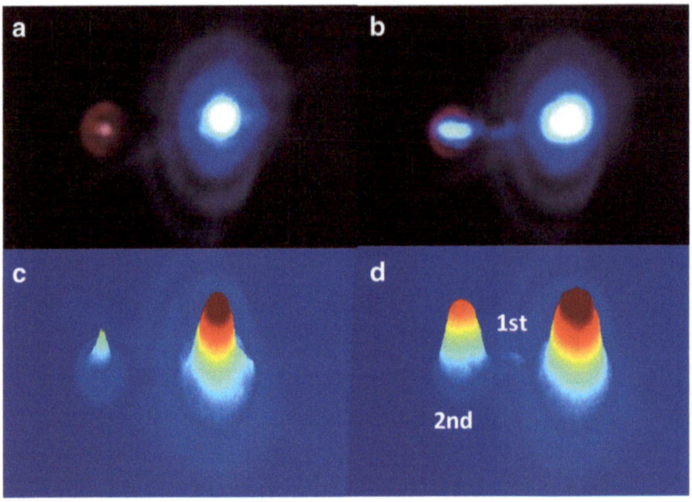

Fig. 10.16 Far-field images of the incident pump and probe pulses (**a**) with and (**b**) without filament interactions and the corresponding 3D plots of the far-field images (**c**, **d**)

the corresponding 3D plot of the pump and probe beams with (b, d) and without (a, c) plasma grating, captured by a digital camera on a paper screen placed at a distance of ~100 cm away from the intersecting point. No SH pulse was observed in the probe beam (Fig. 10.16a) when the pump and probe were not synchronized, while reversely a bright blue spot emerged, collinearly propagated with the probe beam. The diffraction of the SH pulse by the thin plasma grating was determined by the Bragg diffraction condition. In this experiment, we dealt with the situation with small angles θ and φ_m. The Bragg diffraction formula could be reduced to $\varphi_m = m\theta\lambda_{SH}/\lambda_{FW}$. For the case of the incident SH along the pump arm, the first- and second-order diffraction angle should be $\theta/2$ and θ, respectively. The observed results presented in Fig. 10.16b strongly support our explanation on diffraction of the incident SH pulses from the plasma grating.

Interestingly, the first-order diffraction was much weaker than the second-order one. This counterintuitive phenomenon could be explained as defocusing and absorption of the first order by the plasma lattice. The energy of the incident and diffracted SH pulses was measured to be 220 and 5 μJ, with a calculated diffraction efficiency of 2.3%.

We further studied the dependence of the transferred SH pulse along the probe beam upon the time delays between the SH pulse and the grating forming pulse [54]. Two dichroic mirrors were used to separate the SH pulses at the end of the filament interaction. A photodetector and locking-in amplifier were used to further optimize the results. The inset in Fig. 10.17 shows the relative SH pulse energy in the pump and probe arms recorded by scanning the pump–robe delay under the condition of a fixed SH pulse delay of −100 fs. A clear variation of SH energy was

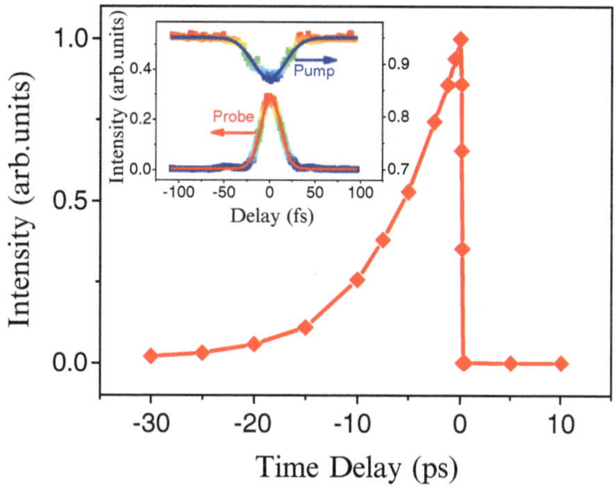

Fig. 10.17 The measured normalized intensity of the transferred SH pulse in the probe arm as a function of the delay of the SH pulse with respect to the FW pulses. *Inset*: SH signal variation in the pump (*blue curve*) and probe (*red curve*) arms for a fixed SH pulse delay of −100 fs

observed only when the pump and probe pulses were synchronized. The pump decrease was accompanied with the probe increase, which clearly indicate that the transferred SH pulse was originated from the pump. As shown in Fig. 10.17, the energy reached a maximum at the zero time delay and only existed when the SH pulse was negatively delayed with respect to the FW pulse. Moreover, the gradually decreased SH diffraction signal as a function of the time delay clearly demonstrated a plasma density decay inside the long-lifetime plasma grating (up to few tens picoseconds).

10.9 Enhanced Third Harmonic Generation with Plasma Gratings in Air

Coherent nonlinear optical frequency conversion is an important aspect of nonlinear optics for its intriguing application in remote sensing [55], single-photon frequency up-conversion [56], and coherent ultraviolet light sources [57]. Much effort has been devoted to improve the energy conversion efficiency by controlling the phase-matching condition between harmonics. In the above, we proclaimed the existence of plasma grating by efficient TH enhancement and diffraction in air. According to the self-guiding model, the clamped filament intensity at a level of 10^{13}–10^{14} W/cm^2 was high enough to produce stable TH generation, whose phase was self-locked with the FW pulse during filamentation [6]. Such self-phase locking may provide a possible way to overcome dispersion restriction to achieve phase matching. However, as the filament was terminated, the balance collapse between the self-focusing and plasma defocusing would result in a rapid decrease of the TH pulses.

In this part, we focus on efficient TH generation assisted by plasma grating in air and compare TH generation with that from a single filament. In the experiments, we

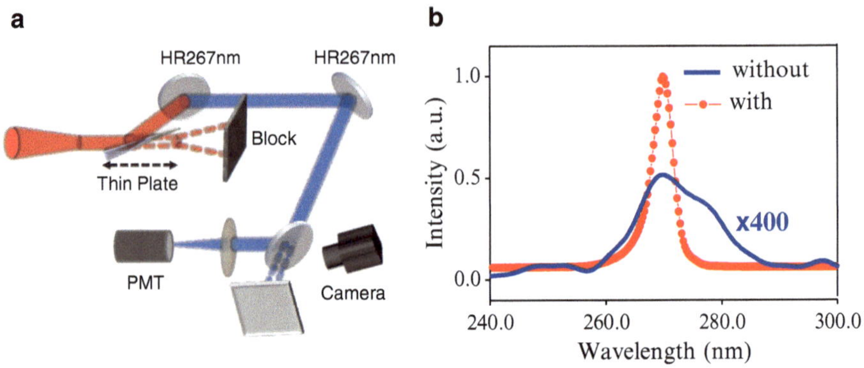

Fig. 10.18 (**a**) The schematic experimental setup for the TH energy and spatial distribution measurement. (**b**) The measured spectra of the TH pulses with and without interaction [41]

explored dependence of the TH enhancement on the crossing angle, polarizations, and the intensity ratios between two intersected filaments. A 1D plasma grating was formed with pump and probe filaments non-collinearly crossed at 3°. In order to study TH enhancement, a fused silica prisms pair at the end of the probe filament was used to separate the TH pulse from the FW pulse. The energy and spatial distributions were, respectively, recorded by a photomultiplier tube (PMT) and a CCD camera, as shown in Fig. 10.18a. Strong spatial and temporal coupling of the interacting filaments was observed when the pump and probe pulses were synchronized in the intersecting point. Strong filament interaction brought about enhanced TH generation in comparison with a single probe filament [41]. The spectra of the generated TH pulses with and without filament interaction are depicted in Fig. 10.18b. The peak TH intensity was enhanced at ~770 times (at the spectral peak around 267 nm) in the presence of filament interaction.

Spatial evolution of the TH enhancement along the filament propagation and the corresponding pulse energy are presented in Fig. 10.19. The TH pulse decreased rapidly with the propagation distance (solid diamond curve), originated from the plasma defocusing and diffraction of the laser beam, while remained almost constant inside the filament owing to the nonlinear phase locking with the FW pulse. The spatial profiles of the generated TH pulse along its propagation direction captured by a CCD camera are also presented in Fig. 10.19, which clearly visualizes the multiring structures of the TH inside the filament, whose divergence angles were measured to be 2.25, 4.09, and 5.85 mrad, respectively. We conclude that the inner rings were generated from the Fraunhofer diffraction of the generated on-axis TH core. By assuming a filament diameter of ~170 μm, we estimated divergence angles of 2.26 and 4.2 mrad for the first and second rings, respectively. Without filament interaction, a sharp decrease of the TH energy was observed at the propagation distances from 102 to 120 cm, where the multiring structures disappeared. In the presence of filament interaction, the generated TH pulses could be

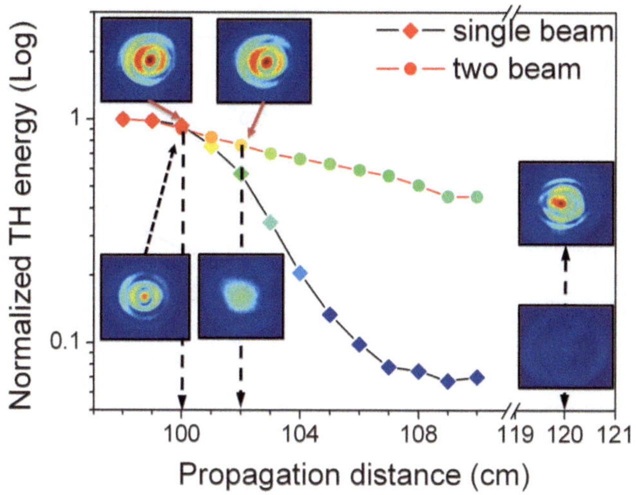

Fig. 10.19 The measured TH pulse energy variation as a function of propagation distance in the cases with and without filament interaction. The TH spatial distributions at three positions of 100, 102, and 120 cm are presented in the *insets*

maintained along a relatively longer distance (solid-circled curve) with sustained ring-shape profiles, indicating the elongation of filament with self-confined diameter due to the filament interaction. From this point of view, two interacting filaments could be used as a method to control filament propagation [17, 41] and increase the nonlinear phase-locking distance, which would be further beneficial for broadband TH generation in filaments.

Since the TH enhancement was based on the filament interaction, which was closely dependent on the experimental parameters such as polarization, crossing angle, driving intensity, and time delay, we systematically investigated the changes in TH conversion energy for a range of parameters. The nonlinear increase of the TH enhancement factor observed at different pump intensities at a non-collinear crossing angle of 9° is shown in Fig. 10.20a. The total TH intensity increased slowly with the control pulses at small intensities. A dramatic TH enhancement was observed with the control peak intensity above 64 GW/cm^2. As the control peak intensity increased higher than 100 GW/cm^2, the TH enhancement was switched to a gradual increase with the control pulses and reached its maximum at 220 GW/cm^2. The phase-matching condition became destroyed above 220 GW/cm^2, leading to a decrease of the TH conversion efficiency.

Figure 10.20b presents a comparison of the TH enhancements driven by pump and probe pulses with different crossing angles. At an increased crossing angle, the requisite control peak intensity should be increased to optimize the TH conversion efficiency. This was mainly caused by non-collinear projection of pump and control pulses into the assembled waveguide, which induced different geometric changes of the TH phase-mismatch under different incidence angles. The geometric effects on

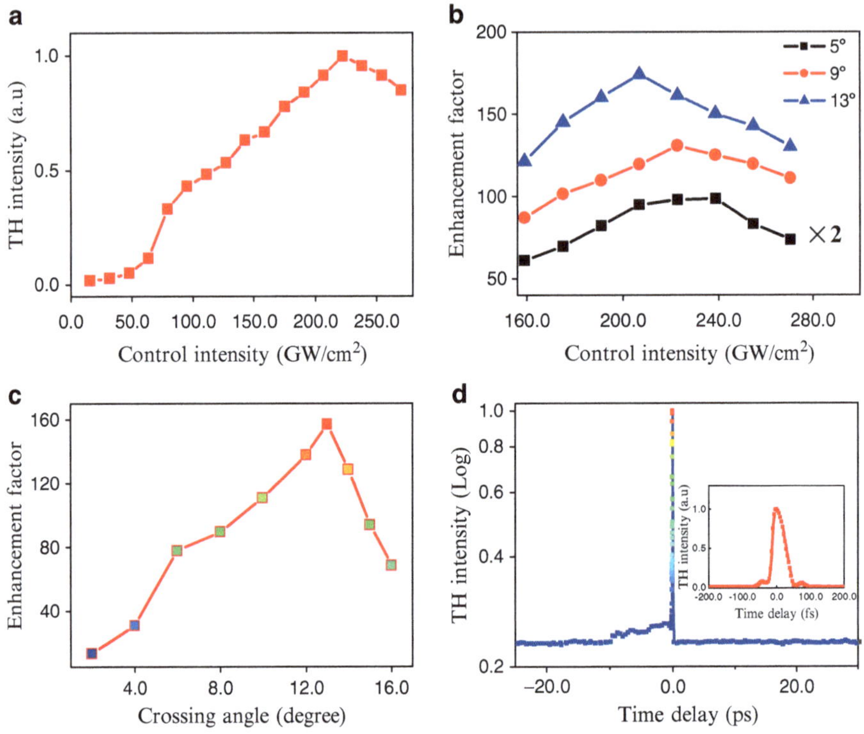

Fig. 10.20 (**a**) Spectrally integrated TH intensity under different pump intensities at a non-collinear crossing angle of 9°. (**b**) Integrated TH enhancement factor as a function of control intensity at different non-collinear crossing angles. (**c**) Integrated TH enhancement factor measured at different non-collinear crossing angles. (**d**) Integrated TH intensity as a function of time delay. *Inset*: the delay dependence of the TH near zero time delay

phase matching were further evidenced by the dependence of the TH enhancement on the crossing angle shown in Fig. 10.20c. Although the assembled waveguide sustained longer lengths at smaller crossing angles, the TH conversion efficiency reached its maximum at a crossing angle ~ 13° under fixed pump and control intensities, and the corresponding enhancement of the generated TH intensity (integrated over the TH spectral range) was further optimized up to 174 times with the control peak intensity around 200 GW/cm^2 (see Fig. 10.20b). The relative polarization of the control and pump pulses also affected the formation of the plasma lattice and the subsequent TH enhancement. As shown in Fig. 10.20d, when the pump and control pulses were temporarily walked off, the enhancement of the TH pulse was still exist for tens of picoseconds, indicating the third-order nonlinearity variation induced by the laser-induced plasma. The inset in Fig. 10.20d presents the detailed TH intensity variation around zero time delay.

10.10 Broadband Third Harmonic Enhancement by Interaction of Intense Few-Cycle and Long Femtosecond Filaments

By using nonlinear interaction between non-collinearly crossed few-cycle and synchronized long filaments, an intensity-enhanced and spectrum-broadened third harmonic (TH) pulses could be generated. We studied the TH enhancement and its dependence on the relative polarization, crossing angle, intensity, and time delay between the two filaments with an experimental setup schematically shown in Fig. 10.21. An output from an amplified Ti:sapphire laser system (50-fs, 800-nm, 1 kHz) was firstly split into two parts after a beam splitter with one of them was further coupled into an argon-filled hollow fiber and then compressed by using chirped mirror pairs to produce few-cycle pulses (probe) with the pulse duration of ~7 fs. A translation stage and a HWP were used in another (pump) arm to precisely control the temporal delay between the pump and probe pulses and the pump pulse polarization, respectively. Both the pump and probe pulses with the corresponding incident pulse energies of 0.55 and 0.50 mJ were focused in air to cross non-collinearly by using two separate lenses with the same focal length of 100 cm. The non-collinear crossing angle was varied from 4.0 to 12.0°. The generated TH pulses were separated from the FW probe pulses by using a pair of silica prisms, which were then measured by a PMT and a spectrometer.

As the long and few-cycle pulses were delayed without temporal overlapping, both pulses propagated in air with no interaction even though they were non-collinearly crossed, and each filament was formed with the typical length of ~3 cm as a result of the counterbalance between the Kerr self-focusing and plasma defocusing. Self-guided filamentation enabled efficient nonlinear frequency conversion since the phase of the FW and the newly component could be locked to fulfill phase-matching condition [6, 41]. However, when the filament of FW disappeared, the TH would give its energy back to the FW due to the phase mismatching, leading to the rapid decrease of the TH energy. This was consistent

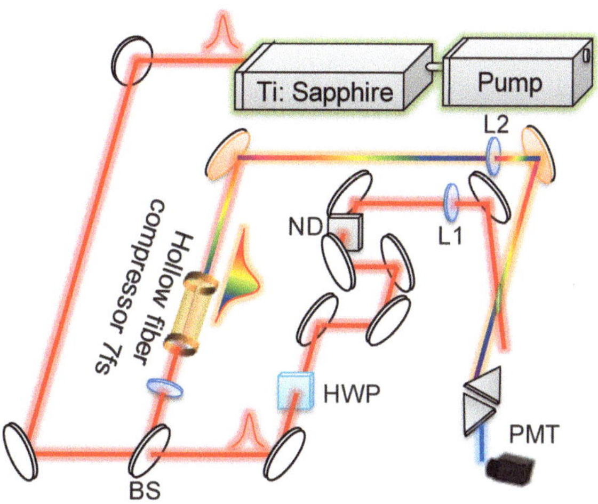

Fig. 10.21 Experimental setup for broadband TH generation in the presence of nonlinear interaction between non-collinearly crossed few-cycle and long-pulse filaments. *BS* beam splitter, *HWP* half-wave plate, *PMT* photomultiplier tube, *L1*, *L2* lens

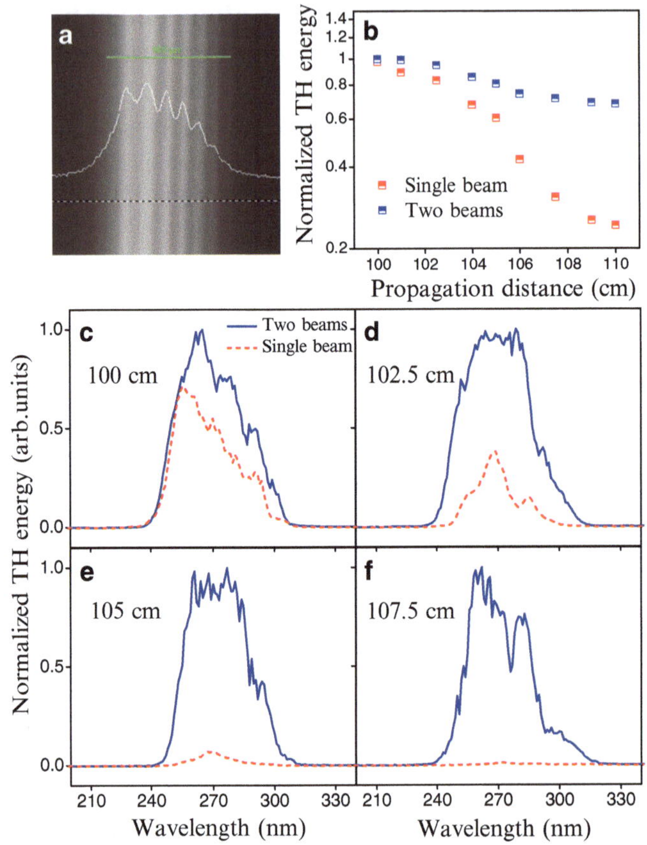

Fig. 10.22 (**a**) The fluorescence image of the interaction region. (**b**) The TH energy versus the propagation distance for the cases with (*blue square*) and without (*red square*) filament interaction. Measured spectra of TH wave generated by few-cycle pulse at different propagation distance of (**c**) 100, (**d**) 102.5, (**e**) 105, and (**f**) 107.5 cm, when the non-collinear filament interaction was turned on (*blue solid curve*) and turned off (*red-dashed curve*). The crossing angle was set to be 5° and the incident pulse energies of the few-cycle and long pulses were, respectively, set to be 0.50 and 0.55 mJ

with the experimental observation shown in Fig. 10.22b (red-squared curve). As the synchronized pump pulse was launched, the energy back conversion from the TH to the FW was suppressed in a long propagation distance [the blue-squared curve in Fig. 10.22b], enhancing TH in the far-field region. All these indicated that the interaction between pump and probe pulses improved the phase-matching condition between the FW and TH pulses, which was further confirmed by the TH spectral distribution at different propagation distances with and without pump pulses (Fig. 10.22c–f). For a single filament generated by few-cycle pulse, the corresponding TH spectrum ranged from 240 to 310 nm at the laser focus around 100 cm [red-dotted curve in Fig. 10.22c]. As the propagation distance increased, the spectral bandwidth was narrowed with a decreased intensity, as shown in

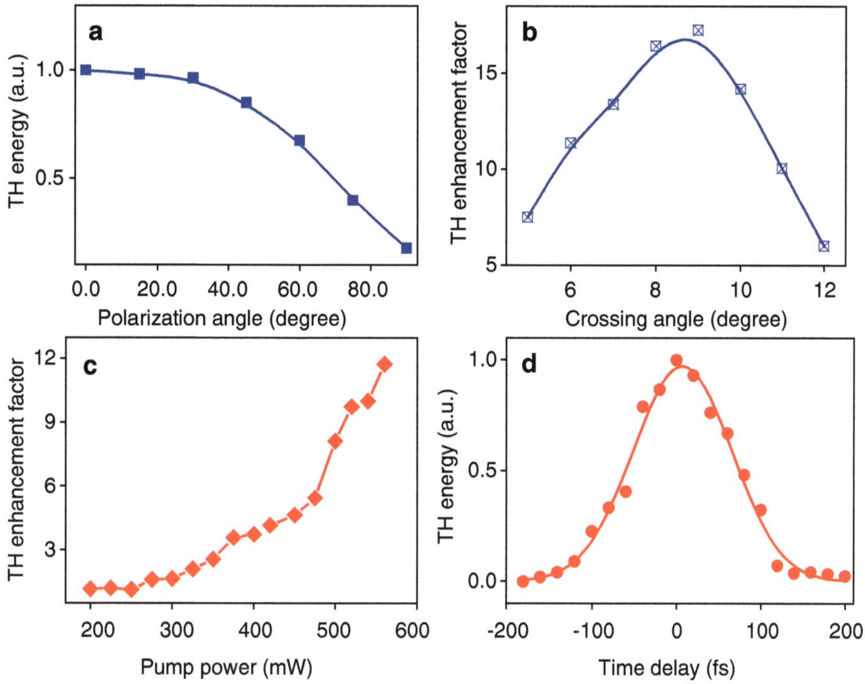

Fig. 10.23 The dependences of the generated TH pulse on (**a**) the relative polarization, (**b**) crossing angle, (**c**) intensity ratio, and (**d**) time delay. The pump (long) and probe (few-cycle) pulse energies were 0.55 and 0.50 mJ, respectively for (**a**, **b**, **d**), and the crossing angle was fixed to be 5° for (**a**, **c**, **d**)

Fig. 10.22d–f. When the synchronized pump pulse was added, the TH spectra at different positions (102.5, 105, and 107.5 cm) showed a similar bandwidth and intensity with that at the focus position.

The evolution of TH energy and spectral distribution confirmed that the TH enhancement in the far-field was originated from the decrease of TH energy back conversion to the FW pulses along their propagations due to phase-matching improvement by adding the synchronized pump pulse. This phenomenon could be understood by considering the filament elongation caused by the nonlinear filament interaction. The synchronized pump pulse significantly increased the intensity in the overlapped region due to the interference between pump and probe pulses. As a result, enhanced nonlinear effects dramatically affected the temporal distribution of the optical field. Meanwhile, the spatial distribution also followed the interference pattern. The modification on the spatial and temporal characteristics of non-collinear filaments could be used to control the filamentation process and lengthen the filamentary propagation.

Figure 10.23a shows the TH energy as a function of the polarization. The maximum and minimum TH enhancement factors were observed as their polarizations were parallel and orthogonal, respectively, where parallel polarizations corresponded to the formation of the plasma lattice. As shown in Fig. 10.23b for TH enhancement at different crossing angles, TH energy increased gradually with the increase of the crossing angle till it reached the maximum enhancement factor of 17 around a crossing angle of 9°. Filamentary propagation as the result of the counterbalance between self-focusing and plasma-defocusing was determined by the combined effects of different pulse characteristics, the optimal filament elongation was achieved with an appropriate crossing angle (9° for pulse duration and energy used in our experiments) to support the maximum of the TH energy measured in the far-field region. Further increase of the non-collinear crossing angle resulted in a reduction of the TH energy. Accordingly, we could adjust the crossing angle to manipulate the filament elongation and then change the TH energy in the far-field region. In addition, the filament interaction could be changed by adjusting the pump intensity and pump–probe time delay. Figure 10.23c and d depicts the generated TH as a function of the pump intensity and relative time delay. The TH enhancement changed under different filament interaction conditions, presenting further evidence that the TH enhancement was determined by the interaction between pump and probe filaments.

10.11 The Formation of an Intense Filament Controlled by Interference of Ultraviolet Femtosecond Pulses

Ultraviolet (UV) femtosecond laser pulses could be generated through nonlinear frequency mixing processes, such as third harmonic generation of the near-infrared femtosecond pulses with cascaded second-order nonlinear processes in the quadratic nonlinear crystals [58–60]. Experimentally, an output from a chirped pulse amplified laser system (800 nm, ~50 fs, 22 mJ, 10 Hz) was applied for TH pulse generation by cascaded nonlinear frequency mixing, as shown in Fig. 10.24. Three successive BBO crystals were used. The first BBO crystal (BBO1, 200 μm, 29.2°-cut, type I) was used for frequency doubling. The group velocity mismatching between the generated SH and the residual FW pulses was compensated by a second BBO crystal (BBO2, 200 μm, 29.2°-cut) with its crystallographic axis rotated 45° with respect to BBO1. The third BBO crystal (BBO3, 200 μm, 44.3°-cut, type I) enabled the TH generation

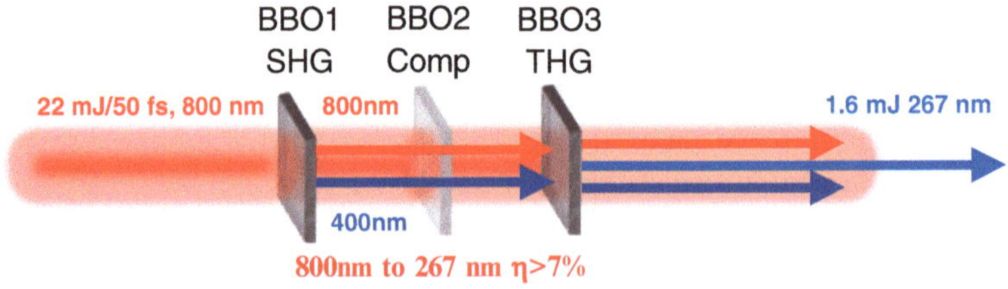

Fig. 10.24 Schematic of efficient TH generation by cascaded nonlinear frequency mixing with three successive BBO crystals. A conversion efficiency of 7% was obtained

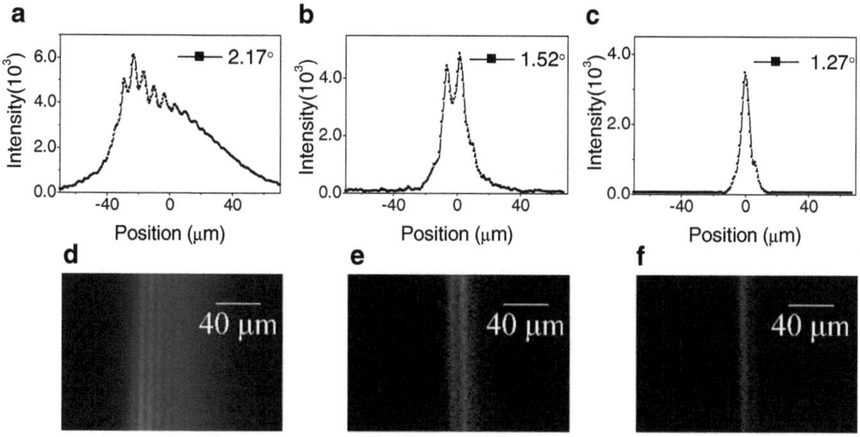

Fig. 10.25 Fluorescence images of the preformed plasma gratings with crossing angles of (**a**) 2.17º, (**b**) 1.52º, and (**c**) 1.27º, and (**d–f**) the corresponding fluorescence profiles

through frequency summing of the FW and SH pulses. The TH pulse was separated from the FW pulse with high-reflection mirrors of the TH pulses.

Similar to non-collinear near-IR filament interaction that induced efficient plasma waveguides [40] and plasma gratings [54], strong nonlinear interaction occurred for two non-collinearly crossed UV femtosecond filaments [61]. Experimentally, we generated an intense TH pulse of 1.6 mJ (~7% efficiency). The TH pulse was then equally split into two beams, and focused by using an $f = 250$ mm curve mirror with a changeable crossing angle. When two UV pulses were synchronized, strong spatiotemporal coupling and great fluorescence enhancement similar to that of the near-IR filament interaction were observed. The formation of plasma grating was directly visualized by the fluorescence detection method. As the interacting UV filaments intersected at decreased crossing angles, the plasma modulation period increased and the number of plasma self-channels in the interaction region decreased. As the UV pulses could be tightly self-guided in filaments of smaller core diameters, only quite a few plasma self-channels were observed in the UV filament overlapping region. Figure 10.25d–f shows the detected fluorescence at three different crossing angles $\theta = 2.17°$, $1.52°$, and $1.27°$, with the corresponding measured fluorescence profiles plotted in Fig. 10.25a–c. Two bright plasma self-channels were observed for $\theta = 1.52°$. Interestingly, a single intense UV filament with a core diameter of 5 μm was produced for $\theta = 1.27°$, as shown in Fig. 10.25f. At sufficiently small crossing angles, the UV pulse interference produced only a few intensity interference peaks in the overlapping region, while UV femtosecond pulses exhibited a large multiphoton ionization probability that caused a serious plasma defocusing. The filament interaction could thus be controlled to have only one central intensity interference peak to sustain filamentary propagation while filamentation was suppressed for all side interference peaks. Accordingly, laser energies were attracted to the central interference peak that assisted filament coalescence into a single filament (Fig. 10.25c, f) [31, 62].

10.12 Conclusions

In summary, non-collinear filament interaction in air was demonstrated to generate plasma lattices of unique wavelength-scale plasma microstructures and periodic density modulation, as a result of Kerr self-focusing at local intensity peaks of the constructive interference followed by its counterbalance with local plasma defocusing. The observed plasma lattices could function as efficient plasma waveguides and plasma gratings that sustained high-peak intensities, fascinating a growing research highlight and stimulating a variety of interesting applications in the field of high-intensity laser physics. We review here some of our recent experimental results based on the periodic plasma density modulation induced by nonlinear interaction of filaments. Our main results are summarized as follows. (a) The formation of self-guided periodic wavelength-scale channels was evidenced by the direct observation of the filament interaction, which revealed wavelength-scale spatial periodicities dependent on the crossing angles and intensity ratios between the non-collinearly overlapped intense femtosecond filaments. We showed that the periodic plasma density modulation could be used to guide the input pulses equivalently as a photonic-crystal waveguide, which enables attractive applications in ultrahigh intense laser optics. (b) Plasma grating assisted diffraction and efficient energy transfer of a delayed second harmonic pulse from one filament to the other were observed in air. The grating dependences on the intensity ratios and relative polarizations of the interacting filaments were studied in details. We anticipate important applications of plasma grating in the field of laser plasma interactions and ultra-intense laser optics. (c) Significant third harmonic enhancement was observed as a result of elongation of laser filament induced by plasma lattices. The experimental results may stimulate further studies for filamentation nonlinear optics, and pave a way for efficient generation of bright ultrashort ultraviolet pulses. (d) UV filaments were coupled to induce interesting fusion and coalesce into a single intense UV filament with a quite small diameter. This could be considered as a new method to control the UV femtosecond filamentation in air. As one of the most significant and interesting phenomena, filament interaction induced periodic plasma microstructures still have plentiful mysterious features and applications, and further experimental and theoretical investigation of the plasma interaction are necessary to understand the dynamics of the involved nonlinear interaction. We expect that such kind of plasma photonic crystal like periodic structure would be vastly applied in the field of ultra-intense laser physics, chemistry, and material science, such as to stretch, compress, and control the propagation of ultra-intense laser pulses, to trap atoms and molecules for chemical reaction in plasma waveguides, to photo-dissociate molecules, to accelerate electrons and ions in the plasma channels, and so on.

Acknowledgements This work was partly supported by National Natural Science Fund (10525416 and 10804032) and National key project for basic research (2011CB808105). Contributions from Heping Zeng's group members and students (Xuan Yang, Yongdong Wang, Liping Shi, Shuai Yuan, Jingxin Ding, Haifeng Pan, Wenxue Li, Jian Wu, and liang'en Ding) are greatly appreciated.

References

1. E. Zeek, K. Maginnis, S. Backus, U. Russek, M. Murnane, G. Mourou, H. Kapteyn, G. Vdovin, Pulse compression by use of deformable mirrors. Opt. Lett. **24**, 493 (1999)
2. C.G. Durfee, A.R. Rundquist, S. Backus, C. Herne, M.M. Murnane, H.C. Kapteyn, Phase matching of high-order harmonics in hollow waveguides. Phys. Rev. Lett. **83**, 2187 (1999)
3. C.G.R. Geddes, C. Toth, J. van Tilborg, E. Esarey, C.B. Schroeder, J. Cary, W.P. Leemans, Guiding of relativistic laser pulses by preformed plasma channels. Phys. Rev. Lett. **95**, 145002 (2005)
4. S.Y. Chen, G.S. Sarkisov, A. Maksimchuk, R. Wagner, D. Umstadter, Evolution of a plasma waveguide created during relativistic-ponderomotive self-channeling of an intense laser pulse. Phys. Rev. Lett. **80**, 2610 (1998)
5. F. Théberge, N. Akozbek, W.W. Liu, A. Becker, S.L. Chin, Tunable ultrashort laser pulses generated through filamentation in gases. Phys. Rev. Lett. **97**, 023904 (2006)
6. N. Aközbek, A. Iwasaki, A. Becker, M. Scalora, S.L. Chin, C.M. Bowden, Third-harmonic generation and self-channeling in air using high-power femtosecond laser pulses. Phys. Rev. Lett. **89**, 143901 (2002)
7. J. Kasparian, R. Sauerbrey, D. Mondelain, S. Niedermeier, J. Yu, J.P. Wolf, Y.B. Andre, M. Franco, B. Prade, S. Tzortzakis, A. Mysyrowicz, M. Rodriguez, H. Wille, L. Woste, Infrared extension of the supercontinuum generated by femtosecond terawatt laser pulses propagating in the atmosphere. Opt. Lett. **25**, 1397 (2000)
8. L. Bergé, S. Skupin, Few-cycle light bullets created by femtosecond filaments. Phys. Rev. Lett. **100**, 113902 (2008)
9. A. Couairon, M. Franco, A. Mysyrowicz, J. Biegert, U. Keller, Pulse self-compression to the single-cycle limit by filamentation in a gas with a pressure gradient. Opt. Lett. **30**, 2657 (2005)
10. V. Kumarappan, K.Y. Kim, H.M. Milchberg, Guiding of intense laser pulses in plasma waveguides produced from efficient, femtosecond end-pumped heating of clustered gases. Phys. Rev. Lett. **94**, 205004 (2005)
11. J. Kasparian, R. Sauerbrey, S.L. Chin, The critical laser intensity of self-guided light filaments in air. Appl. Phys. B. **71**, 877 (2000)
12. L. Bergé, S. Skupin, F. Lederer, G. Mejean, J. Yu, J. Kasparian, E. Salmon, J.P. Wolf, M. Rodriguez, L. Woste, R. Bourayou, R. Sauerbrey, Multiple filamentation of terawatt laser pulses in air. Phys. Rev. Lett. **92**, 225002 (2004)
13. L. Bergé, M.R. Schmidt, J.J. Rasmussen, P.L. Christiansen, K.Ø. Rasmussen, Amalgamation of interacting light beamlets in Kerr-type media. J. Opt. Soc. Am. B **14**, 2550 (1997)
14. W. Królikowski, S.A. Holmstrom, Fusion and birth of spatial solitons upon collision. Opt. Lett. **22**, 369 (1997)
15. V. Antoine, L. Bergé, Femtosecond optical vortices in air. Phys. Rev. Lett. **95**, 193901 (2005)
16. M.-F. Shih, M. Segev, G. Salamo, Three-dimensional spiraling of interacting spatial solitons. Phys. Rev. Lett. **78**, 2551 (1997)
17. T.T. Xi, X. Lu, J. Zhang, Interaction of light filaments generated by femtosecond laser pulses in air. Phys. Rev. Lett. **96**, 025003 (2006)
18. J. Wu, H. Cai, Y. Tong, H. Zeng, Measurement of field-free molecular alignment by cross-defocusing assisted polarization spectroscopy. Opt. Express **17**, 16300 (2009)
19. Y. Feng, H. Pan, J. Liu, C. Chen, J. Wu, H. Zeng, Direct measurement of field-free molecular alignment by spatial (de)focusing effects. Opt. Express **19**, 2852 (2011)
20. H. Cai, J. Wu, X. Bai, H. Pan, H. Zeng, Molecular-alignment-assisted high-energy supercontinuum pulse generation in air. Opt. Lett. **35**, 49 (2010)
21. J. Wu, H. Cai, H. Zeng, A. Couairon, Femtosecond filamentation and pulse compression in the wake of molecular alignment. Opt. Lett. **33**, 2593 (2008)
22. H. Cai, J. Wu, H. Li, X. Bai, H. Zeng, Elongation of femtosecond filament by molecular alignment in air. Opt. Express **17**, 21060 (2009)

23. J. Wu, H. Cai, A. Couairon, H. Zeng, Wavelength tuning of a few-cycle laser pulse by molecular alignment in femtosecond filamentation wake. Phys. Rev. A **79**, 063812 (2009)
24. J. Liu, Y. Feng, H. Li, P. Lu, H. Pan, J. Wu, H. Zeng, Supercontinuum pulse measurement by molecular alignment based cross-correlation frequency resolved optical gating. Opt. Express **19**, 40 (2011)
25. P. Lu, J. Liu, H. Li, H. Pan, J. Wu, H. Zeng, Cross-correlation frequency-resolved optical gating by molecular alignment for ultraviolet femtosecond pulse measurement. Appl. Phys. Lett. **97**, 3478008 (2010)
26. H. Cai, J. Wu, P. Lu, X. Bai, L. Ding, H. Zeng, Attraction and repulsion of parallel femtosecond filaments in air. Phys. Rev. A **80**, 051802 (2009)
27. J. Wu, H. Cai, P. Lu, X. Bai, L. Ding, H. Zeng, Intense ultrafast light kick by rotational Raman wake in atmosphere. Appl. Phys. Lett. **95**, 221502 (2009)
28. P. Russell, Photonic crystal fibers. Science **299**, 358 (2003)
29. J.S. Foresi, P.R. Villeneuve, J. Ferrera, E.R. Thoen, G. Steinmeyer, S. Fan, J.D. Joannopoulos, L.C. Kimerling, H.I. Smith, E.P. Ippen, Photonic-bandgap microcavities in optical waveguides. Nature **390**, 143 (1997)
30. D.G. Ouzounov, F.R. Ahmad, D. Müller, N. Venkataraman, M.T. Gallagher, M.G. Thomas, J. Silcox, K.W. Koch, A.L. Gaeta, Generation of megawatt optical solitons in hollow-core photonic band-gap fibers. Science **301**, 1702 (2003)
31. J.W. Fleischer, M. Segev, N.K. Efremidis, D.N. Christodoulides, Observation of two-dimensional discrete solitons in optically induced nonlinear photonic lattices. Nature **422**, 147 (2003)
32. H.J. Eichler, P. Kuemmel, S. Orlic, A. Wappelt, High-density disk storage by multiplexed microholograms. IEEE J. Sel. Top. Quant. Electron. **4**, 840 (1998)
33. D.C. Meisel, M. Wegener, K. Busch, Three-dimensional photonic crystals by holographic lithography using the umbrella configuration: Symmetries and complete photonic band gaps. Phys. Rev. B **70**, 165104 (2004)
34. H. Zeng, J. Wu, H. Xu, K. Wu, Generation and weak beam control of two-dimensional multicolored arrays in a quadratic nonlinear medium. Phys. Rev. Lett. **96**, 083902 (2006)
35. D.G. Ouzounov, K.D. Moll, M.A. Foster, W.R. Zipfel, W.W. Webb, A.L. Gaeta, Delivery of nanojoule femtosecond pulses through large-core microstructured fibers. Opt. Lett. **27**, 1513 (2002)
36. T. Udem, R. Holzwarth, T.W. Hansch, Optical frequency metrology. Nature **416**, 233 (2002)
37. P. Lodahl, A. Floris van Driel, I.S. Nikolaev, A. Irman, K. Overgaag, D. Vanmaekelbergh, W.L. Vos, Controlling the dynamics of spontaneous emission from quantum dots by photonic crystals. Nature **430**, 654 (2004)
38. A. Mekis, J.C. Chen, I. Kurland, S. Fan, P.R. Villeneuve, J.D. Joannopoulos, High transmission through sharp bends in photonic crystal waveguides. Phys. Rev. Lett. **77**, 3787 (1996)
39. J.K. Ranka, R.S. Windeler, A.J. Stentz, Visible continuum generation in air-silica microstructure optical fibers with anomalous dispersion at 800 nm. Opt. Lett. **25**, 25 (2000)
40. X. Yang, J. Wu, Y. Peng, Y. Tong, P. Lu, L. Ding, Z. Xu, H. Zeng, Plasma waveguide array induced by filament interaction. Opt. Lett. **34**, 3806 (2009)
41. X. Yang, J. Wu, Y. Peng, Y. Tong, S. Yuan, L. Ding, Z. Xu, H. Zeng, Noncollinear interaction of femtosecond filaments with enhanced third harmonic generation in air. Appl. Phys. Lett. **95**, 111103 (2009)
42. H.L. Xu, A. Azarm, J. Bernhardt, Y. Kamali, S.L. Chin, The mechanism of nitrogen fluorescence inside a femtosecond laser filament in air. Chem. Phys. **360**, 171 (2009)
43. G.N. Gibson, R.R. Freeman, T.J. McIlrath, Dynamics of the high-intensity multiphoton ionization of N_2. Phys. Rev. Lett. **67**, 1230 (1991)
44. A. Becker, A.D. Bandrauk, S.L. Chin, S-matrix analysis of non-resonant multiphoton ionisation of inner-valence electrons of the nitrogen molecule. Chem. Phys. Lett. **343**, 345 (2001)

45. R.W.B. Pearse, A.G. Gaydon, *The Identification of Molecular Sepctra*, 4th edn. (Chapman and Hall, New York, 1984)
46. G. Rodriguez, A.R. Valenzuela, B. Yellampalle, M.J. Schmitt, K.Y. Kim, In-line holographic imaging and electron density extraction of ultrafast ionized air filaments. J. Opt. Soc. Am. B **25**, 1988 (2008)
47. M. Centurion, Y. Pu, Z.W. Liu, D. Psaltis, T.W. Hansch, Holographic recording of laser-induced plasma. Opt. Lett. **29**, 772 (2004)
48. D.M. Burland, Applications of holography in the investigation of photochemical reactions. Acc. Chem. Res. **16**, 218 (1983)
49. L. Bergé, S. Skupin, R. Nuter, J. Kasparian, J.P. Wolf, Ultrashort filaments of light in weakly ionized, optically transparent media. Rep. Prog. Phys. **70**, 1633 (2007)
50. M. Mlejnek, M. Kolesik, J.V. Moloney, E.M. Wright, Optically turbulent femtosecond light guide in air. Phys. Rev. Lett. **83**, 2938 (1999)
51. T.K. Gaylord, M.G. Moharam, Thin and thick gratings: terminology clarification. Appl. Opt. **20**, 3271 (1981)
52. C.H. Pai, S.Y. Huang, C.C. Kuo, M.W. Lin, J. Wang, S.Y. Chen, C.H. Lee, J.Y. Lin, Fabrication of spatial transient-density structures as high-field plasma photonic devices. Phys. Plasmas **12**, 070707 (2005)
53. C.C. Kuo, C.H. Pai, M.W. Lin, K.H. Lee, J.Y. Lin, J. Wang, S.Y. Chen, Enhancement of relativistic harmonic generation by an optically preformed periodic plasma waveguide. Phys. Rev. Lett. **98**, 033901 (2007)
54. X. Yang, J. Wu, Y. Tong, L. Ding, Z. Xu, H. Zeng, Femtosecond laser pulse energy transfer induced by plasma grating due to filament interaction in air. Appl. Phys. Lett. **97**, 071108 (2010)
55. K. Stelmaszczyk, P. Rohwetter, G. Méjean, J. Yu, E. Salmon, J. Kasparian, R. Ackermann, J.P. Wolf, L. Wöste, Long-distance remote laser-induced breakdown spectroscopy using filamentation in air. Appl. Phys. Lett. **85**, 3977 (2004)
56. P. Kumar, Quantum frequency conversion. Opt. Lett. **15**, 1476 (1990)
57. J. Kasparian, M. Rodriguez, G. Méjean, J. Yu, E. Salmon, H. Wille, R. Bourayou, S. Frey, Y.-B. André, A. Mysyrowicz, R. Sauerbrey, J.-P. Wolf, L. Wöste, White-light filaments for atmospheric analysis. Science **301**, 61 (2003)
58. A.E. Jailaubekov, S.E. Bradforth, Tunable 30-femtosecond pulses across the deep ultraviolet. Appl. Phys. Lett. **87**, 021107 (2005)
59. R.J. Gehr, M.W. Kimmel, A.V. Smith, Simultaneous spatial and temporal walk-off compensation in frequency-doubling femtosecond pulses in β-BaB_2O_4. Opt. Lett. **23**, 1298 (1998)
60. Y. Wang, X. Dai, J. Wu, L. Ding, H. Zeng, Spectral modulation of ultraviolet femtosecond laser pulse by molecular alignment of CO_2, O_2, and N_2. Appl. Phys. Lett. **96**, 031105 (2010)
61. Y. Wang, Y. Zhang, P. Chen, L. Shi, X. Lu, J. Wu, L. Ding, H. Zeng, The formation of an intense filament controlled by interference of ultraviolet femtosecond pulses. Appl. Phys. Lett. **98**, 111103 (2011)
62. D.N. Christodoulides, R.I. Joseph, Discrete self-focusing in nonlinear arrays of coupled waveguides. Opt. Lett. **13**, 794 (1988)

Chapter 11
The Dawn of Ultrafast Nonlinear Optics in the Terahertz Regime

F. Blanchard, L. Razzari, F.H. Su, G. Sharma, Roberto Morandotti, T. Ozaki, M. Reid, and F.A. Hegmann

11.1 Introduction

The terahertz (THz) frequency range is a specific region of the electromagnetic spectrum also known as the far-infrared (FIR) region. More precisely, THz waves cover the region from 100 GHz to 20 THz, thus bridging the gap between microwaves and infrared light. Physically, 1 THz is equivalent to a wavelength of 300 μm in vacuum, to 33.3 cm^{-1} in terms of wave numbers, to a photon energy of 4 meV, or to a temperature of 48 K. THz waves have the ability to penetrate various materials including non-metallic compounds (papers and plastics), organics, gases, and liquids, thus being a powerful tool for spectroscopic sensing [1]. This portion of the electromagnetic spectrum has been accessible for some time by various means including molecular gas lasers, gyrotrons, and free-electron lasers [2]. Due to complexity, cost,

F. Blanchard (✉)
Institute for Integrated Cell-Material Sciences (iCeMS), Kyoto University, Yoshida-Homachi, Sakyo-ku, Kyoto-shi 606-8501, Japan
e-mail: fblanchard@icems.kyoto-u.ac.jp

L. Razzari
Istituto Italiano di Tecnologia, Genova, Italy

INRS-EMT, Varennes, QC, Canada

F.H. Su
Institute of Solid State Physics, Hefei, China

G. Sharma • R. Morandotti • T. Ozaki
INRS-EMT, Université du Québec, Montreal, QC, Canada

M. Reid
Department of Physics, University of Northern British Columbia, Prince George, BC, Canada

F.A. Hegmann
Department of Physics, University of Alberta, Edmonton, AB, Canada

Z. Chen and R. Morandotti (eds.), *Nonlinear Photonics and Novel Optical Phenomena*, Springer Series in Optical Sciences 170, DOI 10.1007/978-1-4614-3538-9_11,

Table 11.1 Properties of recent high-intensity THz sources

	Large photoconductive switch [24]	Two-color plasma [13]	Optical rectification (OR) in ZnTe [16]	OR by tilted-pulse-front in $LiNbO_3$ [17]	OR by tilted-pulse-front in $LiNbO_3$ [25]	OR by tilted-pulse-front in $LiNbO_3$ [26]
Energy (μJ)	0.8	0.03	1.5	10	50	3
Peak electric field	150 kV/cm	400 kV/cm	230 kV/cm	250 kV/cm		1.0 MV/cm
Bandwidth (THz)	0.1–2	0.3–7	0.1–3	0.1–0.7	0.1–0.5	0.1–2.5
THz spot diameter (FWHM)			1.6 mm			300 μm

and limited frequencies of operation, these sources have traditionally made it difficult to gain full access to the terahertz frequency range. Nevertheless, there were several pioneering works in nonlinear FIR spectroscopy already in the early 1970s, about one decade after the advent of the laser (readers may find a review in [3]). In particular, saturated absorption in the FIR region was first studied in 1970, which led to the optically pumped FIR gas laser [4]. In the 1980s, the first demonstration of THz radiation coherently generated and detected was made. This result coincided with the development of ultrafast lasers and was obtained using a photoconductive antenna emitter [5], where photoexcited carriers induced by an ultrafast laser pulse are accelerated by a biasing electric field. The resulting time varying current $J(t)$ radiates an electromagnetic transient, $E \propto \partial J/\partial t$, whose amplitude and phase depend on various parameters such as carrier mobility, carrier lifetime, bias field, and on the impurity doping concentration [6]. This allowed the birth of coherent time-domain THz spectroscopy (TDTS) [1], which provided unprecedented insights into the nature of molecular vibrations, carrier dynamics in semiconductors, and protein kinetics [7–12]. Even with 30 years of rapid advances in the study of light–matter interactions at THz frequencies, lack of efficient emitters and sensitive detectors in this frequency range has for long time slowed down THz linear and nonlinear spectroscopy.

More recently, innovative generation schemes based on nonlinear optical processes have successfully reached sufficient THz peak power to access the fascinating world of nonlinear optics temporally resolved with subpicosecond resolution [13–20]. In Table 11.1, we show various laser-based techniques to generate high-intensity THz pulses with electric fields greater than 100 kV/cm at the focus. Notice that this table does not describe all the achievements obtained in the field, and one can find a more complete review of THz generation sources in references [21–23]. However, to date and in the context of nonlinear optics at THz frequencies, generating high-intensity THz pulses clearly points toward two promising techniques that are optical rectification and two-color plasma based sources, as shown in Table 11.1.

This book chapter is organized as follows: in Sect. 11.2 we present an overview of the THz generation and detection schemes suitable to perform nonlinear coherent TDTS with subpicosecond time resolution, and we discuss both the advantages and limitations. We then review in Sect. 11.3 a set of absorption bleaching experiments performed on an n-doped semiconductor using a THz Z-scan technique, preceded by a brief description of the recent nonlinear optical phenomena observed at THz frequencies. Finally, in Sect. 11.4, we describe a THz pump–THz probe setup that enables mapping of the conduction band structure in semiconductors.

11.2 Handling THz Waves

11.2.1 Detection Schemes

There are a number of methods used to detect FIR radiation. Thermal detection as well as coherent detection using electro-optic crystals and photoconductors can be used for detecting pulsed terahertz radiation. Traditionally, radiation in the near-infrared through the visible wavelength range at higher frequencies can be detected with great sensitivity with devices such as photomultipliers, photoconductors, and photodiodes. These devices essentially operate on the same principle: a photon has sufficient energy to generate a free electron or a charge–carrier pair, which are detected electronically. These methods work sufficiently well when the photon energies of the radiation to be detected are large enough to generate the charge carriers. The photon energy at 1 THz is about 4 meV, making it very difficult to employ such standard techniques for photon detection. At the same time, very sensitive detection of electromagnetic radiation can be achieved for very long wavelengths by purely electronic means. For example, the detection of radio waves using antennas can be very efficient. However, it becomes more and more difficult to operate antennas, as well as the corresponding electronics, at higher frequencies approaching the THz range. Therefore, a somewhat unique technology is required for detecting radiation in the THz frequency range, lying between the well-developed detection technology for the visible and the well-established detection technology for the microwave portion of the electromagnetic spectrum.

There are basically two types of detection that one would consider—direct detection and coherent detection of THz radiation. Direct detection would likely be considered with incoherent sources of THz radiation or when the THz source has a narrow frequency range. Examples of direct detection would be thermal detectors such as pyroelectric or bolometric detectors. For higher sensitivity, a nonlinear detector such as a Schottky barrier diode or a hot electron bolometer can be used in a heterodyne configuration [27]. Recently, a bolometer-type uncooled THz camera became available [28, 29] allowing direct visualization of the THz radiation beam intensity profile. For the case of coherent detection, the three primary methods are

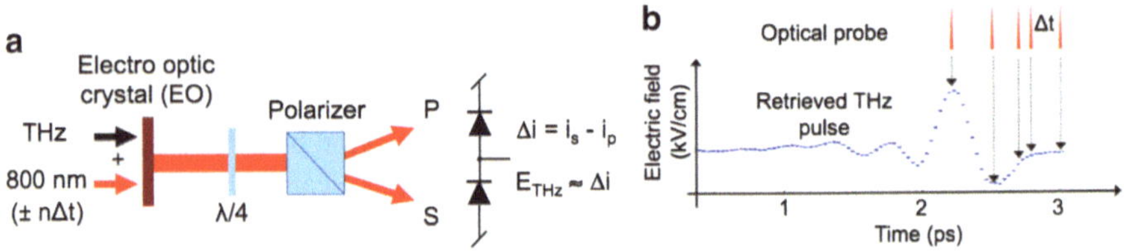

Fig. 11.1 Concept of EO sampling. (**a**) THz-electric-field-induced birefringence into EO crystal by Pockels effect. Polarizer and photodiodes convert phase modulations into amplitude modulations. (**b**) Example of a THz trace recovered by moving the optical delay stage of the probe beam line

photoconductive sampling [5], electro-optic (EO) sampling [30], and more recently laser-plasma based detection [31]. The readers may find a complete description of photoconductive sampling and free-space EO sampling in [32], and a demonstration of broadband THz detection using laser-plasma in [33]. In Fig. 11.1, we show the concept of EO sampling as first introduced by Wu and Zhang [30].

As shown in Fig. 11.1, an optical probe beam is passed co-linearly with a terahertz field inside an EO material. The terahertz field acts as a bias field for the linear electro-optic effect (Pockels effect), which induces a polarization modulation on the probe beam. The polarization modulation is measured using a Wollaston polarizer prism and a pair of photodiodes to monitor the differential polarization signal (S and P), which is balanced (i.e., equal to zero) in the absence of a terahertz field. The differential photodiode signal (Δi) is measured with a lock-in amplifier, and the terahertz signal is mapped out in time by varying the probe delay ($\pm\Delta t$) with respect to the terahertz pulse. In this configuration, a high signal-to-noise ratio (SNR) >10,000 capability is feasible, comparable to photoconductive sampling [34]. Typically, EO sampling is done at the central part of a focused THz spot that gives the peak THz electric field. Limitations in the detected bandwidth mainly arise from the choice of the EO crystal and the temporal pulse width of the sampling probe beam. For detecting THz waves, zinc telluride (ZnTe) [30] and gallium phosphide (GaP) [35] crystals are commonly used, due to the relatively good matching between the THz phase velocity and the optical group velocity at 800 nm (i.e., the emission wavelength of Ti:sapphire lasers). To estimate the peak THz-electric-field-induced birefringence into the EO crystal, the modulation ratio between the two photodiodes signals (i_s and i_p) is used [36]:

$$\frac{i_s - i_p}{i_s + i_p} = \sin\theta \approx \theta = \frac{2\pi}{\lambda} n_0^3 r_{41} E L, \tag{11.1}$$

where n_0, L, and r_{41} are the index of refraction, thickness, and the electro-optical coefficient of the sensor material, respectively. E is the peak THz electric field, λ is the probe beam wavelength, and θ is the modulation detected by the two photodiodes signals i_s and i_p.

11.2.2 Generation by Optical Rectification

Optical rectification (OR) is a well-known second-order nonlinear process that takes place in media lacking inversion symmetry. Already in the 1970s, optical rectification in a nonlinear medium was known to generate FIR light [37]. In spite of this, experimental and theoretical evidence of rectification in (110) crystal semiconductors was only demonstrated in 1994 [38]. This approach is normally more straightforward than the use of a photoconductive antenna, since no external electronic components are required. Other advantages are the wide bandwidth capability associated with optical rectification [39] when matching between the group velocity of the optical pumping wavelength and the phase velocity of the emitted THz radiation is achieved. For efficient generation of THz radiation, Hebling et al. have proposed the idea of using a Cerenkov-type scheme based on a tilted front laser pulse propagation in lithium niobate ($LiNbO_3$) [40]. In this scheme, the large refractive index mismatch between the optical pump and the THz wave can be overcome while taking advantage of the higher nonlinear coefficient of $LiNbO_3$ when compared to a standard GaP or ZnTe crystal. Thanks to this pioneering technique [40], efficiencies greater than 6×10^{-4} were demonstrated using 16 mJ of pump laser energy [17] and 30 μJ of THz energy, with a relatively low peak frequency at 0.35 THz [25, 41].

Physically speaking, optical rectification is a process in which a laser pulse that is traveling through a nonlinear crystal induces a time-dependent polarization change that radiates an electromagnetic wave. The temporal dependence of the radiated field is given by

$$E_{\mathrm{THz}}^{\mathrm{rad}} \propto \frac{\partial^2 P(t)}{\partial t^2}, \tag{11.2}$$

where $E_{\mathrm{THz}}^{\mathrm{rad}}$ is the radiated THz field and $P(t)$ is the polarization change, expressed by

$$P(t) \propto \chi^2(0; -\omega, \omega) E_{\mathrm{opt}}(-\omega) E(\omega). \tag{11.3}$$

Here E_{opt} is the electric field of the optical pump and χ^2 is the second-order electric susceptibility of the material. It is worth mentioning that the polarization of the radiated THz field depends on the crystal orientation as well as on the incident pump polarization. Ideally, the frequency content of the radiated THz field is strictly related to the bandwidth of the incident pump beam. For example, a laser pulse of duration $\tau = 100$ fs can lead to a rectification with $1/\tau \approx 10$ THz bandwidth. However, in typical experimental conditions, one can expect a limitation in the bandwidth of the radiated field due to the imperfect matching between the group velocity of the optical pump and the phase velocity of the radiated THz field, as well as to the presence of phonon absorption lines in the generating crystal.

Matching between the optical group velocity and the THz phase velocity is crucial for an efficient optical rectification as well as for the electro-optic (EO) sampling technique (often used to characterize THz optical pulses) [34]. The phase matching condition for the generation of THz radiation via optical rectification is equivalent to that of difference frequency mixing [39], and is given by:

$$\mathbf{k}\left(\omega_{\mathrm{opt}}+\Omega\right)-\mathbf{k}\left(\omega_{\mathrm{opt}}\right)=\mathbf{k}(\Omega). \tag{11.4}$$

Here $\mathbf{k}$ is the wave vector, ω_{opt} is the angular frequency of the pump beam, and Ω is one particular frequency of the radiated THz wave-packet. When using this expression, we can depict the radiated THz waves as the superposition of all the frequency content in the optical pump beam. Additionally, Ω depends on the second-order dispersion coefficient of the crystal, as well as on the pump beam spectrum and on the phase shift between the pump and the THz waves.

From (11.4), we can write:

$$\frac{k(\Omega)}{\Omega}=\left[\frac{\partial k}{\partial \omega}\right]_{\mathrm{opt}}, \tag{11.5}$$

which states that the phase velocity of the THz waves matches the group velocity of the pump beam [39]. Furthermore, we can introduce the coherence length of an optical rectification process, in which THz waves have a positive buildup [39].

$$l_c=\frac{\pi c}{\omega_{\mathrm{THz}}\left|n_{\mathrm{opt}}-\lambda_{\mathrm{opt}}\left.\frac{\mathrm{d}n_{\mathrm{opt}}}{\mathrm{d}\lambda}\right|_{\lambda_{\mathrm{opt}}}-n_{\mathrm{THz}}\right|}, \tag{11.6}$$

where c is the speed of light, n_{THz} is the refractive index at THz frequencies, n_{opt} is the refractive index of the pump beam, and λ is the wavelength of the pump beam.

Due to the reasons mentioned above, ZnTe is one of the nonlinear materials of choice for the generation and the detection of THz radiation for a collinear phase matching configuration. To better understand the details of THz emission via optical rectification, we will examine the case of a (110) ZnTe crystal. Zincblende crystals have a cubic structure with a $\bar{4}3m$ "point group," thus implying that only the electro-optic tensor components $r_{14}=r_{25}=r_{36}$ are non-zero. In order to simplify our mathematical description, a lab coordinate system, $x'y'z'$, with the $x'y'$ plane on the (110) plane of the crystal can be constructed (see Fig. 11.2 for details) [42].

We define θ as the angle between the y' direction [001] of the lab reference coordinates and the polarization vector of the optical pump beam. The generated THz electric field E_{THz} is projected in the reference plane $x'y'$, where x' is in the direction [−110], according to the formula:

$$\begin{pmatrix} E_{\mathrm{THz}_{x'}} \\ E_{\mathrm{THz}y'} \end{pmatrix} \propto \begin{pmatrix} \sin 2\theta \\ \sin^2(\theta) \end{pmatrix}. \tag{11.7}$$

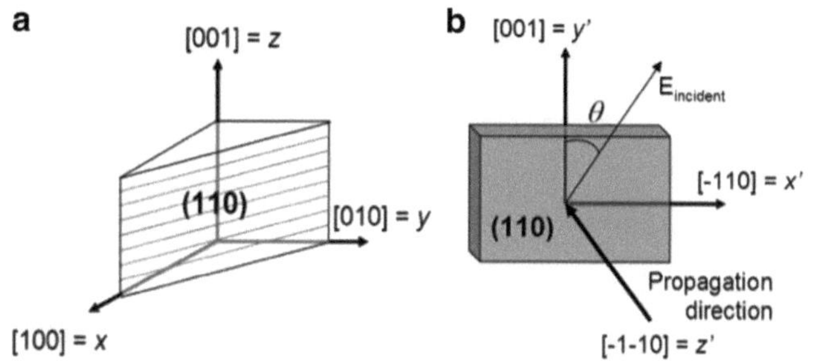

Fig. 11.2 (**a**) (110) plane in *xyz* coordinate. (**b**) Coordinate $x'y'z'$ on the (110) plane

From (11.7), the emitted THz energy is thus proportional to $\sin^2(2\theta) + \sin^4(\theta)$ [38]. According to the newly defined axis given in Fig. 11.2, the polarization orientation α of the irradiated THz waves is given by:

$$\alpha = \tan^{-1}\left(\frac{2}{\tan(\theta)}\right). \tag{11.8}$$

We notice that when $\theta = 90°$, then $\alpha = 0$ and the THz wave is polarized perpendicularly to the optical pump beam. From the above expression for the THz energy, we can easily infer that the maximum emitted THz radiation is found for $\tan(\theta) = \sqrt{2}$, so that $\alpha = \theta = 54.7°$.

The highest THz field ever reported based on optical rectification in ZnTe was obtained at the advanced laser light source (ALLS) laboratory, which is a facility allowing for multiple laser beam interactions, from X-rays to IR, with sufficient power to change significantly the properties of matter and probe its dynamics (http://lmn.emt.inrs.ca/EN/ALLS.htm). The main technology used in this facility is based on ultrafast Ti:sapphire lasers (800 nm) and the THz source is attached to a laser beam line that supplies energies as high as 70 mJ (after a vacuum compressor), delivered by ~ 40-fs-wide laser pulses at a repetition rate of 100 Hz. A schematic of the ALLS THz source is shown in Fig. 11.3 and includes three main parts: a THz generation chamber held under vacuum (~10^{-6} torr), an 800-nm probe beam line propagating in air, and a dry-nitrogen-purged section, where the THz beam travels a distance of over 2 m. The THz source has been built using nonlinear optical rectification for a collinear velocity matching configuration using a large aperture ZnTe crystal, as described and characterized in [16].

In this system, THz pulse waveforms are detected using free-space EO sampling in a second (110) ZnTe crystal of 0.5 mm thickness. The waveform could be reconstructed by using an optical delay stage (A). A lock-in amplifier connected to the output of the balanced photodiodes and referenced to the chopper was used to acquire the THz waveforms. A second optical delay line (B) is also available to set up more complex experiments requiring an auxiliary optical pump or a weak THz probe. The generated THz energy was measured using a pyroelectric detector (Coherent-Molectron J4–05) with a specified sensitivity of 2,624 V/J at 1.06 μm, which we

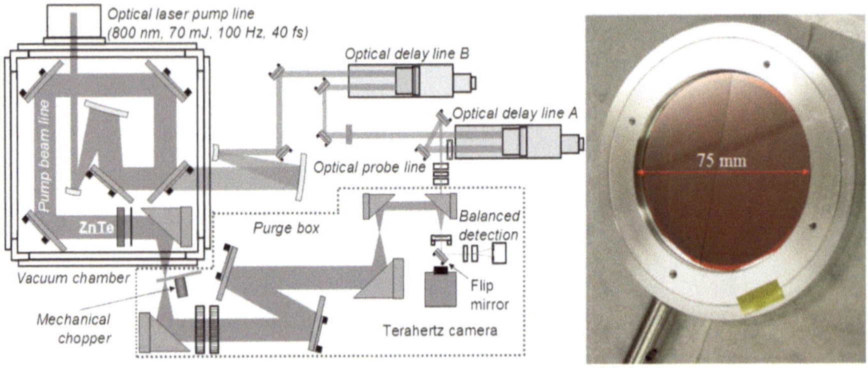

Fig. 11.3 Schematic of the experimental setup for intense THz pulses based on optical rectification in a large aperture ZnTe crystal (*right-hand side*)

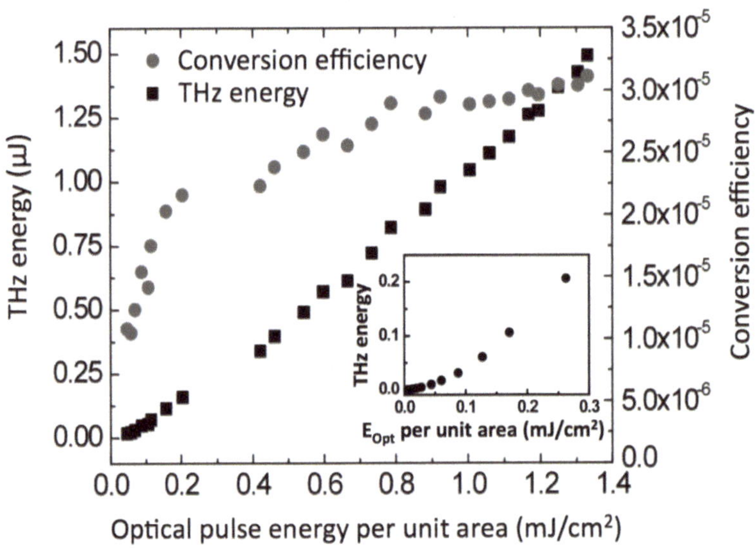

Fig. 11.4 THz pulse energy emitted from the large aperture ZnTe source as a function of the incident 800 nm laser pulse energy per unit area. The corresponding energy conversion efficiency is also shown. The *inset* depicts the magnified portion of the THz energy scaling (below 300 μJ/ cm^2 of pump intensity) obtained using a pyroelectric detector [16]

previously calibrated at THz frequencies using a second pyroelectric detector from Microtech Instruments (based on a $LiTaO_3$ crystal [16]). An interesting result obtained using the large aperture ZnTe source is the scaling of the THz energy versus laser pump energy, shown in Fig. 11.4. The pump beam irradiating the ZnTe crystal has a fluence ranging from 28 μJ/cm^2 (1 mJ) to 1.33 mJ/cm^2 (48 mJ) with a Gaussian shape and a total area of 36 cm^2. At the maximum pump energy of 48 mJ, the THz pulse energy is as high as 1.5 μJ, corresponding to an energy conversion efficiency of 3.1×10^{-5} and an average THz power of 150 μW (for a 100 Hz repetition rate pump pulse train).

The highest THz fields generated to date by laser-based sources in the 0.1–2.5 THz frequency range are achieved using optical rectification by tilted-pulse-front techniques in $LiNbO_3$ crystals [26]. As we mentioned earlier, efficient THz generation needs the group velocity of the optical pump beam to match the phase velocity of the THz wave, that is $v_{\mathrm{vis}}^{\mathrm{gr}} = v_{\mathrm{THz}}^{\mathrm{ph}}$. Many high dielectric constant materials, including $LiNbO_3$ and other ferroelectric materials offer a very high EO coefficient; however, the velocity matching of optical and THz waves in these crystals cannot be achieved collinearly. For these materials, the refractive index in the THz range is more than two times larger than that in the visible range. This problem is overcome by tilting the pulse front of the optical beam using a diffraction grating, as first proposed by Hebling et al. [40] and as subsequently demonstrated by the same group [43, 44]. In particular, recent theoretical predictions show how to fulfill the condition where the grating image coincided with the tilted-pulse-optical-pulse-front [45], and a successful experimental demonstration in a 4-f scheme [26] confirmed the importance of generating a collimated THz beam when a high power laser source is available. These dual successes now give an unprecedented tool (i.e., following the demonstration of a peak electric field as high as 1.0 MV/cm in the 0.1–2.5 THz range [26]) to continuously alter the lattice momenta of the electrons in the solid and thus, to explore their properties within the entire Brillouin zone.

11.3 High-Field Transport and Nonlinear THz Dynamics in Semiconductors

The investigation of high-frequency and high-field transport effects in semiconductors is of great interest in condensed matter physics in order to understand the behavior of fast semiconductor devices operating under extreme conditions. In the last 25 years, intense microsecond to nanosecond (and more rarely, picosecond) terahertz (THz) pulse sources in conjunction with incoherent (i.e., total energy) detection methods have been used to explore the nonlinear optical properties of semiconductors in the FIR region of the spectrum, which has enabled studies of high-field transport, band structure, and carrier–phonon interactions in semiconductors [46, 47]. For example, Mayer and Keilmann [47] first observed third harmonic generation in the FIR in doped semiconductors using 40 ns, 20 cm^{-1} pulses from a FIR laser source, providing new insight into nonlinear electron transport dynamics in semiconductors. More recently, by using coherent detection methods, the electric field of the THz pulse, rather than its intensity, has been mapped in time, giving access to both amplitude and phase information that can be used to extract the real and imaginary parts of the index of refraction (or equivalently, the complex conductivity or dielectric function) without the use of Kramers–Kronig relations. In addition, the brightness of the newly developed THz pulse sources together with the possibility of coherent detection gives orders of magnitude better sensitivity than conventional thermal sources based on bolometric detection. As mentioned previously, the rich information contained in the energy range of THz waves (0.41–12.3 meV) allows one to perform spectroscopy on numerous materials [7–12]. In particular, quasi particle scattering rates in doped semiconductors are in the THz range, and so THz pulses can be used to characterize their overall conductivities [48]. However, since significant advances in high power single-cycle THz pulse generation have occurred relatively recently, only a limited number of experiments investigating the nonlinear optical response of semiconductor materials at THz frequencies and at picosecond timescales have been reported so far. For instance, the nonlinear response of *n*-type GaAs, excited by a THz pulse with an electric field amplitude of 50 kV/cm, has been reported [14]. In addition, THz intensity dependent cross-phase modulation has been observed in electro-optic crystals, in turn leading to spectral shifting, broadening, and modulation of co-propagating laser pulses [15]. Lattice anharmonicity and self-phase modulation in $LiNbO_3$ [19] and THz-electric-field-induced impact ionization in InSb [49] have been reported using intense THz pulses. Moreover, these sources have allowed the observation of a decrease in THz absorption due to intervalley scattering in doped GaAs, Si, and Ge using THz pump–THz probe techniques [50], as well as the observation of ballistic transport of electrons in GaAs across half the Brillouin zone by time-resolved high-field THz measurements [51].

11.3.1 Nonlinear Absorption Bleaching

One of the most common and straightforward nonlinear optical characterization techniques is the open-aperture Z-scan where the transmission of a sample is measured as it passes through the focus of an intense optical beam, thus giving access to nonlinear transmission properties. Widely used in multiphoton absorption studies [52], it has been shown to be effective even for characterizing saturable absorbers [53]. Indeed, THz saturable absorbers have recently been proposed [54]. Generally, since most types of nonlinear phenomena are detectable through the Z-scan technique, one must perform open and closed aperture Z-scan to identify the origin of the transmission change (i.e., index of refraction or nonlinear absorption changes). On the other hand, when a Z-scan technique is used in combination with a THz EO sampling coherent detection scheme, as opposed to incoherent detection, the full temporal waveform evolution is available. This additional feature can directly confirm the presence, or not, of refractive index changes, simply by looking at the phase changes of the temporal waveform. In particular, it is also worth stressing that THz Z-scan gives access to the dynamics of a nonlinear process by simply measuring the change in transmission of a single THz pulse, which is usually not possible at optical frequencies. A schematic of a typical Z-scan technique is shown in Fig. 11.5.

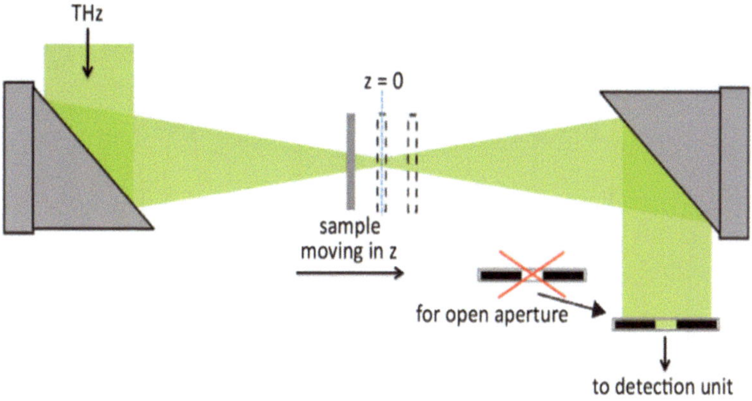

Fig. 11.5 Schematic of the THz Z-scan technique

In order to investigate free-carrier-related nonlinearities at terahertz frequencies, we can take advantage of a well-known material widely used in optoelectronics, namely, indium gallium arsenide (InGaAs). The sample used here is a 500-nm-thick *n*-type $In_{0.53}$ $Ga_{0.47}As$ epilayer (carrier concentration of approximately 2×10^{18} cm^{-3}) grown by metal oxide chemical vapor deposition on a lattice-matched, 0.5-mm-thick semi-insulating indium phosphide (InP) substrate. Its linear transmission characteristic at low THz fields shows that only 3% of the incident THz pulse energy passes through; this strong drop in transmission is mainly due to the high conductivity of the epilayer, where the InP substrate alone has an overall transmission (including absorption and reflection losses) of about 60%.

In Fig. 11.6a, the InGaAs sample is illuminated with 0.8 μJ terahertz pulses with peak electric field of around 200 kV/cm at the focus. In that first experiment, a Z-scan measurement was performed on the sample using an incoherent detection method, measuring the transmitted THz pulse energy with a standard pyroelectric detector. A significant transmission enhancement (red line, Fig. 11.6a) is observed around $z = 0$ mm (focus position) of the Z-scan. In a situation where the same technique is applied to the substrate alone, no effect is observed (as shown by the black line in Fig. 11.6a). The nonlinear transmission becomes clearer when looking at the temporal profile of the transmitted THz electric field at different *z*-positions, as shown in Fig. 11.5b. Taking into account that a linear transmission of the THz pulses is obtained for a position far from focus (i.e., $|z| > 6$ mm), Fig. 11.6b shows no significant temporal shift between each transmitted pulses, thus indicating that the imaginary part of the conductivity is not appreciably changing. Knowing that by calculating the time integral of the modulus squared of electric field, we obtain a quantity proportional to energy, one can corroborate the data obtained in Fig. 11.6a with the one measured in (b). Figure 11.6c shows the evaluation of the transmitted energy calculated using the coherent measurement shown in Fig. 11.6b, which is found to be consistent with the direct energy measurement obtained in Fig. 11.6a. Additional information can be extracted from Fig. 11.6b by looking at the

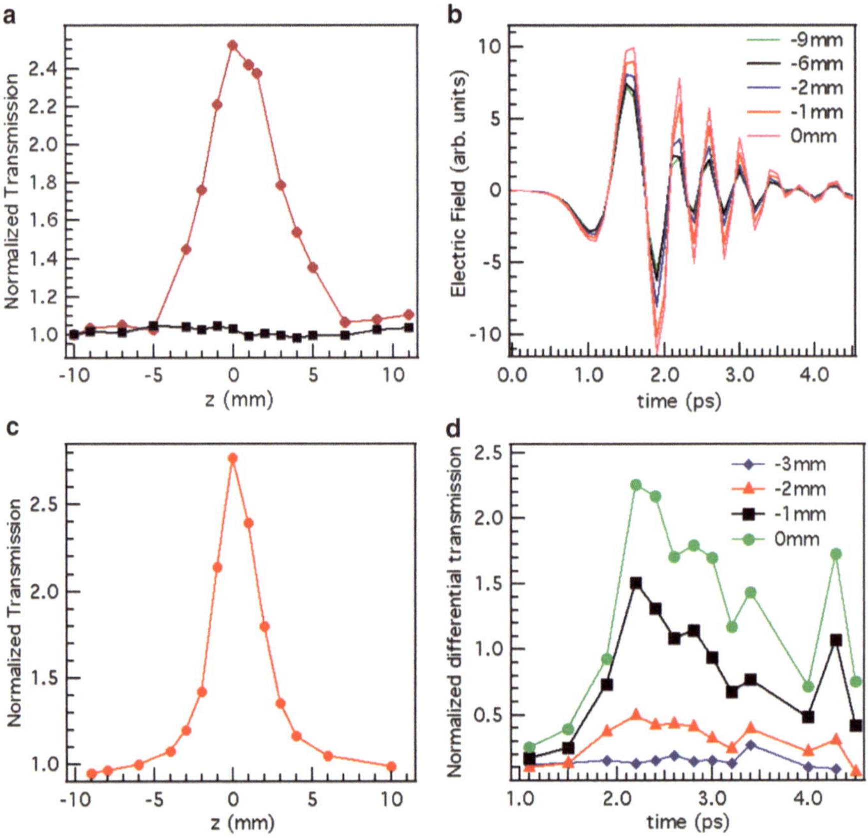

Fig. 11.6 (**a**) Z-scan normalized transmission of the total THz pulse energy measured with a pyroelectric detector after the sample (the *red line* refers to the InGaAs epilayer on an InP substrate and the *black curve* to the InP substrate only). (**b**) Transmitted THz pulse electric field for different positions of the Z-scan. (**c**) Normalized time integral associated with the modulus squared of the transmitted electric field as a function of the *z* position along the scan. (**d**) Normalized electric field differential transmission as a function of time for different *z* position along the scan (adapted from ref. [55])

transmission changes with time, showing dynamical features on a timescale comparable to the THz pulse duration. These subtle changes become more evident in Fig. 11.6d, where the normalized electric field differential transmission is plotted as a function of time for different *z* position along the scan. This operation is obtained by taking the difference between each peak electric field of the transmitted pulses near the focus position with the pulse in the linear transmission regime (i.e., $|z| > 6$ mm). The dynamics of the bleaching process shows an initial increase in transmission over a period of 1 ps (with a peak at $t = 2.2$ ps) followed by a slower decay.

A nonlinear bleaching of terahertz absorption, similar to the one discussed here, was reported some 20 years ago in a pioneering work on FIR nonlinear optics [47]; in

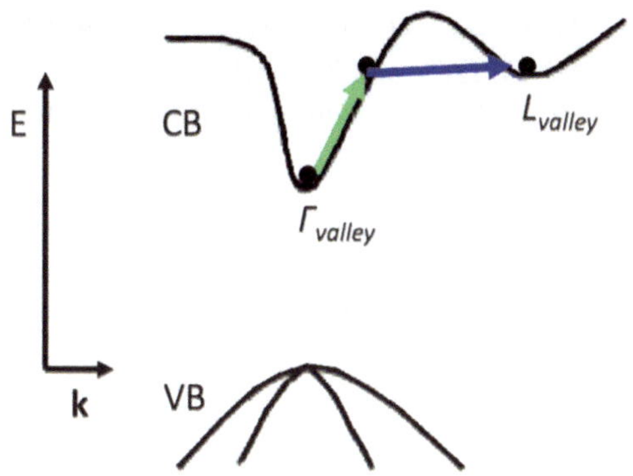

Fig. 11.7 Mechanism of intense THz-pulse-induced intervalley scattering. The electrons in the conduction band (CB) are accelerated by the THz electric field; after acquiring enough kinetic energy they may scatter into a satellite valley (*L* valley in this case)

that case, *n*-doped Germanium and Gallium Arsenide were used. The origin of the effect was tentatively attributed to an electric-field-induced scattering of carriers into upper conduction band valleys. In these valleys, electrons acquire a significantly higher effective mass, reducing the macroscopic conductivity of the sample and thus increasing its transmission. In their case, ns-long THz pulses were used in conjunction with incoherent detection, which does not allow the observation of subpicosecond dynamics. In the case of excitation by few cycle, coherently detected, picosecond THz pulses, free carriers in the Γ valley, are accelerated by the terahertz electric field during each of its oscillations. When they acquire enough kinetic energy to overcome the nearest intervalley separation, they may scatter into an upper valley where, as previously stated, the effective mass is higher and thus the transmission is enhanced. The electrons in this upper valley then scatter back to the Γ valley, so that eventually the transmission starts to drop again, with a time constant related to the intervalley relaxation time. In the particular case described above, as described in Fig. 11.7, the closest upper valley in $In_{0.53}$ $Ga_{0.47}As$ is the *L* valley ($\Delta_{\Gamma-L} = 0.55$ eV), the effective masses in the two valleys are $m_\Gamma = 0.03745 m_e$ and $m_L = 0.26 m_e$, respectively, and the *L*–Γ intervalley relaxation time is known to be about 3.1 ps [56].

11.3.2 Dynamic Intervalley-Electron-Transfer Model

In order to characterize the absorption bleaching phenomenon in doped semiconductors, a simple Drude model incorporating Γ–*L* intervalley scattering can be used to describe the temporal dynamics of the observed nonlinear THz transmission. The THz transmission of our samples, which can be idealized as a thin conducting sheet with thickness *d* on an insulating substrate with index *n*, can be expressed as follows [7]:

$$E_{\text{trans}}(t) = \frac{1}{Y_0 + Y_S}(2Y_0 E_{\text{inc}}(t) - Jd). \tag{11.9}$$

Here, E_{trans} and E_{inc} are the transmitted and incident THz fields, respectively, $Y_0 = (377\Omega)^{-1}$ and $Y_S = nY_0$ are the free-space and substrate admittances, respectively, and J is the current density in the film, which is given by:

$$J = n_\Gamma v_\Gamma e + n_L v_L e, \tag{11.10}$$

where e is the electronic charge, n_Γ and n_L are the electron densities, and v_Γ and v_L are the drift velocities of the Γ and L valley electrons, respectively. The electron velocities driven by the transmitted THz field E_{trans} and the population of electrons in the Γ valley n_Γ can be described by the dynamic equations:

$$\frac{\partial v_i}{\partial t} = \frac{eE_{\text{trans}}}{m_i^*} - \frac{v_i}{\tau_i}, i = \Gamma, L, \tag{11.11}$$

$$\frac{\mathrm{d}n_\Gamma}{\mathrm{d}t} = \frac{n_\Gamma}{\tau_{\Gamma L}} + \frac{n_L}{\tau_{L\Gamma}}, \tag{11.12}$$

$$n_0 = n_L + n_\Gamma. \tag{11.13}$$

Here, τ_Γ, τ_L and m_Γ^*, m_L^* denote the intravalley scattering times and the effective masses in the two valleys, while $\tau_{L\Gamma}^{-1}$ and $\tau_{\Gamma L}^{-1}$ are the scattering rates from one valley to another, and n_0 is the total electron density.

During the absorption bleaching process, the transmitted field E_{trans} accelerates the electrons in the conducting layer of the sample and induces the population transfer between the different valleys of the conduction band. This in turn affects the current density J in (11.9), and hence modifies the transmitted field. This feedback is responsible for the rich and surprising dynamical features associated with the effect under investigation. The change in electron populations in the Γ and L valleys is determined by the intervalley scattering rates. The L–Γ transfer rate $\tau_{L\Gamma}^{-1}$ is kept constant [57], while the Γ–L scattering rate $\tau_{\Gamma L}^{-1}$ is a function of the average kinetic energy of the electrons in the Γ valley [58]:

$$\varepsilon = \frac{1}{2} m_\Gamma^* v_\Gamma^2 + \frac{3}{2} k_B T_L. \tag{11.14}$$

This average kinetic energy is associated to the electrons in the Γ valley with k_B and T_L the Boltzmann constant and the lattice temperature, respectively. The scattering rate $\tau_{\Gamma L}^{-1}$ is zero at low energies but starts to increase rapidly at a threshold value ε_{th} to a maximum value $\tau_{\Gamma L0}^{-1}$ at high energies.

$$\tau_{\Gamma L}^{-1}(\varepsilon_\Gamma) = \begin{cases} 0, & \varepsilon \le \varepsilon_{\text{th}}(1-b) = \varepsilon_1, \\ \text{smooth function for} & \varepsilon_1 < \varepsilon_\Gamma < \varepsilon_2, \\ \tau_{\Gamma L0}^{-1}, & \varepsilon_\Gamma \ge \varepsilon_{\text{th}}(1+b) = \varepsilon_2. \end{cases} \tag{11.15}$$

Since intervalley scattering takes place via emission or absorption of optical phonons, the energy-dependent function of the valley transfer rate, $\tau_{\Gamma L}^{-1}(\varepsilon_\Gamma)$, is tentatively made "smooth." The smooth function is inserted via a seventh-order polynomial section that is continuous up to the third derivative [58]. Here, ε_{th} and b are the threshold energy and smooth width parameter, respectively.

Electric-field-driven intervalley scattering is a well-known mechanism in high-field transport physics, and it can introduce negative differential resistance and Gunn oscillations in direct bandgap semiconductors [59]. The critical DC field required to excite these phenomena in $In_{0.53}$ $Ga_{0.47}As$ is usually in the range of 2.5–4 kV/cm [60]. As one can see in Fig. 11.6, however, at THz frequencies, the effect of absorption bleaching vanishes rapidly, as the peak electric field inside the epilayer drops below 14 kV/cm (at the position $z = \pm 4$ mm in the Z-scan). This indicates that the critical field required to excite the same phenomenon can be higher than its DC counterpart. Figure 11.8 shows how the model presented above can describe both the transmitted energy and the time-domain experimental results first shown in Fig. 11.6, strongly suggesting that the intervalley scattering mechanism is responsible for the observed nonlinear absorption bleaching process (red curves are simulated data in Fig. 11.8). The fitting procedure allows one to quantify the scattering rates of heavily doped InGaAs, as studied here. The Γ–L intervalley scattering rate is found to be about 3.33×10^{13} s^{-1} ($\tau_{\Gamma L} = 30$ fs), close to the value measured in GaAs [61], while the L–Γ intervalley relaxation rate is found to be about 2.50×10^{11} s^{-1} ($\tau_{L\Gamma} = 4$ ps), similar to that found in ref. [56]. Using this model, the rapid transmission change along the Z-scan (Fig. 11.8c) can also be described.

11.4 Mapping of the Conduction Band

Traditionally, the electron effective mass is measured using cyclotron resonance (CR) [60–62]. Pioneering CR studies revealed the anisotropy of the electron effective masses in anisotropic conduction bands, such as in Si and Ge [62]. The nonparabolicity of the conduction band α was first found by CR in the 1960 and also by probing the curvature of the band while varying the excitation energy [63], and the same technique was used to study nonparabolicity in bulk $In_{0.53}$ $Ga_{0.47}As$ [64], showing an increase in the electron effective mass (often called cyclotron mass) as a function of the CR energy (up to 30 meV at 4.2 K). However, CR measurements are obtained only if the carriers are able to complete several closed orbits around the magnetic field prior to suffering collisions with phonons or impurities. Therefore, CR is mainly performed in pure crystals and at low temperature (i.e., liquid helium temperatures below 4 K). Nevertheless, dc field-induced anisotropy of the small-signal microwave conductivity at 35 GHz showed an anisotropic behavior in n-doped InSb samples at low temperatures (77 K) with respect to the polarization of the applied microwave probing field [65]. In those experiments [65], the lower microwave conductivity seen in the parallel

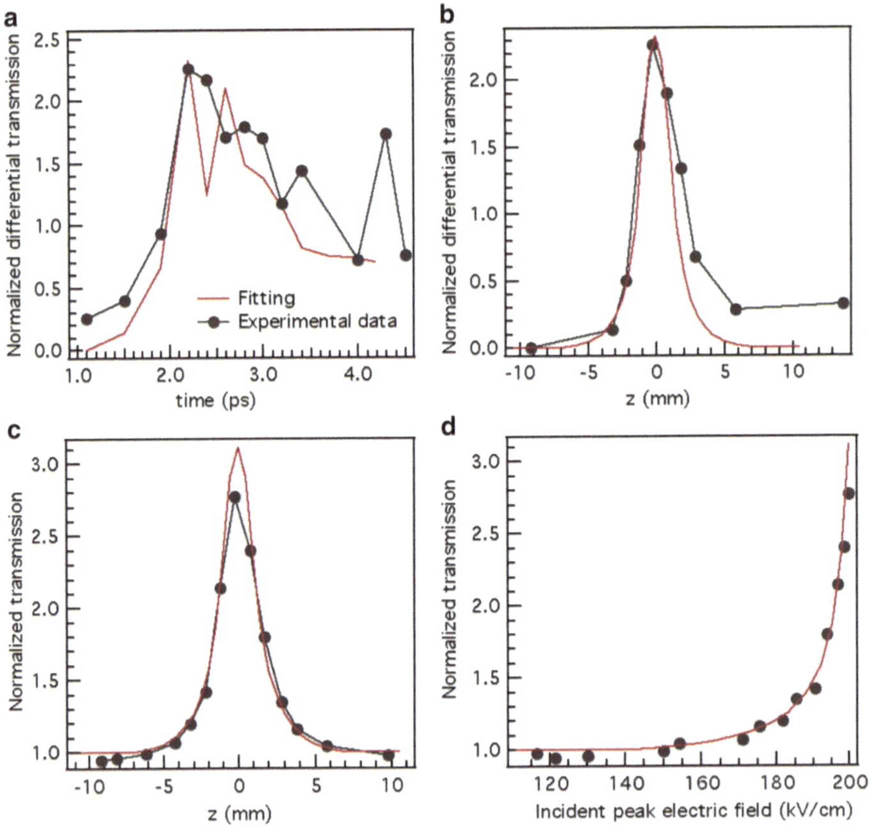

Fig. 11.8 (**a**) Normalized electric field differential transmission as a function of time at the focus of the Z-scan (*Red line*: model; *black line*: experimental data). (**b**) Peak value ($t = 2.2$ ps) of the normalized electric field differential transmission as a function of the z position along the scan. (**c**) Comparison with the data from n. (**d**) Incident electric field dependence of the normalized energy transmission (adapted from ref. [55])

configuration compared to the perpendicular case at high dc fields was attributed to an anisotropy in the intravalley scattering times due to the enhanced electron heating in the parallel direction [65, 66]; however, effects due to band nonparabolicity were not considered. Kaw et al. [67] later showed that the band nonparabolicity was important for understanding the anisotropic microwave conductivity of hot electrons in InSb, but noted a reduction in anisotropy when nonparabolic effects were included, contrary to other models [68]. Further experiments [69] at higher frequencies in the 75 GHz range revealed a microwave anisotropy in InSb at low temperatures opposite to that seen at 35 GHz. It is also important to note that the dc field in these experiments had to be pulsed in order to minimize thermal heating of the sample [69].

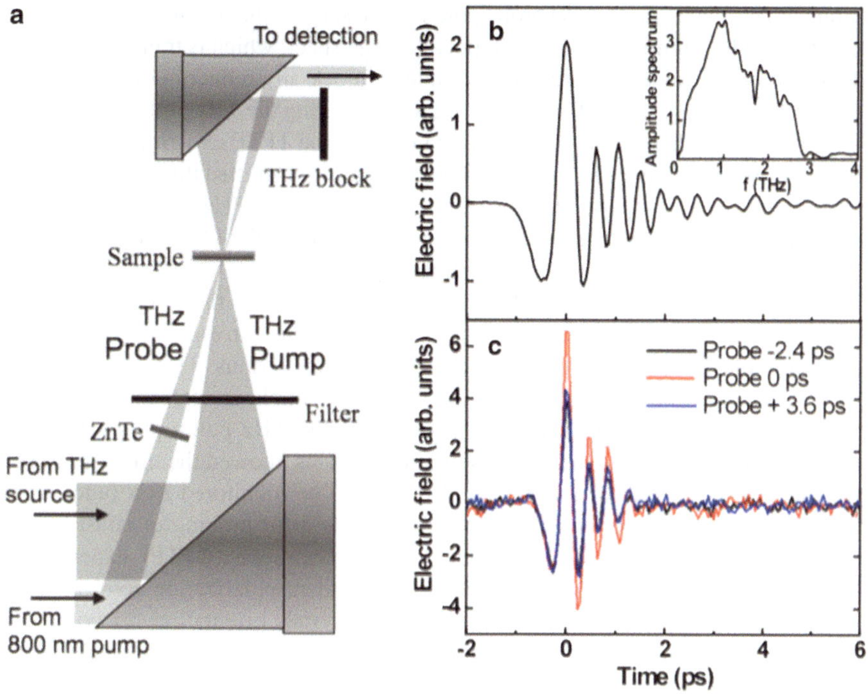

Fig. 11.9 (**a**) Schematic of the TPTP experimental setup. (**b**) Electric field profile of the terahertz pump beam emitted by the ZnTe optical rectification source. *Inset*: amplitude spectrum of the THz pump pulse. (**c**) Electric field profile of the transmitted terahertz probe beam at various times delay between the THz pump and THz probe pulses (adapted from ref. [70])

11.4.1 Polarization Dependence

A novel approach to map the conduction band nonparabolicity in heavily doped semiconductors and at room temperature has been recently proposed using a THz pump–THz probe (TPTP) scheme [70]. This methodology based on TPTP scans is similar to the TDTS technique described previously, with the addition of a THz pump pulse that excites the sample into a non-equilibrium state. In that specific case, the lock-in amplifier is synchronized to detect changes in the THz probe beam at the chopper frequency modulating the THz pump beam. This reads a voltage proportional to the electric field of the THz probe pulse. The temporal scanning between pump and probe beams is obtained by using the delay stage (B) in Fig. 11.3. The TPTP experiment is shown schematically in Fig. 11.9a. In this TPTP scheme, an additional ZnTe crystal 0.5-mm thick was used to generate the THz probe pulses transmitted through the sample and detected by free-space electro-optic sampling. Both the ZnTe crystal for the probe beam and the ZnTe crystal for detection could be rotated to produce (and detect) probe polarization states either parallel or perpendicular to the THz pump beam. A black polyethylene sheet, which is transparent to the THz radiation, was used before the sample to block any remaining 800-nm light transmitted through the ZnTe crystal used for generating the THz probe pulse. It has to be mentioned that the non-collinear geometry of the TPTP experiment presented in Fig. 11.9 allowed the THz pump and probe beams transmitted through the sample to be geometrically separated. Cross-talk between the two THz beams is therefore avoided by simply placing a metallic beam block in the path of the transmitted THz pump beam after the second off-axis parabolic mirror.

Figure 11.9b shows an example of the temporal profile of the THz pump pulses produced by the ZnTe source, and the inset shows the corresponding amplitude spectrum of the THz pump pulse. In Fig. 11.9c, various THz probe pulses waveforms transmitted through the $In_{0.53}$ $Ga_{0.47}As$ sample are presented, showing an increased transmission when overlapped with the THz pump pulse at a zero relative time delay, while the phase is unaffected. This allowed the transmission of the main positive peak of the THz probe pulse to be monitored as a function of pump–probe delay time, which is a common method for probing ultrafast carrier dynamics in semiconductors in optical pump–THz probe (OPTP) experiments [56, 71, 72]. As mentioned previously, an independent polarization control of the probe beam is obtained by changing the ZnTe crystals orientation for both THz probe emitter and detector. This additional feature allows for the detection of anisotropic behavior of samples under investigation.

Figure 11.10a is a schematic of the polarization-dependent TPTP technique and Figs. 11.10b and c are the normalized transmission of the main peak of the THz probe pulse as a function of pump–probe delay time. The presence of the THz pump pulse results in an increase in transmission of the peak electric field of the THz probe pulse. Figure 11.10b shows the transmission change for a probe polarization perpendicular to the pump beam polarization, while Fig. 11.10c shows the same measurement performed for the parallel polarization configuration. In the latter case in Fig. 11.10c, a fast, large amplitude oscillation is observed on top of a slower component similar to that shown in Fig. 11.10b. In particular, when the probe beam is blocked, no residual signal from the pump beam is detected; both beams have to be present inside the sample in order to observe the large amplitude oscillations for the parallel polarizations shown in Fig. 11.10c. Moreover, moving the sample to an off-focus position (or completely removing it from the THz beam) eliminated any modulation signal of the THz probe transmission. This would suggest that the observed signals cannot be attributed to any interference effects between the THz pump and THz probe waveforms at the detector crystal. Furthermore, the presence of small oscillations in the perpendicular polarization case, as shown in Fig. 11.10b, is inconsistent with the requirement of a parallel polarization for the interference effects. As discussed earlier for Fig. 11.9c, the lack of any significant phase shift in the transmitted THz waveform implies that the observed signal is due to the modulation of the peak amplitude of the THz probe pulse. To further corroborate this conclusion, no signal was observed as a function of the pump–probe delay time when the THz detection point was set to a zero-crossing of the THz probe waveform. The increase in

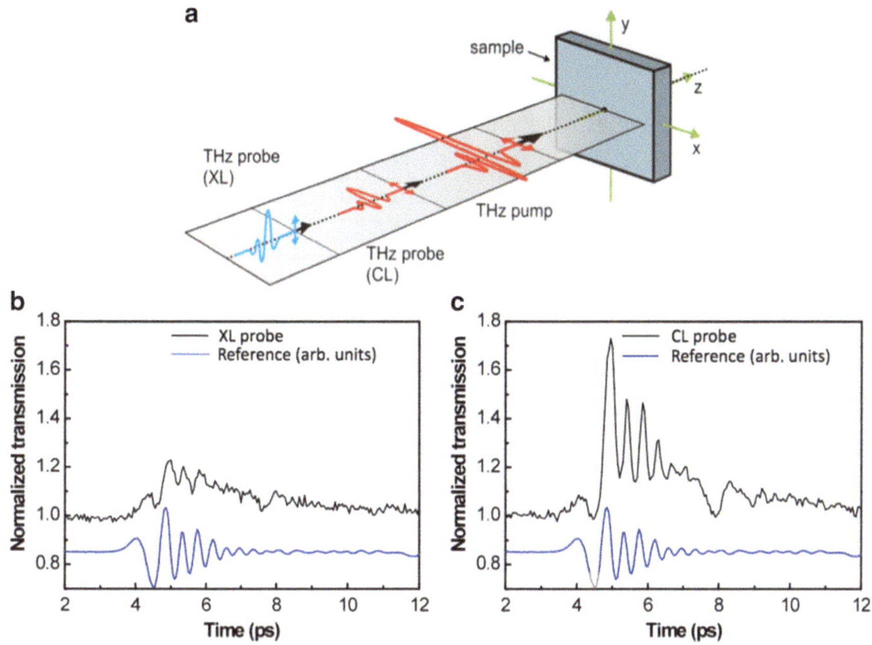

Fig. 11.10 (**a**) Polarization-dependent results for a cross- (XL) and a co- (CL) linear TPTP configuration. Normalized transmission of the THz probe peak electric field as a function of pump–probe delay time for (**b**) XL-TPTP and (**c**) CL-TPTP. The THz pump pulse waveform transmitted through the InGaAs sample (*blue line*) is also shown for reference (adapted from ref. [70])

the transmission of the THz probe beam for cross-linear (XL) polarization, as shown in Fig. 11.10b, is due to a decrease in sample conductivity arising primarily from the THz-pump-pulse-induced intervalley scattering of carriers from the high mobility Γ valley to the lower mobility L valley in InGaAs [55, 73]. As shown in Fig. 11.10b, the decay of the transient absorption bleaching signal observed for perpendicular polarizations is about 3.3 ps, which is consistent with relaxation times of about 3.1 ps for carrier scattering from the L valley back to the central Γ valley reported in the OPTP experiments performed on $In_{0.53}$ $Ga_{0.47}As$ [56]. This dynamics is also consistent with the Z-scan results discussed in the previous section.

11.4.2 Effective Mass Anisotropy

In this study, the interpretation of the anisotropic nature of the large ultrafast modulation of the THz probe transmission observed in the co-linear (CL) TPTP geometry is based on the nonparabolicity of the isotropic conduction band in InGaAs. This anisotropy is depicted in Fig. 11.11a and arises from the different

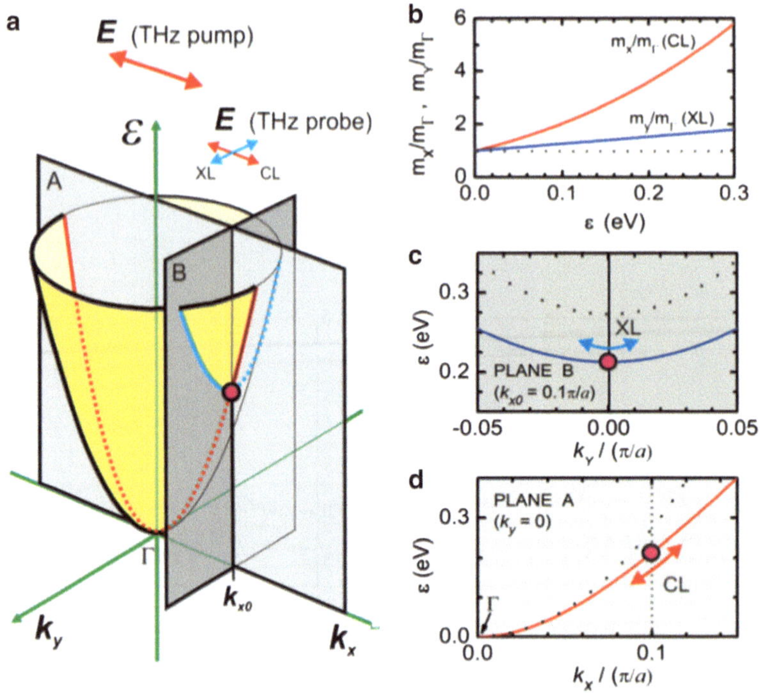

Fig. 11.11 (**a**) Schematic of an isotropic, nonparabolic band for an electron with energy ($\mathbf{k}_x$, $\mathbf{k}_y$). The electric field of the THz pump pulse is polarized along the x-direction and drives the electron high in the band to $\mathbf{k}_{x0}$, where the THz probe pulse senses an anisotropic effective mass due to different band curvatures parallel and perpendicular to the pump field direction, as given by (11.18a) and (11.18b), respectively. (**c**) Electron effective masses as a function of energy normalized to the masses at the bottom of the Γ valley for cross- (**d**) and co-linear (**e**) polarizations (adapted from ref. [70])

orientations in **k** vector that probes the effective mass for electrons high in the band. As shown in Fig. 11.11b, the energy ε of an electron in an isotropic, nonparabolic band with wave vector components $\mathbf{k}_x$ and $\mathbf{k}_y$ along the x and y directions, respectively, is given by [73]:

$$\varepsilon(1 + \alpha\varepsilon) = \frac{\hbar^2}{2m_0}\left(\mathbf{k}_x^2 + \mathbf{k}_y^2\right), \tag{11.16}$$

where α is the nonparabolicity factor of the band and m_0 is the effective mass at the bottom of the band. The polarization of the electric field of the THz pump pulse is along the $\mathbf{k}_x$-axis, such that the parallel and perpendicular TPTP configurations correspond to the electric field of the THz probe pulse polarized along the x and y directions, respectively. If the THz pump field drives an electron to a state with crystal momentum $\hbar\mathbf{k}_{x0}$ in the x-direction, then the band effective masses of the carrier at $(\mathbf{k}_x, \mathbf{k}_y) = (\mathbf{k}_{x0}, 0)$ parallel and perpendicular to the x-direction are given by:

$$m_x^{-1} = \frac{1}{\hbar^2} \frac{\mathrm{d}^2 \varepsilon}{\mathrm{d}\mathbf{k}_x{}^2} \bigg|_{\mathbf{k}_x = \mathbf{k}_{x0}, \mathbf{k}_y = 0}, \tag{11.17a}$$

$$m_y^{-1} = \frac{1}{\hbar^2} \frac{\mathrm{d}^2 \varepsilon}{\mathrm{d}\mathbf{k}_y{}^2} \bigg|_{\mathbf{k}_x = \mathbf{k}_{x0}, \mathbf{k}_y = 0}, \tag{11.17b}$$

for the two components, respectively, which can be expressed as:

$$m_x = m_0 \left(1 + \frac{2\alpha (\hbar \mathbf{k}_{x0})^2}{m_0} \right)^{3/2} = m_0 (1 + 2\alpha\varepsilon)^3, \tag{11.18a}$$

$$m_y = m_0 \left(1 + \frac{2\alpha (\hbar \mathbf{k}_{x0})^2}{m_0} \right)^{1/2} = m_0 (1 + 2\alpha\varepsilon). \tag{11.18b}$$

Figure 11.11c shows the difference in effective masses normalized to the effective mass at the bottom of the band for the perpendicular and parallel cases as a function of the electron energy according to (11.18a) and (11.18b) with $\alpha = 1.33\ \mathrm{eV}^{-1}$ [74]. As one can see, the band masses m_x and m_y are equivalent only at the bottom of the band, but $m_x > m_y$ for energies higher up in the band (see Fig. 11.11d and e). The dotted line in Fig. 11.11b–d is the band masses without the nonparabolicity factor.

The anisotropic electron dynamics induced by the strong electric field of the THz pump pulse can be explained by including the anisotropic effective mass equations (11.18a) and (11.18b) in the dynamic intervalley-electron-transfer model introduced in Sect. 11.3.2 The strong THz pump electric field modulates the electron population in the Γ valley and the electron effective mass in such a manner to drive the electron far away from the Γ point in **k** space. Both pump and probe THz electric fields drive the electron according to the Drude motion equation. However, the weak probe THz electric field is only able to provide the electron with a much lower kinetic energy than the thermal energy. Therefore, the probe THz-field-induced intervalley scattering and effective mass modulation can be neglected. The time-dependent transmission of the main peak of the THz probe pulse with respect to the time delay of the THz pump pulse at the sample can be solved numerically. The calculated peak THz probe transmission as a function of the pump–probe delay time and corresponding evolution of the Γ valley electron density are shown in Fig. 11.12 (red curves). The calculated electron velocities and corresponding normalized effective masses, m_x/m_0 and m_y/m_0, are

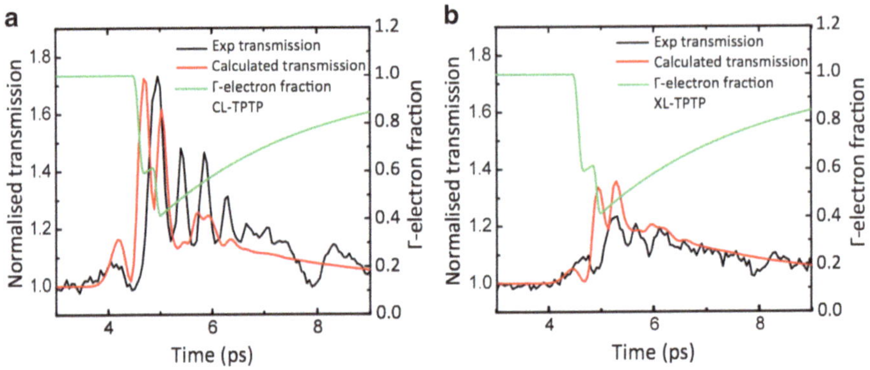

Fig. 11.12 Calculated normalized transmission of the peak THz probe field (*red lines*) and Γ valley electron fraction (*green lines*), compared with the observed THz probe transmission (*black lines*) as a function of the pump–probe delay time for (**a**) the parallel TPTP and (**b**) the perpendicular TPTP configurations, respectively

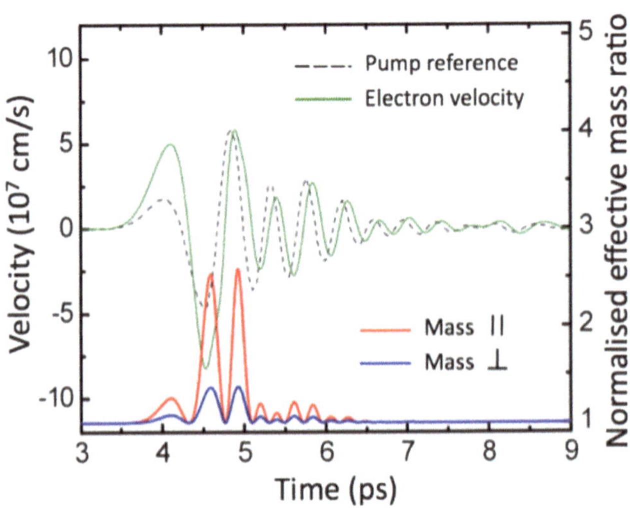

Fig. 11.13 Calculated electron velocity (*green line*) and effective mass normalized to the effective mass at the bottom of the band in the parallel (CL, *red line*) and perpendicular (XL, *blue line*) directions as a function of pump–probe delay time. The THz pump field (*dashed line*) is also shown

shown in Fig. 11.13, and Table 11.2 summarizes the best-fit parameters used in the dynamic intervalley-electron-transfer model for Z-scan and TPTP measurements. As shown in Fig. 11.12, the calculations agree with the experimental data to some extent. The fast oscillations and longer-term response observed in the parallel case, Figs. 11.10b and 11.12a, are thus due to a combination of the change in the electron

Table 11.2 Summary of best-fit parameters used in the model to describe the Z-scan and TPTP results

Parameter	Symbol	Z-scan simulations parameters	TPTP simulation parameters	Units
Nonparabolicity factor [74]	α_Γ	Γ valley: 1.33 *L* valley: 0.59	Γ valley: 1.33 *L* valley: 0.59	eV^{-1}
Γ valley scattering rate	${\tau_\Gamma}^{-1}$	1.00×10^{13}	0.56×10^{13}	s^{-1}
L valley scattering time	${\tau_L}^{-1}$	1.67×10^{13}	1.67×10^{13}	s^{-1}
Threshold energy	ε_{th}	0.13	0.25	eV
Γ–*L* intervalley scattering rate [61]	${\tau_{\Gamma L}}^{-1}$	3.33×10^{13}	3.33×10^{13}	s^{-1}
L–Γ intervalley scattering rate [56]	${\tau_{L\Gamma}}^{-1}$	2.50×10^{11}	2.50×10^{11}	s^{-1}

effective mass (fast oscillations) and the THz-pump-induced intervalley scattering (pedestal with picosecond decay time). As shown in Fig. 11.13, the model is able to reproduce the stronger oscillations seen in the parallel (CL) case compared to the perpendicular (XL) case due to the strong anisotropy in effective mass for electrons high in the band. A more rigorous theoretical approach also shows that the observed TPTP polarization anisotropy arises from effective mass anisotropy of hot carriers due to band nonparabolicity [70].

11.5 Perspectives and Conclusions

In this chapter, we have presented an introduction of THz nonlinear spectroscopy through the absorption bleaching phenomenon in an *n*-doped $In_{0.53}$ $Ga_{0.47}As$ thin film semiconductor. The absorption bleaching was first observed using a THz Z-scan technique and then confirmed using a more sophisticated technique based on a THz-pump–THz-probe experiment. The subpicosecond time resolution of this technique, coupled with the control of probe polarization, reveals the anisotropic nature of the hot electron effective mass inside an isotropic and nonparabolic conduction band. A classical model for the nonlinear conductivity at THz frequencies can be used to understand the origin of the absorption bleaching and confirms the origin of the anisotropic response of the THz probe signal, agreeing with more rigorous models. This new tool may open the way to directly mapping energy bands in semiconductors.

References

1. D. Grischkowsky, S. Keiding, M. van Exter, C. Fattinger, Far-infrared time-domain spectroscopy with terahertz beams of dielectrics and semiconductors. J. Opt. Soc. Am. B **7**, 2006 (1990)
2. P.D. Coleman, Reminiscences on selected millennium highlights in the quest for tunable terahertz-submillimeter wave oscillators. IEEE J. Sel. Top. Quant. Electron. **6**, 1000 (2000)
3. F. Keilmann, Nonlinear far-infrared spectroscopy of solids. Infrared Phys. **31**, 373 (1991)
4. T.Y. Chang, T.J. Bridges, Laser action at 452, 496, and 541 μm in optically pumped CH_3F. Opt. Commun. **1**, 423 (1970)
5. D.H. Auston, P.R. Smith, Generation and detection of millimeter waves by picosecond photonconductivity. Appl. Phys. Lett. **43**, 631 (1983)
6. J.T. Darrow, X.-C. Zhang, D.H. Auston, J.D. Morse, Saturation properties of large-aperture photoconducting antennas. IEEE J. Quant. Electron. **28**, 1607 (1992)
7. M.C. Nuss, J. Orenstein, Terahertz Time-Domain Spectroscopy, in *Millimeter and Submillimeter Wave Spectroscopy of Solids*, ed. by G. Grüner (Springer, Berlin, 1998)
8. D.M. Mittleman, *Sensing with Terahertz Radiation* (Springer, Berlin, 2003)
9. B. Ferguson, X.-C. Zhang, Materials for terahertz science and technology. Nat. Mater. **1**, 26 (2002)
10. P.U. Jepsen, D. Cooke, M. Koch, Terahertz spectroscopy and imaging—Modern techniques and applications. Laser Photon. Rev. **5**, 124 (2011)
11. R. Ulbricht, E. Hendry, J. Shan, T.F. Heinz, M. Bonn, Carrier dynamics in semiconductors studied with time-resolved terahertz spectroscopy. Rev. Mod. Phys. **83**, 543 (2011)
12. J.B. Baxter, G.W. Guglietta, Terahertz spectroscopy. Anal. Chem. **83**, 4342 (2011)
13. T. Bartel, P. Gaal, K. Reimann, M. Woerner, T. Elsaesser, Generation of single-cycle THz transients with high electric-field amplitudes. Opt. Lett. **30**, 2805 (2005)
14. P. Gaal, K. Reimann, M. Woerner, T. Elsaesser, R. Hey, K.H. Ploog, Nonlinear terahertz response of *n*-type GaAs. Phys. Rev. Lett. **96**, 187402 (2006)
15. Y. Shen, T. Watanabe, D.A. Arena, C.-C. Kao, J.B. Murphy, T.Y. Tsang, X.J. Wang, G.L. Carr, Nonlinear cross-phase modulation with intense single-cycle terahertz pulses. Phys. Rev. Lett. **99**, 043901 (2007)
16. F. Blanchard, L. Razzari, H.-C. Bandulet, G. Sharma, R. Morandotti, J.-C. Kieffer, T. Ozaki, M. Reid, H.F. Tiedje, H.K. Haugen, F.A. Hegmann, Generation of 1.5 μJ single-cycle terahertz pulses by optical rectification from a large aperture ZnTe crystal. Opt. Express **15**, 13212 (2007)
17. K.-L. Yeh, M.C. Hoffmann, J. Hebling, K.A. Nelson, Generation of 10 μJ ultrashort terahertz pulses by optical rectification. Appl. Phys. Lett. **90**, 171121 (2007)
18. J. Dai, N. Karpowicz, X.-C. Zhang, Coherent polarization control of terahertz waves generated from two-color laser-induced gas plasma. Phys. Rev. Lett. **103**, 023001 (2009)
19. J. Hebling, K.-L. Yeh, M.C. Hoffmann, K.A. Nelson, High-power THz generation, THz nonlinear optics, and THz nonlinear spectroscopy. IEEE J. Sel. Top. Quant. Electron. **14**, 345 (2008)
20. S. Leinß, T. Kampfrath, K.V. Volkmann, M. Wolf, J.T. Steiner, M. Kira, S.W. Koch, A. Leitenstorfer, R. Huber, Terahertz coherent control of optically dark paraexcitons in Cu_2O. Phys. Rev. Lett. **101**, 246401 (2008)
21. K. Reimann, Table-top sources of ultrashort THz pulses. Rep. Prog. Phys. **70**, 1597 (2007)
22. M.C. Hoffmann, J.A. Fülöp, Intense ultrashort terahertz pulses: generation and applications. J. Phys. D: Appl. Phys. **44**, 083001 (2011)
23. F. Blanchard, G. Sharma, L. Razzari, X. Ropagnol, H.-C. Bandulet, F. Vidal, R. Morandotti, J.-C. Kieffer, T. Ozaki, H.F. Tiedje, H.K. Haugen, M. Reid, F.A. Hegmann, Generation of intense THz radiation via optical methods. IEEE J. Sel. Top. Quant. Electron. **17**, 5 (2011)
24. D. You, R.R. Jones, P.H. Bucksbaum, D.R. Dykaar, Generation of high-power sub-cycle 500-fs electromagnetic pulses. Opt. Lett. **18**, 290 (1993)

25. A.G. Stepanov, S. Henin, Y. Petit, L. Bonacica, J. Kasparian, J.-P. Wolf, Mobile source of high-energy single-cycle terahertz pulses. Appl. Phys. B **101**, 11 (2010)
26. H. Hirori, A. Doi, F. Blanchard, K. Tanaka, Single-cycle terahertz pulses with amplitudes exceeding 1 MV/cm generated by optical rectification in $LiNbO_3$. Appl. Phys. Lett. **98**, 091106 (2011)
27. E. Gerecht, C.F. Musante, Y. Zhuang, K.S. Yngvesson, T. Goyette, J.C. Dickinson, J. Waldman, P.A. Yagoubov, G.N. Goltsman, B.M. Voronav, E.M. Gershenzon, NbN hot electron bolometric mixers—a new technology for low-noise THz receivers. IEEE Trans. Microw. Theory Tech. **47**, 2519 (1999)
28. N. Oda, M. Sano, H. Yoneyama, T. Sasaki, S. Kurashina, I. Hosako, N. Sekine, K. Fukunaga, Y. Ogawa, S. Komatsubara, T. Sudoh, T. Irie, H. Atake, Y. Ikeda, Development of bolometer-type uncooled THz-QVGA sensor and camera, in *34th International Conference on Infrared Millimeter and Terahertz Waves (IRMMW-THz)*, vol. 1, 2009, Busan, Korea (South), 21–25 Sep
29. L. Marchese, M. Bolduc, B. Tremblay, M. Doucet, H. Oulachgar, L. Le Noc, C. Alain, H. Jerominek, A. Bergeron, A microbolometer-based THz Imager, in *Proc. SPIE 7671*, 2010 Monday 5 April 2010, Orlando, Florida, USA
30. Q. Wu, X.-C. Zhang, Free-space electro-optic sampling of terahertz beams. Appl. Phys. Lett. **67**, 3523 (1995)
31. J. Dai, X. Xie, X.-C. Zhang, Detection of broadband terahertz waves with laser-induced plasma in gases. Phys. Rev. Lett. **97**, 103903 (2006)
32. D. Dragoman, M. Dragoman, Terahertz fields and applications. Prog. Quant. Electr. **28**, 1 (2004)
33. N. Karpowicz, J. Dai, X. Lu, Y. Chen, M. Yamaguchi, H. Zhao, X.-C. Zhang, L. Zhang, C. Zhang, M. Price-Gallagher, C. Fletcher, O. Mamer, A. Lesimple, K. Johnson, A coherent heterodyne time-domain spectrometer covering the entire terahertz gap. Appl. Phys. Lett. **92**, 011131 (2008)
34. Q. Wu, M. Litz, X.-C. Zhang, Broadband detection capability of ZnTe electro-optic field detectors. Appl. Phys. Lett. **21**, 2924 (1996)
35. Q. Wu, X.-C. Zhang, 7 terahertz broadband GaP electro-optic sensor. Appl. Phys. Lett. **70**, 1784 (1997)
36. A. Yariv, *Quantum Electronics*, 3rd edn. (Wiley, New York, 1989)
37. H.K. Yang, P.L. Richards, Y.R. Shen, Generation of far-infrared radiation by picosecond light pulses in $LiNbO_3$. Appl. Phys. Lett. **19**, 320 (1971)
38. A. Rice, Y. Jin, X.F. Ma, X.-C. Zhang, D. Bliss, J. Larkin, M. Alexander, Terahertz optical rectification from (110) zincblende crystals. Appl. Phys. Lett. **64**, 1324 (1994)
39. A. Nahata, A.S. Weling, T.F. Heinz, A wideband coherent terahertz spectroscopy system using optical rectification and electro-optic sampling. Appl. Phys. Lett. **69**, 2321 (1996)
40. J. Hebling, G. Almási, I.Z. Kozma, J. Kuhl, Velocity matching by pulse front tilting for large area THz-pulse generation. Opt. Express **10**, 1161 (2002)
41. A.G. Stepanov, L. Bonacina, S.V. Chekalin, J.-P. Wolf, Generation of 30 μJ single-cycle terahertz pulses at 100 Hz repetition rate by optical rectification. Opt. Lett. **33**, 2497 (2008)
42. A. Naaman, Generation and manipulation of THz waves, Ph.D. thesis, Oregon State University, 2006
43. A.G. Stepanov, J. Hebling, J. Kuhl, Efficient generation of subpicosecond terahertz radiation by phase-matched optical rectification using ultrashort laser pulses with tilted pulse fronts. Appl. Phys. Lett. **83**, 3000 (2003)
44. J. Hebling, A.G. Stepanov, G.A. Asi, B. Bartal, J. Kuhl, Tunable THz pulse generation by optical rectification of ultrashort laser pulses with tilted pulse fronts. Appl. Phys. B **78**, 593 (2004)
45. J.A. Fülöp, L. Pálfalvi, G. Almási, J. Hebling, Design of High-energy terahertz sources based on optical rectification. Opt. Express **18**, 12311 (2010)
46. S.D. Ganichev, W. Prettl, *Intense Terahertz Excitation of Semiconductors* (Oxford University Press, Oxford, 2006)

47. A. Mayer, F. Keilmann, Far-infrared nonlinear optics. II. $\chi(3)$ contributions from the dynamics of free carriers in semiconductors. Phys. Rev. B **33**, 6962 (1986)
48. T.I. Jeon, D. Grischkowsky, Nature of conduction in doped silicon. Phys. Rev. Lett. **78**, 1106 (1997)
49. M.C. Hoffmann, J. Hebling, H.Y. Hwang, K.-L. Yeh, K.A. Nelson, Impact ionization in InSb probed by terahertz pump-terahertz probe spectroscopy. Phys. Rev. B **79**, 161201 (2009)
50. M.C. Hoffmann, J. Hebling, H.Y. Hwang, K.-L. Yeh, K.A. Nelson, THz-pump/THz-probe spectroscopy of semiconductors at high field strengths [Invited]. J. Opt. Soc. Am. B **26**, A29 (2009)
51. W. Kuehn, P. Gaal, K. Reimann, M. Woerner, T. Elsaesser, R. Hey, Coherent ballistic motion of electrons in a periodic potential. Phys. Rev. Lett. **104**, 146602 (2010)
52. E. W. Van Stryland, M. Sheik-Bahae, *Z-Scan in Characterization Techniques and Tabulations for Organic Nonlinear Materials*. ed. by M. G. Kuzyk and C. W. Dirk (Marcel Dekker, Inc., 1998), pp. 655–692.
53. B. Gu, Y.X. Fan, J. Wang, J. Chen, J.P. Ding, H.T. Wang, B. Guo, Characterization of saturable absorbers using an open-aperture Gaussian-beam Z scan. Phys. Rev. A **73**, 065803 (2006)
54. M.C. Hoffmann, D. Turchinovich, Semiconductor saturable absorbers for ultrafast terahertz signals. Appl. Phys. Lett. **96**, 151110 (2010)
55. L. Razzari, F.H. Su, G. Sharma, F. Blanchard, A. Ayesheshim, H.C. Bandulet, R. Morandotti, J.C. Kieffer, T. Ozaki, M. Reid, F.A. Hegmann, Nonlinear ultrafast modulation of the optical absorption of intense few-cycle terahertz pulses in n-doped semiconductors. Phys. Rev. B **79**, 193204 (2009)
56. S.E. Ralph, Y. Chen, J. Woodall, D. McInturff, Subpicosecond photoconductivity of $In_{0.53}$ $Ga_{0.47}As$: Intervalley scattering rates observed via THz spectroscopy. Phys. Rev. B **54**, 5568 (1996)
57. K. Blotekjaer, Transport equations for electrons in two-valley semiconductors. IEEE Trans. Electron Dev. **17**, 38 (1970)
58. A.M. Anile, S.D. Hern, Two-valley hydrodynamical models for electron transport in gallium arsenide: simulation of Gunn oscillations. VLSI Des. **15**, 681 (2002)
59. J. Singh, *Electronic and Optoelectronic Properties of Semiconductor Structures* (Cambridge University Press, Cambridge, 2003)
60. M.A. Haase, V.M. Robbins, N. Tabatabaie, G.E. Stillman, Subthreshold electron velocity-field characteristics of GaAs and $In_{0.53}$ $Ga_{0.47}As$. J. Appl. Phys. **57**, 2295 (1985)
61. P.C. Becker, H.L. Fragnito, C.H.B. Cruz, J. Shah, R.L. Fork, J.E. Cunningham, J.E. Henry, C.V. Shank, Femtosecond intervalley scattering in GaAs. Appl. Phys. Lett. **53**, 2089 (1988)
62. G. Dresselhaus, A.F. Kip, C. Kittel, Cyclotron resonance of electrons and holes in silicon and germanium crystals. Phys. Rev. **98**, 368 (1955)
63. E.D. Palik, G.S. Picus, S. Teitler, R.F. Wallis, Infrared cyclotron resonance in InSb. Phys. Rev. **122**, 475 (1961)
64. R.J. Nicholas, C.K. Sarkar, L.C. Brunel, S. Huant, J.C. Portal, M. Razeghi, J. Chevrier, J. Massies, H.M. Cox, Shallow donor spectroscopy and polaron coupling in $Ga_{0.47}In_{0.53}As$. J. Phys. C **18**, L427 (1985)
65. K. Richter, E. Bonek, On the anisotropic microwave conductivity in the hot-electron region of n-InSb. Phys. Stat. Sol. **31**, 579 (1969)
66. D. Mukhopadhyay, B.R. Nag, Microwave incremental conductivity of InSb at 77 K in the presence of high steady electric field. Electron. Lett. **5**, 20 (1969)
67. P.K. Kaw, P.K. Dubey, A.K. Chakravarti, Hot-electron anisotropy of microwave conductivity in nonparabolic semiconductors. J. Appl. Phys. **41**, 3102 (1970)
68. M.S. Sohda, S.K. Sharma, P.K. Dubey, dc field-induced anisotropy of microwave conductivity in nonparabolic semiconductors at low temperatures. J. Appl. Phys. **42**, 5713 (1971)
69. E. Bonek, Millimeter-wave investigation of electronic conduction in semiconducting III-V compounds. J. Appl. Phys. **43**, 5101 (1972)

70. F. Blanchard, D. Golde, F.H. Su, L. Razzari, G. Sharma, R. Morandotti, T. Ozaki, M. Reid, M. Kira, S.W. Koch, F.A. Hegmann, Effective mass anisotropy of hot electrons in nonparabolic conduction bands of n-doped InGaAs films using ultrafast terahertz pump-probe techniques. Phys. Rev. Lett. **107**, 107401 (2011)
71. F.H. Su, F. Blanchard, G. Sharma, L. Razzari, A. Ayesheshim, T.L. Cocker, L.V. Titova, T. Ozaki, J.C. Kieffer, R. Morandotti, M. Reid, F.A. Hegmann, Terahertz pulse induced intervalley scattering in photoexcited GaAs. Opt. Express **17**, 9620 (2009)
72. D.G. Cooke, F.A. Hegmann, E.C. Young, T. Tiedje, Electron mobility in dilute GaAs bismide and nitride alloys measured by time-resolved terahertz spectroscopy. Appl. Phys. Lett. **89**, 122103 (2006)
73. M. Lundstrom, *Fundamentals of Carrier Transport*, 2 edition (Cambridge University Press, 2 edition November 20, 2000), p. 14
74. S.R. Ahmed, B.R. Nag, M. Deb Roy, Hot-electron transport in $In_{0.53}$ $Ga_{0.47}As$. Solid State Electron. **28**, 1193 (1985)

Chapter 12
Generation of Terahertz Radiation via Purcell-Enhanced Nonlinear Frequency Mixing

J. Bravo-Abad and M. Soljačić

12.1 Introduction

As illustrated throughout this book, nonlinear micro- and nanostructured material systems offer unique fundamental ways for tailoring a variety of nonlinear optical processes in a broad spectrum of frequency regimes. In this chapter, we show how the combination of concepts associated with nonlinear optics and those usually ascribed to quantum optics enable identifying novel routes in this general endeavor towards the ultimate control of nonlinear optical phenomena.

The discovery of the quantum Purcell effect more than 60 years ago [1], opened up the fascinating possibility of manipulating, almost at will, the rate of spontaneous emission (SE) of a quantum light emitter by modifying the electromagnetic (EM) density of states of the environment in which the emitter is embedded. With the advent of concepts such as photonic crystals (PhCs) [2–4] and the rapid development of improved nanofabrication techniques, the ability to control the SE of atoms, molecules, or quantum dots, has become of great importance for a broad spectrum of important applications in fields as diverse as illumination, biological and chemical sensing, harvesting of solar energy, and communications [5]. A very promising route that arises in this context consists in revisiting the original inspiration of these concepts in the context of classical electrodynamics [6, 7] and applying them to the realm of nonlinear optics.

In particular, the application of a classical analog of SE enhancement of a radiating system inside a cavity to boost the power radiated by a classical current distribution could be greatly advantageous to solve some of the main challenges [8]

J. Bravo-Abad (✉)
Departamento de Fisica Teorica de la Materia Condensada, Universidad Autonoma de Madrid, Campus de Cantoblanco, 28049 Madrid, Spain
e-mail: jorge.bravo@uam.es

M. Soljačić
Physics Department, MIT, 77 Massachusetts Avenue, Cambridge, MA 02139, USA

Z. Chen and R. Morandotti (eds.), *Nonlinear Photonics and Novel Optical Phenomena*, Springer Series in Optical Sciences 170, DOI 10.1007/978-1-4614-3538-9_12,

arising in the path towards the convergence of photonics and electronics. One of the most remarkable illustrations of this concept resides in the enhancement of the conversion efficiency of nonlinear frequency-mixing processes whose conversion efficiency is intrinsically very low. Of special interest due to its considerable importance for applications [9, 10], is the case of efficient terahertz (THz) generation. In particular, the application of these general principles to develop a compact and powerful source of coherent THz radiation able to operate efficiently at room temperature (RT) shows great promise for a truly practical implementation of wide spectrum of applications in areas so diverse as medicine [11], sensing [12], spectroscopy [13] or security screening [14]. Moreover, the approach described in this chapter establishes a simple alternative to other approaches requiring involved experimental setups, such as the development of room temperature THz quantum-cascade lasers [15] or those approaches requiring more intricate phase-matching setups and powerful lasers [16, 17].

This chapter is organized as follows. In Sect. 12.2, we first review the theoretical foundation of the classical analog of the quantum Purcell effect (Sect. 12.2.1). Then, we show how an approach based on a coupled-mode theory allows generalizing that classical analog to the case of a nonlinear frequency mixing process occurring inside an EM resonant cavity (Sect. 12.2.2). In Sect. 12.3, we show how the application of the general principles obtained from the proposed picture, combined with the unique properties of PhCs, enables low-power efficient THz generation at room-temperature using a compact device based on conventional material systems. Finally, in Sect. 12.4 we provide a set of conclusions for this chapter.

12.2 Theory of Purcell-Enhanced Nonlinear Frequency Mixing

12.2.1 Classical Analog of the Quantum Purcell Effect

To gain physical insight into the approach introduced in this chapter, let us briefly review the theoretical foundation of the analog between the SE process of a two-level atom in a lossless and inhomogeneous dielectric cavity and the power radiated by a classical decaying source placed in the same cavity. We first recall that the enhancement of the SE rate of a two-level atom in a dielectric lossless cavity is given by the so-called Purcell factor $F_p = (3Q/4\pi^2 V_{\text{eff}})(\lambda/n)^3$, where Q and V_{eff} are the quality factor and the effective modal volume of the cavity, respectively. λ stands for the resonant wavelength (in air) of the cavity, and n is the refractive index of the material where the emitter is located. Here, we have assumed that the source is located at the maximum electric-field position inside the cavity, that the frequency of the atomic transition coincides with the resonant frequency of the cavity, and also that there is no polarization mismatch between the electric dipole moment of the emitter and the cavity field. The SE rate (γ) of the considered two-level atom, located at $\mathbf{r} = \mathbf{r}_0$ and interacting with the EM field of the cavity, can be expressed in terms of the classical Green's function of the dielectric structure $\left(G^{\omega}_{\alpha,\beta}(\mathbf{r},\mathbf{r}')\right)$ as [18–20].

$$\gamma = -\frac{4}{hc^3}\frac{e^2}{m^2}p_{12,\alpha}p_{12,\beta}\int d\omega \frac{2\gamma_{12}}{(\omega_0-\omega)^2-\gamma_{12}^2}\mathrm{Im}\left[G^{\omega}_{\alpha,\beta}(\mathbf{r}_0,\mathbf{r}_0)\right], \tag{12.1}$$

where the subindices α and β denote the α-th and β-th Cartesian components of the corresponding vector ($\alpha, \beta = \{x, y, z\}$). $p_{12,\alpha} = \langle 1|\hat{P}|2\rangle$ is the matrix element of the electron momentum operator $\hat{P}$, ω_0 stands for the frequency of the considered atomic transition, whereas γ_{12} represents the corresponding dipole dephasing rate [21].

Next, in order to build the connection between the quantum Purcell effect and its classical counterpart, we take into account that the total energy radiated by a classical dipole current of the form $\mathbf{J}_{\omega}(\mathbf{r}) = (e/m)j(\omega)\,\delta(\mathbf{r}-\mathbf{r}_0)\mathbf{p}_{12}$, can also be written in terms of $G^{\omega}_{\alpha,\beta}(\mathbf{r},\mathbf{r}')$ as

$$E_{\mathrm{rad}} = -\frac{2\pi}{c^2}\frac{e^2}{m^2}p_{12,\alpha}p_{12,\beta}\,\omega_0\int d\omega|j(\omega)|^2\,\mathrm{Im}\left[G^{\omega}_{\alpha,\beta}(\mathbf{r}_0,\mathbf{r}_0)\right]. \tag{12.2}$$

Now, assuming $j(\omega) = 1/(i(\omega_0-\omega)+\gamma_{12})$, by direct comparison of (12.1) and (12.2) we obtain

$$\frac{\gamma_{\mathrm{cav}}}{\gamma_{\mathrm{free}}} = \frac{E_{\mathrm{rad,cav}}}{E_{\mathrm{rad,free}}}, \tag{12.3}$$

where γ_{cav} and γ_{free} stand for the SE rate of the atom in the cavity and in free space, respectively, whereas $E_{\mathrm{rad,cav}}$ and $E_{\mathrm{rad,free}}$ are the total energy radiated by $\mathbf{J}_{\omega}(\mathbf{r})$ in the cavity and free space, respectively.

Noticing that $E_{\mathrm{rad,cav}}/E_{\mathrm{rad,free}} = P_{\mathrm{rad,cav}}/P_{\mathrm{rad,free}}$ (where $P_{\mathrm{rad,cav}}$ and $P_{\mathrm{rad,free}}$ are the average powers radiated by the classical current source inside the cavity and in free space, respectively), the combination of (12.3) with the expression for the Purcell factor F_{p} given above suggests a route for extracting efficiently the power radiated by an arbitrary classical current distribution simply by optimizing ratio Q/V_{eff} characterizing the cavity in which the current is embedded in. We point out that, to our knowledge, (12.3) has been used extensively to compute SE rates of quantum light emitters embedded in complex nanostructures by means of standard computational methods developed for classical electromagnetism (such as the finite-difference-time-domain, FDTD, method), but surprisingly, there has been limited interest in using (12.3) to identify new strategies to enhance conversion efficiencies of nonlinear frequency mixing processes.

12.2.2 Coupled-Mode Theory Analysis

Our goal in this section is to present the theoretical foundation of the nonlinear coupled mode theory (CMT) used throughout this chapter.

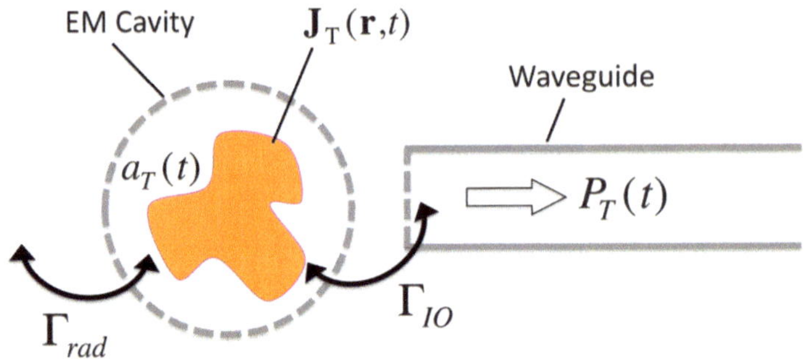

Fig. 12.1 Schematic illustration of the coupled-mode theory applied to obtain the power radiated by an arbitrary classical current distribution embedded in a single-mode cavity and asymmetrically coupled to a waveguide. The main parameters used in this coupled-mode description are also displayed in the sketch

12.2.2.1 Coupled-Mode Theory Approach to Classical Purcell Enhancement

To formulate the generalization of approach discussed above to the case of an arbitrary current distribution $\mathbf{J}_T(\mathbf{r}, t)$, we apply a coupled-mode theory analysis to this problem [22, 23]. To compute the total power emitted by $\mathbf{J}_\mathrm{T}(\mathbf{r}, t)$ inside the cavity, we assume, without loss of generality, that all the radiated power is collected by a waveguide asymmetrically coupled (through evanescent fields) to the cavity (see schematics of Fig. 12.1). If, for simplicity, we further assume that the cavity is single-mode, the electric field inside the cavity $\mathbf{E}_\mathrm{T}(\mathbf{r}, t)$ can be written as

$$\mathbf{E}_\mathrm{T}(\mathbf{r}, t) = a_\mathrm{T}(t)\exp(-i\omega_\mathrm{T}t)\frac{\mathbf{E}_\mathrm{T}^{(0)}(\mathbf{r})}{(1/2)\int d\mathbf{r}\,\varepsilon_0\, n_\mathrm{T}^2(\mathbf{r})\left|\mathbf{E}_\mathrm{T}^{(0)}(\mathbf{r})\right|^2}, \tag{12.4}$$

where $\mathbf{E}_T^{(0)}(\mathbf{r})$ is the electric field profile of the cavity mode. ω_T is the cavity resonant frequency, while $n_\mathrm{T}(\mathbf{r})$ stands for spatially dependent refractive index of the cavity at the corresponding resonance frequency. The function $a_\mathrm{T}(t)$ stands for the slowly varying electric field amplitude. The normalization of $\mathbf{E}_\mathrm{T}(\mathbf{r}, t)$ has been chosen so the time-dependent electric field energy stored in the resonant cavity mode $U(t)$ is simply given by $U(t) = |a(t)|^2$.

Now, by considering the resonant cavity mode as a CMT port [22], one can obtain the following equation of motion governing the temporal evolution of $a_\mathrm{T}(t)$.

$$\frac{\mathrm{d}a(t)}{\mathrm{d}t} = -i\omega_\mathrm{T}a(t) - (\Gamma_\mathrm{rad} + \Gamma_\mathrm{IO} + \Gamma_\mathrm{abs})a(t) + g_\mathrm{in}(t), \tag{12.5}$$

where Γ_rad, Γ_IO, and Γ_abs are the cavity decay rates due to radiative losses, decay into the waveguide, and linear absorption, respectively (the corresponding quality factors are $Q_\mathrm{rad} = \omega_\mathrm{T}/2\Gamma_\mathrm{rad}$, $Q_\mathrm{IO} = \omega_\mathrm{T}/2\Gamma_\mathrm{IO}$, and $Q_\mathrm{abs} = \omega_\mathrm{T}/2\Gamma_\mathrm{abs}$, respectively).

The function $g_{\text{in}}(t)$ is the driving term of the system. In the considered problem, it has a clear physical interpretation: it represents the power radiated by the current distribution $\mathbf{J}_{\text{T}}(\mathbf{r},t)$ inside the cavity. From the Poynting theorem and simple energy conservation arguments, one finds that $g_{\text{in}}(t)$ can be expressed as

$$g_{\text{in}}(t) = \frac{1}{4}\frac{\int d\mathbf{r}\tilde{\mathbf{J}}_{\text{T}}(\mathbf{r},t)\left[\mathbf{E}_{\text{T}}^{(0)}(\mathbf{r})\right]^*}{\sqrt{\int d\mathbf{r}\,\varepsilon_0\, n_{\text{T}}^2(\mathbf{r})\left|\mathbf{E}_{\text{T}}^{(0)}(\mathbf{r})\right|^2}}. \tag{12.6}$$

Here, we have assumed that $\mathbf{J}_{\text{T}}(\mathbf{r},t) = \tilde{\mathbf{J}}_{\text{T}}(\mathbf{r},t)\ \exp(-i\omega_{\text{T}}\, t)$ and we have also introduced the expression for $\mathbf{E}_{\text{T}}(\mathbf{r},t)$ given in (12.4).

Thus, for a given $\tilde{\mathbf{J}}_{\text{T}}(\mathbf{r},t)$ the temporal evolution of the energy inside the cavity, $|a_{\text{T}}(t)|^2$, is obtained by inserting (12.6) into (12.5) and solving the resulting first-order differential equation. Once $a_{\text{T}}(t)$ is known, the total emitted power $P_{\text{T}}(t)$ can be computed simply by using $P_{\text{T}}(t) = 2\Gamma_{\text{IO}}|a_{\text{T}}(t)|^2$.

In particular, after some straightforward algebra, the instantaneous THz emitted power $P_{\text{T,s}}$ can be expressed analytically as

$$P_{\text{T,s}} = 4\left(\frac{n_{\text{T}}}{\pi\, c\, \varepsilon_0\, \lambda_{\text{T}}^2}\right)\left(\frac{Q_{\text{T}}}{Q_{\text{IO}}}\right)\left(\frac{Q_{\text{T}}}{\tilde{V}_{\text{T}}}\right)\left|\int d\,\mathbf{r}\,\mathbf{J}_{\text{T}}(\mathbf{r},t)[\mathbf{E}^{(0)}(\mathbf{r})]^*/E_{\text{T,max}}\right|^2, \tag{12.7}$$

where Q_{T} stands for the total quality factor of the cavity. λ_{T} is the resonant wavelength in air ($\lambda_{\text{T}} = 2\pi c/\omega_{\text{T}}$), whereas $E_{\text{T,max}}$ denotes the maximum value of $|\mathbf{E}_T(\mathbf{r})|$. In (12.7), we have also defined $\tilde{V}_{\text{T}} = V_{\text{T}}/(\lambda_{\text{T}}/n_{\text{T}})^3$, where V_{T} is the effective modal volume of the resonator.

From (12.7), the Purcell enhancement of the power radiated by $\mathbf{J}_{\text{T}}(\mathbf{r},t)$ is apparent through the factor $Q_{\text{T}}/\tilde{V}_{\text{T}}$. As we discuss below, it is precisely this enhancement factor, together with the unprecedentedly large values for $Q_{\text{T}}/\tilde{V}_{\text{T}}$, that can be realized in photonic microresonators [5, 24] that enable increase of the conversion efficiency of optical nonlinear frequency conversion processes to an extent that cannot be achieved by means of any other currently known physical mechanism. Note also that the factor $Q_{\text{T}}/Q_{\text{IO}}$ appearing in (12.7) reflects the necessity to match the intrinsic absorption rate inside the cavity to the decay rate to the waveguide. This Q-matching condition permits the efficient extraction of power generated in the cavity, a general property that appears in a broad range of different contexts (thermal, EM, mechanical, etc.) that require efficient emission of energy from a high-Q system [22, 25].

12.2.2.2 Purcell-Enhanced Second-Order Nonlinear Frequency Mixing

We now turn to the case of a general nonlinear frequency mixing process involving a pump and idler frequencies (denoted hereafter by ω_1 and ω_2, respectively) in a $\chi^{(2)}$ crystal. In this case, the current distribution $\mathbf{J}_{\text{T}}(\mathbf{r},t)$ arises from the temporal variation of the nonlinear polarization vector $\mathbf{J}_{\mathrm{T}}(\mathbf{r},t) = \partial \mathbf{P}^{\mathrm{NL}}(\mathbf{r},t)/\partial t$. In particular, we focus on the case of difference-frequency generation (DFG); the generalization of this approach to other second-order nonlinear frequency mixing process is straightforward. If we assume that the electric fields ω_i (where $i = 1,2$) are given by $\mathbf{E}_i(\mathbf{r},t) = a_i(t)\exp(-i\omega_i t)\mathrm{E}_i^{(0)}(\mathbf{r})/\sqrt{U_i}$ (where U_i is the EM energy stored in the system at frequency ω_i), using (12.7), we find that the instantaneous power $P_{\mathrm{T,s}}$ emitted at the final frequency $\omega_{\mathrm{T}} = \omega_1 - \omega_2$ takes the following form

$$P_{\mathrm{T,s}} = \frac{2^4\, c_0\, \pi n_{\mathrm{T}}}{\varepsilon_0\, \lambda_{\mathrm{T}}^4} \frac{Q_{\mathrm{T}}}{Q_{\mathrm{IO}}} \frac{Q_{\mathrm{T}}}{\tilde{V}_{\mathrm{T}}} |a_1(t)|^2\, |a_2(t)|^2\, |\beta_{\mathrm{eff}}|^2, \tag{12.8}$$

where β_{eff} represents the nonlinear coupling strength between the EM fields involved in the nonlinear difference-frequency mixing. This magnitude can be written as [26]

$$\beta_{\mathrm{eff}} = \frac{\int d\mathbf{r} \sum_{i=1,2,3} \left\{ \left[E_{\mathrm{T}i}^{(0)}(\mathbf{r})\right]^* / E_{\mathrm{T,max}} \right\} \left\{ \sum_{j,k=1,2,3} \chi_{ijk}^{(2)}(\mathbf{r}) E_{1j}^{(0)}(\mathbf{r}) \left[E_{2k}^{(0)}(\mathbf{r})\right]^* \right\}}{\sqrt{\int d\mathbf{r}\, n_1^2 \left|\mathbf{E}_1^{(0)}\right|^2} \sqrt{\int d\mathbf{r}\, n_2^2(\mathbf{r}) \left|\mathbf{E}_2^{(0)}\right|^2}}, \tag{12.9}$$

where $\chi_{ijk}^{(2)}(\mathbf{r})$ stands for the spatial distribution of the second-order nonlinear susceptibility tensor (here, the subindices $\{i,j,k\}$ denote the Cartesian components $\{x,y,z\}$, respectively, of the corresponding electric field vectors).

An important figure of merit for the scheme proposed in this chapter is the enhancement factor (η_{enh}) of the output power $P_{\mathrm{T,s}}$ predicted by (12.8) with respect to the value of $P_{\mathrm{T,s}}$ that one would obtain using traditional approaches to enhance the conversion efficiency of a DFG process. In particular, we find relevant the comparison of our scheme with the case in which the nonlinear coupling coefficient (β_{eff}) is increased simply by reducing the nonlinear interaction area of the modes involved in the frequency mixing process, using for instance a waveguide for the pump, idler and final frequencies (an approach that has been used extensively in the past for enhancing the conversion efficiency of DFG in different wavelength regimes [27, 28]). For simplicity, in this comparison we assume operation in the undepleted regime [i.e., we take $a_1(t) = a_1(0)$ and $a_2(t) = a_2(0)$ in (12.8)]. Additionally, we assume that in our system, both the pump and idler fields are temporally confined, this confinement being characterized by Q-factors Q_1 and Q_2 for ω_1 and ω_2, respectively. As we discuss below for a particular potential implementation of our scheme, this temporal confinement not only permits enhancing $P_{\mathrm{T,s}}$, but also allows introducing a generalization of the canonical phase-matching condition [28], which leads to efficient DFG process even in systems in which the implementation of standard phase matching techniques (such as birefringence) is difficult or even impossible. Keeping these assumptions in mind, after some algebra, one finds that it is possible to write an accurate analytical approximation for η_{enh} as

$$\eta_{\text{enh}} \approx \frac{2^5\,\lambda_1\,\lambda_2}{\pi^3\,\lambda_T^2} \frac{n_T n_{T,\text{eff}} n_{1,\text{eff}} n_{2,\text{eff}}}{n_1^2 n_2^2} \frac{A_{\text{wg}}}{L_{\text{wg}}^2} \frac{Q_T}{Q_{\text{IO}}} \frac{Q_T}{\tilde{V}_T} Q_1 Q_2, \tag{12.10}$$

where λ_1, λ_2, and λ_T stand for the wavelengths corresponding to the pump, idler, and final frequencies, respectively. n_1, n_2, and n_T denote the value of the refractive indexes at ω_1, ω_2, and ω_T, respectively, of the dielectric medium where the frequency mixing takes place. The parameters $n_{1,\text{eff}}$, $n_{2,\text{eff}}$, and $n_{T,\text{eff}}$ represent the effective refractive indices of the waveguide configuration for the corresponding modes at ω_1, ω_2, and ω_T, respectively. A_{wg} is the transversal area of the waveguide system, whereas L_{wg} stands for the corresponding length of the waveguide. Note also that in order to make a meaningful comparison between the cavity system and its waveguide counterpart, when deriving (12.10), we have assumed that in both cases there exists an optimal overlap between $\mathbf{E}_1(\mathbf{r},t)$, $\mathbf{E}_2(\mathbf{r},t)$ and $\mathbf{E}_T(\mathbf{r},t)$. Furthermore, for this comparison we have assumed that perfect phase matching is achieved for the waveguide case.

Equation (12.10) summarizes well the comparison between the different mechanisms that play a role to enhance the conversion efficiency of a DFG process in a waveguide configuration and the scheme to enhance the conversion efficiency described in this chapter. Specifically, we have found that, as we numerically demonstrate below for a particular structure, in the case of THz generation via a $\chi^{(2)}$ DFG process in realistic nonlinear optical material configurations (in which considering the absorption losses at the final THz frequency is a key aspect in determining the ultimate conversion efficiency), it is possible to reach values $\tilde{\eta}_{\text{enh}} 10^3$. We emphasize that this large value for η_{enh} is obtained even if the corresponding phase-matching condition is satisfied in the waveguide system, which is often challenging to implement due to the vast difference between the wavelengths corresponding to the pump and idler modes and the final THz modes. Thus, we believe that the scheme discussed in this chapter could be of a paramount importance for increasing the conversion efficiency of nonlinear frequency conversion processes whose conversion efficiency is intrinsically low due for instance to the lack of a phase-matching mechanism in the considered frequency range, or due to the small value of $\chi^{(2)}$ of the materials of interest. This conclusion is the first important result of this chapter.

12.3 Practical Implementation: Efficient Thz Generation at Room Temperature

In this section we show how the general approach described in Sect. 12.2 can be implemented in practice using the unique properties of PhCs to achieve simultaneous spectral and spatial EM mode engineering. Specifically, in this section we

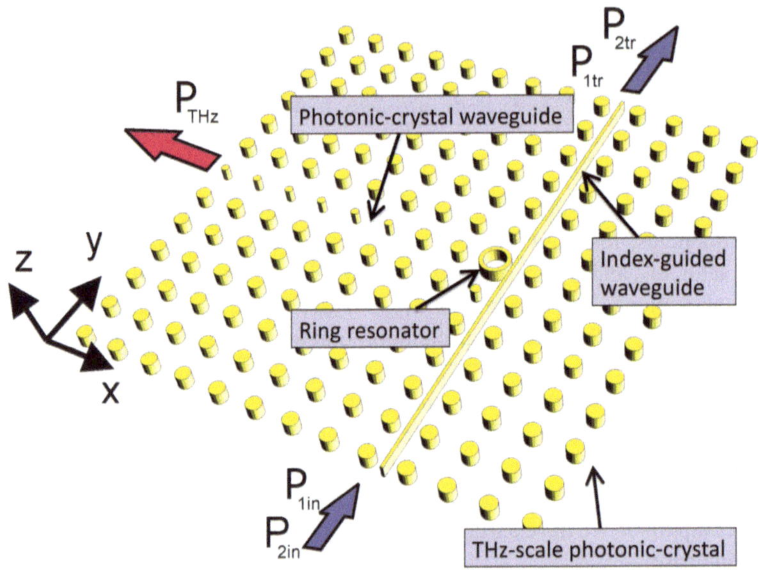

Fig. 12.2 Schematics of the proposed triply resonant photonic structure, which essentially consists of a ring resonator embedded in a THz-scale square-lattice photonic crystal. The ring resonator is side coupled to both an index-guided waveguide transporting the pump and idler powers ($P_{1\text{in}}$ and $P_{2\text{in}}$) and a THz-scale PhC waveguide, which allows to extract efficiently the THz power (P_{THz}) from the system

explore the extent to which the considered approach can be implemented in a realistic triply resonant nonlinear structure [29] that can be applied to solve the current lack of efficient sources and detectors operating at room temperature in the so-called THz frequency gap. We point out that although in this section we have analyzed the THz emission properties of a specific triply resonant structure, we have found that the approach introduced here can be implemented in a variety of different photonic structures and a variety of material systems. To illustrate this fact, similar results as those discussed in this section have been obtained for a PhC configuration different from the one presented here (see ref. [30]).

12.3.1 Triply Resonant Photonic Structure: Linear Mode Analysis

Figure 12.2 displays a schematic of the proposed triply resonant system [29]. The power carried by two NIR beams of wavelengths ω_1 and ω_2 (playing the role of idler and pump beams, respectively, their corresponding powers being $P_{1\text{in}}$ and $P_{2\text{in}}$) is coupled, by means of an index-guided waveguide, to two high-order whispering gallery modes (WGM) supported by a dielectric ring resonator. These WGM at λ_1 and λ_2 are characterized by angular momenta m_1 and m_2, respectively. The ring resonator also acts as a dipole-like defect for λ_T, when embedded in an otherwise

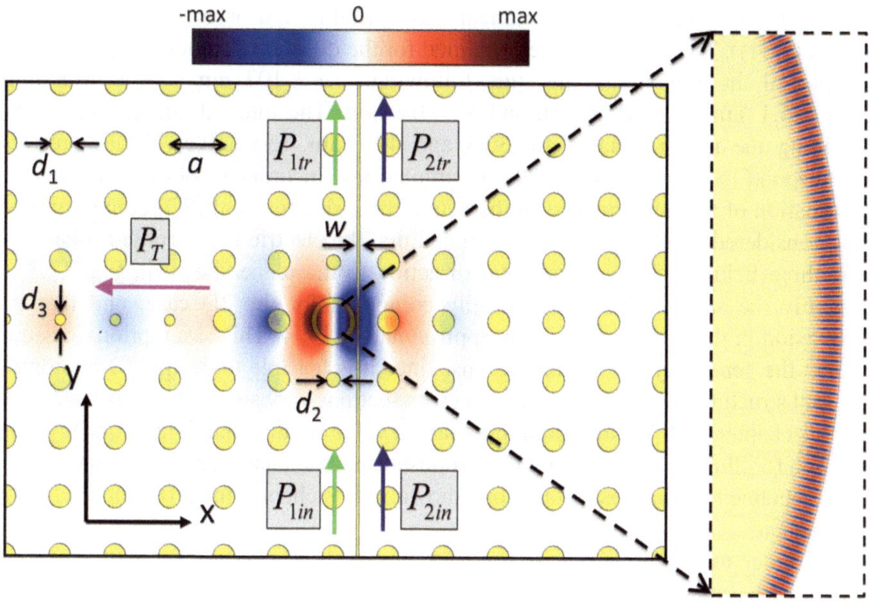

Fig. 12.3 *Left panel*: electric field profile E_z corresponding to the resonant mode appearing at 1 THz in the structure shown in Fig. 12.2. The different geometrical parameters used in the optimization of the structure are also shown. *Right panel* shows an enlarged view of the electric field profile E_x corresponding to a whispering gallery with $m_1 = 572$ circulating inside the dielectric ring shown in the *left panel*. *Shaded areas* in both the main and inset figures represent GaAs regions, while *white areas* represent air

perfectly periodic THz-wavelength scale PhC formed by a square lattice of dielectric rods (see the corresponding electric field profile in Fig. 12.3). Thus, the $\chi^{(2)}$ nonlinear frequency down-conversion interaction that takes place between the two NIR WGM's circulating inside the ring resonator yields a current distribution that radiates inside the PhC cavity at the frequency difference $\omega_T = \omega_1 - \omega_2$; the rate at which the radiation is emitted is strongly enhanced by the PhC environment in which the ring resonator is embedded. In order to extract efficiently the THz output power ($P_{T,s}$) from the PhC cavity, we introduce into the system a PhC waveguide created by reducing the radius of a row of rods (see Fig. 12.2). In addition, in order to break the degeneracy existing between the x- and y-oriented dipole defect modes, the radius of two of the nearest neighbor rods of the ring resonator is reduced with respect to the radius of the other rods in the PhC. This configuration permits having a large value for the Purcell factor ($Q_T/\tilde{V}_T$), along with a high-Q resonant confinement also for the pump and idler frequencies.

Figure 12.3 shows the structure that results from optimizing the geometrical parameters of the system for efficient generation at 1 THz. For this optimization we have assumed that the wavelengths of pump and idler beams are $\lambda_1 = 1,550$ nm and $\lambda_2 = 1,542$ nm, and that the structure is implemented in GaAs (in which the relevant component of the nonlinear susceptibility tensor is $d_{14} = 274\,\mathrm{pm/V}$, see ref. [31]). The optimal values obtained for the different geometrical parameters displayed in Fig. 12.3 are the following, $a = 102\ \mu\mathrm{m}$, $d_1 = 40.8\ \mu\mathrm{m}$, $d_2 = 25.1\ \mu\mathrm{m}$, $d_3 = 18.8\ \mu\mathrm{m}$, and $w = 0.8\ \mu\mathrm{m}$. The internal and external radii defining the ring resonator are 30.5 and 40.1 μm, respectively. These results correspond to two-dimensional (2D) simulations. In order to obtain an accurate estimation of the performance of the actual three-dimensional (3D) counterpart of the considered structure, we have assumed that the electric field profile for each of the three fields in the perpendicular direction to the plane shown in Fig. 12.3 is roughly the same as the mode profile computed for the 2D case; and that its extension in the third dimension is approximately $\lambda_T/4$. This assumption does not affect the generality of our conclusions, since it is feasible to design 3D photonic-crystal structures with 2D electric field cross sections very similar to the ones shown in this chapter [32]. We also point out that the value for the final THz frequency was chosen for illustration purposes. The proposed approach has a general character: we have obtained similar results as those shown in this chapter for different final THz frequencies.

In order to maximize the strength of the nonlinear coupling coefficient that governs the energy transfer between the pump, idler and THz fields, the whole structure must be designed so the dependence of the THz electric field profile on the azimuthal coordinate θ inside the ring cancels the modulation introduced in the nonlinear susceptibility tensor by the local variation of the pump and idler fields with respect to the axes of the nonlinear crystal [33, 34]. This modulation is given by the dependence on θ of the product $\mathbf{E}_1^{(0)}(\mathbf{r})\ \mathbf{E}_2^{(0)*}(\mathbf{r})$, which in the case of the considered WGMs is given by a factor $\exp[i\theta(m_2 - m_1 \pm 2)]$. Note that both the radial and azimuthal components of $\mathbf{E}_i^{(0)}(\mathbf{r})$ are proportional to $\exp(im_i\,\theta)$ (where $i = 1, 2$). Since, in turn, $\mathbf{P}^{\mathrm{NL}}(\mathbf{r}, t)$ is proportional to the product of the Cartesian components of $\mathbf{E}_i^{(0)}(\mathbf{r})$, a factor $\exp(\pm 2\theta)$ must be included, multiplying the phase-matching factor $\exp[i\theta(m_2 - m_1 \pm 2)]$ appearing in the integrand of the numerator in the right-hand side of (12.9). A similar discussion for the case of second-harmonic generation can be found in refs. [33, 34]. For GaAs, and for the above cited values for λ_1 and λ_2, we have found that this condition is fulfilled by two WGMs with $m_1 = 572$ and $m_2 = 575$, and a dipole defect mode ($m_T = 1$) in the THz-scale PhC.

The above discussion can be viewed as a generalization of the canonical phase-matching condition often found in nonlinear optics. Specifically, note that in standard phase-matching techniques the overall efficiency of a DFG process relies entirely on finding a suitable nonlinear material whose dispersion relation permits fulfilling simultaneously frequency and linear momentum conservation in the considered frequency mixing process. However, in the approach introduced here the linear momentum conservation is replaced by the conservation of what is effectively analogous angular momentum.

As shown above, the values of m_1, m_2, and m_T can be tailored almost at will simply by modifying the geometrical parameters that define the THz-scale PhC.

This yields a general and versatile route to phase-matching that does not depend exclusively on the intrinsic properties of naturally existing nonlinear optical materials. This fact could be particularly relevant in those systems in which the canonical phase-matching condition (derived from frequency and linear momenta conservation) cannot be fulfilled.

12.3.2 Numerical Analysis of the Nonlinear Regime

To compute accurately the nonlinear optical dynamics of the structure shown in Fig. 12.2, we have applied a temporal coupled-mode theory (TCMT) formalism similar to that described in refs. [23, 26]. In these references it is shown that this theoretical framework permits characterizing accurately several nonlinear frequency mixing processes, including those in which there exists a large difference between the wavelength corresponding to the pump and the final frequency. Standard numerical methods used in nonlinear nanophotonics (such as the nonlinear finite-difference-time-domain, FDTD, method) are not suitable for being applied to this problem, mainly due to the vast difference between the wavelength corresponding to the pump and the final THz frequency (the THz wavelength is about two orders of magnitude larger than the pump wavelength in the cases of interest).

Figure 12.4 summarizes the results obtained in the continuous-wave (cw) regime for the case in which $P_{1\text{in}} = P_{2\text{in}}$ (the dependence of the results on the ratio $P_{2\text{in}}/P_{1\text{in}}$ is discussed below) and quality factors $Q_1 = Q_2 = 3.5 \times 10^5$ and $Q_\text{T} = 10^3$. These values for Q are compatible with both the absorption coefficient of GaAs at 1 THz ($\alpha = 0.5\,\text{cm}^{-1}$) [35] and the experimental values for the quality factor obtained in similar configurations for the considered ring resonator and the PhC cavity [5, 36]. Notice that in Fig. 12.4, following ref. [26], we have defined $P_0 = \omega_1 \Gamma_1/(4|\beta_\text{eff}|^2 Q_1 Q_2 Q_\text{T})$. The physical interpretation P is the following: it corresponds to the critical power in which the maximum conversion efficiency is reached in the limit $P_{2\text{in}} \rightarrow 0$. For the particular structure analyzed in this work $P_0 = 3.17\,\text{W}$. Notice that to compute this value of P_0 for the actual 3D structure sketched in Fig. 12.2, we have assumed that the electric field profiles in the perpendicular direction to the plane shown in Fig. 12.3 are the same as those obtained in the 2D simulations, and that its extension in the third dimension for each of the three fields is approximately $\lambda_\text{T}/4$.

As seen in Fig. 12.4, for values of $P_{\text{tot,in}} > 0.07\,P_0$, the conversion efficiency (defined here as ratio between the output power at THz and total input power at NIR frequencies) starts departing from the conversion efficiency predicted by the undepleted approximation, eventually reaching the maximum value predicted by the Manley–Rowe quantum limit. Specifically, from the steady-state solution of the TCMT equations, in the considered case when $P_{1\text{in}} = P_{2\text{in}}$, one can find that the

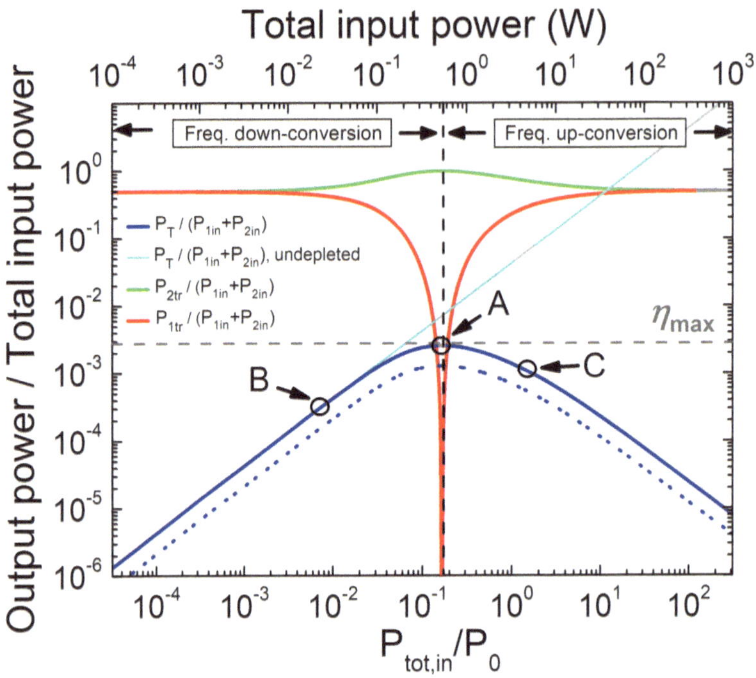

Fig. 12.4 Ratio between total output powers emitted by the system at pump, idler and final THz frequencies (ω_1, ω_2, and ω_T, respectively) versus total input power $P_{tot,in}$. $P_{1in} = P_{2in}$ is assumed in these calculations. *Horizontal dashed line* displays the maximum possible conversion efficiency η_{max} given by the Manley–Rowe quantum limit. For comparison, the conversion efficiency obtained from the undepleted-pump approximation (i.e., by assuming that pump and idler powers remain constant in the considered frequency mixing process) is also shown (see *cyan line*). *Dotted line* displays the effect of linear absorption losses on the conversion efficiency. Labels *A–C* correspond to input powers considered in the time-dependent analysis displayed in Fig. 12.5a–c

maximum conversion efficiency (i.e., the maximum possible value of the ratio $P_{T,s}/(P_{1in} + P_{2in})$) can be written as

$$\eta_{max} = \frac{\Gamma_1 \Gamma_T}{2} \left(\frac{\omega_T}{\omega_1} \right), \tag{12.11}$$

where $\Gamma_T = Q_T/Q_{T,IO}$ and $\Gamma_1 = Q_1/Q_{1,IO}$. This analytical expression is confirmed by our numerical simulations (see Fig. 12.4). As clearly shown in Fig. 12.4, at the critical value of P_{1in} at which this maximum conversion efficiency is reached (in our case $P^c_{tot,in} = 0.17P_0$, or equivalently $P^c_{tot,in} = 0.54\,W$), the pump power that is coupled to the ring resonator is completely down-converted inside the system to power at THz and idler frequencies, giving rise to a sharp minimum in P_{1tr} and a maximum in P_{2tr}. We also point out that the net effect of the absorption losses in the conversion efficiency consists simply in downscaling the results obtained in the lossless case by a factor $\Gamma_1\Gamma_T$ (see the dotted line in Fig. 12.4).

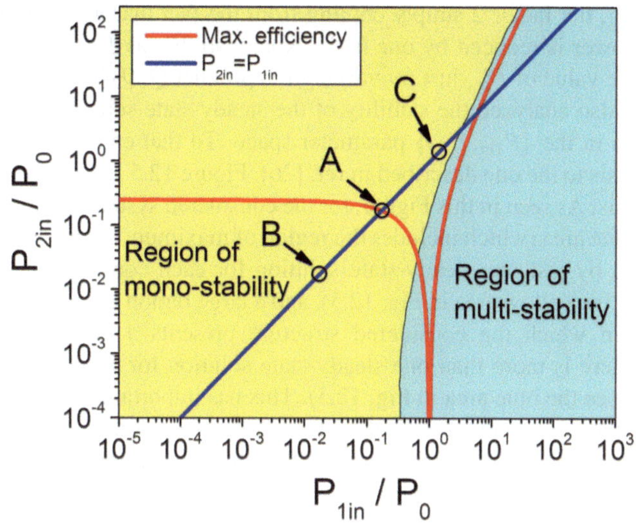

Fig. 12.5 Stability analysis of the steady-state solutions corresponding to the considered nonlinear frequency conversion process as a function of normalized powers of the pump $(P_{1\mathrm{in}})$ and the idler $(P_{2\mathrm{in}})$. *Solid line* renders the dependence between $P_{1\mathrm{in}}$ and $P_{2\mathrm{in}}$ that yields the maximum THz conversion efficiency [as defined in (12.12)] in the analyzed configuration. *Blue line* represents the case $P_{1\mathrm{in}} = P_{2\mathrm{in}}$. *Yellow* and *blue areas* represent the regions of monostability and multistability, respectively, in the space of parameters $\{P_{1\mathrm{in}}, P_{2\mathrm{in}}\}$

We turn now to the case in which $P_{1\mathrm{in}} \neq P_{2\mathrm{in}}$. As shown in ref. [30], for this case, the conversion efficiency (also defined as the maximum possible value of the ratio $P_{\mathrm{T,s}}/(P_{1\mathrm{in}} + P_{2\mathrm{in}})$), it is no longer given by (12.11), but it now depends on the specific ratio between the input pump power $P_{1\mathrm{in}}$ and total input power $P_{1\mathrm{in}} + P_{2\mathrm{in}}$. Specifically, when $P_{1\mathrm{in}} \neq P_{2\mathrm{in}}$ this maximum conversion efficiency can be written in terms of the maximum conversion efficiency $\eta_{\max}$ deduced for the case $P_{1\mathrm{in}} = P_{2\mathrm{in}}$ [see (12.11)], as [30]

$$\tilde{\eta}_{\max} = \eta_{\max} \frac{2P_{1\mathrm{in}}}{P_{1\mathrm{in}} + P_{2\mathrm{in}}} \tag{12.12}$$

In addition, from this analysis one finds that, in this case in which $P_{1\mathrm{in}} \neq P_{2\mathrm{in}}$, the critical powers at which the maximum conversion efficiency takes place ($P^{\mathrm{c}}_{1\mathrm{in}}$ and $P^{\mathrm{c}}_{2\mathrm{in}}$) satisfy the following expression

$$\frac{P^{\mathrm{c}}_{2\mathrm{in}}}{P_0} = \frac{\omega_2}{4\omega_1} \frac{\Gamma_1}{\Gamma_2} \left| 1 - \frac{P^{\mathrm{c}}_{1\mathrm{in}}}{P_0} \right|^2 \tag{12.13}$$

This function is plotted as a solid red line in Fig. 12.5. Note that in the limit $P_{1\mathrm{in}} \to 0$, the value of $\tilde{\eta}_{\max} \to 0$. On the other hand, in the limit $P_{2\mathrm{in}} \to 0$, we obtain

$\tilde{\eta}_{max} \rightarrow 2\eta_{max}$, the factor 2 simply coming from the fact that in that limit the total NIR input power is reduced by one half.[1] Note also that as $P_0 \propto 1/Q_1Q_2Q_T$, one can adjust the value of P^c_{1in} just by varying the product $Q_1Q_2Q_T$.

We have also analyzed the stability of the steady-state solutions of the considered problem in the $\{P_{1in}, P_{2in}\}$ parameter space. To that end, we have applied a similar analysis to the one described in ref. [26]. Figure 12.5 summarizes the results of this analysis. As seen in this Figure, for the considered system there are two areas of stability: one area (which includes the region of maximum conversion efficiency) characterized by just one steady-state solution for each combination $\{P_{1in}, P_{2in}\}$ (displayed as the yellow area in Fig. 12.5), and a large region of the parameter space $\{P_{2in}, P_{1in}\}$ in which the considered structure presents a multistable response, i.e., where there is more than one steady-state solution for a given pair of values $\{P_{1in}, P_{2in}\}$ (see the blue area in Fig. 12.5). This has important consequences when interpreting the experimental data of a possible device based on these ideas.

In order to completely characterize the THz generation process in the analyzed structure, we have also studied the temporal evolution of the response of the system to Gaussian pulse excitations. We consider the case in which $P_{1in}(t) = P_{2in}(t)$. In these calculations, we also assume that the temporal width of the NIR pulses corresponds to the lifetime of the THz-scale cavity ($\tau_{THz} \approx 16$ ns). The value of τ_{THz} is much larger than the lifetime of the WGM modes at the pump and idler frequencies (~ 0.8 ns). Thus, on physical grounds, we expect similar maximum conversion efficiencies as those found in the cw analysis described above. Figure 12.6a–c show the results corresponding to three representative values for the peak power of $P_{tot,in}(t)$ (labeled as A, B, and C, respectively, in Fig. 12.4). As seen in Fig. 12.6a, when the maximum of $P_{tot,in}(t)$ is equal to the critical power, $P^c_{tot,in}$, the pump input pulse is completely consumed after spending approximately 60 ns in the system (i.e., $P_{1tr}(t) \approx 0$ after 60 ns); and, simultaneously, the power of the transmitted idler pulse ($P_{2tr}(t)$) value reaches a peak value that is approximately twice the peak value corresponding to the input idler pulse ($P_{2in}(t)$). For peak values of $P_{tot,in}(t)$ much lower than the critical power, the almost undepleted behavior can be clearly observed (see Fig. 12.6b): the peak powers of the pump and idler pulses are barely modified as they travel through the system. On the other hand, Fig. 12.6c clearly shows how, for input NIR peak powers well beyond the critical power, the pump pulse is completely consumed at $t \approx 40$ ns. However, in contrast to the case displayed in Fig. 12.6a, Fig. 12.6c shows how after $t \approx 40$ ns the up-conversion process that mix ω_T with ω_2 to yield ω_1 starts being relevant, and, consequently $P_{1tr}(t)$ begins increasing with time; which in turn reduces the overall THz conversion efficiency of the process. Finally, Fig. 12.6d displays a summary of our time-dependent simulations in terms of the ratio between the output THz energy and total input NIR energy defined as $E_{THz} = \int_{-\infty}^{\infty} dt\, P_T(t)$

[1] In the limit $P_{2in} \ll P_{1in}$, assuming a lossless triply resonant cavity (i.e., $\Gamma_1 = \Gamma_T = 1$), the maximum conversion efficiency given by (12.12) approaches the conventional Manley-Rowe quantum limit ω_T/ω_1.

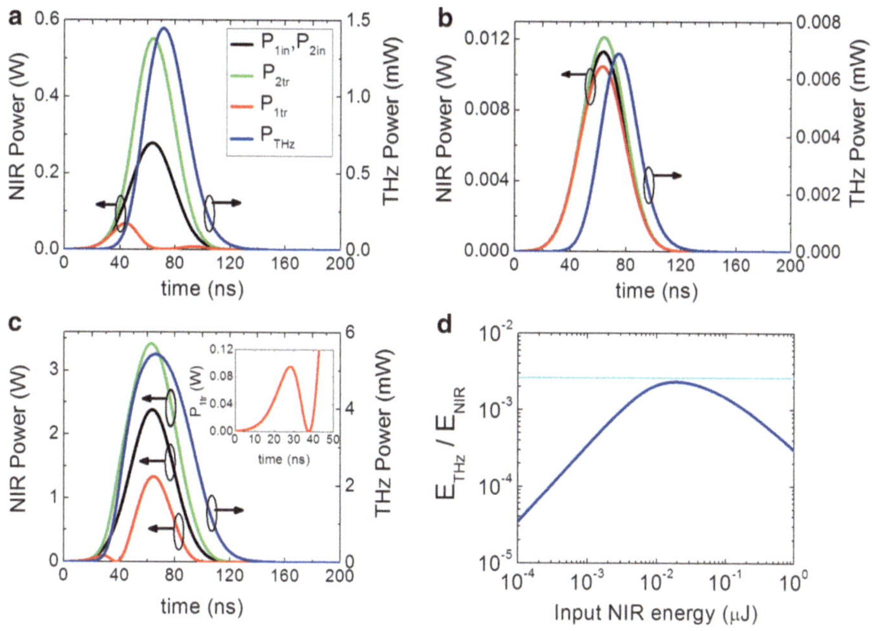

Fig. 12.6 Analysis of the temporal response of the system shown in Fig. 12.1 to Gaussian excitation pulses. Panels (**a**), (**b**), and (**c**) correspond to the peak values for $P_{tot,in}$ shown by labels *A*, *B*, and *C*, respectively in Fig. 12.4. In (**a–c**), the scale on the *left vertical axis* corresponds to the NIR power of the pump and idler frequencies, both for the input and transmitted pulses; whereas the scale on the *right vertical axis* corresponds to the THz output power. *Inset* of (**c**), shows an enlarged view of the temporal dependence of $P_{1tr}(t)$ between $t = 0$ and 50 ns. *Panel* (**d**) displays the ratio between the total output energy (E_{THz}) and the total NIR input energy (E_{NIR}) as a function of E_{NIR}. *Horizontal line* in this panel displays the maximum possible conversion efficiency given by the Manley–Rowe quantum limit

and $E_{NIR} = \int_{-\infty}^{\infty} dt\,[P_{1in}(t) + P_{2in}(t)]$, respectively). As displayed in Fig. 12.6d, the maximum conversion efficiency can be reached for an input energy $E_{NIR}^{c} \approx 0.02$ μJ. Similarly to the calculation of P_0, E_{NIR}^{c} has been computed by assuming that the height of the structure along the *z*-axis (see Fig. 12.2) is $\lambda_T/4$ for all three fields. The value obtained in this way represents a reduction in E_{NIR} of three orders of magnitude with respect to the most efficient schemes for THz generation in nonlinear crystals reported up to date. Furthermore, we emphasize that in addition to powerful lasers, current efficient schemes for THz generation require intricate phase-matching setups, whereas in the system introduced in this manuscript maximum theoretically possible efficiency can be achieved in an integrated structure having a total area of approximately 1 mm^2.

12.4 Conclusions

In conclusion, using a physical picture inspired on a classical analogy of the quantum Purcell effect, we have described a novel route to identify the optimal conditions that enable reaching dramatic enhancements of the conversion efficiency of arbitrary difference-frequency down-conversion process. The approach presented in this chapter has also allowed us to generalize the canonical phase-matching condition found in the considered frequency mixing processes. This generalization enables efficient DFG processes even in the case of nonlinear optical materials where standard phase-matching techniques are difficult or impossible to implement in certain frequency regimes. By means of detailed numerical simulations, we have illustrated the relevance of the proposed scheme by demonstrating complete conversion to THz energy of a 0.02 μJ NIR pump pulse in a realistic 1 mm^2-footprint structure created from GaAs. Alternatively, we have demonstrated that in the continuous-wave regime, the pump powers required to reach quantum-limited conversion efficiency can be reduced up to three orders of magnitude with respect to the conventional approaches for THz generation employed up to date. In contrast to previous high-efficiency THz generation schemes, the concept described in this chapter and ref. [29] opens the way to efficient THz generation from sources that are compact, turn-key, and low-cost, which we believe could enable a broader use of THz sources.

Acknowledgments We acknowledge J.D. Joannopoulos, S.G. Johnson, A.W. Rodriguez, P. Rakich, M. Loncar, I.B. Burgess, Y. Zhang, and M.W. McCutcheon for a fruitful collaboration in this topic. We also thank David Duchesne for helpful discussions. This work was supported by the MRSEC Program of the National Science Foundation under award number DMR-0819762 and by the US Army Research Office through the Institute for Soldier Nanotechnologies under Contract No. W911NF-07-D-0004. J.B.-A. acknowledges financial support by Spanish Spanish Ministry of Science and Innovation through Grant No. RyC-2009-0548.

References

1. E. Purcell, Phys. Rev. **69**, 681 (1946)
2. E. Yablonovitch, Phys. Rev. Lett. **58**, 2059 (1987)
3. S. John, Phys. Rev. Lett. **58**, 2486 (1987)
4. J.D. Joannopoulos, S.G. Johnson, R.D. Meade, J.N. Winn, *Photonic Crystals: Molding the Flow of Light*, 2nd edn. (Princeton University Press, Princeton, NJ, 2008)
5. S. Noda, M. Fujita, T. Asano, Nat. Phot. **1**, 449 (2007)
6. E.U. Condon, J. Appl. Phys. **12**, 129 (1941)
7. C.G. Montgomery, R.H. Dicke, E.M. Purcell (eds.), *Principles of Microwave Circuits* (Institution of Engineering and Technology, London, 1987)
8. E. Ozbay, Science **311**, 189 (2006)
9. B. Ferguson, X.-C. Zhang, Nat. Mater. **1**, 26 (2002)
10. M. Tonouchi, Nat. Phot. **1**, 97 (2007)
11. P.H. Siegel, IEEE Trans. Microw. Theor. Tech. **52**, 2438 (2004)

12. D. Mittleman (ed.), *Sensing with Terahertz Radiation* (Springer, Berlin, 2003)
13. J. Zhang, D. Grischkowsky, Opt. Lett. **29**, 1617 (2004)
14. J.F. Federici, B. Schulkin, F. Huang, D. Gary, R. Barat, F. Oliveira, D. Zimdars, Semicond. Sci. Technol. **20**, 266–280 (2005)
15. M.A. Belkin, F. Capasso, F. Xie, A. Belyanin, M. Fischer, A. Wittmann, J. Faist, Appl. Phys. Lett. **92**, 201101 (2008)
16. K.L. Vodopyanov, M.M. Fejer, X. Yu, J.S. Harris, Y.-S. Lee, W.C. Hurlbut, V.G. Kozlov, D. Bliss, C. Lynch, Appl. Phys. Lett. **89**, 141119 (2006)
17. K.-L. Yeh, M.C. Hoffmann, J. Hebling, K.A. Nelson, Appl. Phys. Lett. **90**, 171121 (2007)
18. G.S. Agarwal, Phys. Rev. A **12**, 1475 (1975)
19. S.M. Barnett, B. Huttner, R. Loudon, Phys. Rev. Lett. **68**, 3698 (1992)
20. Y. Xu, R.K. Lee, A. Yariv, Phys. Rev. A **61**, 033807 (2000)
21. M.O. Scully, M.S. Zubairy, *Quantum Optics* (Cambridge University Press, Cambridge, 1997)
22. H.A. Haus, *Waves and Fields in Optoelectronics* (Prentice-Hall, Englewood Cliffs, NJ, 1984)
23. R. Hamam, M. Ibanescu, E.J. Reed, P. Bermel, S.G. Johnson, E.P. Ippen, J.D. Joannopoulos, M. Soljacic, Opt. Express **16**, 12523 (2008)
24. K.J. Vahala, Nature **424**, 839 (2003)
25. D.L.C. Chan, I. Celanovic, J.D. Joannopoulos, M. Soljacic, Phys. Rev. A **74**, 064901 (2006)
26. I.B. Burgess, A.W. Rodriguez, M.W. McCutcheon, J. Bravo-Abad, Y. Zhang, S.G. Johnson, M. Loncar, Opt. Express **17**, 9241 (2009)
27. A. Ashkin, G.D. Boyd, J.M. Dziedzic, IEEE J. Quantum Electron. **2**, 109124 (1966)
28. R.W. Boyd, *Nonlinear Optics* (Academic, New York, NY, 1992)
29. J. Bravo-Abad, A.W. Rodriguez, J.D. Joannopoulos, P.T. Rakich, S.G. Johnson, M. Soljacic, Appl. Phys. Lett. **96**, 101110 (2010)
30. I.B. Burgess, Y. Zhang, M.W. McCutcheon, A.W. Rodriguez, J. Bravo-Abad, S.G. Johnson, M. Loncar, Opt. Express **17**, 20099 (2009)
31. S. Singh, in *Handbook of Laser Science and Technology*, vol 3, Part I (M.J. Weber CRC, Boca Raton, FL, 1986)
32. M.L. Povinelli, S. Johnson, S. Fan, J.D. Joannopoulos, Phys. Rev. B **64**, 075313 (2001)
33. Y. Dumeige, P. Feron, Phys. Rev. A **74**, 063804 (2006)
34. Z. Yang, P. Chak, A.D. Bristow, H.M. van Driel, R. Iyer, J.S. Aitchison, A.L. Smirl, J.E. Sipe, Opt. Lett. **32**, 826 (2007)
35. J. Hebling, A.G. Stepanov, G. Almasi, B. Bartal, J. Kuhl, Appl. Phys. B **78**, 593 (2004)
36. V.R. Almeida, C.A. Barrios, R.R. Panepucci, M. Lipson, Nature **431**, 1081 (2004)

Chapter 13
Photonic Transition in Nanophotonics

Zongfu Yu and Shanhui Fan

13.1 Introduction

Photonic transition [1] is induced by refractive index modulation. Many photonic structures, including photonic crystals, or waveguides, can be described by a photonic band structure. When these structures are subject to temporal refractive index modulation, photon states can go through interband transitions, in a direct analogy to electronic transitions in semiconductors. Such photonic transitions have been recently demonstrated experimentally in silicon microring resonators [2]. In this chapter, we review two applications using the photonic transitions.

As the first application, we show that based on the effects of photonic transitions, a linear, broadband, and nonreciprocal isolator [3] can be accomplished by spatial–temporal refractive index modulations that simultaneously impart frequency and wavevector shifts during the photonic transition process. This work demonstrates that on-chip isolation can be accomplished with dynamic photonic structures, in standard material systems that are widely used for integrated optoelectronic applications.

In the second application, we show that a high-Q optical resonance can be created dynamically, by inducing a photonic transition between a localized state and a one-dimensional continuum through refractive index modulation [4]. In this mechanism, both the frequency and the external linewidth of a single resonance are specified by the dynamics, allowing complete control of the resonance properties.

This chapter is organized as follows: in Sect. 13.2, we review photonic transition induced by dynamic modulation; in Sects. 13.3 and 13.4, we describe the optical isolator and tunable cavity based on photonic transition, respectively; Sect. 13.5 is the conclusion part.

Z. Yu • S. Fan (✉)
Department of Electrical Engineering, Ginzton Laboratory, Stanford University, 350 Serra Mall (Mail Code: 9505), Stanford, CA 94305-9505, USA
e-mail: shanhui@stanford.edu

Z. Chen and R. Morandotti (eds.), *Nonlinear Photonics and Novel Optical Phenomena*, Springer Series in Optical Sciences 170, DOI 10.1007/978-1-4614-3538-9_13,

13.2 Photonic Transition in a Waveguide

We start by describing the photonic transition process in a silicon waveguide. The waveguide (assumed to be two-dimensional for simplicity) is represented by a dielectric distribution $\varepsilon_s(x)$ that is time-independent and uniform along the z-direction (Fig. 13.1b). Such a waveguide possesses a band structure as shown in Fig. 13.1a, with symmetric and antisymmetric modes located in the first and second band, respectively. An interband transition, between two modes with frequencies and wavevectors $(\omega_1, k_1), (\omega_2, k_2)$ located in these two bands, can be induced by modulating the waveguide with an additional dielectric perturbation:

$$\varepsilon'(x, z, t) = \delta(x)\cos(\Omega t - qz), \quad (13.1)$$

where $\delta(x)$ is the modulation amplitude distribution along the direction transverse to the waveguide. $\Omega = \omega_2 - \omega_1$ is the modulation frequency. Figure 13.1c shows

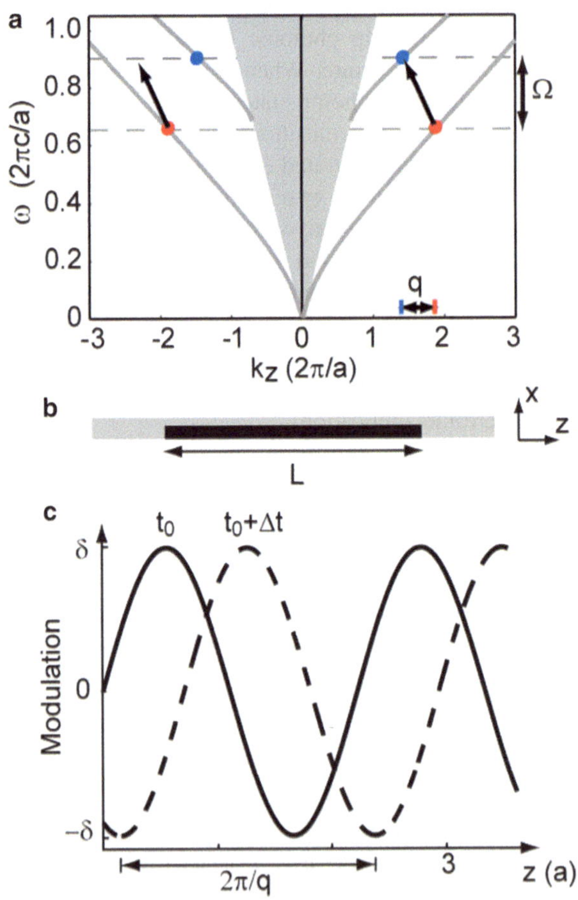

Fig. 13.1 (**a**) Band structure of a slab waveguide. (**b**) Structure of a silicon ($\varepsilon_s = 12.25$) waveguide. Modulation is applied to the dark region. (**c**) The modulation profile at two sequential time steps

the profile of the modulation. Such a transition, with $k_1 \neq k_2$, is referred to as an indirect photonic transition, in analogy with indirect electronic transitions in semiconductors.

We assume that the wavevector q approximately satisfies the phase-matching condition, i.e., $\Delta k = k_2 - k_1 - q \approx 0$. In the modulated waveguide, the electric field becomes:

$$E(x,z,t) = a_1(z)E_1(x)\mathrm{e}^{i(-k_1 z+\omega_1 t)} + a_2(z)E_2(x)\mathrm{e}^{i(-k_2 z+\omega_2 t)}, \tag{13.2}$$

where $E_{1,2}(x)$ are the modal profiles, satisfying the orthogonal condition: (for simplicity, we have assumed the TE modes where the electric field has components only along the y-direction)

$$\frac{v_{gi}}{2\omega_i}\int_{-\infty}^{\infty} \varepsilon(x)E_i^*E_j = \delta_{ij}. \tag{13.3}$$

In (13.3), the normalization is chosen such that $|a_n|^2$ is the photon number flux carried by the nth mode. By substituting (13.2) into the Maxwell's equations, and using slowly varying envelope approximation, we can derive the coupled mode equation:

$$\frac{\mathrm{d}}{\mathrm{d}z}\begin{pmatrix} a_1 \\ a_2 \end{pmatrix} = \begin{pmatrix} 0 & i\frac{\pi}{2l_c}\exp(-i\Delta kz) \\ i\frac{\pi}{2l_c}\exp(i\Delta kz) & 0 \end{pmatrix}\begin{pmatrix} a_1 \\ a_2 \end{pmatrix}, \tag{13.4}$$

where

$$l_c = \frac{4\pi}{\varepsilon_0 \int\limits_{-\infty}^{\infty} \delta(x)E_1(x)E_2(x)\mathrm{d}x}, \tag{13.5}$$

is the coherence length. With an initial condition $a_1(0) = 1$ and $a_2(0) = 0$, the solution to (13.4) is:

$$\begin{aligned} a_1(z) &= \mathrm{e}^{-iz\Delta k/2}\left[\cos\left(\frac{z}{2l_c}\sqrt{\pi^2 + (l_c\Delta k)^2}\right)\right. \\ &\quad \left. + i\frac{l_c\Delta k}{\sqrt{\pi^2 + (l_c\Delta k)^2}}\sin\left(\frac{z}{2l_c}\sqrt{\pi^2 + (l_c\Delta k)^2}\right)\right] \\ a_2(z) &= i\mathrm{e}^{iz\Delta k/2}\frac{\pi}{\sqrt{\pi^2 + (l_c\Delta k)^2}}\sin\left(\frac{z}{2l_c}\sqrt{\pi^2 + (l_c\Delta k)^2}\right) \end{aligned} \tag{13.6}$$

In the case of perfect phase-matching, i.e., $\Delta k = 0$, a photon initially in mode 1 will make a complete transition to mode 2 after propagating over a distance of coherence length l_c (Fig. 13.2a). In contrast, in the case of strong phase-mismatch, i.e., $l_c\Delta k \gg 1$, the transition amplitude is negligible (Fig. 13.2b).

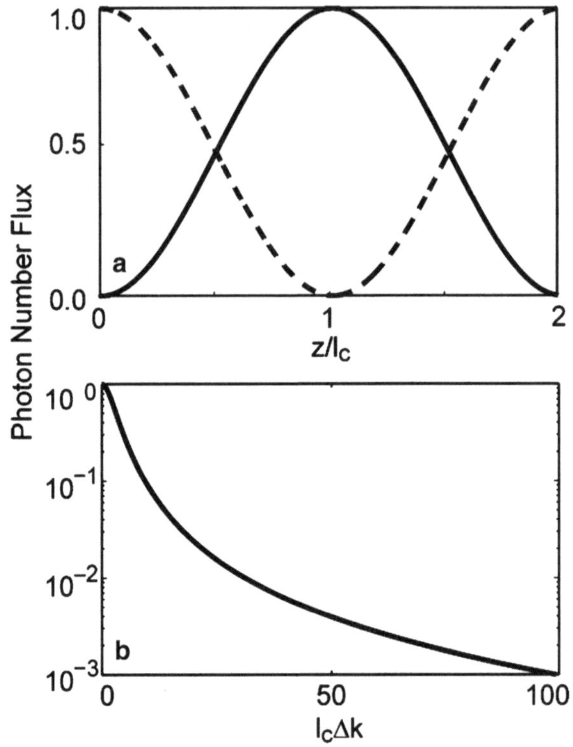

Fig. 13.2 (**a**) Spatial evolution of the photon number flux of two modes (*dashed line* mode 1 and *solid line* mode 2), when a phase-matching modulation is applied to the waveguide. (**b**) Maximum photon flux in mode 2 as a function of phase mismatch. The transition becomes essentially negligible at $l_c\Delta k \gg 1$

13.3 Photonic Transition for Integrated Optical Isolator

In this section, we use the photonic transition described in the previous section to achieve on-chip optical isolation. In an optical network, isolators are an essential component used to suppress back-reflection, and hence interference between different devices. Achieving on-chip optical signal isolation has been a fundamental difficulty in integrated photonics. The need to overcome this difficulty, moreover, is becoming increasingly urgent, especially with the emergence of silicon nanophotonics, which promises to create on-chip optical systems at an unprecedented scale of integration.

To create complete optical signal isolation requires simultaneous breaking of both the time-reversal and the spatial inversion symmetry. In bulk optics, this is achieved using materials exhibiting magneto-optical effects. Despite many efforts [5–8], however, on-chip integration of magneto-optical materials, especially in silicon in a CMOS compatible fashion, remains a great difficulty. Alternatively, optical isolation has also been observed using nonlinear optical processes [9, 10], or in electroabsorption modulators [11]. In either case, however, optical isolation occurs only at specific power ranges, or with associated modulation side bands. In addition, there have been works aiming to achieve partial optical isolation in

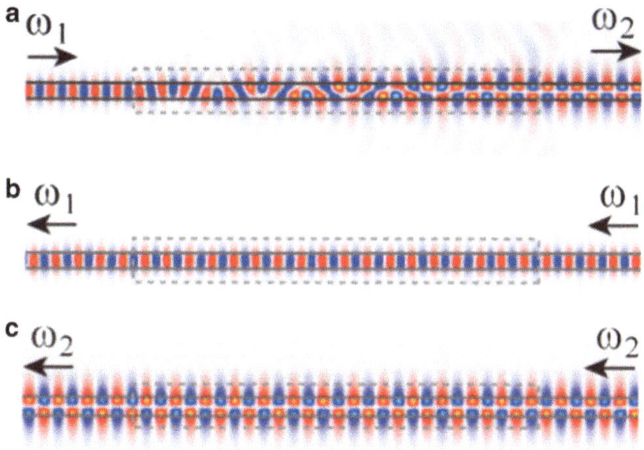

Fig. 13.3 Finite-difference time-domain simulation of an isolator based on photonic transitions. The *box* indicates the regions where the refractive index is modulated. *Blue/red* shows the amplitude of electric fields. *Arrows* indicate propagation directions

reciprocal structures that have no inversion symmetry (for example, chiral structures) [12]. In these systems, the apparent isolation occurs by restricting the allowed photon states in the backward direction, and would not work for arbitrary backward incoming states. None of the above nonmagnetic schemes can provide complete optical isolation.

In this part, we review and expand upon our recent works [3, 13] on creating complete and linear optical isolation using photonic transition. In these works, the temporal profile of the modulation used to induce the transition is chosen to break the time-reversal symmetry, while the spatial profile of the modulation is chosen to break the spatial-inversion and the mirror symmetry. As seen by the finite-difference time-domain simulations, when a silicon waveguide is under a modulation that induces an interband photonic transition, light of frequency ω_1 in forward direction is converted to a higher frequency mode ω_2 by the modulation (Fig. 13.3a). At the same time, light of frequencies ω_1 or ω_2 in the backward direction are not affected by the modulation (Fig. 13.3b, c). Combined with an absorption filter centered at ω_2, this structure can absorb all lights incident from one direction at ω_1, while passing those in the opposite direction, and thus creates a complete isolator behavior. It was also shown that the finite-difference time-domain simulations can also be well reproduced by coupled-mode theory [3].

We use the coupled mode theory as described in Sect. 13.2 to discuss the performance and design considerations for our dynamic isolator schemes. The waveguide system in Sect. 13.2 exhibits strong nonreciprocal behavior: the modulation in (13.1) does not phase-match the mode at $(\omega_1, -k_1)$ with any other mode of the system (Fig. 13.1a). Thus, while the mode at (ω_1, k_1) undergoes a complete photonic transition, its time-reversed counterpart at $(\omega_1, -k_1)$ is not affected at all. Such nonreciprocity arises from the breaking of *both* time-reversal

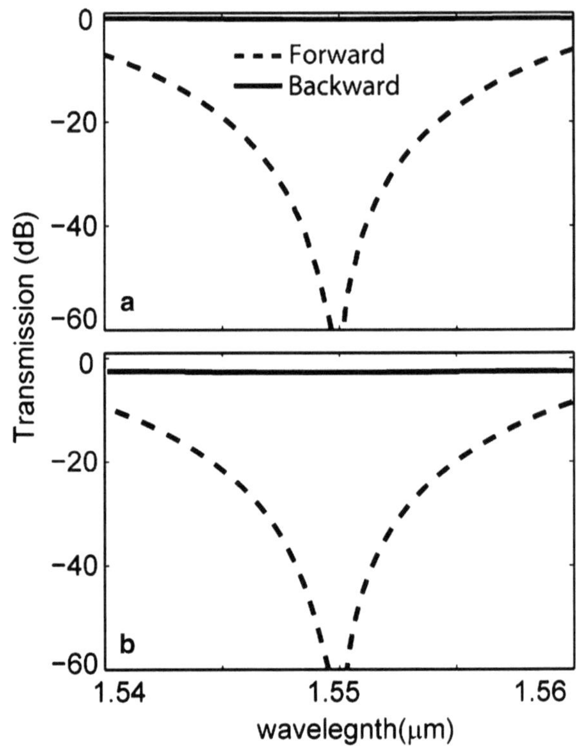

Fig. 13.4 Forward and backward transmission spectra without (**a**) and with (**b**) modulation loss

and spatial-inversion symmetries in the dynamics: The modulation in (13.1) is *not* invariant with either $t \to -t$ or $z \to -z$.

As a specific example, we consider a silicon ($\varepsilon = 12.25$) waveguide of 0.27 μm wide, chosen such that the first and second bands of the waveguide have the same group velocity around wavelength 1.55 μm (or a frequency of 193 THz). The modulation has a strength $\delta_{\max}/\varepsilon_s = 5 \times 10^{-4}$, a frequency $\Omega/2\pi = 20\,\mathrm{GHz}$ and a spatial period $2\pi/|q| = 0.886$ μm. (All these parameters should be achievable in experiments.) The modulation is applied to half of the waveguide width so that the even and odd modes can couple efficiently. The modulation length L is chosen as the coherence length $l_{c0} = 2.19\,\mathrm{mm}$ (Fig. 13.1b) for operation frequency ω_0 at 1.55 μm wavelength. Figure 13.4a shows the transmission for forward and backward directions. The bandwidth is 5 nm with contrast ratio above 30 dB.

For the loss induced by refractive index modulation schemes, e.g., carrier injection modulation, the contrast ratio remains approximately the same as the lossless case, since the modulation loss applies to transmission in both directions. Thus the isolation effect is not affected. As an example, the modulation strength used here $\delta/\varepsilon_s = 5 \times 10^{-4}$ results in a propagating loss of 1.5 cm^{-1} in silicon [14]. This causes an insertion loss about −3.5 dB while the bandwidth remains approximately unchanged (Fig. 13.4b).

In general, nonreciprocal effects can also be observed in intraband transitions involving two photonic states in the same photonic band. However, since typically $\Omega \ll \omega_1$, and the dispersion relation of a single band can typically be approximated as linear in the vicinity of ω_1, cascaded process [2], which generates frequencies at $\omega_1 + n\Omega$ with $n > 1$, is unavoidable, and it complicates the device performance. In contrast, the interband transition here eliminates the cascaded processes.

We would like to emphasize that the modulation frequency can be far smaller than the bandwidth of the signal. This is in fact one of the key advantages of using interband transition. The transition occurs from a fundamental even mode to a second-order odd mode. The generated odd mode can be removed with the use of mode filters that operate based on modal profiles. Examples of such mode filters can be found in [15, 16]. It is important to point out that such mode filters are purely passive and reciprocal, and can be readily implemented on chip in a very compact fashion. Moreover, later in this section, we discuss an implementation of an isolator without the use of modal filters.

13.3.1 *Detailed Analysis of the Isolator Performance*

Below in this section, based on the coupled mode theory, we analyze in details various aspects regarding the performance of the proposed isolator including in particular its operational bandwidth and device size.

13.3.1.1 Bandwidth

The dynamic isolator structure creates contrast between forward and backward propagations by achieving complete frequency conversion only in the forward direction. As discussed above, the modulation is chosen such that, it induces a phase-matched transition from an even mode at the frequency ω_0 to an odd mode at the frequency of $\omega_0 + \Omega$. The length of the waveguide is chosen to be the coherence length $l_c(\omega_0)$ for this transition, such that complete conversion occurs at this frequency ω_0 for the incident light. In order to achieve a broad band operation, one would need to achieve near-complete conversion for all incident light having frequencies ω in the vicinity of ω_0 as well. From (13.6), broad band operation therefore requires that

$$\begin{aligned} \Delta k(\omega) &= 0 \\ l_c(\omega) &= L = l_c(\omega_0). \end{aligned} \tag{13.7}$$

The first condition in (13.7) implies that the phase-matching condition needs to be achieved over a broad range of frequencies, and the second condition implies that the coherence length should not vary as a function of frequency. Deviations from these conditions result in a finite operational bandwidth.

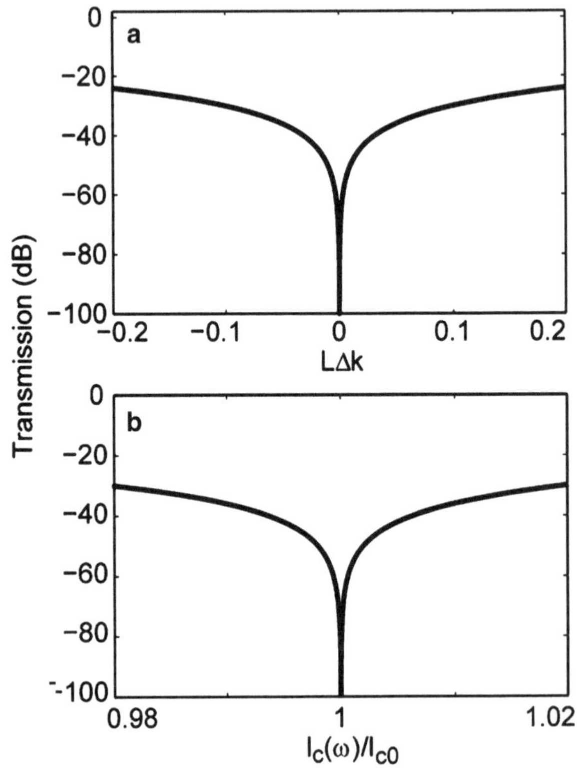

Fig. 13.5 Forward transmission as a function of (**a**) phase mismatch and (**b**) coherence length variation

We consider the phase-matching condition first. In the vicinity of the design frequency ω_0, the wavevector mismatch can be approximated by $\Delta k = k_1(\omega) - k_2(\omega + \Omega) - q \approx ((1/v_{g1}(\omega)) - (1/v_{g2}(\omega + \Omega)))\Delta\omega + 1/2((\mathrm{d}^2k_1(\omega)/\mathrm{d}\omega^2) - (\mathrm{d}^2k_2\omega + \Omega)/\mathrm{d}\omega^2)\Delta\omega^2||_{\omega=\omega_0}$.

Thus, to minimize the phase mismatch, it is necessary, first of all, that the two bands have the same group velocities, i.e., the two bands are parallel to each other. Moreover, it is desirable that the group velocity dispersion of the two bands matches with one another. As a quantitative estimate, assuming that $l_c(\omega) \approx L$ for all frequencies, Fig. 13.5a shows the forward transmission as a function of $L\Delta k$. For a transmission below −30 dB, this requires a phase mismatch of $L\Delta k<0.1$. As a concrete example for comparison purposes, Fig. 13.6a shows the phase mismatch $L\Delta k$ as a function of wavelength for the structure simulated in Fig. 13.4. Notice that $L\Delta k<0.1$ over a bandwidth of 5 nm due to the mismatch of group velocity dispersion in the two guided mode bands. Thus the operating bandwidth of this device for 30 dB contrast is on the order of 5 nm.

For the second condition in (13.7), we note that in most waveguide structures, since the coherence length is determined by the modal profile, it generally varies slowly as a function of frequency. For example, for a waveguide with parameters chosen in Sect. 13.2, the coherence length varies <2% over 20 nm bandwidth

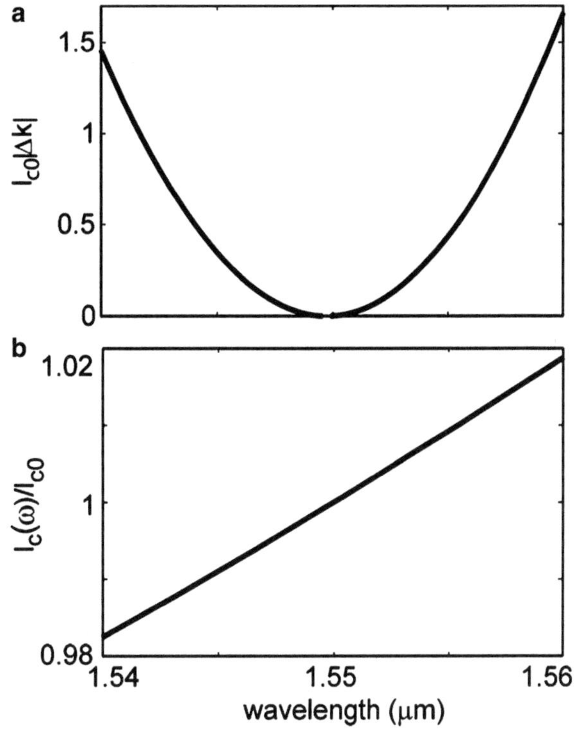

Fig. 13.6 Phase mismatch (**a**) and coherence length (**b**) as a function of wavelength for the device simulated in Fig. 13.4

around 1.55 μm wavelength (Fig. 13.6b). As a simple estimate of how coherence length variation impacts device performance, assuming that $\Delta k(\omega) = 0$ over a broad frequency range, we calculate the forward transmission as a function of coherence length given the modulation length $L = l_c(\omega = \omega_0)$ (Fig. 13.5b). For 2% variation of the coherence length, the forward transmission remains below −30 dB. Comparing Fig. 13.6a, b, therefore, we conclude that for the structure simulated in Fig. 13.4, the 5-nm bandwidth is primarily limited by group velocity dispersion of the two waveguide bands. Since the structure used in Fig. 13.4 is a rather simple, we believe that substantial further enhancement of operating bandwidth is achievable by optimization of waveguide geometry.

13.3.1.2 Device Size

The size of the isolator is determined by the coherence length l_c. Starting from (13.5), and taking into account the normalization of E field (13.3), the coherence length can be written as

$$l_c = \frac{4\pi}{\varepsilon_0 \int\limits_{-\infty}^{\infty} \delta(x) E_1(x) E_2(x) \mathrm{d}x} = \frac{2\pi}{\gamma} \sqrt{\frac{v_{g1} v_{g2}}{\omega_1 \omega_2}} \approx \lambda_0 \cdot \frac{1}{\gamma} \cdot \frac{v_g}{c}, \tag{13.8}$$

where $\gamma = (\int_{-\infty}^{\infty}\delta(x)E_1(x)E_2(x)\mathrm{d}x)/(\sqrt{\int_{-\infty}^{\infty}\varepsilon(x)|E_1|^2\mathrm{d}x\int_{-\infty}^{\infty}\varepsilon(x)|E_2|^2\mathrm{d}x})$ characterizes the effect of modulation. In deriving (13.8), we assume that $\omega_1 \approx \omega_2 \equiv 2\pi c/\lambda_0$, where λ_0 is the wavelength in vacuum, since the modulation frequency is typically far smaller than the optical frequency. Moreover, the two bands are assumed to be parallel to each other, i.e., $v_{g1} \approx v_{g2} \equiv v_g$. Equation (13.8) indicates that the device size is proportional to the group velocity and is inversely proportional to the modulation strength. For a rough estimate, with a modulation strength $\gamma \sim (\delta/\varepsilon) \sim 10^{-4}$, operating at a wavelength of $\lambda \sim 1.5$ μm and $v_g \approx c/3$, the coherence length $l_c \sim 5$ mm. To reduce the size, one can use stronger modulation strength and/or slow light waveguides.

13.3.1.3 Near-Phase-Matched Transition in the Backward Direction

In general, due to energy conservation constraint, a mode with a frequency of ω_1 can only make a transition to modes at $\omega_1 \pm \Omega$. In our design, the modulation is chosen to create a phase-matched transition in the forward direction. However, for most electro-optic or acoustic-optic modulation schemes, the modulation frequency $\Omega \leqslant 100$ GHz is much smaller than the optical frequency. Consequently, as can be seen from Fig. 13.7a, in the backward direction the transition to the mode in the second band with a frequency $\omega_3 = \omega_1 - \Omega$ becomes nearly phase-matched. The wavevector mismatch of this transition is:

$$\Delta k_b = -k_2(\omega_1 - \Omega) + k_1(\omega_1) + q \approx \frac{2\Omega}{v_g}. \tag{13.9}$$

Such a transition results in loss in the backward direction and thus a reduction of contrast between the forward and backward directions.

To calculate such transmission loss in the backward direction, we replace Δk in (13.6) with Δk_b. In general, in order to suppress such backward transmission loss, one needs to have:

$$\Delta k_b \cdot L \geq 1. \tag{13.10}$$

Combining with (13.8), the condition of (13.10) is then transformed to:

$$\frac{2\lambda_0}{c} \cdot \frac{\Omega}{\gamma} \geq 1. \tag{13.11}$$

Remarkably, we note from (13.11) that for electro-optic or acoustic-optic modulation schemes, the effects of weak refractive index modulation γ and low modulation frequency Ω cancel each other out. The use of weak refractive index modulation results in a long coherence length, which helps in suppressing the transition processes that are not phase matched. And it is precisely such a cancelation that enables the construction of dynamic isolators with practical modulation mechanisms.

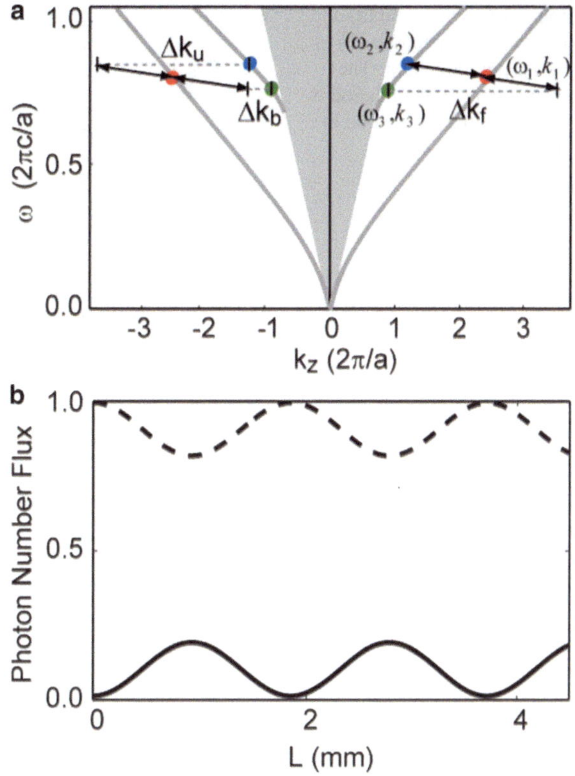

Fig. 13.7 (**a**) The transition diagram for low frequency modulation. (**b**) Spatial evolution of photon flux in the backward direction for an even mode at 1.55 μm wavelength (*dashed line*) and an odd mode (*solid line*) that is 20 GHz lower in frequency. The structure has the same parameters described in Sect. 13.2

For the example shown in Fig. 13.4, the near-phase-matched transition in the backward direction has a $\Delta k_b = 2\pi/2.06\,\text{mm}$ and thus $\Delta k_b L = 6.7$, which results in a loss of -0.22 dB for the backward transmission (Fig. 13.7b).

13.3.2 Design Flexibility

In the previous sections, we have shown that by using interband transition, one can create nonreciprocal mode conversion in a waveguide. Such a waveguide works as an isolator when combined with a modal filter. The performance of such device can be analyzed and optimized using coupled mode theory. In this section, we present two examples to show that such nonreciprocal photon transition can be exploited in a wide range of structures to form nonreciprocal optical devices that satisfy diverse performance requirements. In the first example, we design a four-port isolator/circulator using nonreciprocal phase shift in the interband transitions. In the second example, we use a nonreciprocal ring resonator to demonstrate a compact design for optical isolation.

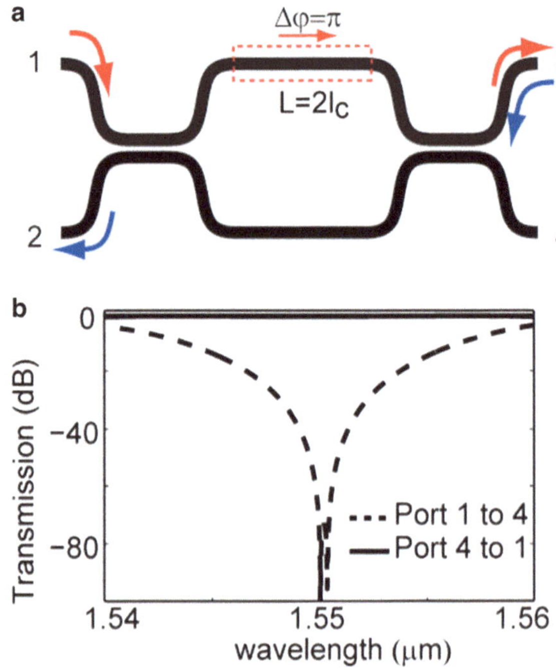

Fig. 13.8 Schematic (**a**) and transmission spectrum (**b**) of a four port circulator. The dynamic index modulation is applied to the waveguide in the *dashed red box*

13.3.2.1 Four-Port Circulator

Figure 13.8a shows the design of a four-port circulator [13]. The structure consists of a Mach–Zehdner interferometer, in which one waveguide arm is subject to the dynamic modulation described above. In contrast to the design in Sect. 13.3, however, here the length of the modulation region is chosen to be twice the coherence length $L = 2l_c$. Thus, light passing through the modulated waveguide in the forward direction will return to the incident frequency (Fig. 13.2a). However, such light experiences a nonreciprocal phase shift due to the photonic transition effect. The use of a Mach–Zehnder interferometer configuration then allows one to construct a circulator. Here no filter is required, which significantly reduces the device complexity.

For concreteness, we assume that the interferometer has two arms with equal length, and uses two 50/50 waveguide couplers. For such an interferometer, the transmission is described by

$$\begin{pmatrix} b_u \\ b_l \end{pmatrix}_{\text{OUT}} = \frac{1}{2}\begin{pmatrix} 1 & i \\ i & 1 \end{pmatrix}\begin{pmatrix} T\exp(i\varphi_p) & 0 \\ 0 & \exp(i\varphi_p) \end{pmatrix}\begin{pmatrix} 1 & i \\ i & 1 \end{pmatrix}\begin{pmatrix} b_u \\ b_l \end{pmatrix}_{\text{IN}}. \quad (13.12)$$

Here, the sub-script "IN" and "OUT" label the input or output. $b_{u/l}$ are the input or output amplitudes in the upper/lower arm. φ_p is the phase acquired due to propagation in the absence of modulation.

In (13.12), the transmission coefficient through the upper arm has an addition contribution from the photon transition:

$$T = e^{-iz\Delta k/2}\left[\cos\left(\frac{z}{2l_c}\sqrt{\pi^2 + (l_c\Delta k)^2}\right) + i\frac{l_c\Delta k}{\sqrt{\pi^2 + (l_c\Delta k)^2}}\sin\left(\frac{z}{2l_c}\sqrt{\pi^2 + (l_c\Delta k)^2}\right)\right], \tag{13.13}$$

which influences both the transmission amplitude and the phase as the wave passes through the upper arm. In our design, we assume a phase-matching modulation with $\Delta k = 0$ for the forward direction, and use a modulated region with $L = 2l_c$. Equation (13.13) shows $T = -1$. In contrast, for the light in the backward direction in the upper arm, in general the phase matching condition is not satisfied. Hence, $T \approx 1$. Thus, in this design, the modulation does not create any frequency conversion. Instead its sole effect is to induce a nonreciprocal phase shift in the upper arm.

The interferometer in Fig. 13.8 exploits such nonreciprocal phase to create a circulator. We have used the coupled mode theory developed in Sect. 13.3, to simulate this structure, assuming the same waveguide parameters as in Fig. 13.4. The results, shown in Fig. 13.8, indicate that lights injected into port 1 completely output through port 3, while in the time reversed case, lights injected into port 3 ends up in port 2. Therefore, this device has exactly the same response function of a four-port circulator. Unlike conventional design, however, no magnetic components are used inside the structure. Alternatively, the device can also function as a two-port isolator. Figure 13.8b shows the transmission spectra in both directions between ports 1 and 4: lights incident from port 4 transmit to port 1 while the reverse transmission is completely suppressed. The contrast ratio for the two directions is above 30 dB for a bandwidth of 5 nm (Fig. 13.8b).

13.3.2.2 Nonreciprocal Ring Resonator

As discussed before, the device size is determined by the coherence length, which typically is above millimeters unless slow light waveguides are used. Substantial reduction of the device footprint can be accomplished using resonator structure at the expense of a smaller operating bandwidth [3]. As an example, we consider a ring resonator (Fig. 13.9a) that supports two anticlockwise rotating resonances, at frequencies ω_1 and ω_2, respectively. Each resonance is further characterized by its wavevector k_1 and k_2 in the waveguide that forms the ring. These two resonances are coupled by applying a dielectric constant modulation along the ring with a profile $\delta(x)\cos[(\omega_1 - \omega_2)t - (k_1 - k_2)z]$, where z measures the propagation distance on the circumference of the ring in counterclockwise direction.

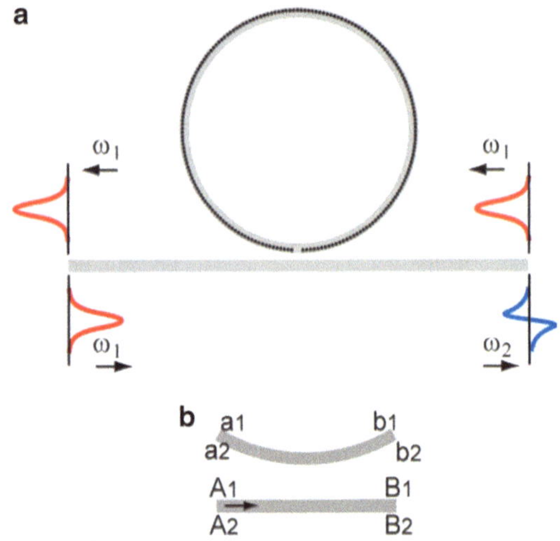

Fig. 13.9 (**a**) Schematic of ring resonator designed for nonreciprocal frequency conversion. The *dark regions* are modulated. (**b**) Schematic of the modes in the ring-waveguide coupling region

To describe the action of this structure, we note that upon completing one round trip, the circulating amplitudes $a_{1,2}$ and $b_{1,2}$ of these two modes (Fig. 13.9b) are related by:

$$\begin{pmatrix} a_1 \\ a_2 \end{pmatrix} = \begin{pmatrix} T_{11} & T_{12} \\ T_{21} & T_{22} \end{pmatrix} \begin{pmatrix} b_1 \\ b_2 \end{pmatrix}, \tag{13.14}$$

where the matrix elements are related to the transition amplitudes for a single round trip, and can be calculated using (13.4). Each of these modes is also coupled to an external waveguide as described by:

$$\begin{pmatrix} b_1 \\ B_1 \\ b_2 \\ B_2 \end{pmatrix} = \begin{pmatrix} r_1 & jt_1 & 0 & 0 \\ jt_1 & r_1 & 0 & 0 \\ 0 & 0 & r_2 & jt_2 \\ 0 & 0 & jt_2 & r_2 \end{pmatrix} \begin{pmatrix} a_1 \\ A_1 \\ a_2 \\ A_2 \end{pmatrix}. \tag{13.15}$$

The external waveguide is also assumed to support two modes with opposite symmetry at the frequencies ω_1, ω_2 respectively. Here, the subscripts label the two frequencies. $A_{1,2}$ and $a_{1,2}$ ($B_{1,2}$ and $b_{1,2}$) are the photon flux amplitudes in the external and ring waveguides before (after) the coupler. The coefficients r, t are taken to be real [17] and $r_{1,2}^2 + t_{1,2}^2 = 1$.

With incident light in mode 1 (i.e., $A_1 = 1, A_2 = 0$) of the external waveguide, combining (13.14) and (13.15), we have

$$B_1 = \frac{r_1 - T_{11} - r_1 r_2 T_{22} + r_2 Det[T]}{1 - r_1 T_{11} - r_2 T_{22} + r_1 r_2 Det[T]}, \tag{13.16}$$

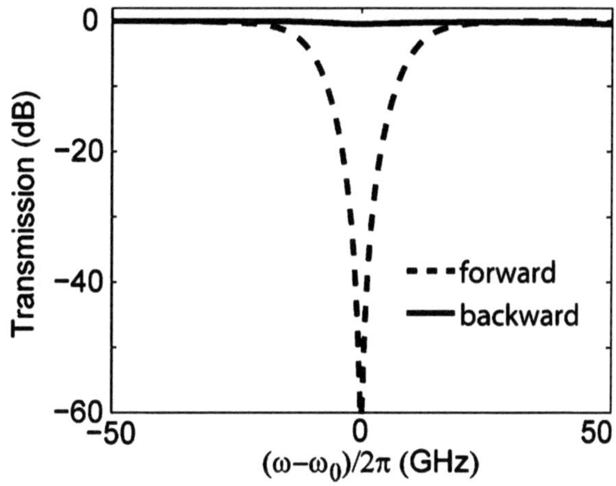

Fig. 13.10 Transmission spectra of a ring-resonator isolator. ω_0 corresponds to 1.55 μm wavelength. The waveguide-ring transmit coefficient is assumed to be $r_{1,2} = 0.95$

where *Det* stands for determinant. Thus, the condition for complete frequency conversion (i.e., $B_1 = 0$) is

$$r_1 - T_{11} - r_1 r_2 T_{22} + r_2 Det[T] = 0. \tag{13.17}$$

In the case that ring is lossless, $Det[T] = 1$ and $T_{11} = T_{22} = \cos((\pi/2)(L/l_c))$, where l_c is the coherence length and L is circumference of the ring. Complete conversion between the two modes can be achieved when the length of the ring is chosen to be

$$\cos\left(\frac{\pi}{2}\frac{L}{l_c}\right) = \frac{r_1 + r_2}{1 + r_1 r_2}. \tag{13.18}$$

With $r_{1,2} \to 1$, $L/l_c \to 0$, the device therefore can provide complete frequency conversion even when its length is far smaller than the coherence length.

As an example, now we use the same waveguide discussed in Fig. 13.4 to form a ring with a radius $r = 12.3$ μm. Such a ring supports two resonant modes: a first band resonant mode at 1.55 μm and a second band mode that is 50 GHz higher in frequency. (This is always achievable by fine tuning the radius and width of the waveguide.) A phase matching modulation is applied to the ring with a coherence length $l_c = 2.37$ mm. At the design wavelength 1.55 μm, the forward transmission is completed suppressed (Fig. 13.10). Here, the complete isolation is achieved with a device size much smaller than the coherence length.

In this section, we have provided some of detailed theoretical considerations for the dynamic isolator structures that we have recently proposed. In contrast to previously considered isolators based on material nonlinearity [9, 10] where isolation is only achievable for a range of incident power, the photonic transition effect studied here is *linear* with respect to the incident light: the effect does not depend upon the amplitude and phase of the incident light. Having a linear process is crucial because the device operation needs to be independent of the format, the timing and the intensity of the pulses used in the system. In conclusion, the structure proposed here shows that on-chip isolation can be accomplished with dynamic modulation, in standard material systems that are widely used for integrated optoelectronic applications.

13.4 Photonic Transition for Tunable Resonance

In this section, we review the tunable resonance based on photonic transition. Resonance appears when a localized state couples to a continuum. In photonics, of particular interest is when the localized state is supported by an optical microcavity, and the continuum is one-dimensional such as in a waveguide. Such waveguide-cavity configurations find applications in filters, sensors, switches, slow-light structures, and quantum information processing devices.

In all applications of resonance, it is essential to accurately control its spectral properties. For the waveguide-cavity resonances, some of the important spectral properties are the resonance frequency, and the *external linewidth* due to waveguide-cavity coupling. The inverse of such linewidth defines the corresponding quality factor (Q) of the cavity.

In this part, we show that a single high-Q resonance can be created by dynamically inducing a photonic transition between a localized state and a one-dimensional continuum. Since the coupling between the continuum and the localized state occurs *solely* through dynamic modulations, both the frequency and the external linewidth of a single resonance are specified by the dynamics, allowing complete control of its spectral properties.

We start by first briefly reviewing the Anderson–Fano model [18, 19], which describes the standard waveguide-cavity systems:

$$H = \omega_c c^+ c + \int \omega_k a_k^+ a_k \mathrm{d}k + V \int (c^+ a_k + a_k^+ c)\mathrm{d}k. \tag{13.19}$$

Here, ω_c is the frequency of a localized state that is embedded inside a one-dimensional continuum of states (Fig. 13.11a) defined by ω_k. $c^+(c)$ and $a_k{}^+(a_k)$ are the bosonic creation (annihilation) operators for localized and continuum states, respectively. V describes the interaction between them. Such a model supports a resonance at $\omega_0 = \omega_c$, with an external linewidth $\gamma = 2\pi(V^2/v_g)$ (defined as the full width at half maximum of the resonance peak). Here, $v_g \equiv \left.\frac{\mathrm{d}\omega_k}{\mathrm{d}k}\right|_{\omega_0}$.

In contrast to the standard Fano-Anderson model, our mechanism is described by the Hamiltonian: (Fig. 13.11b)

$$H = \omega_c c^+ c + \int \omega_k a_k^+ a_k \mathrm{d}k + (V + V_D \cos(\Omega t)) \int (c^+ a_k + a_k^+ c)\mathrm{d}k. \tag{13.20}$$

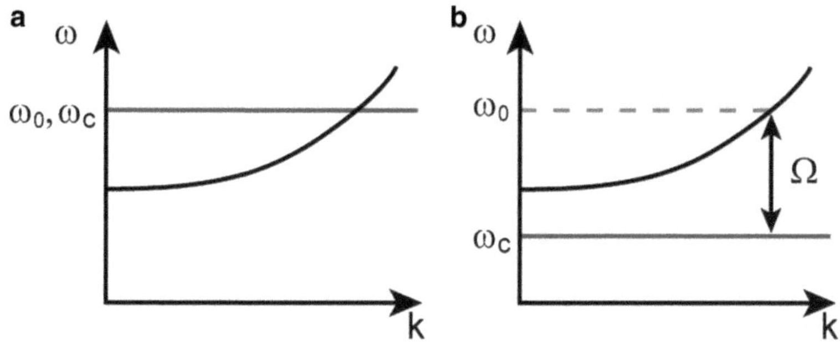

Fig. 13.11 Two different coupling mechanisms between a localized state and a one-dimensional continuum. (**a**) Static case: The frequency ω_c of the localized state lies in the band of the continuum. The static coupling between them results in a resonance at $\omega_0 = \omega_c$. (**b**) Dynamic case: The localized state has its frequency ω_c that falls outside the continuum. A modulation at a frequency Ω creates a photonic transition that couples them, resulting in a resonance at $\omega_0 = \omega_c + \Omega$

Here, unlike in (13.19), we assume that $\omega_k > \omega_c$ for any k. Consequently, the static coupling term $V\int(c^+a_k + a_k^+c)\mathrm{d}k$ no longer contributes to the decay of the resonance. Instead, it only results in a renormalization of ω_c. The localized state decays *solely* through the dynamic term $V_D\cos(\Omega t)\int(c^+a_k + a_k^+c)\mathrm{d}k$, which arises from modulating the system. Such modulation induces a photonic transition between the localized state and the continuum.

For the Hamiltonian of (13.20), one can derive an input–output formalism [20] in the Heisenberg picture, relating $C(t) = c(t)e^{-i\Omega t}$ to the input field operator $a_{\mathrm{IN}}(t)$ as:

$$\frac{\mathrm{d}}{\mathrm{d}t}C = -i(\omega_c + \Omega)C - \frac{\gamma}{2}C + i\sqrt{\gamma}a_{\mathrm{IN}}, \tag{13.21}$$

where $\gamma = 2\pi(((V_D/2)^2)/v_g)$ with $v_g = \left.\frac{\mathrm{d}\omega_k}{\mathrm{d}k}\right|_{\omega=\omega_c+\Omega}$. For an incident wave a_{IN} in the waveguide, the modulated system therefore creates a single resonance at the frequency $\omega_0 = \omega_c + \Omega$. Importantly, unlike the static system in (13.19), here both the frequency ω_0 and the external linewidth γ of the resonance are controlled by the dynamic modulation.

We now realize the Hamiltonian in (13.20) in a photonic crystal heterostructure [21] (Fig. 13.12a). The structure consists of a well and two barrier regions, defined in a line-defect waveguide in a semiconductor ($\varepsilon = 12.25$) two-dimensional photonic crystal. In the barrier regions, the crystal has a triangular lattice of air holes with a radius $r = 0.3a$, where a is the lattice constant. The waveguide supports two TE (H_z, E_x, E_y) modes with even and odd modal symmetry (Fig. 13.12c, light gray lines). In the well region, the hole spacing a' along the waveguide is increased to $1.1a$, which shifts the frequencies of the modes downward (Fig. 13.12c, dark lines) compared to those of the barriers. As a result, the odd modes in the well and the barriers do not overlap in frequencies. Thus, the well can support localized states, which are essentially standing waves formed by two

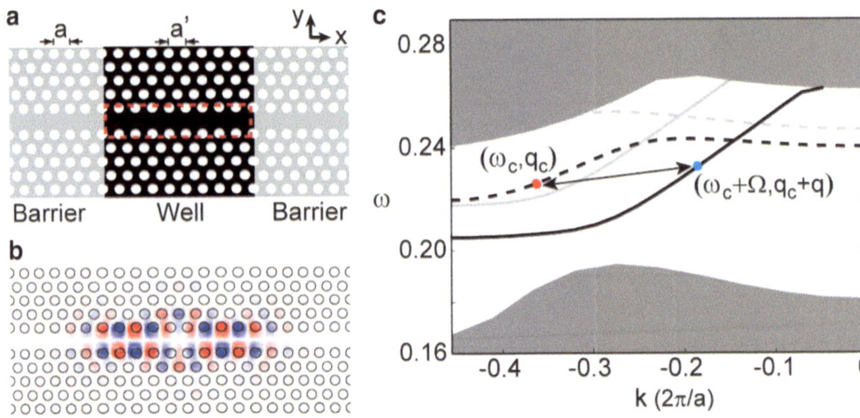

Fig. 13.12 (**a**) A photonic crystal heterostructure. The width of the waveguide measured from the centers of the holes on the two sides is $1.33a$. The *highlighted rectangle* represents the modulated region, which has dimensions of $2a \times 9.7a$. (**b**) Electric field (E_y) profile of a localized state in the well. *Red* and *blue* represent positive and negative maximum amplitudes. (**c**) Dispersion relation of the photonic crystal waveguide modes. The *dark* and *light gray lines* are for modes in the well and barriers, respectively. *Solid* (*dashed*) *lines* represent modes with even (odd) modal symmetry. *Shadowed regions* are the extended modes of the crystal region of the well

counter-propagating odd modes in the well. Figure 13.12b shows one such localized state at the frequency $\omega_c = 0.2252(2\pi c/a)$, with its corresponding waveguide mode at the wavevector $q_c = -0.37(2\pi/a)$ indicated by a red dot in Fig. 13.12c. Without modulation such a localized state cannot leak into the barrier and hence cannot be excited by wave coming from the barrier.

To induce a photonic transition, we modulate the dielectric constant of the well in the form of $\varepsilon_D = \Delta\varepsilon(y)\cos(\Omega t - qx)$. Here, the modulation frequency Ω is chosen such that an even mode in the well at the frequency $\omega_c + \Omega$ can leak into the barriers. The modulation wavevector q is selected to ensure a phase-matched transition between this even mode and the odd mode at (ω_c, q_c) that forms the localized state. Since these two modes have different symmetry, the modulation has an odd transverse profile: $\Delta\varepsilon(y) = \text{sign}(y)\Delta\varepsilon$, with $y = 0$ located at the waveguide center.

In the presence of the modulation, we consider an even mode incident from the left barrier, with a frequency ω in the vicinity of $\omega_c + \Omega$. As it turns out, for the even modes, the transmission coefficients into and out of the well are near unity. Thus, inside the well, the amplitudes of the even mode (Fig. 13.13, blue arrow) at the two edges, $A_{x=0}$ and $A_{x=L}$, are the input and output amplitudes of the system. As the even mode propagates forward from $x = 0$ to $x = L$, the modulation induces a transition to a copropagating odd mode at $\omega - \Omega$ (Fig. 13.13, red arrow). This transition process is described by [3]:

$$\begin{pmatrix} A_{x=L} \\ B_{x=L} \end{pmatrix} = \begin{pmatrix} \exp(iLq_{\omega}) & 0 \\ 0 & \exp(iLq_{\omega-\Omega}) \end{pmatrix} \begin{pmatrix} \sqrt{1-\eta^2} & i\eta \\ i\eta & \sqrt{1-\eta^2} \end{pmatrix} \begin{pmatrix} A_{x=0} \\ B_{x=0} \end{pmatrix}, \tag{13.22}$$

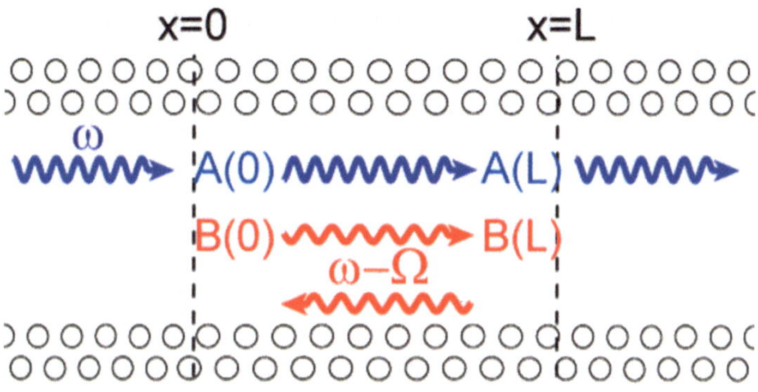

Fig. 13.13 The microscopic theory for photonic transition in the photonic crystal heterostructure. Incident light from the barrier at a frequency ω, as represented by the *blue arrows*, couples to a mode of the well at the frequency $\omega-\Omega$, as represented by the *red arrows*. The *dashed lines* indicate the edges of the well

where $B_{x=0}$ and $B_{x=L}$ are the amplitudes of the copropagating odd mode at $\omega-\Omega$ at the two edges, q_ω and $q_{\omega-\Omega}$ are the wavevectors of the two modes. For weak modulation, the transition rate $\eta=(\Delta\varepsilon/\varepsilon)L\kappa\ll 1$, where κ is the overlap factor between the two modes and the modulation profile.

Once the fields reach $x=L$, the odd mode is completely reflected, and propagates back to $x=0$. We note that no significant photon transition occurs in the backward propagation, since the modulation profile does not phase-match between $(\omega,-q_\omega)$ and $(\omega-\Omega,-q_{\omega-\Omega})$. Consequently,

$$B_{x=0}=\exp(iLq_{\omega-\Omega}+i2\phi)B_{x=L}, \tag{13.23}$$

where ϕ is the reflection phase at the well edge. Also, since there is a localized state at ω_c, the round trip phase at ω_c is $2(Lq_{\omega_c}+\phi)=2\pi n$ where n is an integer. Therefore, the round trip phase for the odd mode at $\omega-\Omega\approx\omega_c$ can be approximated as

$$2(Lq_{\omega-\Omega}+\phi)\approx 2\pi n+(\omega-\Omega-\omega_c)\frac{2L}{v_{gc}}, \tag{13.24}$$

where $v_{gc}=\left.\frac{\mathrm{d}\omega}{\mathrm{d}k}\right|_{\omega=\omega_c}$. Combined (13.22)–(13.24), the transmission spectrum is:

$$T=\frac{A_{x=L}}{\mathrm{e}^{iLq_\omega}A_{x=0}}=\frac{\sqrt{1-\eta^2}-\mathrm{e}^{i(\omega-\omega_0)\frac{2L}{v_{gc}}}}{1-\mathrm{e}^{i(\omega-\omega_0)\frac{2L}{v_{gc}}}\sqrt{1-\eta^2}}\approx\frac{\omega-\omega_0-i\frac{\gamma}{2}}{\omega-\omega_0+i\frac{\gamma}{2}}, \tag{13.25}$$

where $\gamma=(\Delta\varepsilon/\varepsilon)^2(\kappa^2Lv_{gc}/2)$.

The detailed microscopic theory thus predicts all-pass filter response for this dynamic system consisting of a waveguide coupled to a standing-wave localized state. In contrast, in the static system, coupling of a waveguide to a standing-wave localized state always produces either band-pass or band-reflection filters. Moreover, the resonant frequency

$$\omega_0 = \omega_c + \Omega \tag{13.26}$$

and the quality factor

$$Q_e \equiv \frac{\omega_0}{\gamma} = \left(\frac{\varepsilon}{\Delta\varepsilon}\right)^2 \frac{2\omega_0}{\kappa^2 L v_{gc}} \tag{13.27}$$

are completely controlled by the modulation, in agreement with the phenomenological model (13.21).

We numerically test the theory using finite-difference time-domain (FDTD) simulations. We simulate a well with a length of $9.9a$. Such a well supports the localized state shown in Fig. 13.12b. The length of the modulated region $L = 9.7a$ (Fig. 13.12a). We excite the even modes in the left barrier, with a Gaussian pulse centered at $0.235(2\pi c/a)$, and a width of $0.001(2\pi c/a)$. Without the modulation, the transmission coefficient (Fig. 13.14a) is near unity. With the modulation (with a strength $\Delta\varepsilon/\varepsilon = 1.63 \times 10^{-2}$, a frequency $\Omega = 9.8 \times 10^{-3}(2\pi c/a)$, and a wavevector $q = 0.196(2\pi/a)$), the transmission spectrum shows little change (Fig. 13.14b). However, the group delay now exhibits a resonant peak with a quality factor $Q_e = 1.09 \times 10^4$ (Fig. 13.14c, blue line). The structure indeed becomes a high-Q all-pass filter.

The properties of this resonance are controlled by the modulation. The resonant frequency changes linearly with respect to the modulation frequency, as predicted (Fig. 13.14e). (When varying the modulation frequency, we also change the modulation wavevector at the same time to satisfy the phase-matching condition.) The resonance frequency is largely independent of the modulation strength (Fig. 13.14e). The width of the resonance, and the peak delay, can be adjusted by changing the modulation strength (Fig. 13.14d). As a comparison between theory (13.27) and simulations, Fig. 13.14f plots the quality factor as a function of the modulation strength at the fixed modulation frequency $\Omega = 9.8 \times 10^{-3}(2\pi c/a)$. The simulation agrees excellently with the theory. The theory curve is generated with only one fitting parameter: the modal overlap factor $\kappa = 0.99a^{-1}$, which agrees well to a direct and separate calculation of the well waveguide by itself that yields $\kappa = 1.16a^{-1}$. The difference can be attributed to the finite-size effect of the well–barrier interfaces.

We now comment on some of the challenges in the practical implementations. For the simulated structure above, according to (13.27), a modulation strength of $\Delta\varepsilon/\varepsilon = 5 \times 10^{-3}$, which is achievable using carrier injection in semiconductors [14], results in an external quality factor of $Q_e = 1.1 \times 10^5$. In comparison, the radiation quality factors of photonic crystal heterostructure cavities exceeded 10^6 in experiments [22].

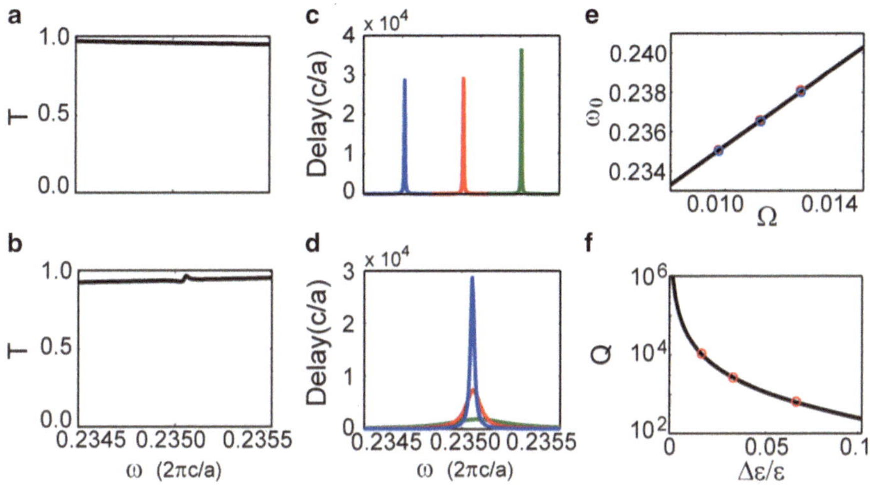

Fig. 13.14 Theory and simulation for the photonic transition process for the structure in Fig. 13.12. (**a**) Transmission spectrum for the unmodulated structure. (**b**) Transmission spectrum in the presence of modulation. The modulation has a frequency $\Omega = 9.8 \times 10^{-3}(2\pi c/a)$ and a strength of $\Delta\varepsilon/\varepsilon = 1.63 \times 10^{-2}$. (**c**) Group delay spectra, with $\Delta\varepsilon/\varepsilon$ fixed at 1.63×10^{-2}. The *blue*, *red* and *green lines* correspond to $\Omega = 9.8 \times 10^{-3}, 11.3 \times 10^{-3}$ and $12.8 \times 10^{-3}(2\pi c/a)$, respectively. (**d**) Group delay spectra, with Ω fixed at $9.8 \times 10^{-3}(2\pi c/a)$. The *blue*, *red*, and *green lines* correspond to $\Delta\varepsilon/\varepsilon = 1.63 \times 10^{-2}, 3.27 \times 10^{-2}$ and 6.53×10^{-2}. (**e**) Resonant frequency as a function of the modulation frequency. The *blue* and *red circles* corresponds to modulation strength of $\Delta\varepsilon/\varepsilon = 1.63 \times 10^{-2}$ and 3.27×10^{-2} respectively. *Circles* are simulation results as determined the peak location of group delay spectra, and the line is from analytical calcualtion. (**f**) Quality factor as a function of modulation strength. *Circles* are simulation results as determined from the peak width in (**d**), the line is from analytic calculation

Regarding the required modulation frequencies, in the simulation, $\Omega = 9.8 \times 10^{-3}(2\pi c/a)$ represents a modulation frequency of 8.1 THz, when the resonance frequency $\omega_0 = 0.235(2\pi c/a)$ corresponds to the wavelength of 1.55 μm. This is in principle achievable, since many index modulation scheme has intrinsic response time below 0.1 ps [23].

As final remarks, in our scheme, the tuning range for the resonant frequency is ultimately limited by the intrinsic response time of the material. Thus, the resonant frequency of the structure have a much wider tuning range, and can be reconfigured with a much higher speed, compared with conventional mechanisms. Moreover, the modulation frequency can typically be specified to a much higher accuracy [24], resulting in far more accurate control of the resonant frequency. Lastly, the localized state here is "dark" since it does not couple to the waveguide in the absence of modulation. Our scheme, which provides a dynamic access to such a dark state, is directly applicable for stopping and storage of light pulses, since the existence of a single dark state is sufficient [25].

13.5 Conclusion

In this chapter, we review the application of photonic transition for optical isolation and tunable resonance. These applications rely on photonic structures that can be dynamically modulated. Experimental techniques to achieve these dynamic structures have undergone fast development. One of the prominent techniques is to use carrier injection to modulate refractive index. Moreover, novel technique based on optical force has also emerged, such as the optomechanical modulation [26]. These developments open exciting opportunities for dynamic photonic structures.

References

1. J.N. Winn, S.H. Fan, J.D. Joannopoulos, E.P. Ippen, Phys. Rev. B **59**(3), 1551–1554 (1999)
2. P. Dong, S.F. Preble, J.T. Robinson, S. Manipatruni, M. Lipson, Phys. Rev. Lett. **100**(3), 033904 (2008)
3. Z. Yu, S. Fan, Nat. Photon. **3**(2), 91–94 (2009)
4. Z. Yu, S. Fan, Appl. Phys. Lett. **96**(1), 011108 (2010)
5. R.L. Espinola, T. Izuhara, M.C. Tsai, R.M. Osgood, Opt. Lett. **29**(9), 941–943 (2004)
6. M. Levy, J. Opt. Soc. Am. B: Opt. Phys. **22**(1), 254–260 (2005)
7. T.R. Zaman, X. Guo, R.J. Ram, Appl. Phys. Lett. **90**(2), 023514 (2007)
8. H. Dotsch, N. Bahlmann, O. Zhuromskyy, M. Hammer, L. Wilkens, R. Gerhardt, P. Hertel, A.F. Popkov, J. Opt. Soc. Am. B: Opt. Phys. **22**(1), 240–253 (2005)
9. M. Soljacic, C. Luo, J.D. Joannopoulos, S.H. Fan, Opt. Lett. **28**(8), 637–639 (2003)
10. K. Gallo, G. Assanto, K.R. Parameswaran, M.M. Fejer, Appl. Phys. Lett. **79**(3), 314–316 (2001)
11. S.K. Ibrahim, S. Bhandare, D. Sandel, H. Zhang, R. Noe, Electron. Lett. **40**(20), 1293–1294 (2004)
12. G. Shvets, Appl. Phys. Lett. **89**, 141127 (2006)
13. Z. Yu, S. Fan, Appl. Phys. Lett. **94**(17), 171116 (2009)
14. B.R. Bennett, R.A. Soref, J.A. Delalamo, IEEE J. Quant. Electron. **26**(1), 113–122 (1990)
15. Y. Jiao, S.H. Fan, D.A.B. Miller, Opt. Lett. **30**(2), 141–143 (2005)
16. B.T. Lee, S.Y. Shin, Opt. Lett. **28**(18), 1660–1662 (2003)
17. H.A. Haus, *Waves and Fields in Optoelectronics* (Prentice-Hall, Englewood Cliffs, NJ, 1984)
18. U. Fano, Phys. Rev. **124**(6), 1866–1878 (1961)
19. P.W. Anderson, Phys. Rev. **109**, 1492 (1958)
20. D.F. Walls, G.J. Milburn, *Quantum Optics* (Springer, Berlin, 1994)
21. B.S. Song, S. Noda, T. Asano, Y. Akahane, Nat. Mater. **4**(3), 207–210 (2005)
22. E. Kuramochi, M. Notomi, S. Mitsugi, A. Shinya, T. Tanabe, T. Watanabe, Appl. Phys. Lett. **88**(4), 041112 (2006)
23. S. Schmittrink, D.S. Chemla, W.H. Knox, D.A.B. Miller, Opt. Lett. **15**(1), 60–62 (1990)
24. T.W. Hansch, Rev. Mod. Phys. **78**(4), 1297 (2006)
25. C.R. Otey, M.L. Povinelli, S.H. Fan, J. Lightwave Technol. **26**(21–24), 3784–3793 (2008)
26. M. Eichenfield, J. Chan, R.M. Camacho, K.J. Vahala, O. Painter, Nature **462**, 78–82 (2009)

Index

Z. Chen and R. Morandotti (eds.), *Nonlinear Photonics and Novel Optical Phenomena*, Springer Series in Optical Sciences 170, DOI 10.1007/978-1-4614-3538-9,

MIX
Papier aus verantwortungsvollen Quellen
Paper from responsible sources
FSC® C105338

If you have any concerns about our products,
you can contact us on
ProductSafety@springernature.com

In case Publisher is established outside the EU,
the EU authorized representative is:
Springer Nature Customer Service Center GmbH
Europaplatz 3, 69115 Heidelberg, Germany

Printed by Libri Plureos GmbH
in Hamburg, Germany